AF413034

Transport in Laser Microfabrication

Emphasizing the fundamentals of transport phenomena, this book provides researchers and practitioners with the technical background they need to understand laser-induced microfabrication and materials processing at small scales. It clarifies the laser materials coupling mechanisms, and discusses the nanoscale confined laser interactions that constitute powerful tools for top-down nanomanufacturing. In addition to analyzing key and emerging applications for modern technology, with particular emphasis on electronics, advanced topics such as the use of lasers for nanoprocessing and nanomachining, the interaction with polymer materials, nanoparticles and clusters, and the processing of thin films are also covered.

Costas P. Grigoropoulos is a Professor in the Department of Mechanical Engineering at the University of California, Berkeley. His research interests are in laser materials processing, manufacturing of flexible electronics and energy devices, laser interactions with biological materials, microscale and nanoscale fluidics, and energy transport.

Transport in Laser Microfabrication: Fundamentals and Applications

COSTAS P. GRIGOROPOULOS

University of California, Berkeley

CAMBRIDGE
UNIVERSITY PRESS

CAMBRIDGE UNIVERSITY PRESS
Cambridge, New York, Melbourne, Madrid, Cape Town, Singapore, São Paulo, Delhi

Cambridge University Press
The Edinburgh Building, Cambridge CB2 8RU, UK

Published in the United States of America by Cambridge University Press, New York

www.cambridge.org
Information on this title: www.cambridge.org/9780521821728

First published 2009

Printed in the United Kingdom at the University Press, Cambridge

A catalog record for this publication is available from the British Library

ISBN 978 0 521 82172 8 hardback

To Mary, Vassiliki, and Alexandra

Contents

Preface

Lasers are effective material-processing tools that offer distinct advantages, including choice of wavelength and pulse width to match the target material properties as well as one-step direct and locally confined structural modification. Understanding the evolution of the energy coupling with the target and the induced phase-change transformations is critical for improving the quality of micromachining and microprocessing. As current technology is pushed to ever smaller dimensions, lasers become a truly enabling solution, reducing thermomechanical damage and facilitating heterogeneous integration of components into functional devices. This is especially important in cases where conventional thermo-chemo-mechanical treatment processes are ineffective. Component microfabrication with basic dimensions in the few-microns range via laser irradiation has been implemented successfully in the industrial environment. Beyond this, there is an increasing need to advance the science and technology of laser processing to the nanoscale regime.

The book focuses on examining the transport mechanisms involved in the laser–material interactions in the context of microfabrication. The material was developed in the graduate course on *Laser Processing and Diagnostics* I introduced and taught in Berkeley over the years. The text aims at providing scientists, engineers, and graduate students with a comprehensive review of progress and the state of the art in the field by linking fundamental phenomena with modern applications.

Samuel S. Mao of the Lawrence Berkeley National Laboratory and the Mechanical Engineering Department of UC Berkeley contributed major parts of Chapters 5, 6, and 9. I wish to acknowledge the contributions of all my former and current students throughout this text. Hee K. Park's, David J. Hwang's, and Seung-Hwang Ko's input extended beyond their graduate studies to post-doctoral stints in my laboratory. I am grateful to Gerald A. Domoto of Xerox Co. for introducing me to an interesting laser topic that evolved into my doctoral thesis at Columbia University. Dimos Poulikakos of the ETH Zürich talked me into starting this book project when I was on sabbatical in Zurich in 2000. His contributions in collaborative work form a key part of the text. I thank Professor Jean M. J. Fréchet of the UC Berkeley College of Chemistry for his contributions as well as Costas Fotakis of the IESL FORTH, Greece, and Dieter Bäuerle of Johannes Kepler University, Austria, for their support and input.

I am indebted to the NSF, DOE, and DARPA for funding work this book benefited from. The expert help of Ms. Ja Young Kim in preparing the artwork was key in completing this book.

Costas P. Grigoropoulos
Berkeley, California, USA

1 Fundamentals of laser energy absorption

1.1 Classical electromagnetic-theory concepts

1.1.1 Electric and magnetic properties of materials

Electric and magnetic fields can exert forces directly on atoms or molecules, resulting in changes in the distribution of charges. Thus, an electric field $\vec{E}$ induces an electric dipole moment or polarization vector $\vec{P}$, while the magnetic induction field $\vec{B}$ drives a magnetic dipole moment or magnetization vector $\vec{M}$. It is convenient to define the electric displacement vector $\vec{D}$ and the magnetic field $\vec{H}$ such that

$$\vec{D} = \varepsilon_0 \vec{E} + \vec{P}, \tag{1.1}$$

$$\vec{H} = \frac{1}{\mu_0} \vec{B} - \vec{M}, \tag{1.2}$$

where ε_0 and μ_0 are the electric permittivity and magnetic permeability, respectively, in vacuum. For isotropic electric materials the vectors $\vec{D}$, $\vec{E}$, and $\vec{P}$ are collinear, while correspondingly for isotropic magnetic materials the vectors $\vec{H}$, $\vec{B}$, and $\vec{M}$ are collinear.

Introducing the electric susceptibility χ, the polarization vector is written as

$$\vec{P} = \chi \varepsilon_0 \vec{E}, \tag{1.3}$$

and, therefore,

$$\vec{D} = \varepsilon_0 (1 + \chi) \vec{E} = \varepsilon_r \varepsilon_0 \vec{E} = \varepsilon \vec{E}, \tag{1.4}$$

where ε is the electric permittivity of the material and $\varepsilon_r = \varepsilon / \varepsilon_0$ the relative permittivity. Analogous expressions are used to describe the magnetic properties of materials:

$$\vec{B} = \mu \vec{H} = \mu_r \mu_0 \vec{H}, \tag{1.5}$$

where μ is the material's magnetic permeability and μ_r the relative magnetic permeability. In a medium where the charge density ρ moves with velocity $\vec{v}$, the free current-density vector $\vec{J}$ is defined as

$$\vec{J} = \rho \vec{v}. \tag{1.6}$$

The magnitude of this current, $|\vec{J}|$, represents the net amount of positive charge crossing a unit area normal to the instantaneous direction of $\vec{v}$ per unit time. The current-density vector is related to the electric field vector via the electric conductivity, σ,

$$\vec{J} = \sigma \vec{E}, \qquad (1.7)$$

which is the continuum form of Ohm's law. In isotropic materials, σ is a scalar quantity, but for crystalline or anisotropic solids σ is a second-order tensor.

1.1.2   Maxwell's equations

The system of Maxwell's equations constitutes the basis for the theory of electromagnetic fields and waves as well as their interactions with materials. For macroscopically homogeneous (uniform) materials, for which ε and μ are constants independent of position, the following relations hold:

(I)

$$\nabla \times \vec{E} = -\mu \frac{\partial \vec{H}}{\partial t}, \qquad (1.8)$$

(II)

$$\nabla \times \vec{H} = \vec{J} + \frac{\partial \vec{D}}{\partial t}, \qquad (1.9)$$

(III)

$$\nabla \cdot \vec{D} = \rho, \qquad (1.10)$$

(IV)

$$\nabla \cdot \vec{B} = 0, \qquad (1.11)$$

where $\vec{D}$, $\vec{B}$, and $\vec{J}$ are related to $\vec{E}$ and $\vec{H}$ through the constitutive Equations (1.4), (1.5), and (1.7). In vacuum where there is no current or electric charge, the Maxwell equations have a simple traveling plane wave solution with the electric and magnetic field orthogonal to one another, and to the direction of propagation.

1.1.2   Boundary conditions

Consider an interface i, separating two media (1) and (2) of different permittivities ε_1, ε_2 and permeabilities μ_1, μ_2 (Figure 1.1). According to Born and Wolf (1999) the sharp and distinct interface is replaced by an infinitesimally thin transition layer. Within this layer ε and μ are assumed to vary continuously. Let $\vec{n}_{12}$ be the local normal at the interface pointing into the medium (2). An elementary cylinder of volume δV and surface area δA is taken within the thin transition layer. The cylinder faces and peripheral wall are normal and parallel to vector $\vec{n}_{12}$, respectively. Since $\vec{B}$ and its derivatives may be assumed continuous over this elementary control volume, the Gauss divergence theorem

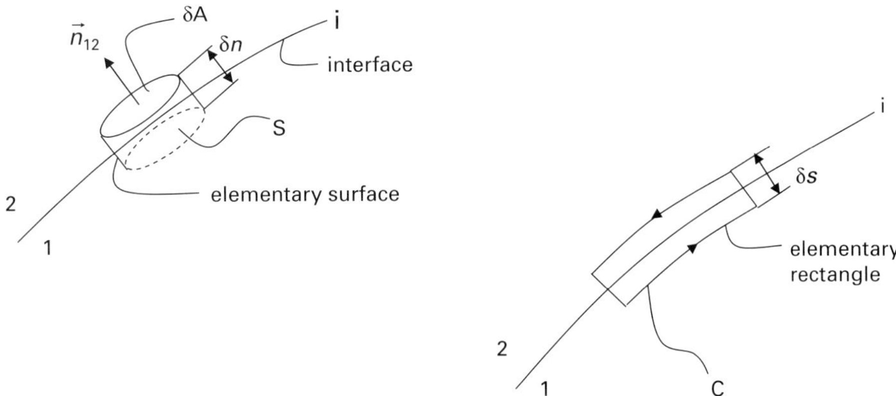

Figure 1.1. Schematics of an elementary volume of height δn and an elementary rectangular contour of width δs across the distinct interface separating media 1 and 2.

is applied to Equation (1.11):

$$\int\!\!\!\int\limits_{\delta V}\!\!\!\int \nabla \cdot \vec{B}\, dV = \int\!\!\!\int\limits_{\delta S} \vec{B} \cdot d\vec{S} = 0. \tag{1.12}$$

The second integral is taken over the surface of the cylinder. In the limit, as the height of the cylinder $\delta h \to 0$, contributions from the peripheral wall vanish and this integral yields

$$(\vec{B}_1 \cdot \vec{n}_1 + \vec{B}_2 \cdot \vec{n}_2)\delta A = 0, \tag{1.13}$$

where $\vec{n}_1 = -\vec{n}_{12}$ and $\vec{n}_2 = \vec{n}_{12}$. Consequently,

$$\vec{n}_{12} \cdot (\vec{B}_2 - \vec{B}_1) = 0. \tag{1.14}$$

The electric displacement vector $\vec{D}$ is treated in a similar manner by applying the Gauss theorem to Equation (1.10):

$$\int\!\!\!\int\limits_{\delta V}\!\!\!\int \nabla \cdot \vec{D}\, dV = \int\!\!\!\int\limits_{\delta S} \vec{D} \cdot d\vec{S} = \int\!\!\!\int\limits_{\delta V}\!\!\!\int \rho\, dV. \tag{1.15}$$

In the limit,

$$\lim_{\delta h \to 0} \int\!\!\!\int\limits_{\delta V}\!\!\!\int \rho\, dV = \int\!\!\!\int\limits_{\delta A} \sigma_s\, dA. \tag{1.16}$$

The above relation defines the surface charge density σ_s. Owing to the vanishing contribution over the peripheral wall as $\delta h \to 0$, Equation (1.16) gives

$$\vec{n}_{12} \cdot (\vec{D}_2 - \vec{D}_1) = \sigma_s. \tag{1.17}$$

These boundary conditions (1.14) and (1.17) can be expressed as

$$B_{2n} = B_{1n}, \tag{1.18a}$$

$$D_{2n} - D_{1n} = \sigma_s, \tag{1.18b}$$

where $B_{2n} = \vec{B}_2 \cdot \vec{n}$, $B_{1n} = \vec{B}_1 \cdot \vec{n}$, $D_{2n} = \vec{D}_2 \cdot \vec{n}$, and $D_{1n} = \vec{D}_1 \cdot \vec{n}$. In other words, the normal components of the magnetic induction vector $\vec{B}$ are always continuous and the difference between the normal components of the electric displacement $\vec{D}$ is equal in magnitude to the surface charge density σ_s.

To examine the behavior of the tangential electric and magnetic field components, a rectangular contour C with two long sides parallel to the surface of discontinuity is considered. Stokes' theorem is applied to Equation (1.8):

$$\iint_{\delta S} \nabla \times \vec{E} \cdot d\vec{S} = \iint_{\delta S} \nabla \times \vec{E} \cdot \vec{s}\, dS = \int_C \vec{E} \cdot d\vec{l} = -\mu \iint_{\delta S} \frac{\partial \vec{H}}{\partial t} \cdot \vec{s}\, dS. \quad (1.19)$$

In the limit as the width of the rectangle $\delta h \rightarrow 0$, the last surface integral vanishes and the contour integral of $\vec{E}$ is reduced to

$$\vec{E}_1 \cdot \vec{t}_1 + \vec{E}_2 \cdot \vec{t}_2 = 0. \quad (1.20)$$

Considering the unit tangent vector $\vec{t}$ along the interface, $\vec{t}_1 = -\vec{t} = -\vec{s} \times \vec{n}_{12}, \vec{t}_2 = \vec{t} = \vec{s} \times \vec{n}_{12}$, Equation (1.20) gives

$$\vec{n} \times (\vec{E}_2 - \vec{E}_1) = 0. \quad (1.21)$$

If a similar procedure is applied to Equation (1.9), then

$$\vec{n} \times (\vec{H}_2 - \vec{H}_1) = \vec{K}, \quad (1.22)$$

where $\vec{K}$ is the surface current density.

The boundary conditions (1.21) and (1.22) are written in the following form:

$$E_{2t} = E_{1t}, \quad (1.23a)$$

$$H_{2t} - H_{1t} = K_t. \quad (1.23b)$$

The subscript t implies the tangential component of the field vector. Thus, the tangential component of the electric field vector $\vec{E}$ is always continuous at the boundary surface and the difference between the tangential components of the magnetic vector $\vec{H}$ is equal to the line current density K, and in radiation problems where $\sigma_s = 0$, $K = 0$. Consequently, the normal components of $\vec{D}$ and $\vec{B}$ and the tangential components of $\vec{E}$ and $\vec{H}$ are continuous across interfaces separating media of different permittivities and permeabilities.

1.1.3   Energy density and energy flux

Light carries energy in the form of electromagnetic radiation. For a single charge q_e, the rate of work done by an external electric field $\vec{E}$ is $q_e \vec{v} \cdot \vec{E}$, where $\vec{v}$ is the velocity of the charge. If there exists a continuous distribution of charge and current, the total rate of work per unit volume is $\vec{J} \cdot \vec{E}$, since $\vec{J} = \rho \vec{v}$. Utilizing (1.9),

$$\vec{J} \cdot \vec{E} = \vec{E} \cdot (\nabla \times \vec{H}) - \vec{E} \cdot \frac{\partial \vec{D}}{\partial t}. \quad (1.24)$$

The following identity is invoked:

$$\nabla \cdot (\vec{E} \times \vec{H}) = \vec{H} \cdot (\nabla \times \vec{E}) - \vec{E} \cdot (\nabla \times \vec{H}), \tag{1.25}$$

and applied to (1.24):

$$\vec{J} \cdot \vec{E} = -\nabla \cdot (\vec{E} \times \vec{H}) - \vec{H} \cdot \frac{\partial \vec{B}}{\partial t} - \vec{E} \cdot \frac{\partial \vec{D}}{\partial t}. \tag{1.26}$$

The above equation is cast as follows:

$$\frac{\partial U}{\partial t} + \nabla \cdot \vec{S} = -\vec{J} \cdot \vec{E}, \tag{1.27a}$$

where

$$U = \frac{1}{2}(\vec{E} \cdot \vec{D} + \vec{B} \cdot \vec{H}), \tag{1.27b}$$

$$\vec{S} = \vec{E} \times \vec{H}. \tag{1.27c}$$

The scalar U represents the energy density of the electromagnetic field and in the SI system has units of $[\mathrm{J/m^3}]$. The vector $\vec{S}$ is called the Poynting vector and has units $[\mathrm{W/m^2}]$. It is consistent to view $|\vec{S}|$ as the power per unit area transported by the electromagnetic field in the direction of $\vec{S}$. Hence, the quantity $\nabla \cdot \vec{S}$ quantifies the net electromagnetic power flowing out of a unit control volume. Equation (1.27a) states the *Poynting vector theorem*.

1.1.4 Wave equations

Recalling the vector identity $\nabla \times (\nabla \times) = \nabla \cdot (\nabla \cdot) - \nabla^2$, Equation (1.8) yields

$$\nabla \times \nabla \times \vec{E} = \nabla \cdot (\nabla \cdot \vec{E}) - \nabla^2 \vec{E} = -\mu \nabla \times \frac{\partial \vec{H}}{\partial t}. \tag{1.28}$$

Invoking (1.9), the right-hand side of the above is

$$-\mu \nabla \times \frac{\partial \vec{H}}{\partial t} = -\mu \frac{\partial \vec{J}}{\partial t} - \mu \varepsilon \frac{\partial^2 \vec{E}}{\partial t^2},$$

and (1.28) gives

$$\nabla^2 \vec{E} - \nabla \left(\frac{\rho}{\varepsilon} \right) = \mu \left(\frac{\partial \vec{J}}{\partial t} + \varepsilon \frac{\partial^2 \vec{E}}{\partial t^2} \right),$$

or

$$\nabla^2 \vec{E} - \nabla \left(\frac{\rho}{\varepsilon} \right) = \mu \sigma \frac{\partial \vec{E}}{\partial t} + \mu \varepsilon \frac{\partial^2 \vec{E}}{\partial t^2}. \tag{1.29a}$$

Similarly, it can be shown that

$$\nabla^2 \vec{H} = \mu \sigma \frac{\partial \vec{H}}{\partial t} + \mu \varepsilon \frac{\partial^2 \vec{H}}{\partial t^2}. \tag{1.29b}$$

For propagation in vacuum, $\rho = 0$, $\sigma = 0$, $\mu = \mu_0$, and $\varepsilon = \varepsilon_0$, and Equations (1.29a) and (1.29b) give

$$\nabla^2 \vec{E} = \mu_0 \varepsilon_0 \frac{\partial^2 \vec{E}}{\partial t^2}, \tag{1.30a}$$

$$\nabla^2 \vec{H} = \mu_0 \varepsilon_0 \frac{\partial^2 \vec{H}}{\partial t^2}. \tag{1.30b}$$

The above are wave equations indicating a speed of wave propagation $c_0 = 1/\sqrt{\mu_0 \varepsilon_0}$, i.e. the speed of light in vacuum. For propagation in a perfect dielectric, $\rho = 0$, $\sigma = 0$, and the following apply

$$\nabla^2 \vec{E} = \mu \varepsilon \frac{\partial^2 \vec{E}}{\partial t^2}, \tag{1.31a}$$

$$\nabla^2 \vec{H} = \mu \varepsilon \frac{\partial^2 \vec{H}}{\partial t^2}. \tag{1.31b}$$

The propagation speed in this case is $c = 1/\sqrt{\mu \varepsilon}$. The index of refraction then is defined:

$$n = \frac{c_0}{c} = \sqrt{\frac{\mu \varepsilon}{\mu_0 \varepsilon_0}}. \tag{1.32}$$

Since, at optical frequencies, $\mu_0 \cong \mu$, the refractive index is approximated as

$$n \cong \sqrt{\frac{\varepsilon}{\varepsilon_0}}. \tag{1.33}$$

Equations (1.30) and (1.31) can be satisfied by monochromatic plane-wave solutions with a constant amplitude A and of the general form

$$\psi = A e^{i(\omega t - \vec{r} \cdot \vec{s})}, \tag{1.34}$$

where $\vec{r}$ and $\vec{s}$ are the position vector and the wavevector, respectively.

The angular frequency ω and the magnitude of the wavevector $\vec{s}$ are related by

$$|\vec{s}| = \omega \sqrt{\mu \varepsilon}. \tag{1.35}$$

According to (1.34), the field has the same values at locations $\vec{r}$ and times t that satisfy

$$\omega t - \vec{r} \cdot \vec{s} = \text{const.} \tag{1.36}$$

The above prescribes a plane normal to the wavevector $\vec{s}$ at any time instant t (Figure 1.2). The plane is called a *surface of constant phase*, often referred to as a *wavefront*.

The plane-wave electromagnetic fields are expressed by

$$\vec{E} = \vec{u}_1 E_0 e^{i(\omega t - \vec{r} \cdot \vec{s})}, \tag{1.37a}$$

$$\vec{H} = \vec{u}_2 H_0 e^{i(\omega t - \vec{r} \cdot \vec{s})}, \tag{1.37b}$$

where $\vec{u}_1$ and $\vec{u}_2$ are constant unit vectors and E_0 and H_0 are the constant-in-space complex amplitudes.

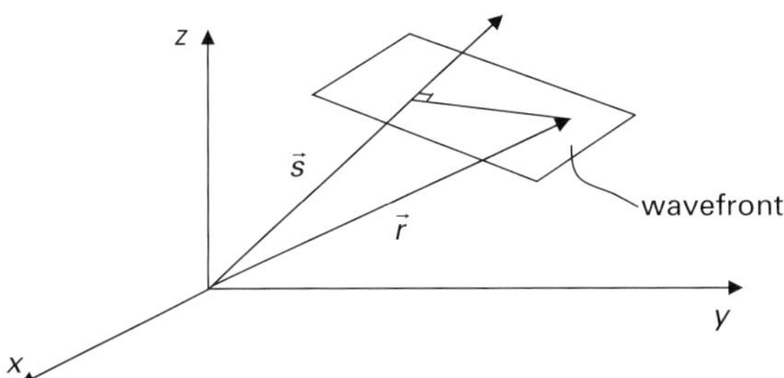

Figure 1.2. A schematic diagram depicting a plane wave propagating normal to the direction $\vec{s}$.

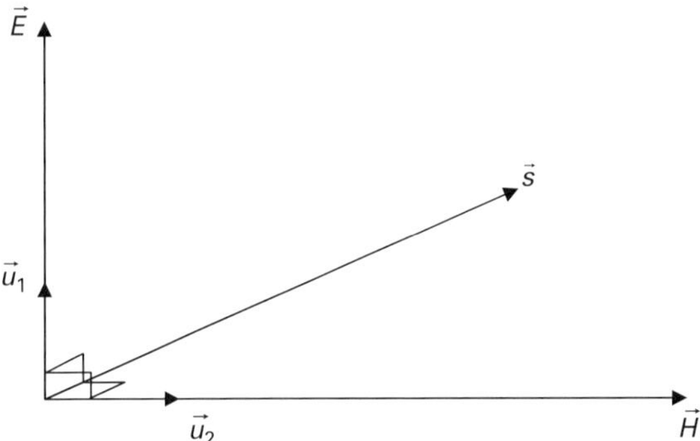

Figure 1.3. A schematic diagram depicting the instantaneous vectors $\vec{E}$ and $\vec{H}$ that form a right-hand triad with the unit vector $\vec{s}$ along the propagation direction.

In a homogeneous, charge-free medium, $\nabla \cdot \vec{E} = \nabla \cdot \vec{H} = 0$. Hence,

$$\vec{u}_1 \cdot \vec{s} = \vec{u}_2 \cdot \vec{s} = 0, \tag{1.38}$$

meaning that $\vec{E}$ and $\vec{H}$ are both perpendicular to the direction of propagation (Figure 1.3). For this reason, electromagnetic waves in dielectrics are said to be *transverse*.

The curl Maxwell equations impose further restrictions on the field vectors. By applying (1.38) in (1.8), it can be shown that

$$\vec{u}_2 = \frac{\vec{s} \times \vec{u}_1}{|\vec{s}|}. \tag{1.39}$$

The triad $(\vec{u}_1, \vec{u}_2, \vec{s})$ therefore forms a set of orthogonal vectors, and $\vec{E}$ and $\vec{H}$ are in phase with amplitudes in constant ratio, provided that ε and μ are both real (Figure 1.4). The plane wave described is a transverse wave propagating in the direction $\vec{s}$ with a time-averaged flux of energy

$$\vec{S} = \frac{|E_0|^2}{2\omega\mu}\vec{u}_3 = \frac{\vec{E}^* \cdot \vec{E}}{2\omega\mu}\vec{u}_3, \tag{1.40}$$

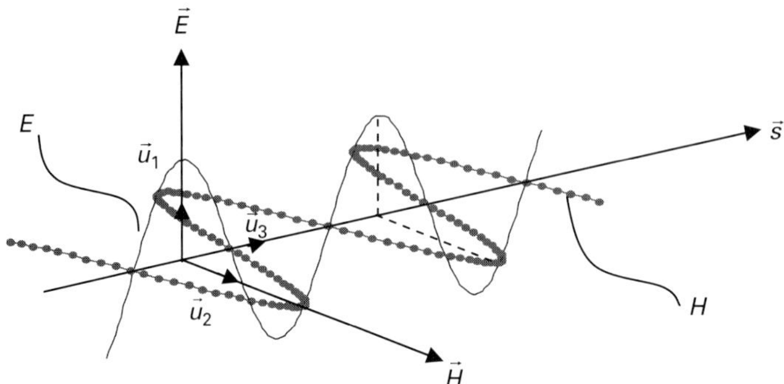

Figure 1.4. In a medium of real refractive index, the electric and magnetic fields are always in phase.

where $\vec{E}^*$ is the conjugate of the complex electric field vector. The time-averaged energy density

$$U = \frac{1}{2}\varepsilon |E_0|^2 = \frac{1}{2}\varepsilon \vec{E} \cdot \vec{E}^*. \tag{1.41}$$

1.1.5   Electromagnetic theory of absorptive materials

The optical properties of perfect dielectric media are completely characterized by the real refractive index. In such media, it is assumed that electromagnetic radiation interacts with the constituent atoms with no energy absorption. In contrast, especially for metals, very little light penetrates to a depth beyond $1\,\mu$m at visible wavelengths. Consider then media with nonzero electric conductivity that absorb energy but do not redirect a collimated light beam. Let $\vec{E}$ and $\vec{H}$ be the real parts of periodic variations:

$$\vec{E}(x, y, z, t) = \mathrm{Re}[\vec{E}^{\mathrm{c}}(x, y, z)\mathrm{e}^{\mathrm{i}\omega t}], \tag{1.42a}$$

$$\vec{H}(x, y, z, t) = \mathrm{Re}[\vec{H}^{\mathrm{c}}(x, y, z)\mathrm{e}^{\mathrm{i}\omega t}]. \tag{1.42b}$$

The superscript c indicates a complex quantity. Utilizing Maxwell's equations (1.8)–(1.11),

$$\nabla \times \vec{E}^{\mathrm{c}} = -\mathrm{i}\omega\mu\vec{H}^{\mathrm{c}}, \tag{1.43a}$$

$$\nabla \times \vec{H}^{\mathrm{c}} = (\sigma + \mathrm{i}\omega\varepsilon)\vec{E}^{\mathrm{c}}, \tag{1.43b}$$

$$\nabla \cdot \vec{E}^{\mathrm{c}} = 0, \tag{1.43c}$$

$$\nabla \cdot \vec{H}^{\mathrm{c}} = 0. \tag{1.43d}$$

Taking the curl of (1.43a) and combining this with (1.43b) gives

$$\nabla \times \nabla \times \vec{E}^{\mathrm{c}} = -\mathrm{i}\omega\mu(\sigma + \mathrm{i}\omega\varepsilon)\vec{E}^{\mathrm{c}}. \tag{1.44}$$

Utilizing the identity $\nabla \times \nabla \times \vec{E}^{\mathrm{c}} = \nabla \cdot (\nabla \cdot \vec{E}^{\mathrm{c}}) - \nabla^2\vec{E}^{\mathrm{c}}$ combined with (1.43c),

$$\nabla \times \nabla \times \vec{E}^{\mathrm{c}} = -\nabla^2\vec{E}^{\mathrm{c}}. \tag{1.45}$$

Combining the above with (1.43a) and (1.43b) gives

$$\nabla^2 \vec{E}^c = i\omega\mu(\sigma + i\omega\varepsilon)\vec{E}^c, \tag{1.46}$$

or

$$\nabla^2 \vec{E}^c = -\omega^2\mu\left(\varepsilon - i\frac{\sigma}{\omega}\right)\vec{E}^c, \tag{1.47}$$

which is written as

$$\nabla^2 \vec{E}^c + (k^c)^2 \vec{E}^c = 0. \tag{1.48}$$

The complex wavenumber k^c satisfies

$$(k^c)^2 = \omega^2\mu\left(\varepsilon - i\frac{\sigma}{\varpi}\right) = \omega^2\mu\varepsilon^c.$$

The quantity

$$\varepsilon^c = \varepsilon - i\frac{\sigma}{\omega}$$

is the complex dielectric constant.

A complex velocity v^c and a complex refractive index n^c can then be defined:

$$v^c = \frac{1}{\sqrt{\mu\varepsilon^c}}, \tag{1.49}$$

$$n^c = \frac{c_0}{v^c} = \sqrt{\frac{\mu\varepsilon^c}{\mu_0\varepsilon_0}}. \tag{1.50}$$

Let $n^c = n - ik$, where n is the real part of the complex refractive index and k the imaginary part, the so-called attenuation index:

$$(n^c)^2 = n^2 - k^2 - 2ink = \mu c_0^2\left(\varepsilon - i\frac{\sigma}{\omega}\right). \tag{1.51}$$

Equating the real and the imaginary parts, and solving for n^2 and k^2, gives

$$n^2 = \frac{c_0^2}{2}\left[\sqrt{\mu^2\varepsilon^2 + \left(\frac{\mu\sigma}{2\pi v}\right)^2} + \mu\varepsilon\right], \tag{1.52a}$$

$$n^2 = \frac{c_0^2}{2}\left[\sqrt{\mu^2\varepsilon^2 + \left(\frac{\mu\sigma}{2\pi v}\right)^2} - \mu\varepsilon\right]. \tag{1.52b}$$

Equation (1.48) implies wave propagation. The simplest solution is that of a plane, time-harmonic wave:

$$\vec{E}^c(\vec{r}, t) = \vec{E}_0^c e^{-i[k^c(\vec{r}\cdot\vec{s}) - \omega t]}, \tag{1.53}$$

where $\vec{s}$ is a unit vector along the direction of propagation. Since

$$k^c = \frac{\omega n^c}{c_0} = \frac{\omega(n - ik)}{c_0},$$

the above can be written as

$$\vec{E}^{c} = \vec{E}_{0}^{c} e^{-\frac{\omega}{c_0} k(\vec{r}\cdot\vec{s})} e^{i\omega[-\frac{n}{c_0}(\vec{r}\cdot\vec{s})+t]}. \tag{1.54}$$

The real part of this expression represents the electric vector:

$$\vec{E} = \vec{E}_{0} e^{-\frac{\omega}{c_0} k(\vec{r}\cdot\vec{s})} \cos\left\{\omega\left[-\frac{n}{c_0}(\vec{r}\cdot\vec{s})+t\right]\right\}. \tag{1.55}$$

A similar expression can be developed for the magnetic field vector. The energy flux per unit area is given by the Poynting vector, $\vec{S} = \vec{E} \times \vec{H}$, which is then

$$\vec{S} = \mathrm{Re}(\vec{E}_{0}^{c} e^{i\omega t}) \times \mathrm{Re}(\vec{H}_{0}^{c} e^{i\omega t}), \tag{1.56}$$

and then

$$\vec{S} = \frac{1}{4}\left[(\vec{E}_{0}^{c} e^{i\omega t} + \vec{E}_{0}^{c*} e^{-i\omega t})\right] \times \left[(\vec{H}_{0}^{c} e^{i\omega t} + \vec{H}_{0}^{c*} e^{-i\omega t})\right],$$

$$\vec{S} = \frac{1}{4}\left[\vec{E}_{0}^{c} \times \vec{H}_{0}^{c} e^{2i\omega t} + \vec{E}_{0}^{c*} \times \vec{H}_{0}^{c*} e^{-2i\omega t} + \vec{E}_{0}^{c*} \times \vec{H}_{0}^{c} + \vec{E}_{0}^{c} \times \vec{H}_{0}^{c*}\right]. \tag{1.57}$$

Consider a time interval $[-T', T']$ large compared with the fundamental wave period, $T = 2\pi/\omega$, which is $O(10^{-15}\text{ s})$:

$$\frac{1}{2T'} \int_{-T'}^{T'} e^{2i\omega t}\, \mathrm{d}t = \frac{1}{4i\omega T'}[e^{2i\omega t}]_{-T'}^{T'} = \frac{1}{4i\frac{2\pi}{T}T'} 2\cos(\omega T') = \frac{T}{2\pi i T'}\cos(\omega T'). \tag{1.58}$$

Evidently, if the EM energy flux given in (1.57) is averaged over $[-T', T']$ with $T' \gg T$, the first two periodic terms will contribute very little. Hence, the averaged energy is

$$\vec{S}_{\mathrm{av}} = \frac{1}{2}\mathrm{Re}(\vec{E}_{0}^{c} \times \vec{H}_{0}^{c*}) = \vec{s}\frac{1}{2}\frac{\mathrm{Re}(n^{c})}{\mu c_0}|\vec{E}_{0}^{c}|^{2}. \tag{1.59}$$

The above expressions indicate that the energy flux carried by a wave propagating in an absorbing medium is proportional to the squared modulus of its complex amplitude and to the real part of the complex refractive index of the medium. The modulus of the Poynting vector, i.e. the monochromatic radiative intensity, I'_{λ}, is

$$I'_{\lambda} = |\vec{S}_{\mathrm{av}}| = I'_{\lambda,0} e^{-\frac{2\omega}{c_0}k(\vec{r}\cdot\vec{s})} = I'_{\lambda,0} e^{-\gamma(\vec{r}\cdot\vec{s})}, \tag{1.60}$$

where γ is the absorption coefficient of the medium:

$$\gamma = \frac{2\omega k}{c_0} = \frac{4\pi k}{\lambda_0}. \tag{1.61}$$

In the above, λ_0 is the wavelength in vacuum. As shown in Figure 1.5, the energy flux drops to $1/e$ of $I'_{\lambda,0}$ after traveling a distance d, the so-called absorption penetration depth:

$$d = \frac{1}{\gamma} = \frac{\lambda_0}{4\pi k}. \tag{1.62}$$

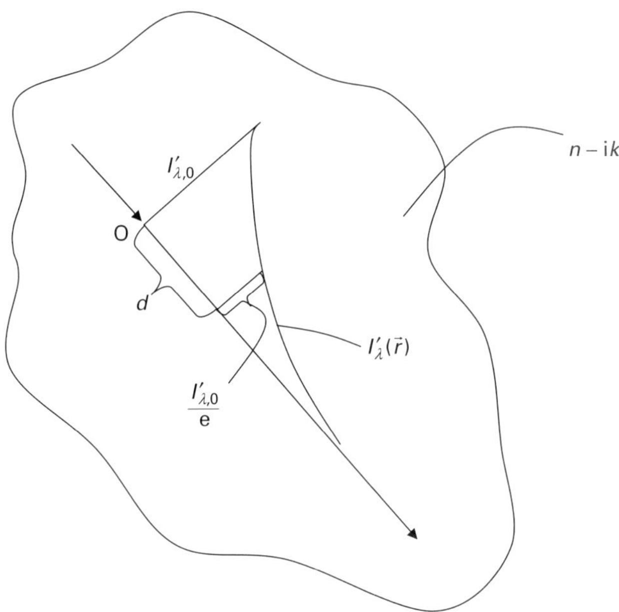

Figure 1.5. A schematic diagram depicting the exponential decay of a monochromatic beam in a medium of complex refractive index $n - ik$.

1.1.7   Refraction and reflection at a surface

Perfect dielectric media

Consider a plane light wave of electric field $\vec{E}_0^+$ incident at the plane interface of two perfect dielectric media, characterized by the real refractive indices n_0 and n_1 (Figure 1.6). The electric field is decomposed to the two polarized components E_{0p}^+ and E_{0s}^+, where p and s indicate polarizations parallel and normal to the plane of incidence (xOz). The superscripts $+$ and $-$ indicate forward and backward wave propagation. Let θ^+ and θ_t be the angles of incidence and refraction. The phase factors associated with the incident and refracted waves are of the form

$$\exp\left\{i\omega\left[t - \frac{n_0}{c_0}(\vec{r}\cdot\vec{s}^{+})\right]\right\} \text{ (incident)},$$

$$\exp\left\{i\omega\left[t - \frac{n_0}{c_0}(\vec{r}\cdot\vec{s}^{-})\right]\right\} \text{ (reflected)},$$

$$\exp\left\{i\omega\left[t - \frac{n_1}{c_0}(\vec{r}\cdot\vec{s}_t)\right]\right\} \text{ (transmitted)}.$$

At the boundary separating the two media ($z = 0$), the phase-factor arguments have to match. Consequently,

$$t - \frac{n_0}{c_0}(\vec{r}\cdot\vec{s}^{+}) = t - \frac{n_0}{c_0}(\vec{r}\cdot\vec{s}^{-}) = t - \frac{n_1}{c_0}(\vec{r}\cdot\vec{s}_t). \tag{1.63}$$

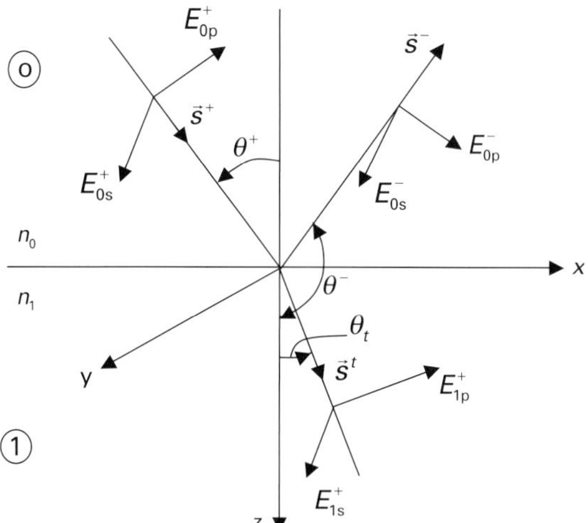

Figure 1.6. A plane wave is incident on the flat interface ($z = 0$) separating two semi-infinite media of refractive indices n_0 and n_1. The electric field vector is decomposed to the parallel (p) and normal (s) polarizations.

At $\vec{r} = (x, y, 0)$,

$$n_0(xs_x^+ + ys_y^+) = n_0(xs_x^- + ys_y^-) = n_1(xs_{x,t} + ys_{y,t}). \tag{1.64}$$

Since the above has to hold for all (x, y),

$$n_0 s_x^+ = n_0 s_x^- = n_1 s_{x,t}, \tag{1.65}$$

$$n_0 s_y^+ = n_0 s_y^- = n_1 s_{y,t}. \tag{1.66}$$

On the other hand,

$$s_x^+ = \sin\theta^+, \qquad s_y^+ = 0, \qquad s_z^+ = \cos\theta^+ \geq 0, \tag{1.67a}$$

$$s_x^- = \sin\theta^-, \qquad s_y^- = 0, \qquad s_z^- = \cos\theta^- \leq 0, \tag{1.67b}$$

$$s_{x,t} = \sin\theta_t, \qquad s_{y,t} = 0, \qquad s_{z,t} = \cos\theta_t \geq 0. \tag{1.67c}$$

Equation (1.65) therefore gives

$$n_0 \sin\theta^+ = n_0 \sin\theta^- = n_1 \sin\theta_t. \tag{1.68}$$

Since $\sin\theta^+ = \sin\theta^-$ and $\cos\theta^+ = -\cos\theta^-$, it is inferred that

$$\theta^+ = \theta = \pi - \theta^-. \tag{1.69}$$

Equation (1.68) and the statement that the directional vector $\vec{s}^-$ of the reflected wave lies on the plane of incidence (xOy), as expressed by (1.67b), constitute the *law of reflection*. By considering Equations (1.66) and (1.67c), we find that

$$n_0 \sin\theta^+ = n_1 \sin\theta_t. \tag{1.70}$$

The above equation, combined with the fact that the directional vector $\vec{s}_t$ of the refracted wave lies on the plane of incidence as deduced from (1.67c), expresses *Snell's law of reflection*. The tangential components of the electric and magnetic vectors have also to be continuous across the interface:

$$
\begin{aligned}
E_{0x} &= (E_{0p}^{+} + E_{0p}^{-})\cos\theta = E_{1x} = E_{1p}^{+}\cos\theta_t, \\
E_{0y} &= E_{0s}^{+} + E_{0s}^{-} = E_{1y} = E_{1s}^{+}, \\
H_{0x} &= n_0(-E_{0s}^{+} + E_{0s}^{-})\cos\theta = H_{1x} = -n_1 E_{1s}^{+}\cos\theta_t, \\
H_{0y} &= n_0(E_{0p}^{+} - E_{0p}^{-}) = H_{1y} = n_1 E_{1p}^{+}.
\end{aligned}
$$

From the above, the Fresnel coefficients for reflection, $r_{F,1}$, and transmission, tr_1, are derived for the p and s polarizations:

$$
\frac{E_{0p}^{-}}{E_{0p}^{+}} = \frac{n_0\cos\theta_t - n_1\cos\theta}{n_0\cos\theta_t + n_1\cos\theta} = r_{F,1p},
\tag{1.71a}
$$

$$
\frac{E_{1p}^{+}}{E_{0p}^{+}} = \frac{2n_0\cos\theta}{n_0\cos\theta_t + n_1\cos\theta} = tr_{1p},
\tag{1.71b}
$$

$$
\frac{E_{0s}^{-}}{E_{0s}^{+}} = \frac{n_0\cos\theta - n_1\cos\theta_t}{n_0\cos\theta + n_1\cos\theta_t} = r_{F,1s},
\tag{1.71c}
$$

$$
\frac{E_{1s}^{+}}{E_{0s}^{+}} = \frac{2n_0\cos\theta}{n_0\cos\theta + n_1\cos\theta_t} = tr_{1s}.
\tag{1.71d}
$$

The monochromatic, directional reflectivities for the two polarizations are

$$
\rho_{\lambda,p}' = \frac{(E_{0p}^{-})^2}{(E_{0p}^{+})^2} = r_{F,1p}^2,
\tag{1.72a}
$$

$$
\rho_{\lambda,s}' = \frac{(E_{0s}^{-})^2}{(E_{0s}^{+})^2} = r_{F,1s}^2,
\tag{1.72b}
$$

and the respective transmissivities are

$$
\tau_{\lambda,p}' = \frac{n_1(E_{1p}^{+})^2}{n_0(E_{0p}^{+})^2} = \frac{n_1}{n_0}tr_{1p}^2,
\tag{1.73a}
$$

$$
\tau_{\lambda,s}' = \frac{n_1(E_{1s}^{+})^2}{n_0(E_{0s}^{+})^2} = \frac{n_1}{n_0}tr_{1s}^2.
\tag{1.73b}
$$

For normal incidence on an isotropic medium the reflectivity and transmissivity become

$$
\rho_{\lambda,p}' = \rho_{\lambda,s}' = \left(\frac{n_0 - n_1}{n_0 + n_1}\right)^2,
\tag{1.74a}
$$

$$
\tau_{\lambda,p}' = \tau_{\lambda,s}' = \frac{4n_0 n_1}{(n_0 + n_1)^2}.
\tag{1.74b}
$$

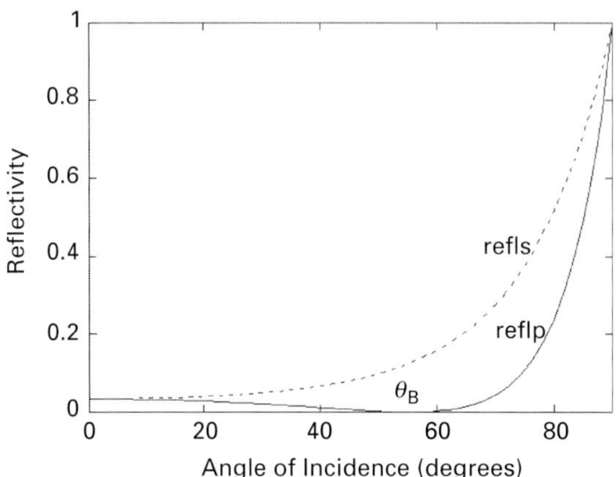

Figure 1.7. The angular dependence of the surface reflectivity for a bulk dielectric sample having refractive index $n_1 = 1.45$ for radiation incident through a medium of refractive index $n_0 = 1$. The curve labeled "reflp" is for parallel polarization and the curve marked "refls" is for normal polarization. The Brewster angle is indicated by "θ_B."

For $\theta > 0$, Equations (1.73), combined with (1.72) and Snell's law of refraction, are used to obtain the reflectivities and transmissivities. Equations (1.72) can be simplified:

$$r_{F,1p} = \frac{\tan(\theta - \theta_t)}{\tan(\theta + \theta_t)}, \tag{1.75a}$$

$$tr_{1p} = \frac{2\sin\theta_t \cos\theta}{\sin(\theta + \theta_t)\cos(\theta - \theta_t)}, \tag{1.75b}$$

$$r_{F,1s} = -\frac{\sin(\theta - \theta_t)}{\sin(\theta + \theta_t)}, \tag{1.75c}$$

$$tr_{1s} = \frac{2\sin\theta_t \cos\theta}{\sin(\theta + \theta_t)}. \tag{1.75d}$$

The denominators are finite, except in the case of p-polarization, when $\tan(\theta + \theta_t) \to \infty$, for $\theta + \theta_t = \pi/2$. In this case,

$$\sin\theta_t = \sin\left(\frac{\pi}{2} - \theta\right) \Rightarrow \frac{n_0}{n_1}\sin\theta = \cos\theta \Rightarrow \tan\theta = \frac{n_1}{n_0}.$$

At this angle, the polarizing or Brewster angle is $\rho'_{\lambda,p} = 0$. Figure 1.7 gives an example of the angular reflectivity variation for a perfectly dielectric (transparent) bulk material for the parallel and normal polarizations. It is noted that for unpolarized light, the reflectivity is

$$\rho'_\lambda(\theta) = \frac{1}{2}(\rho'_{\lambda,p}(\theta) + \rho'_{\lambda,s}(\theta)). \tag{1.76}$$

Reflection at the surface of an absorbing medium

The equations for light propagation in a transparent medium can be modified for the case of an absorbing medium, by replacing the real refractive index by its complex

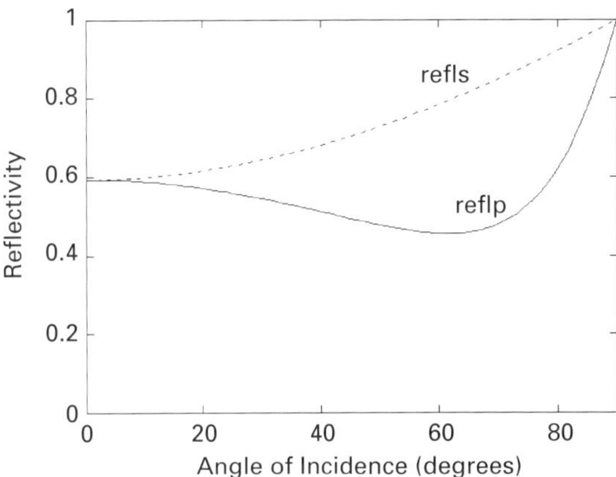

Figure 1.8. The angular dependence of the surface reflectivity for a bulk absorbing sample having refractive index $n_1^c = 0.7 - 2i$ for radiation incident through a medium of refractive index $n_0 = 1$. The curve labeled "reflp" is for parallel polarization and the curve marked "refls" is for normal polarization.

counterpart:

$$n_1 \to n_1^c = n_1 - ik_1.$$

In this case the angle of refraction is complex, and is identified by the generalized version of Snell's law of refraction:

$$\sin \theta_t^c = \frac{n_0 \sin \theta}{n_1 - ik_1}. \tag{1.77}$$

The complex Fresnel coefficients for normal incidence are

$$r_{F,1p} = r_{F,1s} = \frac{n_0 - n_1 + ik_1}{n_0 + n_1 + ik_1}. \tag{1.78}$$

The normal incidence surface reflectivity is

$$\rho'_{\lambda,p} = \rho'_{\lambda,s} = (r_{F,1})^2 = \frac{(n_0 - n_1)^2 + k_1^2}{(n_0 + n_1)^2 + k_1^2}. \tag{1.79}$$

Figure 1.8 gives an example of the angular reflectivity variation for an absorbing medium. For parallel polarization, the reflectivity exhibits a minimum at the so-called "pseudo-Brewster" angle, although it does not vanish there.

Consider light impinging on the surface of an absorbing medium at an oblique angle of incidence. The complex electric field in the medium (1) is

$$\vec{E}_1^c(\vec{r}, t) = \vec{E}_{t,0}^c e^{-i[k_1^c(\vec{r}\cdot\vec{s}_t)-\omega t]}, \tag{1.80}$$

where $k_1^c = \omega(n_1 - ik_1)/c_0$.

The complex unit vector of propagation in medium (1) is defined by

$$s_{t,x} = \sin\theta_t^c = \frac{\sin\theta}{n_1 - ik_1} = \frac{\sin\theta(n_1 + ik_1)}{n_1^2 + k_1^2}, \tag{1.81a}$$

$$s_{t,y} = 0, \tag{1.81b}$$

$$s_{t,z} = \cos\theta_t^c = \sqrt{1 - \sin^2\theta_t^c} = \sqrt{1 - \left[\frac{\sin\theta(n_1 + ik_1)}{n_1^2 + k_1^2}\right]^2}$$

$$= \sqrt{1 - \frac{\left(n_1^2 - k_1^2\right)\sin^2\theta}{\left(n_1^2 + k_1^2\right)^2} - i\frac{2n_1 k_1 \sin^2\theta}{\left(n_1^2 + k_1^2\right)^2}}. \tag{1.81c}$$

According to Born and Wolf (1999), it is convenient to set

$$s_{t,z} = qe^{i\delta}, \tag{1.82}$$

where q and δ are real. By taking the squares and equating the real and imaginary parts of (1.80c) and (1.81), it can be shown that

$$q^2 \cos(2\delta) = 1 - \frac{\left(n_1^2 - k_1^2\right)\sin^2\theta}{\left(n_1^2 + k_1^2\right)^2}, \tag{1.83a}$$

$$q^2 \sin(2\delta) = -\frac{2n_1 k_1 \sin^2\theta}{\left(n_1^2 + k_1^2\right)^2}. \tag{1.83b}$$

The phase factor is then given by

$$k_1^c(\vec{r}\cdot\vec{s}_t) = \frac{\omega}{c_0}(n_1 - ik_1)(x s_{t,x} + z s_{t,z})$$

$$= \frac{\omega}{c_0}(n_1 - ik_1)\left[\frac{x\sin\theta(n_1 + ik_1)}{n_1^2 + k_1^2} + z(q\cos\delta + iq\sin\delta)\right]$$

$$= \frac{\omega}{c_0}[x\sin\theta + zq(n_1\cos\delta + k_1\sin\delta) + izq(n_1\sin\delta - k_1\cos\delta)]. \tag{1.84}$$

On examining (1.80) and (1.84), it is evident that the surfaces of constant amplitude are given by

$$z = \text{const}, \tag{1.85}$$

i.e. they are planes parallel to the surface $z = 0$. On the other hand, the surfaces of constant phase are given by

$$x\sin\theta + zq(n_1\cos\delta + k_1\sin\delta) = \text{const}. \tag{1.86}$$

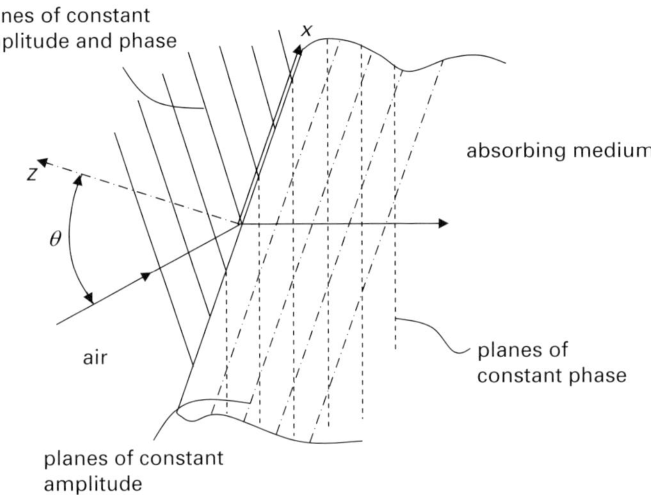

Figure 1.9. In an absorbing medium the planes of constant phase and constant amplitude coincide only in the case of normal incidence.

These are planes whose normal vectors are at an angle θ_t' with respect to the outward normal to the boundary:

$$\cos\theta_t' = \frac{q(n_1\cos\delta + k_1\sin\delta)}{\sqrt{\sin^2\theta + q^2(n_1\cos\delta + k_1\sin\delta)^2}}, \tag{1.87a}$$

$$\sin\theta_t' = \frac{\sin\theta}{\sqrt{\sin^2\theta + q^2(n_1\cos\delta + k_1\sin\delta)^2}}. \tag{1.87b}$$

Since the planes of constant amplitude and constant phase do not coincide, as is the case in lossless media, light propagation in absorbing materials takes the form of *inhomogeneous waves* (Figure 1.9).

1.1.8 Laser light absorption in multilayer structures

A laser beam is incident on a multilayer stack of films, $z_0 \geq \ldots \geq z_{j-1} \geq z \geq z_j \geq \ldots z_N$, which is stratified in the z-direction, as shown in Figure 1.10. The laser-beam propagation axis is on the x–y plane. A wave of unit strength, and wavelength, λ_0, is considered incident on the stratified structure at the angle θ_0. The case of arbitrary polarization can be treated as a superposition of TE (transverse electric) and TM (transverse magnetic) polarized waves. For TE-polarized light

$$E_y = U(z)\exp\left[i\left(\frac{2\pi}{\lambda_0}n_0\sin\theta_0\, y - \omega t\right)\right], \tag{1.88a}$$

$$H_y = V(z)\exp\left[i\left(\frac{2\pi}{\lambda_0}n_0\sin\theta_0\, y - \omega t\right)\right], \tag{1.88b}$$

$$H_z = W(z)\exp\left[i\left(\frac{2\pi}{\lambda_0}n_0\sin\theta_0\, y - \omega t\right)\right]. \tag{1.88c}$$

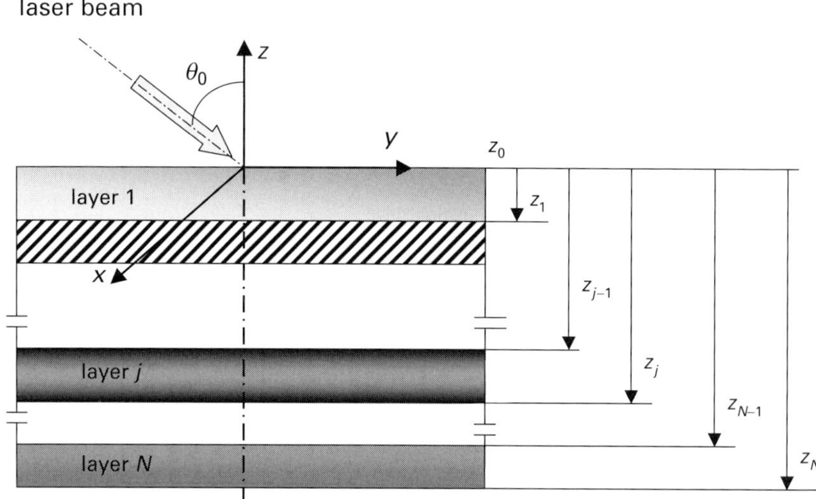

Figure 1.10. A schematic diagram of a laser beam incident on a multilayer structure.

Using Maxwell's equations, it is found that W is linearly dependent on U, and that the solution can be expressed in the form of a characteristic transmission matrix, defined by

$$\begin{Bmatrix} U_0 \\ V_0 \end{Bmatrix} = M^c(z) \begin{Bmatrix} U(z) \\ V(z) \end{Bmatrix}, \tag{1.89a}$$

where

$$M^c(z) = \begin{bmatrix} \cos\left(\dfrac{2\pi}{\lambda_0} n^c z \cos\theta^c\right) & -\dfrac{i}{p^c} \sin\left(\dfrac{2\pi}{\lambda_0} n^c z \cos\theta^c\right) \\ -i p^c \sin\left(\dfrac{2\pi}{\lambda_0} n^c z \cos\theta^c\right) & \cos\left(\dfrac{2\pi}{\lambda_0} n^c z \cos\theta^c\right) \end{bmatrix}. \tag{1.89b}$$

In the above expression, $p^c = n^c \cos\theta^c$ for a TE wave, and $p^c = \cos\theta^c/n^c$ for a TM wave. The angle θ^c is complex for absorbing films and is defined by the generalized version of Snell's law of refraction:

$$n^c \sin\theta^c = n_0 \sin\theta_0. \tag{1.90}$$

The multilayer system transmission matrix is

$$M^c = \prod_{j=1}^{N} M^c(z_j - z_{j-1}). \tag{1.91}$$

The lumped structure reflectivity and transmissivity can be obtained. The reflection and transmission Fresnel coefficients, r_F and tr, are

$$r_F = \frac{\left[M^c(1,1) + M^c(1,2)n^c_{ss}\right]n^c_a - \left[M^c(2,1) + M^c(2,2)n^c_{ss}\right]}{\left[M^c(1,1) + M^c(1,2)n^c_{ss}\right]n^c_a + \left[M^c(2,1) + M^c(2,2)n^c_{ss}\right]}, \tag{1.92a}$$

$$tr = \frac{2n^c_a}{\left[M^c(1,1) + M^c(1,2)n^c_{ss}\right]n^c_a + \left[M^c(2,1) + M^c(2,2)n^c_{ss}\right]}, \tag{1.92b}$$

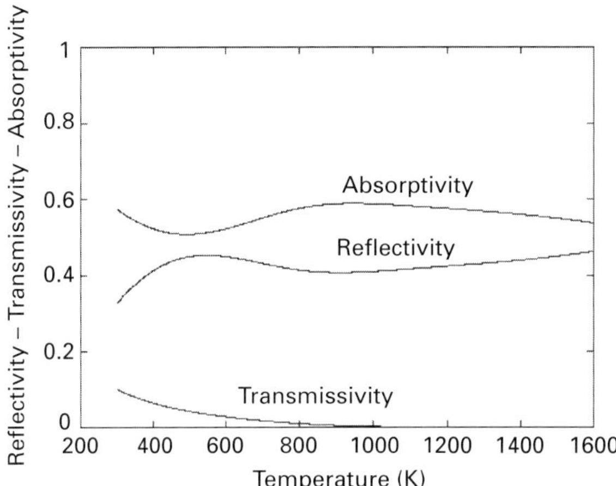

Figure 1.11. The temperature dependence of the optical properties for a 0.5-μm silicon film on a glass substrate at the 514.5-nm Ar$^+$-laser wavelength.

where the subscripts ss and a indicate the substrate $(Z > Z_N)$ and ambient $(Z < Z_0)$, respectively.

The structure reflectivity and transmissivity, in terms of r_F and tr, follow:

$$R = \frac{|E_a^-|^2}{|E_a^+|^2} = |r_F|^2, \tag{1.93a}$$

$$\tau = \frac{n_{ss}|E_{ss}^+|^2}{n_a|E_a^+|^2} = \frac{n_{ss}}{n_a}|tr|^2. \tag{1.93b}$$

As an example, Figure 1.11 shows the calculated temperature dependences of the reflectivity, transmissivity, and absorptivity, for normal incidence of Ar$^+$ laser light, of wavelength $\lambda = 0.5145$ μm, upon a silicon layer of thickness $d_{Si} = 0.5$ μm, deposited on a glass substrate. The variation with temperature of the complex refractive index of silicon (Sun *et al.*, 1997) generates distinct changes in the optical properties of the film due to interference effects, even though the bulk-silicon normal-incidence reflectivity varies slowly with temperature. When silicon melts, it exhibits a metallic behavior with an abrupt rise in reflectivity and drop in absorptivity.

Returning to the stratified multilayer structure, the time-averaged power flow per unit area that crosses the plane perpendicular to the z-axis is given by the magnitude of the Poynting vector,

$$S(z) = \frac{1}{2}\,\mathrm{Re}[\vec{E}(z) \times \vec{H}^*(z)]. \tag{1.94}$$

A plane wave is assumed incident on the structure, with electric field amplitude E_a^+. The corresponding energy flow along the z-direction is

$$\vec{S} = \frac{n_a}{2\mu c}|E_a^+|^2\vec{k}. \tag{1.95}$$

The electric field amplitudes of the reflected and transmitted waves, E_a^- and E_{ss}^+, are obtained using the above expressions. The electric and magnetic fields in the mth layer,

$m = 1 \ldots, N$, are given by

$$E_m(z) = E_m^+ \exp\left[-\mathrm{i}k_m^{\mathrm{c}}(z - z_m)\right] + E_m^- \exp\left[+\mathrm{i}k_m^{\mathrm{c}}(z - z_m)\right], \qquad (1.96\mathrm{a})$$

$$H_m(z) = \frac{n_m^{\mathrm{c}}}{c\mu}\left\{E_m^+ \exp\left[-\mathrm{i}k_m^{\mathrm{c}}(z - z_m)\right] - E_m^- \exp\left[+\mathrm{i}k_m^{\mathrm{c}}(z - z_m)\right]\right\}, \qquad (1.96\mathrm{b})$$

where

$$k_m^{\mathrm{c}} = \frac{2\pi}{\lambda}n_m^{\mathrm{c}}, \qquad z_m = \sum_{j=1}^{m-1} d_j.$$

Continuity of the electric and magnetic field at the interfaces is applied to obtain a recursive formula for the amplitudes of the electric field:

$$E_m^+ = \frac{1}{2}\left[E_{m-1}^+\left(1 + \frac{n_{m-1}^{\mathrm{c}}}{n_m^{\mathrm{c}}}\right)\mathrm{e}^{-\mathrm{i}k_{m-1}^{\mathrm{c}}d_{m-1}} + E_{m-1}^-\left(1 - \frac{n_{m-1}^{\mathrm{c}}}{n_m^{\mathrm{c}}}\right)\mathrm{e}^{+\mathrm{i}k_{m-1}^{\mathrm{c}}d_{m-1}}\right], \qquad (1.97\mathrm{a})$$

$$E_m^- = \frac{1}{2}\left[E_{m-1}^+\left(1 - \frac{n_{m-1}^{\mathrm{c}}}{n_m^{\mathrm{c}}}\right)\mathrm{e}^{-\mathrm{i}k_{m-1}^{\mathrm{c}}d_{m-1}} + E_{m-1}^-\left(1 + \frac{n_{m-1}^{\mathrm{c}}}{n_m^{\mathrm{c}}}\right)\mathrm{e}^{+\mathrm{i}k_{m-1}^{\mathrm{c}}d_{m-1}}\right]. \qquad (1.97\mathrm{b})$$

Calculation of the amplitudes E_m^+ and E_m^- starts from the first layer, for which $d_{m-1} = 0$. Once the electric field has been determined, the power flow is evaluated everywhere in the structure. At a location z within the mth layer

$$\vec{S}_m(z) = \frac{1}{2\mu c}\,\mathrm{Re}\left[\left(n_m^{\mathrm{c}}\right)^*\left(E_m^1(z) + E_m^2(z)\right) \times \left(E_m^1(z) - E_m^2(z)\right)^*\right]\vec{k}, \qquad (1.98)$$

where

$$E_m^1(z) = E_m^+ \exp\left[-\mathrm{i}k_m^{\mathrm{c}}(z - z_m)\right], \qquad (1.99\mathrm{a})$$

$$E_m^2(z) = E_m^- \exp\left[+\mathrm{i}k_m^{\mathrm{c}}(z - z_m)\right]. \qquad (1.99\mathrm{b})$$

The local energy flow is normalized by the energy flux incident on the structure:

$$\vec{S}_m(z) = \frac{1}{n_{\mathrm{a}}|E_{\mathrm{a}}^+|^2}\,\mathrm{Re}\left[\left(n_m^{\mathrm{c}}\right)^*\left(E_m^1(z) + E_m^2(z)\right) \times \left(E_m^1(z) - E_m^2(z)\right)^*\right]. \qquad (1.100)$$

The local laser-energy absorption per unit volume is

$$Q_{\mathrm{abs}}(x, y, z, t) = Q_{\mathrm{las}}(x, y, t)\frac{\mathrm{d}S(z)}{\mathrm{d}z}, \qquad (1.101)$$

where Q_{las} is the incident laser intensity distribution.

1.2 Optical properties of materials

1.2.1 Classical theories of optical constants

Consider a medium characterized by a complex refractive index $n^c = n - ik$. The complex dielectric constant (or complex permittivity) is defined by

$$\varepsilon^c = \varepsilon_0(1 + \chi) - i\frac{\sigma}{\omega} = \varepsilon_0(\varepsilon' - i\varepsilon''), \tag{1.102}$$

where

$$\varepsilon' = \frac{\mathrm{Re}(\varepsilon^c)}{\varepsilon_0} = 1 + \mathrm{Re}(\chi) = n^2 - k^2 \tag{1.103a}$$

and

$$\varepsilon'' = \frac{\mathrm{Im}(\varepsilon^c)}{\varepsilon_0} = \mathrm{Im}(\chi) - \frac{\sigma}{\omega\varepsilon_0} = 2nk. \tag{1.103b}$$

The above equation implies that both the electric conductivity and the susceptibility contribute to the imaginary part of the complex permittivity and thereby to the absorption characteristics of the material. The part $\mathrm{Im}[-\sigma/(\omega\varepsilon_0)]$ represents the contribution due to the free current density, while the part $\mathrm{Im}(\chi)$ is caused by the current density associated with bound charges. For a nonmagnetic material (i.e. $\mu = \mu_0$) the components of the complex refractive index are derived from (1.103a) and (1.103b):

$$n = \frac{\sqrt{\sqrt{\varepsilon'^2 + \varepsilon''^2} + \varepsilon'}}{2}, \tag{1.104a}$$

$$k = \frac{\sqrt{\sqrt{\varepsilon'^2 + \varepsilon''^2} - \varepsilon'}}{2}. \tag{1.104b}$$

The Lorentz model for nonconductors

According to this model, polarizable matter is represented as a collection of identical, independent, and isotropic harmonic oscillators of mass m and charge e, whereas electrons are permanently bound to the core and immobile atoms. In response to a driving force produced by the local (effective) field, the oscillators undergo a displacement from equilibrium $\vec{x}$ and are acted upon by a linear restoring force $K_s\vec{x}$, where K_s is the spring stiffness, and a damping force $b\dot{\vec{x}}$, where b is the damping constant. The equation of motion is

$$m\ddot{\vec{x}} + b\dot{\vec{x}} + K_s\vec{x} = e\vec{E}_{\mathrm{local}}. \tag{1.105}$$

The excitation is assumed of periodic form,

$$\vec{E}_{\mathrm{local}} = \vec{E}_{0,\mathrm{local}}e^{i\omega t}. \tag{1.106}$$

Of interest is the resulting periodic response:

$$\vec{x} = \vec{x}_0 e^{i\omega t}. \tag{1.107}$$

Equation (1.105) yields

$$\vec{x} = \frac{(e/m)\vec{E}_{\text{local}}}{\omega_0^2 - \omega^2 - i\zeta\omega}. \tag{1.108}$$

where the resonance frequency $\omega_0 = \sqrt{K_s/m}$ and $\zeta = b/m$. For $\zeta \neq 0$, the displacement and the electric field are not in phase. Equation (1.108) is rewritten

$$\vec{x} = \frac{(e/m)\vec{E}_0 e^{i\omega t} e^{i\phi}}{\sqrt{\left(\omega_0^2 - \omega^2\right)^2 + \zeta^2\omega^2}}, \tag{1.109}$$

or

$$\vec{x} = A e^{i\phi}, \tag{1.110}$$

where

$$A = \frac{1}{\sqrt{\left(\omega_0^2 - \omega^2\right)^2 + \zeta^2\omega^2}}, \qquad \phi = \arctan\left(\frac{\zeta\omega}{\omega_0^2 - \omega^2}\right).$$

The induced dipole moment of an oscillator is $e\vec{x}$. If N is the number of oscillators per unit volume, the polarization $\vec{P}$ (dipole moment per unit volume), neglecting local effects, is

$$\vec{P} = N e\vec{x} = \frac{\omega_p^2}{\omega_0^2 - \omega^2 - i\zeta\omega}\varepsilon_0\vec{E}, \tag{1.111}$$

where the plasma frequency $\omega_p = \sqrt{Ne^2/(m\varepsilon_0)}$ and ω_0 is the resonance frequency.

The complex permittivity is then extracted:

$$\frac{\varepsilon^c}{\varepsilon_0} = 1 + \chi = 1 + \frac{\omega_p^2}{\omega_0^2 - \omega^2 - i\zeta\omega}, \tag{1.112a}$$

$$\varepsilon' = 1 + \chi' = 1 + \frac{\omega_p^2\left(\omega_0^2 - \omega^2\right)}{\left(\omega_0^2 - \omega^2\right)^2 + \zeta^2\omega^2}, \tag{1.112b}$$

$$\varepsilon'' = \chi'' = \frac{\omega_p^2\zeta\omega}{\left(\omega_0^2 - \omega^2\right)^2 + \zeta^2\omega^2}. \tag{1.112c}$$

The maximum value of ε'' occurs approximately at ω_0, under the condition that $\zeta \ll \omega_0$. For frequencies close to resonance, Equations (1.112b) and (1.112c) yield

$$\varepsilon' = 1 + \frac{\omega_p^2(\omega_0 - \omega)/(2\omega_0)}{(\omega_0 - \omega)^2 + (\zeta/2)^2}, \tag{1.113a}$$

$$\varepsilon'' = \frac{\zeta\omega_p^2/(4\omega_0)}{(\omega_0 - \omega)^2 + (\zeta/2)^2}. \tag{1.113b}$$

According to (1.113b), the maximum value of ε'' is approximately $\omega_p^2/(\zeta\omega_0)$ and the full-width-at-half-maximum points are at $\omega = \omega_0 \pm \zeta/2$.

In the ideal case of no absorption, i.e. $\zeta = 0$, the real refractive index goes to infinity, $n_{\omega \to \omega_0^{\pm}} \Rightarrow \pm\infty$. The regime of anomalous dispersion is the only region in the radiation spectrum of decreasing n with increasing frequency. Taking the limit of Equation (1.112) for high frequencies, $\omega \gg \omega_0$:

$$\varepsilon' \cong 1 - \frac{\omega_{\mathrm{p}}^2}{\omega^2} \Rightarrow n \cong \sqrt{\varepsilon'} \cong 1 - \frac{\omega_{\mathrm{p}}^2}{2\omega^2}, \tag{1.114a}$$

$$\varepsilon'' \cong \frac{\zeta \omega_{\mathrm{p}}^2}{\omega^3} \Rightarrow k \cong \frac{\varepsilon''}{2} \cong \frac{\zeta \omega_{\mathrm{p}}^2}{2\omega^3}. \tag{1.114b}$$

Consequently, the reflectivity at normal incidence exhibits the following trend:

$$\rho' \cong \left(\frac{\omega_{\mathrm{p}}}{2\omega}\right)^4. \tag{1.115}$$

The behavior of the dielectric constant in the infrared (IR) regime, i.e. for $\omega \ll \omega_0$, is as follows:

$$\varepsilon' \cong 1 + \frac{\omega_{\mathrm{p}}^2}{\omega_0^2}, \tag{1.116a}$$

$$\varepsilon'' \cong \frac{\zeta \omega_{\mathrm{p}}^2 \omega}{\omega_0^4}. \tag{1.116b}$$

Hence, the real part of the electric permittivity asymptotically approaches a constant value in the far IR, while the imaginary part tends to vanish. The trends predicted by the single Lorentz model are exhibited in Figure 1.12. Measured refractive-index components of glassy SiO_2 can be found, for example, in Palik (1985).

In general, the dielectric function of a collection of oscillators is given by

$$\frac{\varepsilon^{\mathrm{c}}}{\varepsilon_0} = 1 + \sum_j \frac{\omega_{\mathrm{p}j}^2}{\omega_{0j}^2 - \omega^2 - \mathrm{i}\zeta_j \omega}, \tag{1.117}$$

where $\omega_{\mathrm{p}j}$, ω_{0j}, and ζ_j are the plasma frequency, the resonance frequency, and the damping constant assigned to the jth harmonic oscillator. This multiple-oscillator model can be used for fitting the radiation properties of materials over a broad spectral range. The most important resonance arises from *interband transitions* of valence-band electrons to the conduction band. To induce an interband transition, the photon energy has to exceed the band-gap energy E_{g}. Insulators have band-gap energies in the deep ultraviolet range and concentrations of free carriers (electrons and holes) are very small in the visible range. The band-gap energy of semiconductors is in the visible or near-IR range. It is therefore possible for free carriers to contribute to the optical response spectra in the visible range. In addition to electronic transitions, resonant coupling to high-frequency optical phonons usually occurs at near-IR frequencies.

The Drude model for conductors

In conducting media, not all electrons are bound to atoms. The optical response of metals is dominated by *free electrons* in states close to the Fermi level. If an external field is applied, their motion will become more orderly. Since there are no resonance frequencies, the optical response of a collection of free electrons can be obtained by

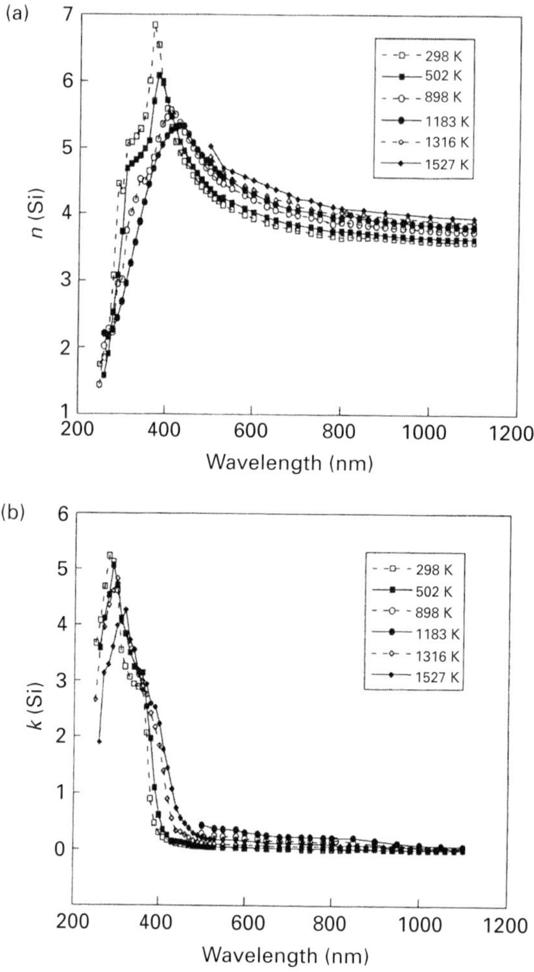

Figure 1.14. Spectral temperature dependences (a) of the refractive index, n, of pure single-crystalline silicon; and (b) of the extinction coefficient, k, of pure single-crystalline silicon. From Sun *et al.* (1997), reproduced with permission from Elsevier.

reflectivity:

$$n(\lambda, T) = n_0(\lambda) + a_n(\lambda)(T - T_{0n}), \qquad (1.130\text{a})$$

$$k(\lambda, T) = a_k(\lambda)\exp\!\left(\frac{T}{T_{0k}}\right), \qquad (1.130\text{b})$$

$$R(\lambda, T) = R_0(\lambda) + a_R(\lambda)(T - T_{0R}). \qquad (1.130\text{c})$$

where $T_{0n} = 25\,^\circ\mathrm{C}$, $T_{0k} = 498\,^\circ\mathrm{C}$, and $T_{0R} = 25\,^\circ\mathrm{C}$ are reference temperatures for fitting. The components of the complex dielectric function, together with the absorption coefficient, are shown in Figure 1.15. The peak E_0 is considered to arise primarily from a M_0 critical point in the joint density of states for the $\Gamma_{25}^{v} \rightarrow \Gamma_{15}^{c}$ transition. The E_1 peak in ε'' near 3.4 eV corresponds to either an M_0 or an M_1 critical point for $\Lambda_3^{v} \rightarrow \Lambda_1^{c}$. The

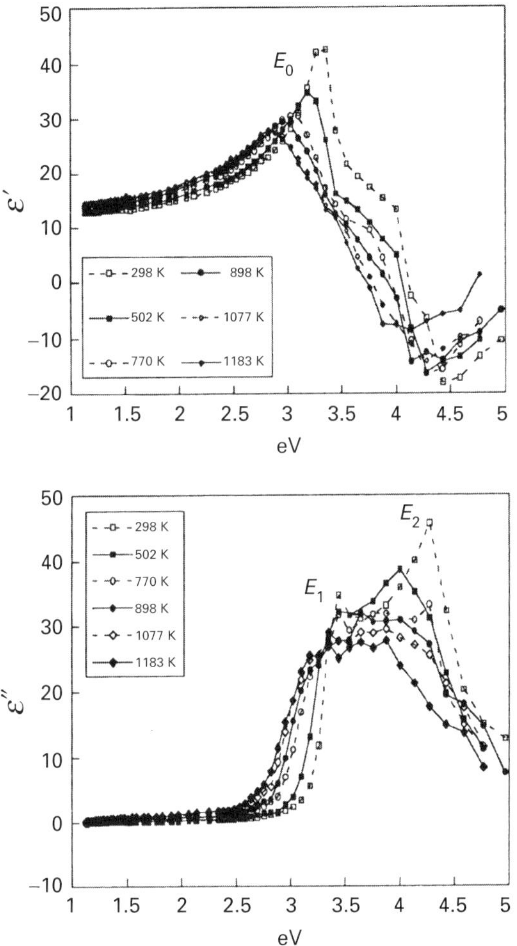

Figure 1.15. Spectral temperature dependences of the components of the complex dielectric function $\varepsilon^c = \varepsilon_0(\varepsilon' - i\varepsilon'')$ of pure single-crystalline silicon. From Sun *et al.* (1997), reproduced with permission from Elsevier.

E_2 peak in ε'' near 4.4 eV is considered to be due to several critical points, including the transitions $\Sigma_2^v \to \Sigma_3^c$. All these features become broader as the temperature increases.

Photon absorption in semiconductor material strongly depends on the interaction of the incident photon flux, the electronic structure, and the lattice dynamics of the semiconductor. Owing to the collective electronic interaction in crystalline silicon, an electronic indirect band gap of 1.12 eV exists, with the Fermi level usually placed between the conduction band and the valence band. On the other hand, the lattice phonon spectrum is generated by the finite-temperature field experienced by the crystalline silicon. Incident photons interact with the crystalline silicon via three routes: photon–phonon interaction, photon–electron interaction (including both conducting and valence electrons), and phonon–electron–photon interaction. Light absorption is essentially a result of these interactions in solid silicon. These interactions are often influenced by each other during the absorption process. As the incident photon energy, $h\nu$, increases, i.e. the wavelength decreases, an increasing number of valence electrons can be excited to the conduction band. More interestingly, electrons in the deep valence band or the lower valence band surface that are shifted from the center of the $E\!-\!k$ diagram can be excited directly to the other conduction band without phonon-assisted excitation in cases of sufficiently high photon energies. This phenomenon can be seen for all temperatures from 298 K to 1183 K. However, the relative decrease in the absorption coefficient observed for photon energies higher than 4.0 eV may be related to the existence of multiple conduction bands in k-space. The other important issue in photon absorption is the dependence on temperature. As stated earlier, the nature of the indirect band gap in silicon requires phonon-assisted electron excitation by the photon from the valence band to the conduction band, in order to conserve both energy and momentum. Therefore, the phonon spectrum in silicon at a fixed lattice temperature greatly influences the absorption process. Higher lattice temperature corresponds to higher-momentum phonons, thereby helping the excitation of more electrons to different conduction bands in the k-space. This is consistent with the experimental observation that the absorption coefficient increases with lattice-temperature elevation up to 4.0 eV photon energy. The opposite absorption trend with temperature above 4.0 eV was discussed above. The decrease of the band gap at elevated temperatures also enhances the electron excitation, resulting in a higher absorption coefficient. At very high lattice temperature, direct phonon absorption of photons can be realized. This provides an additional channel for absorption enhancement in crystalline silicon.

Detailed analysis of the fine structure of the absorption near the absorption edge has been given by McFarlane *et al.* (1958).

The optical band gap is also a function of temperature, given by Thurmond (1975):

$$E_{\mathrm{g}}(T) = E_{\mathrm{g}}^0 - \frac{AT^2}{\beta + T}, \tag{1.131}$$

where the value of the optical band gap at 0 K is taken at 1.17 eV, and the fitting parameters are $A = 4.73 \times 10^{-4}$ eV K^{-1} and $\beta = 635$ K.

Amorphous silicon and polysilicon

The optical properties of a-Si are sensitive to preparation conditions, surface conditions, surface oxides, hydrogen content, and also the degree of disorder in the sample. Figure 1.16 shows the temperature-dependent properties of 50-nm-thick a-Si films that were deposited by low-pressure chemical vapor deposition from silane (SiH_4) at 550 °C on quartz substrates (Moon *et al.*, 2000). The hydrogen content in these films was lower than 2%. Spectroscopic ellipsometry was performed over the range 250–1520 nm and up to a temperature of 671 K, in order to prevent crystallization. The measured spectra do not exhibit the sharp features of c-Si, but instead show a rather broad and smooth shoulder.

The structure and consequently the complex refractive index of silicon films deposited by chemical vapor deposition are in general dependent on the deposition conditions and

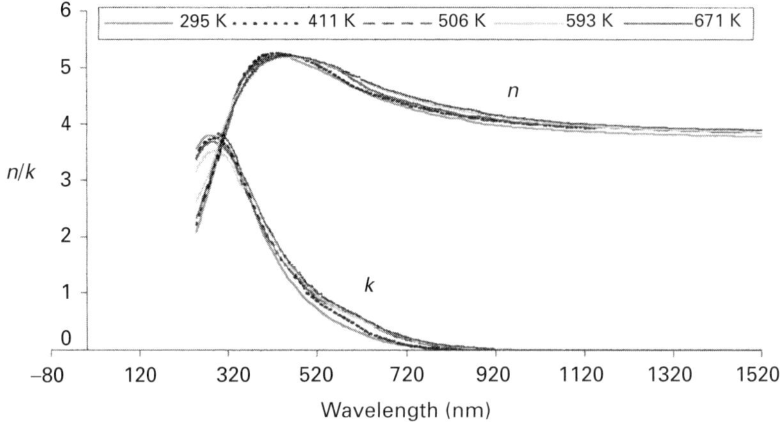

Figure 1.16. Temperature-dependent optical properties of a-Si. From Moon *et al.* (2000), reproduced with permission from Taylor and Francis.

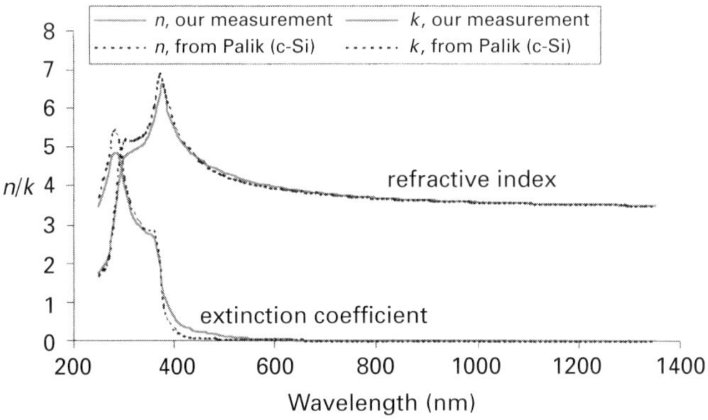

Figure 1.17. Comparison of room-temperature optical properties of laser-annealed poly-Si with those of crystalline silicon. From Moon *et al.* (2000), reproduced with permission from Taylor and Francis.

the post-processing annealing procedures. It is believed that polysilicon films can be modeled as mixtures of void fractions and amorphous and single-crystalline components using effective-medium theory (Montaudon *et al.*, 1985). Polysilicon films were grown from 50-nm-thick a-Si films by line-scanning a XeCl excimer laser beam of nanosecond pulse duration at wavelength $\lambda = 308$ nm. Thirty excimer laser pulses at the laser energy density (fluence) of 380 mJ/cm^2 were used to produce an average grain size of about 80 nm and maximum roughness less than 10 nm. As shown in Figure 1.17, the measured results are similar to data given in Palik (1985) for crystalline silicon.

Liquid silicon

Spectroscopic ellipsometry was applied by Shvarev *et al.* (1974) to measure the optical properties of liquid silicon in the wavelength range 0.4–1.0 µm at temperatures of 1450 and 1600 °C (the equilibrium melting temperature is 1412 °C). Within the range 0.4–1 μm, the measured components of the complex refractive index can be fitted by linear relations to $n(\lambda) = 1.8 + 5(\lambda - 0.4)$ and $k(\lambda) = 4 + 5(\lambda - 0.4)$, where λ is in micrometers. The conductivity, $\sigma(\omega) = nk(c/\lambda)$, increases with wavelength, reaching the value of $9.7 \times 10^{15}\,\mathrm{s}^{-1}$ at 1 μm, which is close to the dc conductivity of $10.9 \times 10^{15}\,\mathrm{s}^{-1}$, implying a short relaxation time. No evidence of interband absorption was revealed in these data that produced an excellent match with the Drude-model predictions.

References

Born, M., and Wolf, E., 1999, *Principles of Optics*, 7th edn, Cambridge, Cambridge University Press.

Jellison, G. E. Jr., and Modine, F. A., 1983, "Optical Functions of Silicon between 1.7 and 4.7 eV at Elevated Temperatures," *Phys. Rev. B*, **27**, 7466–7472.

Macfarlane, G. G., McLean, T. P., Quarrington, J. E., and Roberts, V., 1958, "Fine Structure in the Absorption-Edge Spectrum of Si," *Phys. Rev.*, **111**, 1245–1254.

Montaudon, P., Debroux, M. H., Ferrieu, F., and Vareille, A., 1985, "Optical Characterization of Polycrystalline Silicon Films before and after Oxidation," *Thin Solid Films*, **125**, 235–241.

Moon, S., Lee, M., Hatano, M., and Grigoropoulos, C. P., 2000, "Interpretation of Optical Diagnostics for the Analysis of Laser Crystallization of Amorphous Silicon Films," *Micro. Therm. Eng.*, **4**, 25–38.

Palik, E. D., 1985, *Handbook of Optical Constants of Materials*, Vols. I and II, London, Academic Press.

Prokhorov, A. M., Konov, V. I., Ursu, I., and Mihailescu, I. N., 1990, *Laser Heating of Metals*, Bristol, Adam Hilger, p. 17.

Shvarev, K. M., Baum, B. A., and Geld, P. V., 1974, "Optical Properties of Liquid Silicon," *Fiz. Tverd. Tela*, **16**, 3246–3248.

Sun, B. K., Zhang, X., and Grigoropoulos, C. P., 1997, "Spectral Optical Functions of Silicon in the Range of 1.13 to 4.96 eV at Elevated Temperatures," *Int. J. Heat Mass Transfer*, **40**, 1591–1600.

Thurmond, C. D., 1975, "Standard Thermodynamic Functions for Formation of Electrons and Holes in Si, GaAs and GaP," *J. Electrochem. Soc.*, **122**, 1133.

2 Lasers and optics

2.1 Lasers for materials processing

Lasers (the acronym from light amplification by stimulated emission of radiation), with their unique coherent, monochromatic, and collimated beam characteristics, are used in ever-expanding fields of applications. Different applications require laser beams of different pulse duration and output power. Lasers employed for materials processing range from those with a high peak power and extremely short pulse duration to lasers with high-energy continuous-wave output.

2.1.1 Continuous-wave – millisecond – microsecond lasers

Continuous-wave (CW) and long-pulsed lasers are typically used to process materials either at a fixed spot (penetration material removal) or in a scanning mode whereby either the beam or the target is translated. Millisecond- and microsecond-duration pulses are produced by chopping the CW laser beam or by applying an external modulated control voltage. Fixed Q-switched solid-state lasers with pulse durations from tens of microseconds to several milliseconds are often used in industrial welding and drilling applications. Continuous-wave carbon dioxide lasers (wavelength $\lambda = 10.6\,\mu\mathrm{m}$ and power in the kilowatt range) are widely employed for the cutting of bulk and thick samples of ceramics such as SiN, SiC, and metal-matrix ceramics (e.g. Duley, 1983). Continuous-wave laser radiation allows definition of grooves and cuts. On the other hand, low-power CO_2 lasers in the 10–150-W range are used for marking of wood, plastics, and glasses. Argon-ion lasers operating in the visible range ($\lambda = 419$–514 nm) are utilized for trimming of thick and thin resistors. In the biomedical field various CW lasers have been used. For example, the CO_2 laser radiation is absorbed in the tissue within a layer of depth about 20 μm, thus achieving a continuous ablation front. On the other hand, the Nd : YAG ($\lambda = 1064$ nm) and argon-ion radiation penetration is of the order of millimeters, giving rise to explosive ablation events. Srinivasan *et al.* (1995) used an argon-ion laser operating in the UV ranges of 275–305 and 350–380 nm to ablate polyimide Kapton films. By chopping the laser beam to produce millisecond and microsecond pulses, they showed that the ablation process scales with an intensity threshold rather than the commonly used fluence threshold. This is certainly not surprising, since, if the laser energy is spread over a long pulse, the beam intensity weakens and the induced temperature and structural and chemical response of the target may be of different nature. In fact, Srinivasan (1992) showed that the etching of polymer films with long, millisecond–microsecond pulses leaves evidence of molten material and carbonization of the walls, although not indicative of the ablation process that characterizes nanosecond-pulse UV laser ablation. Microfabrication applications involving direct writing can be effected by CW Ar^+ and Kr^+ lasers, utilizing frequency-doubled lines. High-power CW Nd : YAG lasers operating on the fundamental wavelength and on frequency-doubled and -tripled harmonics are often used for various cutting and microprocessing applications.

2.1.2   Nanosecond lasers

Technological development in the manufacturing of gas and solid-state lasers has greatly advanced in terms of reliability and in many cases has enabled the transition from the laboratory environment to industrial applications. Many pulsed laser ablation experiments have used excimer lasers (usually KrF at $\lambda = 248$ nm, but also XeCl at $\lambda = 308$ nm and ArF at $\lambda = 193$ nm) with pulse duration in the range 20–30 ns, maximum pulse energy in the range of 0.25–1 J, and pulse repetition rate typically 5–300 Hz. Since most materials are strong absorbers of UV-wavelength radiation, the excimer-laser light is absorbed in a very shallow region near the irradiated material surface. On the other hand, the very short duration of the laser pulse brings the peak power up to 10^{10} W/cm^2. These two features make the excimer laser a successful tool for initiating photochemical and/or photothermal ablation. Thus, the excimer laser is the most efficient ablation tool operating in the UV range for precision micromachining and surface patterning (Brannon, 1989; Horiike *et al.*, 1987; Patzel and Endert, 1993), chemical or physical modification (Rothschild and Ehlrich, 1988; Phillips *et al.*, 1993; Srinivasan and Braren, 1989), and via-hole formation in electronic-circuit packaging (Lankard and Wolbold, 1992). On the other hand, pulsed laser deposition (PLD) using excimer lasers has enabled fabrication of novel thin-film materials of high quality and superior properties than those obtainable with conventional manufacturing techniques. This method is comprehensively reviewed in Chrisey and Hubler (1994). *Q*-switched Nd : YAG lasers with pulse duration of about 7–10 ns, pulse energies in the near-IR wavelength of $\lambda = 1064$ nm, of power typically from 10 mJ to about 1 J, and with repetition rates of 10 Hz are versatile ablation tools since they can operate at various wavelengths. Frequency-doubled pulses at $\lambda = 532$ nm, tripled pulses at $\lambda = 355$ nm, and quadrupled pulses at $\lambda = 266$ nm carry respectively lower energies. Pulsed laser deposition can be effected with solid-state lasers; for example Davanloo *et al.* (1992) reported the production of amorphic diamond (diamondlike) films using *Q*-switched Nd : YAG laser radiation at $\lambda = 1064$ nm. A relatively inexpensive ablation tool is the transversely excited atmospheric pressure (TEA) CO_2 laser, that generates low-repetition-rate high-energy pulses in the kilojoule range, while it provides low-energy ($\sim$1 J) pulses at the 1 kHz repetition rate. The pulse has a short high-energy spike 100–200 ns wide and a longer trailing component of duration 1–10 ms; the two parts may carry comparable energies. Another cost-effective ablation laser for applications requiring relatively low energies is the N_2 laser that operates at $\lambda = 337$ nm, with pulse duration from 7–10 ns, pulse energies in the 100-mJ range to <10 mJ, and repetition rates of about 10 Hz. Diode-pumped solid-state lasers such as Nd : YLF and Nd : YAG and having pulse energies of hundreds of millijoules and operating on the fundamental or frequency-doubled wavelengths are attractive for micromachining applications because of their small size, flexibility, and high repetition rates (tens of kilohertz). For ablation of biological materials, free-running solid-state lasers are often used. They have long pulse duration of hundreds of milliseconds, which can be shortened by Q-switching to tens of nanoseconds, with corresponding energies in the 1-J range and tens of millijoules. Typical crystals are Ho : YAG ($\lambda = 2.1\,\mu\text{m}$) and Er : YAG ($\lambda = 2.94\,\mu\text{m}$) modules, with respective radiation-penetration depths in water of about 40 μm and 1 μm, thus achieving different absorption characteristics. For reference, it is noted that the wavelength 2.94 μm is located right at the peak absorption in water. In general, the nature of the ablation process in terms of angular distributions and energies of the ejected particles depends on the laser wavelength. Even though the structure and properties of the target material obviously affect the outcome, Sappey and Nogar (1994) note that, for comparable energies, long wavelengths usually produce thermal behavior, whereas UV ablation exhibits nonthermal characteristics and intermediate visible wavelengths yield results whose interpretation may be ambiguous.

2.1.3 Picosecond lasers

Whereas the nanosecond time scale is much longer than the characteristic relaxation times in metallic systems, invoking the thermal picture, the picosecond regime is still longer, but comparable. It has been claimed that picosecond-laser ablation of multicomponent targets offers the distinct advantage of preserving the target's stoichiometry in the chemistry of the ejected plume. It is noted that collisional and chemical-reaction effects in the target phase may introduce departures, since conflicting evidence has been presented. Most of the ablation work with picosecond lasers is done with pulsed solid-state lasers. For example, a 35-ps Nd : YAG laser producing 15 mJ at $\lambda = 1064\,\text{nm}$ and 10 mJ at $\lambda = 266\,\text{nm}$ was used to ablate Cu by Mao *et al.* (1993). A mode-locked Nd : YAG laser with pulse duration 50 ps, operating at $\lambda = 532\,\text{nm}$, was used by Bostanjoglo *et al.* (1994) to ablate free-standing metal films of thickness 50–90 nm under fluences in the range 0.6–8 J/cm^2. Fundamental studies on the picosecond-laser–plasma interactions were for example conducted using a Nd : glass laser system based on the chirped pulse amplification and compression (CPAC) technique that yielded 1.3-ps, 1.05-μm pulses with an average energy of 10 mJ (Chen *et al.*, 1993). For pulsed-laser processing in the IR range, the free-electron laser (FEL) offers tunability and high power as has been demonstrated for advanced materials-science (Brau, 1988) and medical (Danly *et al.*, 1987) applications.

2.1.4 Femtosecond lasers

In the sub-picosecond or femtosecond regime, the laser pulse is shorter than the relaxation times, and the equilibrium assumption is no longer valid, necessitating treatment of the microscopic mechanisms of energy transfer via quantum mechanics. One notable characteristic of femtosecond lasers is the high radiation intensity that has the ability to create high-density plasmas. On the other hand, by beating the thermal diffusion time scale, femtosecond-laser radiation can in principle be used for micromachining with minimal thermal damage to the surrounding area. In the UV range, KrF ($\lambda = 248$ nm) excimer lasers with typical pulse duration 500 fs and pulse energies in the range several to tens of millijoules have been demonstrated in the processing of Al and glassy C (Sauerbrey *et al.*, 1994), Ni, Cu, Mo, In, Au, W (Preuss *et al.*, 1995), fused silica (Ihlemann, 1992), and ceramics such as Al_2O_3, MgO, and ZrO_2 (Ihlemann *et al.*, 1995), and for polymer ablation (Bor *et al.*, 1995; Wolff-Rottke *et al.*, 1995). KrF excimer lasers have also been used in studies of high-density-gradient Al and Au plasmas (Fedosejevs *et al.*, 1990) and for production of soft X-rays from Al (Teubner *et al.*, 1993). The latter was also accomplished from Cu and Ta targets by near-IR Ti : sapphire-laser irradiation at $\lambda = 807$ nm, with pulse duration 120 fs and pulse energy 60 mJ (Kmetec *et al.*, 1992). A Ti : sapphire system with pulse duration of 150 fs and $\lambda = 770$ nm was used in studies of gold ablation (Pronko *et al.*, 1995), while pulse durations of 170–200 fs at wavelength $\lambda = 798$ nm and energy 4 mJ ablated polymers through a multiphoton ablation mechanism. Ti : sapphire-laser technology is often utilized in the laboratory environment and has recently made inroads into industrial applications, for example in the repair of lithographic masks. Amplified systems typically deliver 1-mJ near-IR pulses at maximum frequency 1 kHz. By utilizing stronger pumping lasers it is possible to extract pulses in the range tens of millijoules, albeit at reduced frequencies. On the other hand, the laser system could be configured to deliver microjoule or picojoule pulses at higher frequencies. Intense, visible dye-laser radiation (pulse duration 160 fs, $\lambda = 616$ nm, energy 5 mJ) generated Si plasmas of high-energy-X-ray-emitting density (Murnane *et al.*, 1989). In a biomedical application (Kautek *et al.*, 1994), a dye laser (pulse duration 300 fs, $\lambda = 615$ nm, pulse energy >0.18 mJ) produced high-quality ablation in human corneas, characterized by damage zones less than 0.5 μm wide. An interesting development in high-repetition-rate femtosecond-laser systems has been the development of (fiber) lasers that may give megahertz repetition rates with hundreds of nanojoules per pulse or correspondingly pulse rates of $\sim$100 kHz at tens of microjoules per pulse at $\lambda = 1,064$ nm with pulse duration $\sim$200 fs. Similar specifications have been achieved with disk or directly diode-pumped high-energy femtosecond oscillators. These lasers are utilized for film micromachining and the internal writing of optical waveguide structures in transparent materials.

2.2 Some specific laser systems

2.2.1 CO_2 lasers

The CO_2 laser has very high conversion efficiency of 20%–30%. The emission is due to rotational lines in the 9.4–10.4-μm bands of the CO_2 molecule. Dispersive elements, such as gratings, are used to select the desired wavelength. With no wavelength selection, operation occurs with peak intensity in the neighborhood of 10.6 μm. Pumping is accomplished either by electron impact or via collisions with vibrationally excited background N_2 gas. Continuous-wave lasing is triggered by longitudinal electrical discharge through flowing gas in an axial tube configuration. Electrical pulsing can produce trains of millisecond- or microsecond-long pulses, while electro-optical or acousto-optical intensity modulation can yield trains of $O(ns)$-long pulses. Because of practical limits in the length of the laser cavity, commercial CW CO_2 lasers in the 500–1000-W range use a folded-tube configuration and axial flow conditions. High CW CO_2 laser powers in the range of tens of kilowatts can be obtained by fast flow transport. Conversely, low powers can be obtained by discharge confined within a waveguide. Sealed-off cavities can be used for laser powers in the range of 10 W. Pulsed operation is triggered in transversely excited atmospheric-pressure (TEA) lasers, where the discharge is applied transverse to the optical axis. After an elapsed time of milliseconds a strong spike pulse of width 100–200 ns is emitted, followed by a longer and lower-amplitude tail that lasts for tens of microseconds.

The quality, symmetry, beam profile, and stability of the CO_2 laser are affected by gradients of the refractive index within the cavity. These gradients are caused by temperature and pressure gradients in the flow. Slow-axial-flow lasers exhibit large power and wavelength instabilities, whereas fast-flow lasers have more stable performance but are more complicated mechanically.

2.2.2 Argon-ion lasers

Continuous-wave Ar^+ lasers emit a bunch of wavelengths (more than 15) in the spectral range 351–514.5 nm. The stronger lines are in the visible at 514.5, 488, and 465.8 nm. Single-line operation at 514.5 nm can produce power of tens of watts. However, such power levels are reached at the cost of a requirement for significant electrical energy, since the efficiency is about 0.1%. Continuous-wave ion lasers have collisionally broadened linewidths of ~ 10 GHz, meaning that many longitudinal modes will appear for a resonator length of 1 m. Selection of a single lasing frequency can then be done using an internal Brewster prism that may be integrated with a reflective resonator mirror. The large linewidth of the Ar^+-laser transitions implies that mode-locking can be done with pulse lengths of ~ 100 ps.

2.2.3 Excimer lasers

Excimers (excited dimers) are diatomic molecules that exist only in an electronically excited state and dissociate in the ground state. This fundamental characteristic is particularly useful, since inversion between the upper bound state and the dissociating lower state is automatically maintained. Figure 2.1 shows a simplified schematic diagram with the general features of the transition process. Electron excitation of the neutral atoms $A + B$ leads to $A^* + B$ and the bound state $(AB)^*$ having energy level of several electron-volts. The bound state then falls to an unstable ground state that cannot survive and breaks apart to neutral atoms.

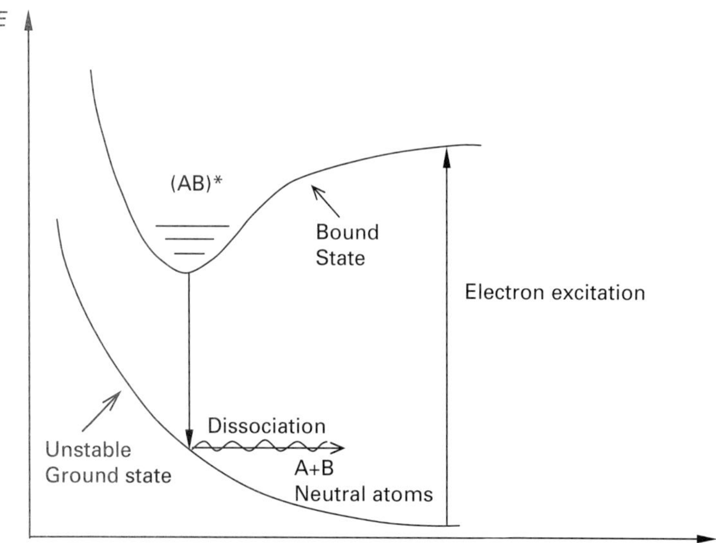

Figure 2.1. A schematic diagram of the transition process in an excimer laser.

The excimer molecules used are rare-gas halides (RGHs), including ArF (193 nm), KrCl (222 nm), KrF (248 nm), XeBr (282 nm), XeBr (282 nm), XeCl (308 nm), and XeF (351 nm). The pulse lengths are typically in the range of tens of nanoseconds. A fast switching device regulates the high-voltage discharge. The KrF excimer-laser beam output yields the strongest emission, with pulse energies of the order of 1 J, and maximum repetition rates of 500 Hz. The temporal shape of the laser intensity profile varies with the laser system. Concerns limiting widespread implementation of excimer lasers stem from the toxic nature of the halogen gases and the limited gas lifetime due to accumulation of impurities in the laser cavity. The latter has been significantly improved with installation of gas processors. One has to recognize that excimer lasers produce powerful pulses in the UV range that may be advantageous for precision micromachining, especially of organic materials. As will be later discussed in a subsequent chapter, interaction of UV pulses with polymers triggers unique photo-physico-chemical mechanisms.

2.2.4   Nd : YAG lasers

The crystal rods are a neodymium (Nd^{3+})-doped yttrium aluminum garnett, $(Y_3Al_5O_{12})_{13}$, matrix. This material has excellent thermal stability and properties (thermal conductivity) suitable for both short-pulse and CW operation. Pumping of these lasers is done by arc lamps, such as krypton or xenon lamps, or diode lasers. The output of the Nd : YAG laser is relatively insensitive to temperature, lending itself to wide use for medium- to high-power applications. The dominant laser emission is in the near-IR spectrum at the fundamental 1ω wavelength of 1064 nm. Higher harmonics can be generated via nonlinear crystals, namely 2ω at 532 nm, 3ω at 355 nm, and 4ω at 266 nm.

However, the efficiency of the lasing process tends to decrease rapidly for the higher harmonics, while the pulse-to-pulse stability also deteriorates.

2.2.5 Ti : sapphire lasers (femtosecond lasers)

Generation of femtosecond laser pulses requires a medium with a broadband gain spectrum together with an active or passive mode-locking mechanism. Since the late 1980s, titanium (Ti)-doped sapphire crystals have been a primary medium of choice owing to its broad gain bandwidth to support femtosecond pulses. Through a Kerr-lens mode-locking (KLM) mechanism, laser pulses with only two optical cycles at full-width half-maximum (FWHM) with center wavelength 800 nm have been generated with Ti : sapphire laser oscillators (Sutter *et al.*, 1999; Ell *et al.*, 2001). Kerr-lens mode-locking, which was developed in 1990, applies the nonlinear optical Kerr effect in the laser gain medium (refractive index increases with intensity) to lock the phase between different cavity modes. The pulse energy obtained from mode-locked femtosecond laser oscillators is typically a few nanojoules, which is insufficient to induce most nonlinear optical and electronic processes involved in laser–material interactions. However, direct amplification of femtosecond-laser pulses could lead to damage of the laser gain medium as well as optical components due to their high peak intensity. The chirped pulse amplification (CPA) scheme invented in the 1980s (Strickland and Mourou, 1985) avoids such damage by inducing a positive chirp that stretches the femtosecond pulses in time before amplification, followed by compressing the amplified pulses back to the initial short pulse. For instance, a 100-fs laser pulse with energy of 1 nJ can be stretched to 100 ps through a grating stretcher with large positive group-velocity dispersion, before being amplified to typically 1 mJ in a regenerative amplifier consisting of an optical cavity with an optical relay to regulate the number of passes through the amplifying medium. The amplified pulse is eventually compressed back to the initial 100 fs (1 mJ) in a grating compressor with matching negative group-velocity dispersion. Applications of CPA to femtosecond lasers have enabled the construction of multi-terawatt photon sources.

2.3 Basic principles of laser operation

2.3.1 Light amplification

The basic laser structure consists of an active optical gain medium that amplifies electromagnetic waves, a pumping source that pumps energy into the active medium, and an optical resonator that is composed of two highly reflective mirrors. The pump source could be a flash lamp, a gas discharge, or an electrical current source. Using radiative transition between atomic energy levels as an example, under thermodynamic equilibrium at temperature T, the population number density N_i, i.e., the number of atoms per unit volume at the ith energy level (Demtröder, 1996) is, according to Boltzmann statistics,

$$N_i = N \frac{g_i}{Z} \mathrm{e}^{-\frac{E_i}{k_B T}}. \tag{2.1}$$

The statistical weights g_i satisfy

$$Z = \sum_i g_i \mathrm{e}^{-\frac{E_i}{k_B T}}, $$

and

$$N = \sum_i N_i.$$

The pumping action boosts the population $N(E_k)$ corresponding to the kth energy level to exceed the population $N(E_i)$ at the energy level E_i, thereby achieving inversion:

$$N_k = N(E_k) > N(E_i) = N_i. \tag{2.2}$$

The transition from higher to lower energy level $E_k \rightarrow E_i$ is accompanied by the release of a photon of frequency v:

$$\frac{E_k - E_i}{h} = v. \tag{2.3}$$

The corresponding absorption coefficient is

$$\gamma(v) = \left[N_i - \left(\frac{g_i}{g_k} \right) N_k \right] \sigma_c(v), \tag{2.4}$$

where $\sigma_c(v)$ is the cross section for the radiative transition. The axial intensity dependence in the active medium is

$$I(v, z) = I(v, 0) \mathrm{e}^{-\gamma(v) z}. \tag{2.5}$$

If $N_k > (g_k/g_i) N_i$, the absorption coefficient becomes negative, implying amplification rather than attenuation. The intensity after a round-trip between the resonator mirrors that are spaced at the distance L_{cav} would be

$$I(v, 2L_{\mathrm{cav}}) = I(v, 0) \exp(-2\gamma(v) L_{\mathrm{cav}} - \gamma_{\mathrm{loss}}), \tag{2.6}$$

where the factor γ_{loss} lumps together the losses induced by the partial reflectivity of the mirrors, attenuation by impurities and contamination in the active medium, diffraction and scattering of the propagating laser beam, etc.

For amplification, $-2\gamma(v) L_{\mathrm{cav}} - \gamma_{\mathrm{loss}} > 0$. Considering (2.4), a population threshold for lasing is defined by

$$\Delta N = N_k \frac{g_i}{g_k} - N_i > \Delta N_{\mathrm{thr}} = \frac{\gamma_{\mathrm{loss}}}{2\sigma_c(v) L_{\mathrm{cav}}}. \tag{2.7}$$

2.3.2   Circular Gaussian beams in a homogeneous medium

The spatial profile of a laser beam at the laser exit aperture is determined by the geometry of the laser cavity. When the cross section of the cavity is symmetrical, as in the case of cylindrically or rectangular shaped cavities, the spatial profiles become simple. Transverse electromagnetic modes are characterized by TEM_{mn}, where m and n indicate the number of modes in two orthogonal directions. The mode of highest symmetry is the TEM_{00} mode. It can be shown (Yariv, 1971, 1989) that the electric field intensity for a TEM_{00} Gaussian laser beam that is propagating in the z-direction and has a circular profile is

$$E(x, y, z) = E_0 \frac{w_0}{w(z)} \exp\left\{ -i(k_w z - \eta(z)) - r^2 \left[\frac{1}{w^2(z)} + \frac{ik_w}{2R_w(z)} \right] \right\}, \qquad (2.8)$$

where $r = \sqrt{x^2 + y^2}$, $k_w = 2\pi n / \lambda$, and

$$w^2(z) = w_0^2 \left(1 + \frac{z^2}{z_R^2} \right), \qquad (2.9a)$$

$$R_w(z) = z \left(1 + \frac{z_R^2}{z^2} \right), \qquad (2.9b)$$

$$\eta(z) = \tan^{-1} \left(\frac{z}{z_R} \right), \qquad (2.9c)$$

$$z_R = \frac{\pi w_0^2 n}{\lambda}. \qquad (2.9d)$$

The parameter $w(z)$ represents the distance at which the field amplitude drops by a factor $1/e$ compared with its value on the beam axis. At that location, the beam intensity drops by a factor of $1/e^2$. The parameter w_0 is the minimum beam-spot size, located at the plane $z = 0$. The analogy to radiation emitted from a point source located at the origin of the coordinate system may be drawn. The complex electric field is

$$E \propto \frac{e^{-ik_w R}}{R} = \frac{1}{R} \exp\left(-ik_w \sqrt{x^2 + y^2 + z^2} \right). \qquad (2.10)$$

Far away from the point source and for $z \gg \sqrt{x^2 + y^2}$, the above is approximated by

$$E \propto \frac{1}{R} \exp\left(-ik_w z - ik_w \frac{x^2 + y^2}{2z} \right) \simeq \frac{1}{R} \exp\left(-ik_w z - ik_w \frac{x^2 + y^2}{2R} \right). \qquad (2.11)$$

Comparison of (2.8) with (2.11) indicates that the parameter R in (2.8) may be viewed as the radius of curvature of the wavefronts at z. The quantity z_R signifies the distance at which the diameter of the laser beam changes by a factor of $\sqrt{2}$ and is called the Rayleigh length. Furthermore, $dR/dz|_{z=z_R} = 0$, indicating that the absolute value of the radius of curvature assumes an infinite value at $z = 0$, passes through a minimum at $z = z_R$, and then increases linearly with z for $z = z_R$. It is noted that the wavelength λ and the minimum focal beam waist w_0 completely specify the shape of the propagating laser beam.

The depth of focus is given by the following relation:

$$d_{\text{dof}} = \frac{\pi w_0^2}{\lambda} \sqrt{\chi_{\text{dof}}^2 - 1}. \qquad (2.12)$$

On setting the acceptable focus to be within 2%, i.e. $\chi_{\text{dof}} = w(z = d_{\text{dof}})/w_0 = 1.02$, the depth of focus is estimated to be about 60 μm for $w_0 = 10$ μm and $\lambda = 1.064$ μm. For large z, the hyperboloids $x^2 + y^2 = \text{const} \times w^2(z)$ are asymptotic to the cone of half-apex angle $\theta_{\text{beam}} = \lambda/(\pi w_0)$. This angle indicates the beam half-divergence in radians. At long distances, $z \gg z_{\text{R}}$, the beam size is therefore given by $w(z) = z\theta_{\text{beam}}$.

2.3.3   Elliptic Gaussian beams in a homogeneous medium

The wave equation allows solutions in which the variation in the x and y directions is given by

$$E \propto \exp\left(-\frac{x^2}{w_x^2} - \frac{y^2}{w_y^2}\right) \tag{2.13}$$

with $w_x \neq w_y$. Specifically, the spatial beam-intensity profile will be given by

$$E(x, y, z) = E_0 \frac{\sqrt{w_{0x} w_{0y}}}{\sqrt{w_x(z)w_y(z)}} \exp\Bigg[- \mathrm{i}(k_w z - \eta(z))$$
$$- x^2\left(\frac{1}{w_x^2(z)} + \frac{\mathrm{i}k_w}{2R_{wx}(z)}\right) - y^2\left(\frac{1}{w_y^2(z)} + \frac{\mathrm{i}k_w}{2R_{wy}(z)}\right)\Bigg]. \tag{2.14}$$

The elliptically shaped beam can be viewed as a composition of two independent circular Gaussian beams in the x- and y-directions. The x-beam waist and radius of curvature are

$$w_x^2(z) = w_{0x}^2\left[1 + \left(\frac{\lambda(z - z_x)}{\pi w_{0x}^2 n}\right)^2\right], \tag{2.15a}$$

$$R_{wx}(z) = (z - z_x)\left[1 + \left(\frac{\pi w_{0x}^2 n}{\lambda(z - z_x)}\right)^2\right]. \tag{2.15b}$$

Analogous expressions apply for the y-beam. The phase delay is

$$\eta(z) = \frac{1}{2}\tan^{-1}\left(\frac{\lambda(z - z_x)}{\pi w_{0x}^2 n}\right) + \frac{1}{2}\tan^{-1}\left(\frac{\lambda(z - z_y)}{\pi w_{0y}^2 n}\right). \tag{2.16}$$

2.3.4   Higher-order beams

If the condition of azimuthal symmetry is removed, the spatial distribution of the electric field allows higher-order modes:

$$E_{l,m}(x, y, z) = E_0 \frac{w_0}{w(z)} H_l\left(\sqrt{2}\frac{x}{w(z)}\right) H_m\left(\sqrt{2}\frac{y}{w(z)}\right)$$
$$\times \exp\Bigg[-\frac{x^2 + y^2}{w^2(z)} - \frac{\mathrm{i}k_w(x^2 + y^2)}{2R_w(z)} - \mathrm{i}k_w z + \mathrm{i}(l + m + 1)\eta\Bigg], \tag{2.17}$$

where $\eta = \tan^{-1}(z/z_{\text{R}})$ and $z_{\text{R}} = \pi w_0^2 n/\lambda$. The functions H_l and H_m are Hermite polynomials of orders l and m.

The phase-shift along the beam-propagation axis z is

$$\vartheta(z) = k_w z - (l + m + 1)\tan^{-1}\left(\frac{z}{z_{\mathrm{R}}}\right). \tag{2.18}$$

2.3.5 Optical resonators with spherical mirrors

Consider a circular Gaussian beam formed in a laser cavity with the spatial profile given by Equation (2.9). Suppose that a resonator is formed by two reflectors, placed at the locations z_1 and z_2, so that their radii of curvature R_{w1} and R_{w2} are the same as those of the wavefronts at the respective locations:

$$R_{w1} = z_1 + \frac{z_{\mathrm{R}}^2}{z_1}, \qquad z_1 = +\frac{R_{w1}}{2} \pm \frac{1}{2}\sqrt{R_{w1}^2 - 4z_{\mathrm{R}}^2}, \tag{2.19a}$$

$$R_{w2} = z_2 + \frac{z_{\mathrm{R}}^2}{z_2}, \qquad z_2 = +\frac{R_{w2}}{2} \pm \frac{1}{2}\sqrt{R_{w2}^2 - 4z_{\mathrm{R}}^2}. \tag{2.19b}$$

If the mirrors are separated by the distance L_{cav}, the Rayleigh length is determined by

$$z_{\mathrm{R}}^2 = \frac{L_{\mathrm{cav}}(-R_{w1} - L_{\mathrm{cav}})(R_{w2} - L_{\mathrm{cav}})(R_{w2} - R_{w1} - L_{\mathrm{cav}})}{(R_{w2} - R_{w1} - 2L_{\mathrm{cav}})^2},$$

$$w_0 = \left(\frac{\lambda z_{\mathrm{R}}}{\pi n}\right)^{\frac{1}{2}}. \tag{2.20}$$

A symmetrical resonator is formed if $R_{w2} = -R_{w1} = R_w$. In this case, the Rayleigh length is

$$z_{\mathrm{R}}^2 = \frac{(2R_w - L_{\mathrm{cav}})L_{\mathrm{cav}}}{4}. \tag{2.21}$$

For a confocal resonator, $R_w = L_{\mathrm{cav}}$ and the beam waist is

$$w_0|_{\mathrm{conf}} = \left(\frac{\lambda L_{\mathrm{cav}}}{2\pi n}\right)^{\frac{1}{2}}, \tag{2.22}$$

and the beam size at the mirrors is

$$w_{1,2} = \sqrt{2}w_0|_{\mathrm{conf}}. \tag{2.23}$$

2.3.6 Resonance frequencies of optical resonators

Consider a spherical resonator with mirrors at the locations z_1 and z_2 (Yariv, 1971). The resonance condition for the (l, m) mode is

$$\vartheta_{l,m}(z_2) - \vartheta_{l,m}(z_1) = \xi\pi, \tag{2.24}$$

where ξ is any integer, and the phase $\vartheta_{l,m}(z)$ is given by Equation (2.18). For a resonator length $L_{\mathrm{cav}} = z_2 - z_1$, the above equation gives

$$k_{w\xi}L_{\mathrm{cav}} - (l + m + 1)\left[\tan^{-1}\left(\frac{z_2}{z_{\mathrm{R}}}\right) - \tan^{-1}\left(\frac{z_1}{z_{\mathrm{R}}}\right)\right] = \xi\pi, \tag{2.25}$$

and consequently, for a fixed (l, m) mode, the wavenumber and frequency separation are

$$k_{w\xi+1} - k_{w\xi} = \frac{\pi}{L_{\text{cav}}}, \tag{2.26a}$$

$$\nu_{\xi+1} - \nu_{\xi} = \frac{c}{2n L_{\text{cav}}}. \tag{2.26b}$$

For a fixed ξ, higher-order modes will resonate as

$$k_{w1} L_{\text{cav}} - (l + m + 1)_1 \left[\tan^{-1}\left(\frac{z_2}{z_R}\right) - \tan^{-1}\left(\frac{z_1}{z_R}\right) \right] = \xi\pi, \tag{2.27a}$$

$$k_{w2} L_{\text{cav}} - (l + m + 1)_2 \left[\tan^{-1}\left(\frac{z_2}{z_R}\right) - \tan^{-1}\left(\frac{z_1}{z_R}\right) \right] = \xi\pi. \tag{2.27b}$$

The wavenumber spacing is

$$k_{w1} - k_{w2} = \frac{1}{L_{\text{cav}}}[(l + m + 1)_1 - (l + m + 1)_2]\left[\tan^{-1}\left(\frac{z_2}{z_R}\right) - \tan^{-1}\left(\frac{z_1}{z_R}\right) \right]. \tag{2.28}$$

The corresponding frequency separation, $\Delta\nu|_{l+m}$, caused by the change $\Delta(l + m)$ is

$$\Delta\nu|_{l+m} = \frac{c}{2\pi L_{\text{cav}} n}\left[\tan^{-1}\left(\frac{z_2}{z_R}\right) - \tan^{-1}\left(\frac{z_1}{z_R}\right) \right]. \tag{2.29}$$

For the confocal resonator, $z_2 = -z_1 = z_R$ and

$$\tan^{-1}\left(\frac{z_2}{z_R}\right) - \tan^{-1}\left(\frac{z_1}{z_R}\right) = \frac{\pi}{4}.$$

Hence, the frequency spacing becomes

$$\Delta\nu|_{l+m} = \frac{1}{2}\Delta(l + m)\frac{c}{2n L_{\text{cav}}}. \tag{2.30}$$

The frequencies generated by the mode-order change either coincide with the resonance frequencies or fall half-way between them (Figure 2.2).

2.3.7   Spectral characteristics of laser emission

The design of the laser resonator can be optimized to allow amplification of only the fundamental TEM_{00} mode, since the higher-order modes suffer severe scattering losses due to high divergence. The spectral content of the laser emission depends on the spectral width of the absorption line of the active medium. As shown in Figure 2.3, only frequencies exceeding the population threshold and contained within the frequency span $(\nu_0 - \delta\nu_{\text{thr}}, \nu_0 + \delta\nu_{\text{thr}})$ will be amplified. It is also apparent that these surviving frequencies will experience a different amplification gain as they are removed from the central resonance frequency. For a gas laser, the typical Doppler broadening width is $O(\text{MHz})$, limiting the number of lasing frequencies. In contrast, for solid-state or liquid

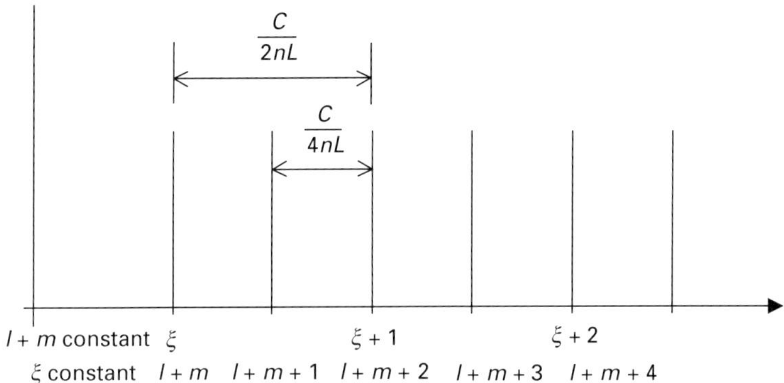

Figure 2.2. Location of the resonance frequencies in a confocal resonator as a function of the longitudinal mode ξ and the beam-profile modes m and n.

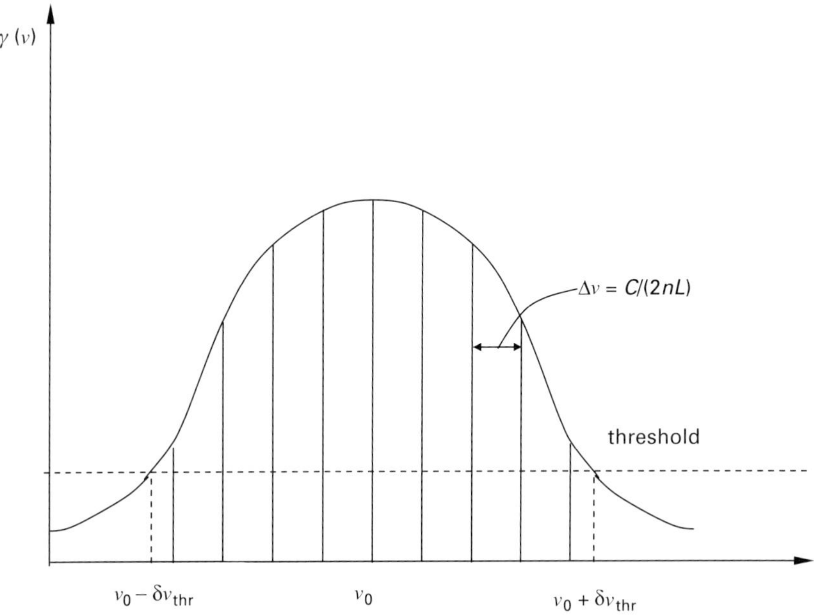

Figure 2.3. Surviving frequencies within the frequency span of a resonator.

(dye) lasers, the line width is larger, $O(\mathrm{GHz})$, thus allowing more longitudinal modal frequencies.

Emission of a single longitudinal mode can be forced by reducing the length of the resonator, so that $c/(2nL_{\mathrm{cav}}) > \delta\nu_{\mathrm{thr}}$. However, this method might not be practical, since the laser output intensity scales with the volume and length of the active medium. Selection of a single mode could then be done externally by a spectrometer or interferometer, but preferably internally by inserting a spectral selective device such as a filter, an etalon, a Michelson-type interferometer, or a grating into the laser cavity.

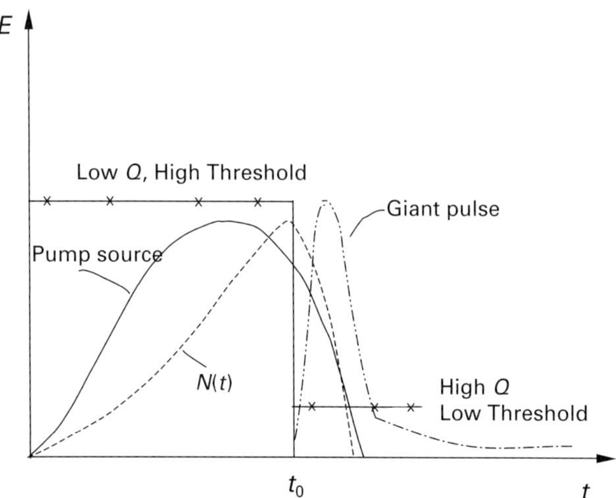

Figure 2.4. A schematic diagram of a Q-switching process.

2.3.8   *Q*-Switching

Short pulses can be produced by the Q-switching method schematically outlined in Figure 2.4. The pump pulse builds, on a rather slow time scale, a transient population density $N(t)$, but no amplification can happen because the losses and hence the population threshold are kept high (low quality factor, Q). At the time t_0, the losses are drastically reduced, the quality factor Q is switched to a high value, and the energy built into the high value of $N(t_0)$ is released to produce the so-called "giant" pulse.

The switching device may be an electro-optical (E-O) Q-switch based on the birefrigence effect of an electro-optical element that is characterized by different refractive indices for light polarized in two orthogonal directions. Electro-optical Q-switches used in practice employ either the Pockels-cell effect (longitudinal or transverse) or the Kerr effect. In longitudinal E-O Q-switches, the electric voltage is applied parallel to the cell. When the voltage is "off," the crystal is uni-axial, but, when the voltage is turned "on," the crystal becomes bi-axial, producing birefringence. The most frequently used crystals are KDP (0.4–0.8 μm), KD*P (0.25–1.2 μm), CDP, LiNbO$_3$ (1–4 μm), CdTe (2–4 μm), and GaAs (2–4 μm) with rise times of $\approx$1 ns. A drawback is that the operating voltage has to be in the kilovolt range, adding bulk to the device.

Acoustic waves propagating through elasto-optical solids modify the refractive index of the material. Consequently, a standing acoustic-wave pattern generated by a temporally modulated RF voltage source creates an optical grating that diffracts the laser beam away from the original direction of propagation. Common materials for the acousto-optical (A-O) Q-switches are quartz in the visible and near-IR ranges, and Te and GaAs in the IR range. Typical rise times are about 100 ns per millimeter thickness.

Mechanical Q-switches are formed by rotating, translating, or oscillating an optical element to interrupt the beam path. For example, loss of parallelism by only part of a milliradian inhibits lasing. Since the motion is mechanical, the switching action cannot be very fast, and the minimum rise time is tens of microseconds.

Saturable absorbers that normally absorb at the laser wavelength generate passive Q-switching. As the electromagnetic (EM) intensity increases, the absorption coefficient decreases due to depletion of ground-state molecules, and the material becomes transparent or bleached. Passive Q-switches in the visible and near-IR spectral ranges are usually based on organic-dye solutions, pigmented plastics, or colored glasses. Passive Q-switches in the far-IR are based on gas saturation absorbers.

2.3.9 Mode-locking

As outlined previously, laser oscillations take place at a number of frequencies, separated by

$$\omega_{\xi+1} - \omega_{\xi} = \frac{\pi c}{L_{\mathrm{cav}}} = \omega. \tag{2.31}$$

At any point inside the laser cavity, the electric field resulting from the multimode oscillation is

$$E(t) = \sum_n E_n e^{i[(\omega_0+n\omega)t+\vartheta_n]}. \tag{2.32}$$

It is therefore inferred that $E(t)$ is periodic with period

$$T_{\mathrm{p}} = \frac{2\pi}{\omega} = \frac{2L_{\mathrm{cav}}}{c},$$

i.e. the round-trip time in the oscillator. While the different modes generally oscillate randomly, useful results are achieved if the phases ϑ_n are fixed, say $\vartheta_n = 0$. For simplicity, it is assumed that there exist N_{mod} oscillating modes having equal amplitudes. The corresponding electric field is

$$E(t) = E_0 \sum_{-\frac{N_{\mathrm{mod}}-1}{2}}^{\frac{N_{\mathrm{mod}}+1}{2}} e^{i(\omega_0+n\omega)t} = E_0 e^{i\omega_0 t} \frac{\sin(N_{\mathrm{mod}}\omega t/2)}{\sin(\omega t/2)}. \tag{2.33}$$

The laser intensity is

$$I(t) \approx E(t)E^*(t) \approx \frac{\sin^2(N_{\mathrm{mod}}\omega t/2)}{\sin^2(\omega t/2)}. \tag{2.34}$$

The above expression represents a train of pulses with period T_{p} and peak power N_{mod} times the average power distributed over all the locked modes (Figure 2.5). If the spectral width of the laser line above threshold is $\Delta\omega_{\mathrm{thr}}$, the number of the locked modes is $N_{\mathrm{lock}} \approx \Delta\omega_{\mathrm{thr}}/\omega$. The individual pulse width is then

$$t_{\mathrm{pulse}} = \frac{T_{\mathrm{p}}}{N_{\mathrm{lock}}} \approx \frac{2\pi/\omega}{\Delta\omega_{\mathrm{thr}}/\omega} = \frac{2\pi}{\Delta\omega_{\mathrm{thr}}} = \frac{1}{\Delta\nu_{\mathrm{thr}}}.$$

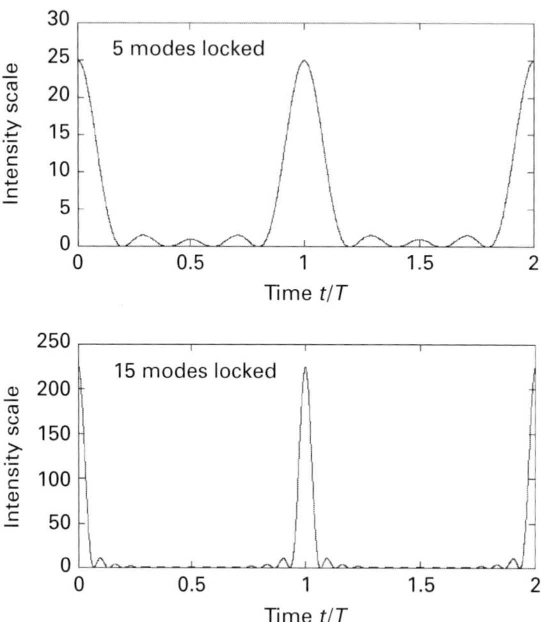

Figure 2.5. Plots of optical field amplitude resulting from phase locking of 5 and 15 laser modes.

For Nd : YAG lasers, the line frequency at $\lambda = 1064$ nm is $\Delta\nu_{\mathrm{thr}} = 1.2 \times 10^{10}$ s^{-1}, yielding a pulse length of $t_{\mathrm{pulse}} \sim 80$ ps. Gas lasers may have a lower line frequency width above the lasing threshold, in the range of megahertz, and can therefore produce mode-locked lengths only in the nanosecond range, whereas dye lasers can in principle produce mode-locked pulses as short as femtoseconds. Active mode-locking by an A-O or E-O device entails periodic modulation of the EM field at a frequency equal to the intermodal frequency ω.

2.4 Definition of laser intensity and fluence variables

2.4.1 Gaussian beams

Fundamental TEM$_{00}$ operation implies that the instantaneous distribution of laser intensity across the beam is Gaussian:

$$I(r, t) = I_{\mathrm{pk}}(t)\exp\left(-\frac{2r^2}{w^2}\right), \qquad (2.35)$$

where w is the radius of the point where the intensity drops by a factor of $1/e^2$ with respect to the peak intensity I_{pk} at $r = 0$ and r is the radial coordinate. Frequently, the radius at which the laser-beam intensity drops by a factor of $1/e$ is specified as $w_{1/e} = w/\sqrt{2}$.

2.4.2   Continuous-wave laser beams

Continuous-wave (CW) laser beams are those for which the laser intensity is constant with time, $I_{pk}(t) = I_{pk}$ except for transient fluctuations and the long-term drift:

$$I(r, t) = I_{pk} \exp\left(-\frac{2r^2}{w^2}\right). \tag{2.36}$$

The total laser power is defined by

$$P = \int_0^{+\infty} I(r)2\pi r \, dr = \frac{\pi w^2}{2} I_{pk}. \tag{2.37}$$

2.4.3   Pulsed laser beams

The dimensionless temporal pulse profile is characterized by

$$p(t) = \begin{cases} \dfrac{I_{pk}(t)}{I_{pk;max}}, & t < t_{pulse}, \\ 0, & t > t_{pulse}, \end{cases} \tag{2.38}$$

where $I_{pk;max}$ is the peak intensity at $t = t_{max}$. It is common to characterize pulse lengths with the FWHM pulse length, t_{FWHM}. This is the temporal width of the pulse evaluated at the intensity $I_{FWHM} = I_{pk;max}/2$.

The *local transient laser fluence* may be defined as follows:

$$F_{tr}(r, t) = \int_{-\infty}^{t} I(r, t')dt'. \tag{2.39}$$

This quantity represents the energy per unit area incident at a specific location, until the elapsed time t. The *total energy* carried by the laser pulse is given by

$$E_{pulse} = \int_{-\infty}^{+\infty} \int_0^{+\infty} I(r, t')2\pi r \, dr \, dt'. \tag{2.40}$$

The *pulse fluence*, F, is defined as the pulse energy divided by an area corresponding to a circular disk of radius w:

$$F = \frac{E_{pulse}}{\pi w^2}. \tag{2.41}$$

The intensity distribution $I(r, t)$ is

$$I(r, t) = \frac{2E_{pulse}}{\pi w^2} \frac{p(t)}{\int_{-\infty}^{+\infty} p(t') \, dt'} \exp\left(-\frac{2r^2}{w^2}\right). \tag{2.42}$$

For a triangular temporal profile,

$$p(t) = \begin{cases} 0, & t < 0, \\ \dfrac{t}{t_{max}}, & 0 < t < t_{max}, \\ \dfrac{t_{pulse} - t}{t_{pulse} - t_{max}}, & t_{max} < t < t_{pulse}. \end{cases} \tag{2.43}$$

Smooth pulses whose peak intensities lie at $t = t_{\text{max}}$ can be fitted by

$$p(t) = \left(\frac{t}{t_{\text{max}}}\right)^{\xi} \exp\left[\rho_{\text{temp}}\left(1 - \frac{t}{t_{\text{max}}}\right)\right], \tag{2.44}$$

where ρ_{temp} is a parameter characterizing the temporal profile. For a sinusoidally modulated CW laser beam

$$p(t) = 1 - \cos(\omega_{\text{mod}}t), \tag{2.45}$$

where ω_{mod} is the modulation frequency.

If a Gaussian beam of circular cross section is focused by a cylindrical lens, or if the beam is incident on the target at an oblique angle of incidence, the resulting profile is elliptical. On the other hand, the output of excimer, nitrogen, TEA CO_2, and solid-state lasers may have a roughly flat-topped cross section. In general, the laser intensity profile is non-Gaussian and the intensity is a function of the spatial coordinates (x, y) on the irradiated target plane and time: $I(x, y, t)$. The local transient laser fluence may be defined in a manner analogous to Equation (2.36):

$$F_{\text{tr}}(x, y, t) = \int_{-\infty}^{t} I(x, y, t')\mathrm{d}t'. \tag{2.46}$$

If A is the area of the irradiated spot, the pulse fluence is defined simply as

$$F = \frac{E}{A}. \tag{2.47}$$

The laser parameters regulating laser processing are the wavelength, λ, the polarization, and the intensity distribution $I(x, y, t)$ on the target surface. The temporal and spatial dependence of the intensity distribution depends on the mode structure and on the external modulation through the beam-delivery system.

2.5   Optical components

Spherical lenses are most commonly used in ablation systems. If beam expansion along one direction is needed, cylindrical lenses can be used. For example, the raw excimer-laser beam usually is of rectangular-elliptical cross section, with an aspect ratio of $\sim$3–5; a cylindrical lens can be used for expanding the shorter dimension to give a square cross section. The lens performance, with regard to the theoretical prediction, depends on lens aberrations: spherical aberration, coma, astigmatism, field curvature, and distortion. To reduce spherical aberration, apertures can be used to attenuate the beam rays diverging from the optical axis. Alternatively, a condensing plano-convex lens can be used, with the convex surface facing the incoming laser beam. Coma, which results from imaging light rays from off-axis points as ring-like structures, can be eliminated by control over the lens shape, as well as by utilizing apertures, again at the expense of some power loss. For demanding applications, custom-made lenses providing the necessary corrections may be necessary. In any case, the choice of the materials for optical components depends on the laser wavelength, energy level, and pulse-repetition rate.

Operation in the wavelength range from 350 to 1000 nm can be handled with quartz, pyrex, or other glasses. In the UV range, from 190 to 350 nm, UV-grade fused silica is adequate for relatively low repetition rates, but color formation and a significant transmission loss have been observed under prolonged operation at high repetition rates (Krajnovich $et\ al.$, 1992). In this case, single-crystalline quartz, CaF_2, or MgF_2 lenses must be used. For the near-IR Nd : YAG, Nd : YLF, Nd : glass lasers and the like, anti-reflection-coated quartz and glass lenses are normally adequate. However, these materials are not transparent farther into the infrared. The crystalline alkali halides NaCl and KCl, and various semiconductor materials, such as ZnSe, CdTe, GaAs, and Ge, are highly transmitting in the far IR, in the 10-μm spectral range of CO_2 laser radiation. While transmission is a major concern, high thermal conductivity, hardness, smoothness, and chemical resistance are also desirable when coatings need to be applied. Thin-film antireflective (AR) coatings typically reduce the surface reflectance to about 0.01.

The selection of mirrors must also be done carefully, to avoid damage by the incident laser radiation. Multilayer dielectric coatings are designed to enhance the reflectivity to 0.99 for the particular wavelength and laser-beam incidence angle. It should be cautioned that use of these mirrors at other wavelengths and incidence directions might have adverse effects, since the reflectance is decreased. Beamsplitters are also used to sample part of the beam for temporal profile measurement, to divert a portion of the beam for pump–probe schemes, or to share the laser beam among separate experiments.

For controlled ambient pressure and composition, but also to provide the appropriate chemical-reaction environment, for example in pulsed-laser-deposition systems, experiments are being conducted in vacuum chambers. The laser windows through which the laser enters the deposition chamber have to be made from high-quality optical materials. To avoid losses by scattering, the window surfaces must be optically smooth ($\lambda/10$ flatness). Deposition of ablated particles onto the windows may lead to gradual transmission loss and potentially to local damage.

2.6 Beam delivery

2.6.1 Gaussian beam focusing

Consider focusing of a laser beam by a thin lens of focal length, f (Figure 2.6). Let the incoming beam's half-divergence angle be θ_1, and assume that the beam waist w_{01} is formed at a distance z_1 from the lens, with $x_1 = z_1 - f$. The focused beam's focal waist w_{02} is formed at a distance z_2 ahead of the lens with $x_2 = z_2 - f$, and the half-divergence angle is θ_2. The thin-lens equation in Newtonian form is (O'Shea, 1985)

$$x_1 x_2 = (z_1 - f)(z_2 - f) = f^2 - f_0^2, \tag{2.48}$$

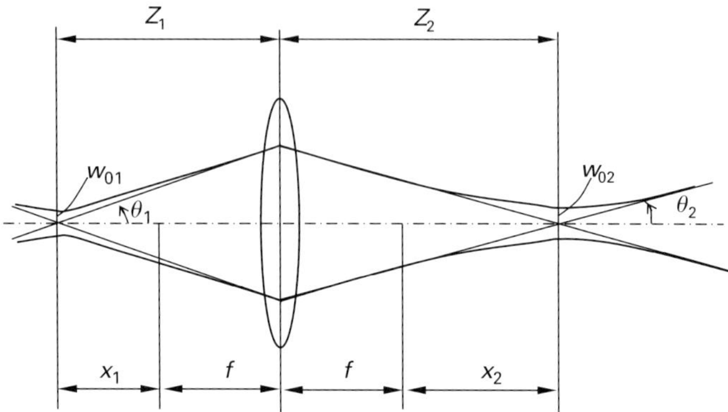

Figure 2.6. Focusing of a circular Gaussian beam with a thin spherical lens.

where the term f_0 accounts for diffraction effects and is given by

$$f_0^2 = z_{R1} z_{R2}. \tag{2.49}$$

The focused beam's parameters are

$$w_{02} = \alpha_{\text{foc}} w_{01}, \tag{2.50}$$

where

$$\alpha_{\text{foc}} = \frac{|f|}{\sqrt{(z_1 - f)^2 + z_{R1}^2}}. \tag{2.51}$$

The location of the focused beam's waist, the Rayleigh range, and the half-beam divergence are

$$z_2 = f + \alpha_{\text{foc}}^2 (z_1 - f), \tag{2.52}$$

$$z_{R2} = \alpha_{\text{foc}}^2 z_{R1}, \tag{2.53}$$

$$\theta_2 = \frac{w_{02}}{z_{R2}}. \tag{2.54}$$

Suppose that the incoming beam is collimated, i.e. $z_1 \gg f$. In this case,

$$\alpha_{\text{foc}} \cong \frac{f}{\sqrt{z_1^2 + z_{R1}^2}}. \tag{2.55}$$

Two limiting cases may be distinguished. When the incoming beam's focal waist is well within the Rayleigh range, $z_1 \ll z_{R1}$,

$$\alpha_{\text{foc}} = \frac{f}{z_{R1}}, \tag{2.56a}$$

$$w_{02} \cong f \theta_1, \tag{2.56b}$$

$$z_2 \cong f. \tag{2.56c}$$

In the more common case, if the lens is positioned far outside the Rayleigh range, $z_1 \gg z_{R1}$,

$$\alpha_{\text{foc}} \cong \frac{f}{z_1}, \tag{2.57a}$$

$$w_{02} \cong \frac{f w_{01}}{z_1}, \tag{2.57b}$$

$$\theta_2 \cong \frac{z_1 \theta_1}{f}, \tag{2.57c}$$

$$z_2 \cong f. \tag{2.57d}$$

Recalling that

$$z_1 \cong \frac{w_l}{\theta_1} = \frac{w_l}{\lambda/(\pi w_{01})},$$

where w_l is the incoming beam's size at the lens, Equation (2.57b) yields

$$w_{02} \cong \frac{f \lambda}{\pi w_l} = \frac{2\lambda}{\pi} f^{\#}, \tag{2.58}$$

where $f^{\#} = f/(2w_l)$ is the f-number of the focusing system.

In laser microprocessing systems where tight focusing is required, the beam divergence is first reduced via collimating and expanding systems that may be either converging or diverging. In principle, if a beam were perfectly collimated, i.e. $\theta_2 = 0$, the minimum beam radius w_{02} would have to be infinite. Hence, it makes sense to look for either a minimum divergence or a maximum distance of the beam waist from the lens element. In the first case, a minimum of θ_2 according to Equation (2.54) would correspond to a maximum z_{R2}, which occurs for $z_2 = f$. In the second case, according to (2.52) z_2 is a maximum when $z_1 = f + z_{R1}$, implying that the beam is considered collimated when a positive focal-length element displaces the focal waist by one Rayleigh range ahead of the focal point. Consequently,

$$\alpha_{\text{foc}} = \frac{|f|}{\sqrt{(z_1 - f)^2 + z_{R1}^2}} = \frac{|f|}{\sqrt{2} z_{R1}}, \tag{2.59a}$$

$$z_2 = f + \alpha_{\text{foc}}^2(z_1 - f) = f + \frac{f^2}{2z_{R1}^2}(z_{R1}) = f + \frac{f^2}{2z_{R1}}. \tag{2.59b}$$

Equation (2.59b) shows that good collimation can be obtained by utilizing a long focal length accepting a tightly focused (and hence of small Rayleigh range) laser beam. Focusing systems with f-numbers $f^{\#} < 1$ require special design and high-quality materials. Thus, in practice, the achievable focal radius $w_0 \sim \lambda$. When the beam divergence is decreased and the spot size is decreased, the depth of focus is also decreased. This might not be advantageous in ablation applications for which a relatively large focal depth is required, such as in the processing of nonplanar specimens. Equation (2.48) shows that, when $w_0 \sim \lambda$, the depth of focus d_{dof} becomes a fraction of λ, requiring high-precision positioning.

If laser processing is performed in an ambient gas environment, consideration must be given to gas breakdown, the probability of which is increased by the presence of dust particles and impurities that are first removed from the target. As discussed by Von Allmen (1987), air-breakdown thresholds in the vicinity of absorbing targets are of the order of 10^7 W/cm^2 for CO_2 lasers and 10^9 W/cm^2 for Nd lasers. Direct writing is achieved by focusing the Gaussian laser beam at normal incidence onto the specimen. For preserving the optical alignment in patterning operations, it is customary to translate the substrate using precision micropositioning stages. Another practical limiting factor when using short-focal-length objectives is that ablation products may be deposited at the lens surfaces. Another major problem in the processing of electronic components is the redeposition of debris onto the target surface. To avoid debris accumulation, nozzles are sometimes used to blow an inert gas such as nitrogen over the surface.

Whereas direct ablation of the target by irradiation from the top is the usual ablation mode, in the laser-induced-forward-transfer (LIFT) technique, a target film is transferred to the receiving wafer (Kantor *et al.*, 1995). The irradiated film is deposited onto a transparent substrate, through which the laser beam is focused. The thickness of the ablated film is of the order of 100 nm, and the gap between the receiving wafer and the film is in the tens-of-micrometers range.

2.6.2   Beam shaping and homogenization

In several applications, it is necessary to produce uniform irradiation at the target surface through homogenization. In the case of coherent laser beams that have Gaussian profiles, this is accomplished by diffractive optics using gratings (Veldkamp, 1982), phase plates (Possin *et al.*, 1983), or holographic techniques. Diffractive optical elements may also be used to obtain, for example, ring-shaped beams that are markedly free of depth-of-focus restrictions. The shaping of beams resulting from unstable laser resonators that typically have strong ring ripples is not straightforward. It is, however, possible to achieve shaping of beams from stable solid-state laser resonator configurations to nearly flat-top profiles via two-dimensional lens arrays.

The raw beam emerging from the excimer laser is incoherent and spatially nonuniform. To avoid hot spots for quantitative experiments, it is necessary to provide means of homogenizing. The simplest homogenizer (Figure 2.7) is a tunnel type, whose internal surfaces are finely polished. The tunnel material of preference is aluminum, since it can be polished easily, yielding a reflectance in the UV range of about 90%. The excimer-laser beam is focused slightly outside the tunnel opening. Very tight focusing may lead to air-plasma and/or tunnel-material damage at the entrance. The size of the opening is determined by the desired laser fluence range and shape from the cross-sectional profile of the laser beam. On the other hand, the length of the homogenizer is designed to provide an adequate number of bounces and beam mixing at the tunnel exit. Use of diffuser plates upstream of the homogenizer entrance may improve beam quality. If the beam fluence needs to be characterized accurately, one may try to eliminate the falling crests of the laser beam by placing an aperture after the exit of the homogenizer. The

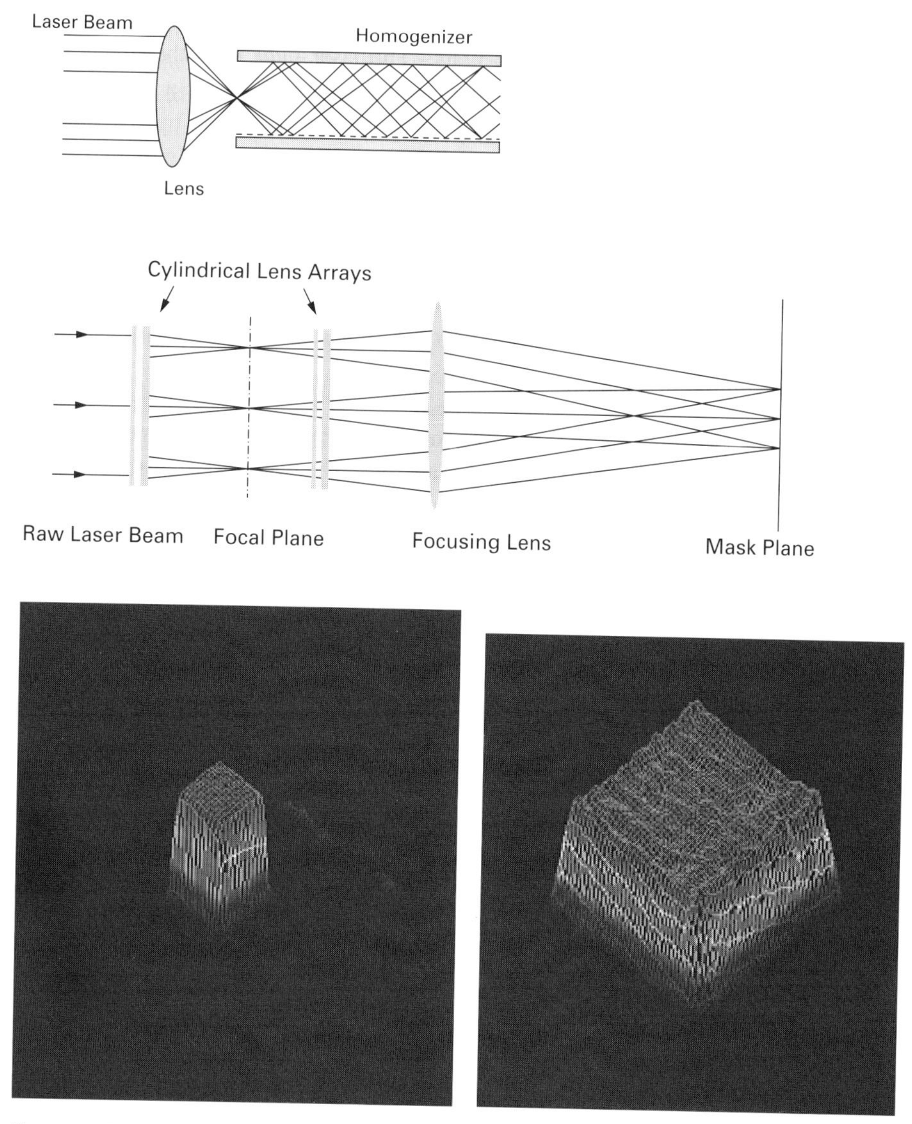

Figure 2.7. The top and middle parts show the schematics of a tunnel-type beam homogenizer and a fly's-eye homogenizer, respectively. For the excimer-laser beam profile, the bottom left shows a homogenized-field (11 mm × 11 mm) image captured using a beam profiler, CCTV lens, and fluorescence plate. The bottom right shows a YAG beam profile (7.5 mm × 7.5 mm) captured by a CCD camera combined with a beam profiler.

beam homogenizer provides a simple and flexible means for improving the beam quality at the expense of a power reduction that can reach 40% in the case of tight focusing and multiple bounces on the tunnel walls.

Another alternative is to use a fly's-eye-type beam homogenizer, as is often done in mask-projection lithography. Shown in Figure 2.7 is a homogenizer consisting of two arrays of cylindrical lenses that are parallel to each other. The spatially nonuniform

where

$$G_X(x, t | x', t') = \frac{1}{\sqrt{4\pi\alpha(t-t')}} e^{-\frac{(x-x')^2}{4\alpha(t-t')}}, \tag{3.16a}$$

$$G_Y(y, t | y', t') = \frac{1}{\sqrt{4\pi\alpha(t-t')}} e^{-\frac{(y-y')^2}{4\alpha(t-t')}}, \tag{3.16b}$$

$$G_Z(z, t | z', t') = \frac{2}{d} \sum_{m=1}^{\infty} e^{-\zeta_m^2 \alpha(t-t')/d^2} \cos\left(\zeta_m \frac{z}{d}\right) \cos\left(\zeta_m \frac{z'}{d}\right), \tag{3.16c}$$

where the eigenvalues are $\zeta_m = \pi(m - \frac{1}{2})$.

Some simple solutions are given in the following discussions.

3.2.2   One-dimensional heat conduction

For a spatially uniform laser-beam distribution, $I(x, y, t) = I(t)$, if the radius or characteristic dimension of the laser beam is much larger than d_{abs} and d_{th}, the temperature profile in the material is one-dimensional.

(i) For $d_{abs}/d_{th} \ll 1$, negligible heat losses from the surface, and assuming a laser source incident on an infinite target, the one-dimensional solution to Equation (3.1) yields a surface temperature $T_{su}(t)$ as follows:

$$T_{su}(t) = \frac{(1-R)\sqrt{\alpha/\pi}}{K} \int_0^t \frac{I(t')dt'}{\sqrt{t-t'}}. \tag{3.17}$$

(ii) For a surface source and triangular temporal laser pulse profile that would, for example, be fitted to an excimer-laser pulse,

$$T_{su}(t) = \frac{8\sqrt{\alpha/\pi}}{K} F(1-R)\left[t^{3/2} - \left(1 + \frac{t_{max}}{t_{pulse} - t_{max}}\right)(t - t_{max})^{3/2}\right],$$
$$t_{max} < t < t_{pulse}, \tag{3.18}$$

where F is the pulse fluence and t_{max} is the time instant of the peak temporal intensity.

The peak surface temperature $T_{su,pk}$ occurs at a time $t_{pk} = t_{pulse}^2/(2t_{pulse} - t_{max})$ and is given by

$$T_{su,pk} = \frac{8F(1-R)}{3K} \sqrt{\frac{\alpha/\pi}{2t_{pulse} - t_{max}}}. \tag{3.19}$$

(iii) For a beam of constant intensity impinging on a slab of thickness d with skin absorption and insulated surfaces:

$$T(z, t) = \frac{(1-R)I_{pk}t}{\rho C_p d} + \frac{(1-R)I_{pk}d}{K}\left[\frac{3z^2 - d^2}{6d^2} - \frac{2}{\pi^2}\sum_{n=1}^{\infty}\frac{(-1)^n}{n^2}\right.$$
$$\left. \times \exp\left(-\frac{\alpha n^2 \pi^2 t^2}{d^2}\right)\cos\left(\frac{n\pi z}{d}\right)\right]. \tag{3.20}$$

(iv) For a pulse with temporal shape $I(t) = Bt^{m/2}$, where $m = -1, 0, 1, 2, \ldots$ incident on a medium of thickness d:

$$
\begin{aligned}
T(z, t) = {} & \frac{2^{m+1}(1 - R)Bt^{m/2}\sqrt{\alpha t}\,\Gamma(m/2 + 1)}{K} \\
& \times \sum_{n=0}^{\infty}\left[i^{m+1}\operatorname{erfc}\left(\frac{(2n+1)d - z}{\sqrt{4\alpha t}}\right) + i^{m+1}\operatorname{erfc}\left(\frac{(2n+1)d + z}{\sqrt{4\alpha t}}\right)\right],
\end{aligned}
$$

$$(3.21)$$

where erfc is the complementary error function, while the Γ function is defined as

$$
\Gamma(\upsilon) = \int_0^{\infty} \xi^{\upsilon-1}\exp(-\xi)\,\mathrm{d}\xi,
$$

and the repeated integrals of the error function are

$$
i^m\operatorname{erfc}(\upsilon) = \int_{\upsilon}^{\infty} i^{m-1}\operatorname{erfc}(\xi)\,\mathrm{d}\xi, \qquad m = 0, 1, 2, \ldots
$$

(v) Considering the laser-beam attenuation in a semi-infinite medium subjected to constant laser intensity, the temperature distribution is

$$
\begin{aligned}
T(z, t) = {} & \frac{2(1 - R)I_{\mathrm{pk}}}{K}\sqrt{\alpha t}\,i\operatorname{erfc}\left(\frac{z}{2\sqrt{\alpha t}}\right) - \frac{(1 - R)I_0}{\gamma K}\mathrm{e}^{-\gamma z} \\
& + \frac{(1 - R)I_{\mathrm{pk}}}{2\gamma K}\exp(\gamma^2\alpha t - \gamma z)\operatorname{erfc}\left(\gamma\sqrt{\alpha t} - \frac{z}{2\sqrt{\alpha t}}\right) \\
& + \frac{(1 - R)I_{\mathrm{pk}}}{2\gamma K}\exp(\gamma^2\alpha t + \gamma z)\operatorname{erfc}\left(\gamma\sqrt{\alpha t} + \frac{z}{2\sqrt{\alpha t}}\right). \qquad (3.22)
\end{aligned}
$$

Accordingly, the surface temperature is

$$
T_{\mathrm{su}}(t) = \frac{I_{\mathrm{pk}}(1 - R)}{K}\left\{\sqrt{\frac{\alpha t}{\pi}} - \frac{1}{\gamma}\left[1 - \mathrm{e}^{\gamma^2\alpha t}\operatorname{erfc}(\gamma\sqrt{\alpha t})\right]\right\}. \qquad (3.23)
$$

(vi) For a beam of constant intensity incident on a slab of thickness d with finite absorption coefficient

$$
\begin{aligned}
T(z, t) = {} & \frac{(1 - R)I_{\mathrm{pk}}}{2\gamma\alpha\rho C_p}\sum_{n=-\infty}^{+\infty}\sigma_n A_n + \frac{(1 - R)I_{\mathrm{pk}}}{\rho C_p}\left(\frac{t}{\alpha}\right)^{\frac{1}{2}} \\
& \times \sum_{n=-\infty}^{+\infty}\sigma_n B_n - \frac{(1 - R)I_{\mathrm{pk}}\mathrm{e}^{-\gamma z}}{\gamma\alpha\rho C_p}, \qquad (3.24)
\end{aligned}
$$

where $\sigma_n = (-1)^n$,

$$
A_n = \exp[\gamma^2 \alpha t \pm \gamma(z - 2nd)] \left[\mathrm{erfc}\left(\gamma\sqrt{\alpha t} \pm \frac{z - 2nd}{\sqrt{4\alpha t}} \right) \right.
$$

$$
\left. - \mathrm{erfc}\left(\gamma\sqrt{\alpha t} \pm \frac{z - 2nd}{\sqrt{4\alpha t}} + \frac{d}{\sqrt{4\alpha t}} \right) \right],
$$

$$
B_n = 2\mathrm{i}\,\mathrm{erfc}\left(\frac{|z - 2nd|}{\sqrt{4\alpha t}} \right) - \exp(-\gamma d)\mathrm{i}\,\mathrm{erfc}\left[\frac{|z - (2n \pm 1)d|}{\sqrt{4\alpha t}} \right].
$$

(vii) In the case of a large absorption depth compared with the thermal penetration depth, $1/\gamma \gg \sqrt{\alpha t_{\mathrm{pulse}}}$, the above equation yields the peak surface temperature at the end of the laser pulse:

$$
T_{\mathrm{su,pk}} = \frac{\gamma F(1 - R)}{\rho C_p}. \tag{3.25}
$$

It is noted that the peak surface temperature in this case is in direct proportionality to the laser pulse energy and does not depend on the pulse duration.

(viii) Consider surface skin-depth absorption in a layer of thickness d, with thermal conductivity and diffusivity K_1 and α_1, respectively, on a semi-infinite medium of thermal conductivity and diffusivity K_2 and α_2. The temperature in the layer is

$$
T_1(z, t) = \frac{(1 - R)I_{\mathrm{pk}}}{K_1} \sqrt{4\alpha_1 t} \left[\sum_{n=-\infty}^{+\infty} \xi^{|n|} \mathrm{i}\,\mathrm{erfc}\left(\frac{|z - 2nd|}{\sqrt{4\alpha_1 t}} \right) + 2\mathrm{i}\,\mathrm{erfc}\left(\frac{z}{\sqrt{4\alpha_1 t}} \right) \right],
$$

$$
\tag{3.26}
$$

where

$$
\xi = \frac{K_1\sqrt{\alpha_2} - K_2\sqrt{\alpha_1}}{K_1\sqrt{\alpha_2} + K_2\sqrt{\alpha_1}}.
$$

The temperature in the substrate is

$$
T_2(z', t) = \frac{2T_L}{\Lambda + 1} \sum_{n=0}^{+\infty} \xi^n I_1(z^*), \tag{3.27}
$$

where $z' = z - d$,

$$
T_L = \frac{2(1 - R)I_{\mathrm{pk}}\sqrt{\alpha_2 t}}{\sqrt{\pi}\,K_2}, \qquad \Lambda = \frac{K_1}{K_2}\sqrt{\frac{\alpha_2}{\alpha_1}},
$$

$$
z^* = \frac{z'}{\sqrt{4\alpha_2 t}} + \frac{(2n + 1)d}{\sqrt{4\alpha_1 t}}, \qquad I_1(x) = \sqrt{\pi}\,\mathrm{i}\,\mathrm{erfc}(x).
$$

3.2.2 Beams of circular cross section

In the case of radial symmetry it is more convenient to set the problem in cylindrical coordinates (r, z). The energy equation for constant properties is

$$\frac{1}{\alpha}\frac{\partial T}{\partial t} = \frac{\partial^2 T}{\partial r^2} + \frac{1}{r}\frac{\partial T}{\partial r} + \frac{\partial^2 T}{\partial z^2} + \frac{Q_{ab}(r, z, t)}{K}. \tag{3.28}$$

The energy-absorption term is

$$Q_{ab}(r, z, t) = (1 - R)\gamma I_{pk} \exp\left(-\frac{r^2}{w^2} - \gamma z\right). \tag{3.29}$$

Consider the beam incident on a semi-infinite medium whose surface experiences negligible losses. The respective initial and boundary conditions are

$$T(r, z, 0) = 0, \tag{3.30a}$$

$$\left.\frac{\partial T}{\partial z}\right|_{z=0} = 0, \tag{3.30b}$$

$$T(r, z \to \infty) \to 0. \tag{3.30c}$$

The solution is written as

$$T(r, z, t) = \frac{1}{\rho C_p} \int_{t'=0}^{t} \int_{r'=0}^{+\infty} \int_{z'=0}^{+\infty} Q_{ab}(r', z', t')G(r, z, t|r', z', t')2\pi r'\, dr'\, dz'\, dt'. \tag{3.31}$$

The Green function is

$$G(r, z, t|r', z', t') = G_R(r, t|r', t') \cdot G_Z(z, t|z', t'), \tag{3.32}$$

where

$$G_R(r, t|r', t') = \frac{1}{4\pi\alpha(t - t')} \exp\left[-\frac{(r - r')^2}{4\alpha(t - t')}\right]$$

$$\times \exp\left[-\frac{rr'}{2\alpha(t - t')}\right] I_0\left[\frac{rr'}{2\alpha(t - t')}\right], \tag{3.33a}$$

$$G_Z(z, t|z', t') = \frac{1}{\sqrt{4\pi\alpha(t - t')}}\left\{\exp\left[-\frac{(z - z')^2}{4\alpha(t - t')}\right] + \exp\left[-\frac{(z + z')^2}{4\alpha(t - t')}\right]\right\}. \tag{3.33b}$$

In the above, I_0 is the modified Bessel function of the first kind and zeroth order.

Consider a beam of power P over a disk of radius w, incident on a semi-infinite substrate. The intensity is uniform over the irradiated area, $I(t) = P(t)/(\pi w^2)$, and the laser energy is absorbed over a skin depth.

(i) A pulse of energy E_{pulse} is instantaneously released on the surface, i.e. $I(t) = I_{pk}\delta(t)$, where $\delta(t)$ is the Kronecker delta function:

$$T(r, z, t) = \frac{(1 - R)E_{pulse}}{2\rho C_p \pi w^2(\pi \alpha^3 t^3)^{1/2}}$$

$$\times \int_0^w \exp\left(-\frac{r^2 + r'^2 + z^2}{4\alpha t}\right)I_0\left(\frac{rr'}{2\alpha t}\right)r'\, dr'. \tag{3.34}$$

(ii) A pulse of arbitrary temporal distribution $I(t)$ is incident on the surface:

$$T(r, z, t) = \frac{1 - R}{4\pi\rho C_p \alpha^{3/2}} \int_0^t \int_0^w \frac{I(t')}{(t - t')^{3/2}} \exp\left(-\frac{r^2 + r'^2 + z^2}{4\alpha(t - t')}\right)$$

$$\times I_0\left(\frac{rr'}{2\alpha(t - t')}\right) r'\, \mathrm{d}r'\, \mathrm{d}t'. \tag{3.35}$$

3.2.3　Gaussian laser beams

A variety of analytical expressions for the temperature rise induced by laser beams of Gaussian intensity cross-sectional profiles incident on semi-infinite substrates, finite slabs, and thin films, are given in Bäuerle (1996) and Prokhorov *et al.* (1990). For negligible heat-transfer losses the following expressions are derived on the basis of the $1/e$ intensity radius, $w = w_{1/e}$.

(i) For surface absorption by an arbitrary temporal distribution of peak intensity $I_{pk}(t)$

$$T(r, z, t) = \frac{(1 - R)w^2}{\rho C_p (\pi\alpha)^{1/2}} \int_0^t \frac{I_{pk}(t')}{(t - t')^{1/2}[4\alpha(t - t') + w^2]}$$

$$\times \exp\left(-\frac{z^2}{4\alpha(t - t')} - \frac{r^2}{4\alpha(t - t') + w^2}\right) \mathrm{d}t'. \tag{3.36}$$

(ii) An approximate expression for the rise in surface temperature at the origin ($r = 0$, $z = 0$) induced by a rectangular laser pulse is

$$T(0, 0, t) = \frac{I_{pk}(1 - R)\gamma w^2}{\rho C_p} \int_0^t \frac{1}{w^2 + 4\alpha t'} \, \mathrm{erfc}\left(\gamma\sqrt{\alpha t'}\right) \exp(\gamma^2 \alpha t') \mathrm{d}t'. \tag{3.37}$$

(iii) The following expression can be derived for the temperature induced by a laser beam of time-varying peak intensity $I_{pk}(t)$ and for volumetric absorption:

$$T(r, z, t) = \frac{2(1 - R)\gamma}{K} \int_{\mu=0}^{\sqrt{\alpha t}} \left[\frac{\mu \exp\left(\gamma^2\mu^2 - \dfrac{r^2}{w^2 + 4\mu^2}\right)}{w^2 + 4\mu^2}\right]$$

$$\times \left\{\exp(-\gamma z)\left[1 - \mathrm{erf}\left(\gamma\mu - \frac{z}{2\mu}\right)\right]\right.$$

$$\left. + \exp(\gamma z)\left[1 - \mathrm{erf}\left(\gamma\mu + \frac{z}{2\mu}\right)\right]\right\} I_{pk}(\mu)\mathrm{d}\mu, \tag{3.38}$$

where the variable μ is defined as $\mu = \sqrt{\alpha(t - t')}$.

If a laser beam of Gaussian cross section is considered, the heat flow is essentially perpendicular to the target surface direction if the beam radius is much greater than the thermal penetration depth, i.e. $w \gg \sqrt{\alpha t_{pulse}}$. In the other extreme case, if $w \ll \sqrt{\alpha t_{pulse}}$ the laser heat source may be considered a point source and the isotherms concentric hemispherical surfaces:

$$T(r, 0, t) = \frac{(1 - R)w^2}{\rho C_p \sqrt{\pi \alpha}} \int_0^t \frac{I(t')}{(t - t')^2 [4\alpha(t - t') + w^2]} \exp\left(-\frac{r^2}{4\alpha(t - t') + w^2}\right) dt'.$$

(3.39)

More-complicated expressions for the linear heat conduction, accounting for the absorption of laser energy in the material, can be derived using Green-function methods (Özisik, 1993). For the purpose of a quick estimate, it is useful to recall simple relations for the temperature rise of the target, such as

$$T \approx \frac{(1 - R)E_{\text{pulse}}}{\rho C_p V_{\text{HAZ}}},$$

(3.40)

where E_{pulse} is the laser pulse energy and V_{HAZ} is an estimate of the heat-affected material volume:

$$V_{\text{HAZ}} = \begin{cases} \pi w^2 \sqrt{\alpha t_{\text{pulse}}}, & w \gg \sqrt{\alpha t_{\text{pulse}}}, \\ \dfrac{4}{3}\pi (\alpha t_{\text{pulse}})^{3/2}, & w \ll \sqrt{\alpha t_{\text{pulse}}}. \end{cases}$$

(3.41)

3.2.4 Beam motion–quasi-static thermal field

Equation (3.6) is written as

$$\frac{\partial^2 T}{\partial x^2} + \frac{\partial^2 T}{\partial y^2} + \frac{\partial^2 T}{\partial z^2} + \frac{Q_{\text{ab}}(x, y, z, t)}{K} = \frac{1}{\alpha}\frac{\partial T}{\partial t}.$$

(3.42)

Consider now scanning of the laser beam with a constant velocity U along the positive x-axis. A new ξ-coordinate is introduced so that the coordinate system (ξ, y, z) is anchored to the laser beam:

$$\xi = x - Ut.$$

(3.43)

The heat-conduction equation (3.42) is transformed to the new set of coordinates:

$$\frac{\partial^2 T}{\partial \xi^2} + \frac{\partial^2 T}{\partial y^2} + \frac{\partial^2 T}{\partial z^2} + \frac{Q_{\text{ab}}(\xi, y, z, t)}{K} = \frac{1}{\alpha}\left(\frac{\partial T}{\partial t} - U\frac{\partial T}{\partial \xi}\right).$$

(3.44)

In an alternative point of view, the solid appears as moving at a velocity U and in the negative ξ-direction toward an observer situated at the origin of the coordinate system (ξ, y, z). The negative sign in front of the velocity term in Equation (3.44) is due to this apparent motion.

If the solid target is long enough in the x-direction compared with the thermal penetration depth, the temperature field becomes stationary with respect to the coordinate system (ξ, y, z). In other words, the temperature field appears to be independent of time for an observer stationed at the origin of the (ξ, y, z) coordinate system. For this quasi-static temperature field, the transient term $\partial T/\partial t$ vanishes:

$$\frac{\partial^2 T}{\partial \xi^2} + \frac{\partial^2 T}{\partial y^2} + \frac{\partial^2 T}{\partial z^2} + \frac{Q_{\mathrm{ab}}(\xi, y, z, t)}{K} = -\frac{U}{\alpha}\frac{\partial T}{\partial \xi}. \tag{3.45}$$

Equation (3.29) can be transformed to a more convenient form by introducing a new dependent variable $T(\xi, y, z) = \Theta(\xi, y, z)\mathrm{e}^{-\frac{U\xi}{2\alpha}}$:

$$\frac{\partial^2 \Theta}{\partial \xi^2} + \frac{\partial^2 \Theta}{\partial y^2} + \frac{\partial^2 \Theta}{\partial z^2} - \left(\frac{U}{2\alpha}\right)^2 \Theta + \frac{Q_{\mathrm{ab}}(\xi, y, z, t)\mathrm{e}^{\frac{U\xi}{2\alpha}}}{K} = 0. \tag{3.46}$$

Assuming a point source of continuous power P_1 scanning over a semi-infinite medium, the above equation yields the following elementary solution:

$$T(r, \xi) = \frac{(1-R)P_l}{2\pi K r} \exp\left(-\frac{Ur}{2\alpha} - \frac{U\xi}{2K}\right), \tag{3.47}$$

where $r = \sqrt{\xi^2 + y^2 + z^2}$.

Quasi-static solution for an elliptic beam profile

Consider a beam of elliptic Gaussian intensity distribution scanning with constant velocity U over a semi-infinite target. The absorption of laser light is considered a skin-depth event. The beam intensity profile is

$$I(\xi, y) = I_{\mathrm{pk}} \exp\left[-\left(\frac{\xi^2}{r_x^2} + \frac{y^2}{r_y^2}\right)\right]. \tag{3.48}$$

Following the derivation by Moody and Hendel (1982), the characteristic radius r and eccentricity β are defined as follows:

$$r = \sqrt{r_x r_y}, \qquad \beta = r_y/r_x.$$

The quasi-static temperature distribution (fixed with respect to the (ξ, y) coordinate system) is described by

$$\frac{\partial^2 T}{\partial \xi^2} + \frac{\partial^2 T}{\partial y^2} + \frac{\partial^2 T}{\partial z^2} - \frac{U}{\alpha}\frac{\partial T}{\partial \xi} = 0, \tag{3.49a}$$

$$-K\left.\frac{\partial T}{\partial z}\right|_{z=0} = (1-R)I(\xi, y), \tag{3.49b}$$

$$T(\xi, y, z \to \pm\infty) = 0. \tag{3.49c}$$

The solution to the above system is given by

$$T(\xi, y, z) = \frac{(1-R)I_{\mathrm{pk}}r}{\sqrt{\pi}K} \int_0^\infty \phi(\chi)\mathrm{d}\chi, \tag{3.50}$$

where

$$\phi(\chi) = \frac{\exp\left\{-\left[\dfrac{(\xi' + U'\chi^2)^2}{\chi^2 + 1/\beta} + \dfrac{y'^2}{\chi^2 + \beta} + \dfrac{z'^2}{\chi^2}\right]\right\}}{\sqrt{(\chi^2 + 1/\beta)(\chi^2 + \beta)}}. \tag{3.51}$$

In the above the parameters are defined as follows:

$$\xi' = \frac{\xi}{r}, \qquad y' = \frac{y}{r}, \qquad z' = \frac{z}{r}, \qquad U' = \frac{Ur}{4\alpha r}. \tag{3.52}$$

Scanning beam over a substrate of finite thickness

Consider a Gaussian beam of elliptic cross section. The beam is scanned over a slab that is extensive in the x- and y-directions but of finite thickness d in the z-direction. Losses represented by a linearized heat-transfer coefficient, $h_{\mathrm{conv},u}$, are allowed from the top surface, while the bottom surface at $z = d$ is kept at the ambient temperature, which is taken to be zero for convenience. The initial condition and boundary conditions complementing the field equation (3.28) are

$$T(\xi, y, z, 0) = 0, \tag{3.53}$$

$$K \left. \frac{\partial T}{\partial z} \right|_{z=0} = h_{\mathrm{conv},u} T(\xi, y, 0, t), \tag{3.54a}$$

$$T(\xi, y, d) = 0, \tag{3.54b}$$

$$T(\xi, y \to \pm\infty) = 0. \tag{3.54c}$$

The following transformation may be applied:

$$T(\xi, y, z, t) = W(\xi, y, z, t)\exp\left(-\frac{U\xi}{2\alpha} - \frac{U^2 t}{4\alpha}\right). \tag{3.55}$$

The field equation and initial and boundary conditions on the transformed variable W are written as follows:

$$\frac{\partial W}{\partial t} = \alpha \left(\frac{\partial^2 W}{\partial \xi^2} + \frac{\partial^2 W}{\partial y^2} + \frac{\partial^2 W}{\partial z^2} \right) + \frac{Q_{\mathrm{ab},W}(\xi, y, z, t)}{\rho C_p}. \tag{3.56}$$

The absorption term $Q_{\mathrm{ab},W}$ is

$$Q_{\mathrm{ab},W}(\xi, y, z, t) = (1 - R)\gamma I_{\mathrm{pk}} \exp\left[-\left(\frac{\xi}{w_x}\right)^2 - \left(\frac{y}{w_y}\right)^2 - \gamma z + \frac{U\xi}{2\alpha} + \frac{U^2}{4\alpha} t \right]. \tag{3.57}$$

The initial condition is

$$W(\xi, y, z, 0) = 0. \tag{3.58}$$

The boundary conditions are

$$K \left. \frac{\partial W}{\partial z} \right|_{z=0} = h_{\mathrm{conv},u,W} W(\xi, y, 0), \tag{3.59a}$$

where $h_{\mathrm{conv},u,W} = h_{\mathrm{conv},u} + KU/(2\alpha)$,

$$W(\xi, y, d) = 0, \tag{3.59b}$$

$$W_{\xi, y \to \pm\infty} \to 0. \tag{3.59c}$$

The temperature distribution is given by

$$W(x, y, z, t) = \frac{1}{\rho C_p} \int_{t'=0}^{t} \int_{\xi'=-\infty}^{+\infty} \int_{y'=-\infty}^{+\infty} \int_{z'=0}^{+\infty} Q_{\mathrm{ab},W}(\xi', y', z', t')$$
$$\times \, G(\xi, y, z, t \,|\, \xi', y', z', t') \, \mathrm{d}x' \, \mathrm{d}y' \, \mathrm{d}z' \, \mathrm{d}t'. \tag{3.60}$$

The Green-function solution is

$$G(x, y, z, t \,|\, x', y', z', t') = \frac{1}{4\pi\alpha(t - t')} \mathrm{e}^{-\frac{(\xi-\xi')^2 + (y-y')^2}{4\alpha(t-t')}} \frac{2}{d} \sum_{m=1}^{\infty} \mathrm{e}^{-\zeta_m^2 \alpha(t-t')/d^2}$$
$$\times \frac{(\zeta_m^2 + B^2)\sin\left[\zeta_m\left(1 - \dfrac{z}{d}\right)\right]\sin\left[\zeta_m\left(1 - \dfrac{z'}{d}\right)\right]}{\zeta_m^2 + B^2 + B},$$
$$\tag{3.62}$$

where the eigenvalues are given by the transcendental equation $\zeta_m \cot \zeta_m = -B$ and $B = h_{\mathrm{conv},u,W} d / K$.

3.3   Melting

3.3.1   Interface boundary conditions

In pure-element materials, the transition to the melting phase normally occurs at a specified temperature. The propagation of the solid/liquid interface is prescribed by the energy balance, which may be thought of as a kinematic boundary condition. The moving interface is assumed to be isothermal at the equilibrium melting temperature, T_{m}, if no overheating or undercooling is assumed:

$$T_s(\vec{r} \in S_{\mathrm{int}}) = T_l(\vec{r} \in S_{\mathrm{int}}) = T_{\mathrm{m}}, \tag{3.63}$$

$$k_s \left.\frac{\partial T_s}{\partial n}\right|_{\vec{r}\in S_{\mathrm{int}}} - k_l \left.\frac{\partial T_l}{\partial n}\right|_{\vec{r}\in S_{\mathrm{int}}} = L_{\mathrm{sl}} U_{\mathrm{int};n}, \tag{3.64}$$

where L_{sl} is the latent heat for melting, $\partial/\partial n$ indicates the derivative of the interface along the normal direction vector $\vec{n}$ at any location of the interface, $\vec{r} \in S_{\mathrm{int}}$, and pointing into the liquid region, while $U_{\mathrm{int};n}$ is the velocity of the interface along $\vec{n}$. The above interfacial boundary conditions, together with the heat-conduction equations in the solid and liquid regions, specify the "Stefan problem." Analytical solutions are scarce and limited to the one-dimensional phase change in materials with constant properties that is driven by a surface temperature differential, the so-called "Neumann solution" (Carslaw and Jaeger, 1959). Numerical solutions implementing the exact boundary conditions are nontrivial in multiple dimensions, requiring front-fixing or front-tracking techniques.

3.3.2   The enthalpy formulation

An alternative way of modeling phase change is the enthalpy method, which circumvents the need for exact tracking of the transient interface motion. The enthalpy function is used to account for phase change:

$$H(T) = \int_{T_0}^{T} \rho_s(T)C_{p,s}(T)\mathrm{d}T, \qquad T < T_\mathrm{m}, \tag{3.65}$$

$$H(T) = \int_{T_0}^{T_\mathrm{m}} \rho_s(T)C_{p,s}(T)\mathrm{d}T + \int_{T_\mathrm{m}}^{T} \rho_l(T)C_{p,l}(T)\mathrm{d}T + L_{sl}, \qquad T > T_\mathrm{m}, \tag{3.66}$$

where the density ρ and specific heat C_p vary differently with temperature in the solid and liquid phases. For $T = T_\mathrm{m}$ the enthalpy function assumes values between H_s and H_l,

$$H_s = \int_{T_0}^{T_\mathrm{m}} \rho_s(T)C_{p,s}(T)\mathrm{d}T, \tag{3.67a}$$

$$H_l = \int_{T_0}^{T_\mathrm{m}} \rho_s(T)C_{p,s}(T)\mathrm{d}T + L_{sl}. \tag{3.67b}$$

The enthalpy value $H = H_s$ is assigned to solid material at the melting temperature, while $H = H_l$ corresponds to pure liquid at the same temperature. Thus, there exists a region in which the melting is partial, which is defined by

$$H_s < H < H_l; \qquad T = T_\mathrm{m}. \tag{3.68}$$

Each point within this region can be assigned a solid fraction $f_{\mathrm{ph,s}}$ and a liquid fraction $f_{\mathrm{ph,l}}$, for which

$$f_{\mathrm{ph,s}} + f_{\mathrm{ph,l}} = 1. \tag{3.69}$$

The enthalpy function during melting at $T = T_\mathrm{m}$ is given by

$$H = H_s + f_{\mathrm{ph,l}}L_{sl}. \tag{3.70}$$

Using the enthalpy as dependent variable, together with the temperature, Equation (3.3) is written

$$\frac{\partial H}{\partial t} = \nabla \cdot (K(T)\nabla T) + Q_{\mathrm{ab}}(x, y, z, t). \tag{3.71}$$

The above scheme can be readily implemented numerically utilizing either explicit or implicit schemes in multi-dimensional domains (e.g. Grigoropoulos *et al.*, 1993a). Because of the relative simplicity of the numerical approach, enthalpy-based schemes are usually preferred, unless a more accurate specification of the motion of the boundary and the driving temperature field gradients is required, as for example in crystal growth.

3.3.3   Approximations

The time for the inception of melting, t_m, can be calculated using expressions for the temperature rise:

$$T_{su}(t_m) = T_m. \tag{3.72}$$

For a rectangular large-area pulse incident on a bulk surface absorber, the threshold fluence, $F_{th,m}$, necessary for melting is given from Equation (3.17) assuming $t = t_{pulse}$:

$$F_{th,m} = \frac{1}{2(1-R)}\sqrt{\frac{\pi}{\alpha}}K(T_m - T_0)\sqrt{t_{pulse}}. \tag{3.73}$$

Approximate expressions for the maximum melt depth, $d_{l,max}$, are given in Bäuerle (1996):

(a) for fluences close to $F_{th,m}$

$$d_{l,max} \approx \sqrt{\frac{\alpha t_{pulse}}{\pi}}\frac{F - F_{th,m}}{F_{th,m}}; \tag{3.74}$$

(b) for $F > F_{th,m}$

$$d_{l,max} \approx \sqrt{\alpha t_{pulse}}\left[\ln\left(\frac{F - F_{th,m}}{F_{th,m}}\right)\right]^{\frac{1}{2}} \approx \sqrt{\alpha t_{pulse}}\left(\frac{F - F_{th,m}}{F_{th,m}}\right)^{\frac{1}{2}}; \tag{3.75}$$

(c) and for $F \gg F_{th,m}$ (but still below the vaporization threshold) energy balance gives

$$d_{l,max} \approx \frac{(1-R)F - q_{loss}}{\rho C_p(T_m - T_0) + L_{sl}}, \tag{3.76}$$

where the losses via conduction, convection, and radiation are lumped together in q_{loss}.

The sensible heat, $\rho C_p T_m$, or more accurately $\int_{T_0}^{T_m}\rho C_p T\,dT$, is in general of comparable order of magnitude to the latent heat. In the case of one-dimensional nanosecond laser melting of metals, the temperature in the molten layer becomes uniform rather rapidly, and most of the solidification process is driven by the thermal gradient across the solid–liquid interface into the solid material.

3.3.4   Departures from equilibrium at the melt interface

The Stefan statement of the phase-change problem assumes that the interface dynamics is governed by the heat flow rather than the phase-transition kinetics. This assumption is true only for low melting speeds. According to the quasi-chemical formulation of crystal growth from the melt (Jackson and Chalmers, 1956; Jackson, 1975), for a flat interface the velocity of recrystallization ($U_{int}(T_{int}) > 0$), or melting ($U_{int}(T_{int}) < 0$), is

$$U_{int}(T_{int}) = C\exp\left(-\frac{Q_{act}}{k_B T_{int}}\right)\left[1 - \exp\left(-\frac{L_{sl}\,\Delta T}{k_B T_{int} T_m}\right)\right], \tag{3.77}$$

where

$$C = R_{\mathrm{M}}^{0}\,\exp\!\left(-\frac{L_{\mathrm{sl}}}{k_{\mathrm{B}}T_{\mathrm{m}}}\right); \qquad \Delta T = T_{\mathrm{m}} - T_{\mathrm{int}}.$$

In the above, k_{B} is Boltzmann's constant, T_{int} is the interface temperature, Q_{act} is the activation energy for viscous or diffusive motion in the liquid, and R_{M}^{0} is the rate of melting at equilibrium. For moderate ΔT, Equation (3.77) is linearized:

$$U_{\mathrm{int}}(T_{\mathrm{int}}) = c_{\mathrm{int}}\,\Delta T. \tag{3.78}$$

In the above, c_{int} is the slope of the interface velocity response function near T_{m}. On the basis of the above arguments, the melt-front temperature is higher than T_{m}, while undercooling is observed in resolidification. To calculate the motion of the phase boundary, it is necessary to solve the heat-conduction equation in the solid and liquid phases and apply (3.64) and (3.78) as boundary conditions at the interface. The classical theory implies symmetry for c_{int} in the melting and recrystallization processes. Evaluation of the X-ray diffraction studies by Larson *et al.* (1982, 1986) has challenged this argument (Peercy *et al.*, 1987) by showing that there is asymmetry in the interface response function, yielding significant undercooling in the recrystallization process. However, although departures from equilibrium are important for determining the recrystallization process, they usually do not affect severely the overall energy balance and therefore are of relatively minor consequence to ablation.

3.4   Ablative material removal

There are several ablative mechanisms by which material, either atomic or bulk, can be released from the surface of the target. References to "thermal" or "photothermal" ablation generally embrace a model in which laser light is converted to lattice vibrational energy before bond breaking liberates atomic material from the bulk surface. The thermal mechanism is distinct from a "photochemical" or "electronic" processes, in which laser-induced electronic excitations lead directly to bond breaking before an electronic to vibrational energy transfer has occurred. Both thermal and electronic sputtering mechanisms lead to the liberation of atomic-size material from the surface. This is distinct from two other ablation mechanisms, identified in the literature as "hydrodynamical" and "exfoliational," which introduce bulk material into the ablation plume. The hydrodynamical mechanism is ascribed to the liberation of micrometer-sized droplets, following motion in the molten phase. Exfoliation refers to an erosive-like mechanism by which material is removed from the surface in solid flakes. Separation of flakes from the surface is thought to occur along energy-absorbing defects in the material. It should be pointed out that these mechanisms are not necessarily distinguishable in a specific laser-ablation system, in which more than one mechanism can occur, either simultaneously or in different phases of the ablation process, depending upon the range of the laser parameters.

At first sight, one can invoke the classical picture of thermal vaporization from a heated surface through transition to the liquid phase to describe material removal from metallic targets for moderate laser fluences on time scales that allow establishment of LTE. That would be the case for nanosecond laser irradiation, since relaxation times in metals are in the sub-picosecond regime. Experiments have shown that laser sputtering of metals can occur at very low fluences, well below the perceived melting threshold. Experiments on sputtering from the molten phase also unveiled evidence that cannot be explained through the classical thermal-vaporization model. At higher laser fluences, the nascent metal vapor is photo-ionized, leading to further heating of the plasma through a cascade process as discussed for example by Dreyfus (1991). Even though laser ablation is a complex phenomenon defying a unified treatment, it is worth recalling some relevant thermal considerations.

3.4.1 Surface vaporization

Experimentally, it is usually easier to achieve vaporization than to control melting without significant material loss to the vapor phase. For moderate laser intensities, the laser-induced peak target surface temperature is below the thermodynamic critical point, and a sharp interface separates the vapor from the liquid phase. Both the sensible heat and the latent heat of melting are typically much smaller than the latent heat of vaporization, implying that evaporation is dominant in the energy balance. For a surface absorber, $1/\gamma \ll \sqrt{\alpha t_{\mathrm{pulse}}}$, simple energy-balance considerations give the following estimate for the material-removal depth, d_{abl}:

$$d_{\mathrm{abl}} = \frac{(1 - R)(F - F_{\mathrm{sh}}) - q_{\mathrm{loss}}}{\rho C_p(T_{\mathrm{bp}} - T_0) + L_{\mathrm{sl}} + L_{\mathrm{lv}}}, \qquad (3.79)$$

where F_{sh} represents the fluence loss due to plasma shielding. This estimate is more appropriate for short pulses, since conduction losses become more significant for longer pulses. For laser energy intensities $I < 10^8$ W/cm^2, energy absorption by the evaporated particles is insignificant, so the vapor phase may be considered transparent.

According to the thermal surface-vaporization picture, the material-removal rate is limited by the surface temperature. Neglecting recondensation of vapor onto the surface, the rate of evaporation from the liquid surface, j_{ev}, can be described using kinetic theory:

$$j_{\mathrm{ev}} = N_{\mathrm{l}}\left(\frac{k_{\mathrm{B}}T_{\mathrm{l}}}{2\pi m_{\mathrm{a}}}\right)^{\frac{1}{2}} \exp\left(-\frac{L_{\mathrm{lv}}}{k_{\mathrm{B}}T_{\mathrm{l}}}\right) - \vartheta_{\mathrm{s}}N_{\mathrm{v}}\left(\frac{k_{\mathrm{B}}T_{\mathrm{v}}}{2\pi m_{\mathrm{a}}}\right)^{\frac{1}{2}}, \qquad (3.80)$$

where N_{l} and N_{v} are the numbers of atoms per unit volume for liquid and vapor, h_{lv} and T_{v} are the latent heat of vaporization and temperature of vapor, m_{a} is the atomic mass, and k_{B} is the Boltzmann constant. The first term on the right represents the evaporation rate from the liquid surface at temperature T_{l}. The second term represents a damping of this evaporation rate due to the return of liquid molecules to the liquid surface. The

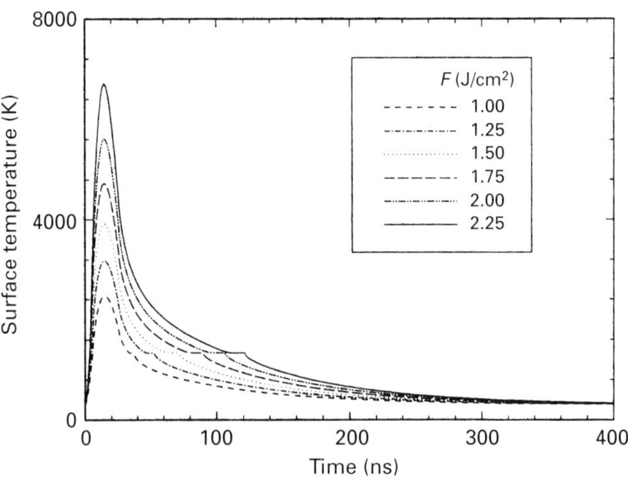

Figure 3.1. Time histories of the surface temperature of a gold substrate subjected to excimer-laser pulses of duration 26 ns and various fluences. The ambient pressure is $P_\infty =$ 1 atm and the laser-spot radius is 1 mm. From Ho *et al.* (1995), reprinted with permission from the American Institute of Physics.

parameter ϑ_s, called the sticking coefficient, represents the probability that a vapor atom returning to the liquid surface is finally adsorbed on the liquid surface.

The total ablation depth, d_{abl}, due to surface evaporation is

$$d_{abl} = \int_0^\infty j_{ev}(t)m_a/\rho \, dt. \tag{3.81}$$

If it is assumed that the thermodynamic path of the process rides the saturation curve, calculated surface temperatures (Ho *et al.*, 1995) may exceed the nominal atmospheric boiling temperature, as shown in Figure 3.1. Often the concept of a surface temperature fixed at the nominal atmospheric boiling temperature is adopted, leading to the unfounded prediction of subsurface heating and explosive material removal. The previously outlined thermal model implies that the material removal from the melt is a continuous process. However, as the liquid surface temperature increases, the ablation rate also increases sharply. It is reasonable to assume that most of the material is removed near the peak surface temperature and that an ablation threshold ascribed to substantial removal rates (e.g. >1 Å/pulse) can be defined.

For nanosecond laser pulses, the duration of melting is of the order of a few tens of nanoseconds. Hydrodynamic motion due to the melt instability caused by the acceleration of the molten phase following the volumetric expansion upon melting may thus develop over hundreds of pulses (Kelly and Rothenberg, 1985; Bennett *et al.*, 1995). In the case of thin metal films on poorly conducting substrates, the melt duration is substantially longer, in the microsecond range, thus allowing sufficient time for the development of fluid flow and the removal of metal film material hundreds of nanometers thick, in sharp contrast to the expectations arising from the thermal-vaporization model. Figure 3.2 shows the dependence of the average ablation depth on laser fluence

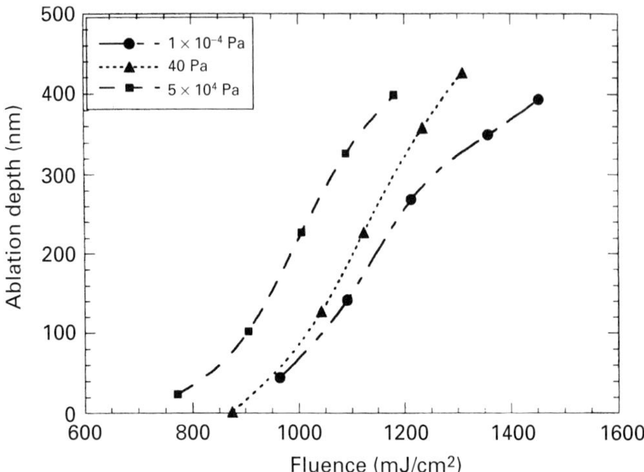

Figure 3.2. The dependence of the ablation rate on laser fluence and argon background gas pressure for excimer-laser ablation of a 0.5-μm-thick gold film on a quartz microbalance. From Zhang *et al.* (1997), reprinted with permission from Springer-Verlag.

and background pressure for a 0.5-μm-thick gold film deposited on a quartz crystal microbalalance (QCM) (Benes, 1984) and ablated by low-energy excimer-laser pulses at the wavelength $\lambda = 248$ nm (Zhang *et al.*, 1997).

Besides surface vaporization and material removal due to melt instabilities, boiling could be thought of as providing another material-removal mechanism. For ideally absorbing media that are free from impurities, voids, and structural microdefects, and for laser intensities $I < 1\,\mathrm{GB/cm^2}$, with corresponding sub-micrometer laser-radiation penetration depths, Rykalin *et al.* (1988) argued that volumetric vaporization could be significant only at temperatures exceeding tens of thousands of kelvins. Such high temperatures exceed the critical point. Further thermodynamic considerations showed that volumetric vaporization is important for large radiation-penetration depths.

3.4.2   Knudsen-layer effects

Assuming a thermally activated process, vapor molecules escaping from the free surface possess a half-Maxwellian velocity distribution:

$$f_s^{\text{forw}} \propto \exp\left[-\frac{2e_{\mathrm{I}} + m\left(u_x^2 + u_y^2 + u_z^2\right)}{2k_{\mathrm{B}}T_{\mathrm{su}}}\right], \qquad u_z \geq 0, \quad -\infty < u_x, \quad u_y < +\infty,$$

(3.82)

where the direction z is normal to the target surface, e_{I} is the accessible internal energy, and T_{su} is the surface temperature. The average u_z in the above equation is

$$\langle u_z \rangle = \sqrt{\frac{2k_{\mathrm{B}}T_{\mathrm{su}}}{\pi m}}.$$

(3.83)

Table 3.1. Flow-property ratios across the Knudsen layer for $\gamma_{ad} = 5/3$

M_K	ρ_K/ρ_s	T_K/T_s	P_K/P_s	β	T_{0K}/T_s	$\dfrac{\rho_K w_K}{\rho_s\sqrt{RT_s/(2\pi)}}$
0	1	1	1	1	1	0
0.005	0.927	0.980	0.908	1.007	0.981	0.148
0.1	0.861	0.960	0.827	1.017	0.964	0.273
0.2	0.748	0.922	0.690	1.051	0.935	0.465
0.4	0.576	0.851	0.490	1.215	0.896	0.688
0.6	0.457	0.785	0.358	1.682	0.879	0.786
0.8	0.371	0.725	0.269	2.947	0.879	0.817
1.0	0.308	0.669	0.206	6.287	0.892	0.816

The evaporated molecules experience collisions that generate backward flow toward the surface. As few as 2–3 collisions per emitted particle suffice to establish an equilibrium isotropic Maxwellian distribution at the edge of the so-called *Knudsen layer* (KL). The resulting distribution is of the form

$$f_K \propto \exp\left[-\frac{2e_1 + m\left(u_x^2 + u_y^2 + (u_z - u_K)^2\right)}{2k_B T_K} \right]. \tag{3.84}$$

In the above, the subscript K indicates conditions at the edge of the KL. In addition, the flow velocity u_K is considered equal to the sonic velocity at the KL boundary,

$$u_K \approx \sqrt{\frac{\gamma_{ad} k_B T_K}{m}}, \tag{3.85}$$

and the temperature ratio T_K/T_{su} is given by

$$\sqrt{\frac{T_K}{T_{su}}} = -\frac{\sqrt{\pi \gamma_{ad}/2}}{2(\eta + 4)} + \sqrt{1 + \frac{\pi \gamma_{ad}}{8(\eta + 4)^2}}. \tag{3.86}$$

The adiabatic exponent γ_{ad} is the ratio of the specific heats:

$$\gamma_{ad} = \frac{C_p}{C_v} = \frac{\eta + 5}{\eta + 3}, \tag{3.87}$$

where η is the number of accessible internal degrees of freedom, with $\eta = 0$ for an atom, 2 for a rotating diatomic molecule, and 4 for a rotating and vibrating diatomic molecule. The KL may be treated numerically by solving the distributions (3.84) and enforcing conservation of mass, momentum, and energy. Assuming that all back-scattered particles recondense on the surface, the following expressions were derived by Knight (1979):

$$\frac{T_K}{T_{su}} = \left[\sqrt{1 + \pi\left(\frac{\gamma_{ad} - 1}{\gamma_{ad} + 1}\frac{m}{2}\right)^2} - \sqrt{\pi}\frac{\gamma_{ad} - 1}{\gamma_{ad} + 1}\frac{m}{2} \right]^2, \tag{3.88a}$$

$$\frac{\rho_K}{\rho_{su}} = \sqrt{\frac{T_{su}}{T_K}}\left[\left(m^2 + \frac{1}{2}\right)e^{m^2}\,\mathrm{erfc}(m) - \frac{m}{\sqrt{\pi}} \right] + \frac{1}{2}\frac{T_{su}}{T_K}[1 - \sqrt{\pi}\,m e^{m^2}\,\mathrm{erfc}(m)], \tag{3.88b}$$

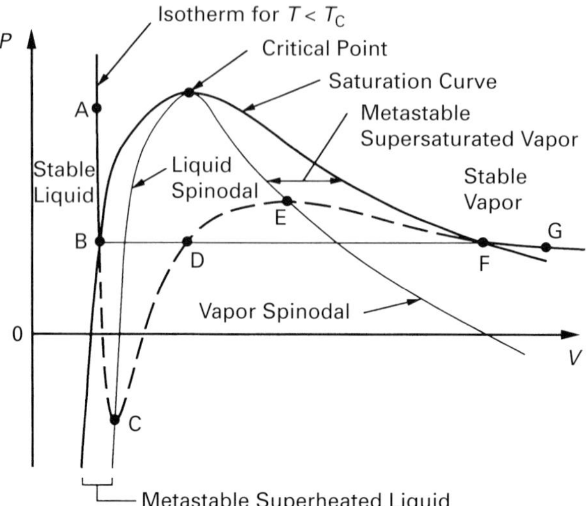

Figure 3.3. Spinodal lines and metastable regions on a P–v diagram. From Carey (1992), reproduced with permission from Taylor & Francis.

According to Knight (1979), the KL analysis provides no information about the value of m, which is defined through the Mach number, M_K, just outside the KL:

$$M_K = \frac{u_K}{\sqrt{\gamma_{ad} R T_K}} = m\sqrt{\frac{2}{\gamma_{ad}}}. \tag{3.89}$$

Anisimov (1968), assumed sonic conditions at the outer edge of the KL. In contrast, the analysis described above allows free specification of M_K, whose value can be determined by the flow state away from the KL. In a more detailed analysis, Knight (1976) showed that application of jump conditions should be restricted to $M_K \leq 1$. Table 3.1 gives flow-property ratios across the KL for atomic material ejection. Also the pressure, $P_K = \rho_K R T_K$, and the stagnation temperature of the gas leaving the Knudsen layer, $T_{0K} = [1 + (\gamma_{ad} - 1)M_K^2/2]T_K$, are given.

If the Knudsen–Hertz vaporization relation is assumed, the forward flow of vaporized particles will be

$$j^{forw} = N_s\sqrt{\frac{k_B T_{su}}{2\pi m_a}}, \tag{3.90}$$

where N_s is the particle number density at the vaporizing surface. By assuming atomic material ejection and recognizing that the recess surface velocity is much lower than the material-ejection velocity, it can be shown that 18% of all evaporating particles return to the surface ($j^{back} \approx 0.18\,j^{forw}$), where recondensation is nearly certain to occur. Hence, the net evaporation rate is lower than the equilibrium vaporization prediction:

$$j_{ev} \approx 0.82\,j^{forw}. \tag{3.91}$$

3.4.3 Explosive phase-change

The criterion for mechanical stability of a pure substance is that the density increases with pressure at constant temperature:

$$\left(\frac{\partial P}{\partial v}\right)_T < 0, \tag{3.92}$$

where P is pressure, v specific volume, and T temperature. A liquid that is superheated above its equlibrium saturation temperature exists in a nonequilibrium condition referred to as a *metastable state*.

Figure 3.3 depicts an isotherm traversing the vaporization dome as v is increased. While the metastable liquid is not in equlibrium, it is mechanically stable, i.e. it Satisfies Equation (3.99). The situation changes between points C and E, where $(\partial P/\partial v)_T > 0$, signifying a violation of the stability criterion. The locus of the *spinodal-limit* or *intrinsic-stability states* across which the coefficient $(\partial P/\partial v)_T$ changes from negative to positive is called the *spinodal curve*. Either the van der Waals or the Berthelot equation of state may be assumed to hold under the nonequilibrium conditions of interest:

$$(P + aT^{-c}v^{-2})(v - b) = RT, \tag{3.93}$$

where the parameter c is 0 for the van der Waals and 1 for the Berthelot equation of state. By enforcing both $(\partial P/\partial v)_T = 0$ and $(\partial^2 P/\partial v^2)_T = 0$ at the critical point (P_c, T_c, v_c) the constants a and b can be deduced in terms of the critical-point coordinates and the equation of state is written as

$$\left(P_r + 3T_r^{-c}v_r^{-2}\right)\left(v_r - \frac{1}{3}\right) = \frac{8}{3}T_r, \tag{3.94}$$

where the reduced thermodynamic variables are

$$P_r = \frac{P}{P_c}, \qquad T_r = \frac{T}{T_c}, \qquad v_r = \frac{v}{v_c}. \tag{3.95}$$

In terms of the reduced properties, the spinodal limit is found at

$$\left(\frac{\partial P_r}{\partial v_r}\right)_{T_r} = 0, \tag{3.96}$$

and the following relation is obtained for the spinodal limit:

$$(T_r)_{sp} = \left[\frac{(3v_r - 1)^2}{4v_r^3}\right]^{1/(c+1)}. \tag{3.97}$$

The van der Waals equation of state predicts that a liquid can be superheated to over 80% of the critical point before reaching the spinodal limit. Slightly higher values are predicted by the Berthelot equation of state. These arguments can be extended to metals subjected to pulsed heating (Miotello and Kelly, 1995; Kelly and Miotello, 1996) following the work by Martynyuk (1974, 1976).

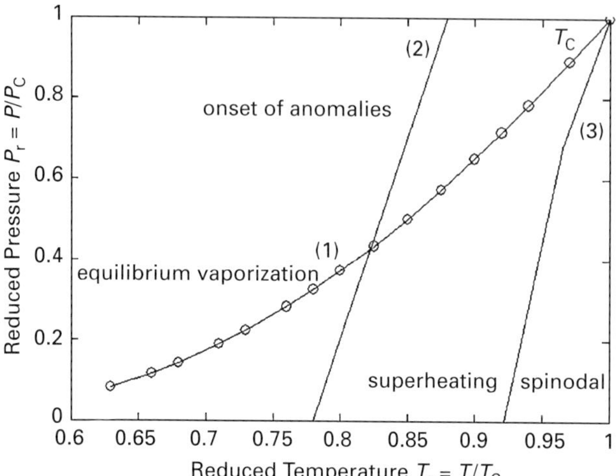

Figure 3.4. The phase diagram of a metal in the neighborhood of the critical point T_c (Kelly and Miotello, 1996). Equilibrium vaporization refers to experimental equilibrium vapor pressures obtained for Cs when both liquid and vapor are present.

Figure 3.4 shows the P–T diagram for a typical metal (Martynyuk, 1983). The curve marked by (1) is the experimental equilibrium binodal; curve (2) is the spinodal calculated by use of the Berthelot equation of state; and line (3) marks the onset of anomalies in the density and specific heat. Upon approaching the critical point, T_c (for $T \gtrsim 0.80\,T_c$) the specific heat exhibits a rapid rise, while the density drop deviates from the essentially linear dependence followed until that range. Under conditions of slow heating, the liquid pressure is expected to increase, riding the saturation curve, until $P_{sat}(T) = P_\infty$, where P_∞ is the ambient pressure. The transition temperature T_{boil} is the equilibrium boiling temperature. However, under fast heating rates the system is forced into the metastable (superheated) regime.

Homogeneous nucleation is expected in a superheated liquid. The nucleation rate for stationary (steady-state) conditions follows the classical relation

$$I_{nuc} \approx N_l B \exp\left(-\frac{\Delta G_c}{k_B T}\right), \tag{3.98}$$

where N_l is the number of atoms per unit volume of the liquid, ΔG_c is the work of formation of a critical vapor nucleus at the temperature T, and B is a function of rather weak dependence with respect to temperature and pressure. Homogeneous nucleation is possible only at relatively high levels of superheating. Martynyuk (1974) calculated that for cesium $I_{nuc} = 1$ cm^{-3} s^{-1} at $T = 0.874T_c$. However, $I_{nuc} = 10^{26}$ cm^{-3} s^{-1} at just $T = 0.905T_c$. The homogeneous nucleation following the instantaneous conversion of a liquid into the metastable state is in effect a transient process whose temporal dependence is given by

$$I_{\text{nuc}} \approx I_{\text{nuc},0} \exp\left(-\frac{\tau}{t}\right). \tag{3.99}$$

In the above, τ is the characteristic time scale for the establishment of steady-state nucleation. The range of τ for many metals is 10^{-9}–10^{-8} s. Consequently, transient effects will manifest themselves for heating times $<10^{-7}$ s and corresponding heating rates $>10^{11}$ K s^{-1}. Typical absorption depths in metals are about $10\,\text{nm} = 10^{-8}$ m. According to Miotello and Kelly (1995) it may be assumed that this skin depth is instantaneously raised to the critical-point neighborhood. The volume per unit area of this superheated zone is $V^* = 10^{-8}$ cm^3. A reasonable estimate of the critical vapor nucleus radius is $O(10^{-6}$ cm), leading to a number of critical nuclei in V^* of $O(10^{11})$. The number of critical nuclei formed within the duration of the laser pulse would then be $I_0 V^* t_{\text{pulse}}$, yielding $O(10^{11}$–$10^{12})$ critical nuclei under the assumption of $I_0 = 10^{26}$ cm^{-3} s^{-1} as mentioned above and for pulse durations t_{pulse} in the range 1–10 ns. On the basis of this simplified qualitative analysis, it is therefore argued that the homogeneous-nucleation theory is relevant for the length and temporal scales of nanosecond pulsed laser heating.

Latent-heat effects vanish in the vicinity of T_{c}. The advancement of the vaporization front can then be traced with the temporal progress of the spinodal isotherm. It is also pertinent to note that the pressure rises sharply near T_{c}. This may result in liquid expulsion in the form of droplets, thereby further enhancing the material-removal rate.

It is noted that, although explosive boiling may be an inevitable process when the liquid is superheated, there are limitations according to kinetic theory. When the liquid is superheated, homogeneous bubble nucleation occurs and the liquid experiences large density fluctuation. Only if these bubbles reach a critical radius, r_{c}, will they grow spontaneously. Bubbles with radius less than r_{c} are likely to collapse, and it takes the bubble a time τ_{c} to grow to the critical radius r_{c}. The expressions for r_{c} and τ_{c} are (Carey, 1992)

$$r_{\text{c}} = \frac{2\sigma_{\text{ST}}}{P_{\text{sat}}(T_1)\exp[v_l(P_l - P_{\text{sat}}(T_1))/(RT_1)] - P_1}, \tag{3.100}$$

$$\tau_{\text{c}} = r_{\text{c}}\left[\frac{2}{3}\left(\frac{T_1 - T_{\text{sat}}(P_1)}{T_{\text{sat}}(P_1)}\right)\frac{L_{\text{v}}\rho_{\text{v}}}{\rho_1}\right]^{-\frac{1}{2}}, \tag{3.101}$$

where σ_{ST} is the surface tension of the liquid, L_{v} and R are the latent heat of vaporization and gas constant, respectively, ρ_1 and ρ_{v} are the densities of superheated liquid and vapor, with $v_1 = 1/\rho_1$. T_1 is the temperature of the superheated liquid, which can be taken as $0.85T_{\text{c}}$ when explosive boiling occurs, P_{sat} and T_{sat} are the saturation pressure and temperature at the temperature of the superheated liquid, which can be obtained from the Clausius–Clapeyron relation, and P_1 is the pressure of the superheated liquid, which can be approximated by the recoil pressure of the evaporating vapor (Von Allmen, 1987), which is $0.54 p_{\text{sat}}(T_1)$. According to the power-law relation of surface tension σ for a liquid metal (Yoshida, 1994), the surface tension drops about 80% at the assumed T_l. Using these parameters, it can be estimated that r_c and τ_c are approximately 0.6 μm and 70 ns for a nanosecond laser ablation of silicon.

These calculations indicate that it would take bubbles about 70 ns to grow to the critical radius of 0.6 μm. Subsequently, the superheated liquid will undergo a transition into a mixture of vapor and liquid droplets, followed by explosive boiling of the liquid–vapor mix. However, during the nanosecond laser pulse as studied by Lu *et al.* (2002), the bubble doesn't have enough time to reach the critical radius during the laser pulse. As a result, without efficient energy dissipation, the liquid temperature can exceed the critical temperature if the laser irradiance is sufficiently high.

The thermal penetration depth $d_{th} = 0.969(\alpha\tau)^{1/2}$ during a laser pulse of duration τ is much larger than the optical penetration $1/\gamma$ in the case of 3-ns laser irradiation of silicion, where α is the thermal diffusivity of the liquid silicon, which is about 0.75 cm²/s, and γ is the absorption coefficient. Therefore, the thermal penetration depth is calculated to be about 0.47 μm. The critical diameter of the bubble is $d_c = 2r_c$, or 1.2 μm, which is larger than the thermal penetration depth; the bubble cannot grow to its critical radius during the laser pulse.

On the basis of the above analysis, a refined delayed-phase-explosion model (Lu *et al.*, 2002) was proposed to describe nanosecond ablation where explosive boiling plays a role in material removal. Explosive boiling will not occur during the laser pulse. However, without significant bubble formation, a high-temperature layer will form below the target surface during the laser pulse, with a depth equal to about the thermal penetration depth. At the same time the target undergoes normal vaporization from the extreme outer surface. Mass ablation below the laser-irradiance threshold is generated by this normal-vaporization mechanism, with the vaporization flux governed by the Hertz–Knudsen equation. At high laser irradiance, after the laser pulse has been completed, the high-temperature liquid layer will propagate into the target with thermal diffusion. Part of the liquid layer in the target may approach the critical temperature and, therefore, new bubbles will emerge inside the superheated liquid, eventually leaving the target.

References

Anisimov, S. I., 1968, "Vaporization of Metal Absorbing Laser Radation," *Sov. Phys. JETP*, **27**, 182–183.

Bäuerle, D., 1996, *Laser Processing and Chemistry*, Heidelberg, Springer-Verlag.

Benes, E., 1984, "Improved Quartz Microbalance Technique," *J. Appl. Phys.*, **56**, 608.

Bennett, T. D., Grigoropoulos, C. P., and Krajnovich, D. J., 1995, "Near-Threshold Laser Sputtering of Gold," *J. Appl. Phys.*, **77**, 849–864.

Carey, V. P., 1992, *Liquid–Vapor Phase-Change Phenomena*, Bristol, PA, Taylor & Francis.

Carslaw, H. S., and Jaeger, J. C., 1959, *Conduction of Heat in Solids*, 2nd edn, Oxford, Oxford University Press.

Dreyfus, R. W., 1991, "Cu⁰, Cu⁺, and Cu₂ from Excimer-Ablated Copper," *J. Appl. Phys.*, **69**, 1721–1729.

Fucke, W., and Seydel, U., 1980, "Improved Experimental Determination of Critical-Point Data for Tungsten," *High Temp. – High Pressures*, **12**, 419–432.

Grigoropoulos, C. P., Rostami, A. A., Xu, X., Taylor, S. L., and Park, H. K., 1993a, "Localized Transient Surface Reflectivity Measurements and Comparison to Heat Transfer Modeling in Thin Film Laser Annealing," *Int. J. Heat Mass Transfer*, **36**, 1219–1229.

Ho, J.-R., Grigoropoulos, C. P., and Humphrey, J. A. C., 1995, "Computational Model for the Heat Transfer and Gas Dynamics in the Pulsed Laser Evaporation of Metals," *J. Appl. Phys.*, **78**, 4696–4709.

Jackson, K. A., 1975, "Theory of Melt Growth," in *Crystal Growth and Characterization*, edited by R. Ueda and J. B. Mullin, Amsterdam, North-Holland.

Jackson, K. A., and Chalmers, B., 1956, "Kinetics of Solidification," *Can. J. Phys.*, **34**, 473–490.

Kelly, R., 1990, "On the Dual Role of the Knudsen Layer and Unsteady, Adiabatic Expansion in Pulse Sputtering Phenomena," *J. Chem. Phys.*, **92**, 5047–5056.

Kelly, R., and Miotello, A., 1996, "Comments on Explosive Mechanisms of Laser Sputtering," *Appl. Surf. Sci.*, **96–98**, 205–215.

Kelly, R., and Rothenberg, J. E., 1985, "Laser Sputtering: Part III. The Mechanism of the Sputtering of Metals at Low Energy Densities," *Nucl. Instrum. Meth. Phys. Res.*, **B7/8**, 755–763.

Knight, C. J., 1976, "Evaporation from a Cylindrical Surface into Vacuum," *J. Fluid Mech.*, **75**, 469–486.

1979, "Theoretical Modeling of Rapid Surface Vaporization with Back Pressure," *AIAA J.*, **17**, 519–523.

Larson, B. C., Tischler, J. Z., and Mills, D. M., 1986, "Nanosecond Resolution Time-Resolved X-Ray Study of Silicon during Pulsed-Laser Irradiation," *J. Mater. Res.*, **1**, 144–154.

Larson, B. C., White, C. W., Noggle, T. S., and Mills, D., 1982, "Synchrotron X-Ray Diffraction Study of Silicon during Pulsed Laser-Annealing," *Phys. Rev. Lett.*, **48**, 337–340.

Lu, Q. M., Mao, S. S., Mao, X. L., and Russo, R. E., 2002, "Delayed Phase Explosion during High-Power Nanosecond Laser Ablation of Silicon," *Appl. Phys. Lett.*, **80**, 3072–3074.

Martynyuk, M. M., 1974, "Vaporization and Boiling of Liquid Metal in an Exploding Wire," *Sov. Phys. Techn. Phys.*, **19**, 793–797.

1976, "Mechanism for Metal Damage by Intense Electromagnetic Radiation," *Sov. Phys. Techn. Phys.*, **21**, 430–433.

1983, "Critical Constants of Metals," *Russ. J. Phys. Chem.*, **57**, 494–501.

Miller, J. C., 1969, "Optical Properties of Liquid Metals at High Temperatures" *Phil. Mag.*, **20**, 1115–1132.

Miotello, A., and Kelly, R., 1995, "Critical Assessment of Thermal Models for Laser Sputtering at High Fluences," *Appl. Phys. Lett.*, **67**, 3535–3537.

Moody, J. E., and Hendel, R. H., 1982, "Temperature Profile Induced by a Scanning CW Laser Beam," *J. Appl. Phys.*, **53**, 4364–4371.

Özisik, M. N., 1993, *Heat Conduction*, 2nd edn, New York, John Wiley.

Palik, E. D., 1985, *Handbook of Optical Constants of Solids*, Vol. I, London, Academic.

1991, *Handbook of Optical Constants of Solids*, Vol. II, London, Academic.

Park, H. K., Grigoropoulos, C. P., Poon, C. C., and Tam, A. C., 1996, "Optical Probing of the Temperature Transients during Pulsed-Laser Induced Vaporization of Liquids," *Appl. Phys. Lett.*, **68**, 596–598.

Peercy, P. S., Thompson, M. O., and Tsao, J. Y., 1987, "Dynamics of Rapid Solidification in Silicon," *Proc. Mater. Res. Soc.*, **5**, 15–30.

Prokhorov, A. M., Konov, V. I., Ursu, I., and Mihailescu, I. N., 1990, *Laser Heating of Metals*, New York, Adam Hilger.

Rykalin, N., Uglov, A., Zuev, I., and Kokora, A., 1988, *Laser and Electron Beam Materials Processing Handbook*, Moscow, Mir Publishers, pp. 194–232.

Von Allmen, M., 1987, *Laser-Beam Interactions with Materials*, Heidelberg, Springer-Verlag.

Yoshida, A., 1994, "Critical Phenomenon Analysis of Surface-Tension of Liquid-Metals," *J. Jap. Inst. Metals*, **58**, 1161–1168.

Zhang, X., Chu, S. S., Ho, J. R., and Grigoropoulos, C. P., 1997, "Excimer Laser Ablation of Thin Gold Films on a Quartz Crystal Microbalance at Various Argon Background Pressures," *Appl. Phys. A*, **64**, 545–552.

4 Desorption at low laser energy densities

4.1 Vapor kinetics

4.1.1 Statistical mechanics

For a single molecule the partition function q should result from a summation of the energies corresponding to the different microstates over all possible generalized momenta $\hat{p}$ and position coordinates $\hat{q}$:

$$q = C_q \int \int \int \cdots \int \mathrm{e}^{\frac{-\hat{H}(\hat{p}_1, \hat{p}_2, \dots, \hat{p}_\xi, \hat{q}_1, \hat{q}_2, \dots, \hat{q}_\xi)}{k_{\mathrm{B}} T}} \, \mathrm{d}\hat{p}_1 \, \mathrm{d}\hat{p}_2 \cdots \mathrm{d}\hat{p}_\xi \, \mathrm{d}\hat{q}_1 \, \mathrm{d}\hat{q}_2 \cdots \mathrm{d}\hat{q}_\xi, \quad (4.1)$$

where $\hat{H}$ is the Hamiltonian corresponding to the particular microstate and C_q is a proportionality constant (e.g. Carey, 1999). In the specific case of the classical limit of a monatomic gas, energy is stored only by translation. In this case, the Hamiltonian $\hat{H}$ and the constant C_q are

$$\hat{H} = \frac{1}{2m}\left(\hat{p}_x^2 + \hat{p}_y^2 + \hat{p}_z^2\right), \qquad C_q = \frac{1}{h^3}, \quad (4.2)$$

where m is the atom mass and h Planck's constant.

The translational partition function is then

$$q_{\mathrm{tr}} = \frac{V}{h^3} \int \int \int \mathrm{e}^{-\frac{\hat{p}_x^2 + \hat{p}_y^2 + \hat{p}_z^2}{2mk_{\mathrm{B}}T}} \, \mathrm{d}\hat{p}_x \, \mathrm{d}\hat{p}_y \, \mathrm{d}\hat{p}_z, \quad (4.3)$$

where V is the gas volume and the integration limits extend from $-\infty$ to $+\infty$. Evaluation of Equation (4.3) gives

$$q_{\mathrm{tr}} = \frac{V(2\pi m k_{\mathrm{B}} T)^{3/2}}{h^3}. \quad (4.4)$$

The probability of finding an atom with momenta between $\langle \hat{p}_x, \hat{p}_y, \hat{p}_z \rangle$ and $\langle \hat{p}_x + \mathrm{d}\hat{p}_x, \hat{p}_y + \mathrm{d}\hat{p}_y, \hat{p}_z + \mathrm{d}\hat{p}_z \rangle$ is

$$f_{\mathrm{prob}}(\hat{p}_x, \hat{p}_y, \hat{p}_z) = \frac{\dfrac{1}{h^3} \mathrm{e}^{-\frac{\left(\hat{p}_x^2 + \hat{p}_y^2 + \hat{p}_z^2\right)}{2mk_{\mathrm{B}}T}} \, \mathrm{d}\hat{p}_x \, \mathrm{d}\hat{p}_y \, \mathrm{d}\hat{p}_z}{q_{\mathrm{tr}}}$$

$$= \left(\frac{1}{2\pi m k_{\mathrm{B}} T}\right)^{\frac{3}{2}} \exp\left[-\frac{\hat{p}_x^2 + \hat{p}_y^2 + \hat{p}_z^2}{2mk_{\mathrm{B}}T}\right] \mathrm{d}\hat{p}_x \, \mathrm{d}\hat{p}_y \, \mathrm{d}\hat{p}_z. \quad (4.5)$$

By transforming from the momentum to the velocity phase space we obtain

$$
f_{\text{prob}}(u_x, u_y, u_z) = \frac{\dfrac{1}{h^3} e^{-\frac{m\left(u_x^2 + u_y^2 + u_z^2\right)}{2k_B T}} \, \mathrm{d}u_x \, \mathrm{d}u_y \, \mathrm{d}u_z}{q_{\text{tr}}}
$$

$$
= \left(\frac{m}{2\pi k_B T}\right)^{\frac{3}{2}} \exp\left[-\frac{m\left(u_x^2 + u_y^2 + u_z^2\right)}{2k_B T}\right] \mathrm{d}u_x \, \mathrm{d}u_y \, \mathrm{d}u_z.
$$

If a spherical polar coordinate system is adopted, the above expression is written

$$
f_{\text{prob}}(u) = \left(\frac{m}{2\pi k_B T}\right)^{\frac{3}{2}} \exp\left(-\frac{mu^2}{2k_B T}\right) u^2 \sin\theta \, \mathrm{d}\theta \, \mathrm{d}\phi \, \mathrm{d}u. \tag{4.6}
$$

Integrating the above over a full sphere gives the probability distribution for the absolute magnitude of the velocity:

$$
f_{\text{prob}}(u) \, \mathrm{d}u = \left(\frac{m}{2\pi k_B T}\right)^{\frac{3}{2}} \exp\left(-\frac{mu^2}{2k_B T}\right) u^2 \, \mathrm{d}u \int_0^{\pi} \sin\theta \, \mathrm{d}\theta \int_0^{2\pi} \mathrm{d}\phi
$$

$$
= 4\pi u^2 \left(\frac{m}{2\pi k_B T}\right)^{\frac{3}{2}} \exp\left(-\frac{mu^2}{2k_B T}\right) \mathrm{d}u. \tag{4.7}
$$

The number of molecules per unit volume of the gas with speed in the range of u to $u + \mathrm{d}u$ is

$$
N_u = N f_{\text{prob}}(u) \mathrm{d}u, \tag{4.8}
$$

where N is the number of molecules per unit volume.

Consider now the efflux of molecules from a plane surface of area S. The number of molecules crossing S in time $\mathrm{d}t$ with speed in the range u to $u + \mathrm{d}u$ within a pencil of a solid angle $\mathrm{d}\Omega$ along the direction θ with respect to the normal to S is

$$
N_u \frac{\mathrm{d}\Omega}{4\pi} u S \cos\theta \, \mathrm{d}t.
$$

Since $\mathrm{d}\Omega = \sin\theta \, \mathrm{d}\theta \, \mathrm{d}\phi$, the number of molecules crossing unit area per unit time is

$$
N \left(\frac{m}{2\pi k_B T}\right)^{\frac{3}{2}} u^3 \exp\left(-\frac{mu^2}{2k_B T}\right) \mathrm{d}u \sin\theta \cos\theta \, \mathrm{d}\theta \, \mathrm{d}\phi. \tag{4.9}
$$

The total number of molecules released from a unit area of the surface can be found by integrating (4.9) over all speeds through a hemisphere:

$$
j = N \left(\frac{m}{2\pi k_B T}\right)^{\frac{3}{2}} \int_0^{\infty} u^3 \exp\left(-\frac{mu^2}{2k_B T}\right) \mathrm{d}u \int_0^{\frac{\pi}{2}} \sin\theta \cos\theta \, \mathrm{d}\theta \int_0^{2\pi} \mathrm{d}\phi
$$

$$
= N \left(\frac{m}{2\pi k_B T}\right)^{\frac{3}{2}} \frac{\pi}{2} \left(\frac{2k_B T}{m}\right)^2. \tag{4.10}
$$

The probability then of a molecule leaving the surface possessing a speed between u and $u + \mathrm{d}u$ is found by integrating (4.9) over a hemisphere and scaling with (4.10):

$$
f_{\mathrm{prob}}(u)\mathrm{d}u = \frac{N\left(\dfrac{m}{2\pi k_{\mathrm{B}}T}\right)^{\frac{3}{2}} u^3 \exp\left(-\dfrac{mu^2}{2k_{\mathrm{B}}T}\right)\mathrm{d}u\,\pi}{N\left(\dfrac{m}{2\pi k_{\mathrm{B}}T}\right)^{\frac{3}{2}} \dfrac{\pi}{2}\left(\dfrac{2k_{\mathrm{B}}T}{m}\right)^2}
$$

$$
= \frac{m^2 u^3 \exp\left(-\dfrac{mu^2}{2k_{\mathrm{B}}T}\right)\mathrm{d}u}{2(k_{\mathrm{B}}T)^2}. \tag{4.11}
$$

The translational energy probability is found by imposing $f_{\mathrm{prob}}(E_{\mathrm{tr}})\mathrm{d}E_{\mathrm{tr}} = f(u)\mathrm{d}u$ with $E_{\mathrm{tr}} = mu^2/2$ and $\mathrm{d}E_{\mathrm{tr}} = mu\,\mathrm{d}u$:

$$
f_{\mathrm{prob}}(E_{\mathrm{tr}}) = \frac{E_{\mathrm{tr}}\exp\left(-\dfrac{E_{\mathrm{tr}}}{k_{\mathrm{B}}T}\right)}{(k_{\mathrm{B}}T)^2}. \tag{4.12}
$$

The mean translational energy is

$$
\bar{E}_{\mathrm{tr}} = \int_0^\infty E_{\mathrm{tr}} f_{\mathrm{prob}}(E_{\mathrm{tr}})\mathrm{d}E_{\mathrm{tr}} = 2k_{\mathrm{B}}T. \tag{4.13}
$$

Correspondingly, the mean translational speed is

$$
\bar{u} = \int_0^\infty u f_{\mathrm{prob}}(u)\mathrm{d}u = 3\left(\frac{\pi k_{\mathrm{B}}T}{2m}\right)^{\frac{1}{2}}. \tag{4.14}
$$

4.1.2 Collisional processes in the plume

The above relations were derived under the assumption that the nascent distribution is preserved, i.e. in the limit of collisionless sputtering. If, for a hard-sphere model, the scattering cross section σ_{c} is considered independent of the absolute and relative velocities, the differential rate of near-surface collisions is

$$
\mathrm{d}j_{\mathrm{col}} = \kappa_{\mathrm{c}} N_1 N_2 \sigma_{\mathrm{c}} \vec{u}_{\mathrm{r}} f_{\mathrm{prob}}(\vec{u}_1) f_{\mathrm{prob}}(\vec{u}_2)\mathrm{d}\vec{u}_1\,\mathrm{d}\vec{u}_2, \tag{4.15}
$$

where $\vec{u}_{\mathrm{r}}$ is the relative velocity:

$$
\vec{u}_{\mathrm{r}} = \vec{u}_1 - \vec{u}_2 = \vec{u}_1^{\mathrm{col}} - \vec{u}_2^{\mathrm{col}}. \tag{4.16}
$$

κ_{c} is a symmetry factor equal to $\frac{1}{2}$ if $f_{\mathrm{prob}}(\vec{u}_1)$ and $f_{\mathrm{prob}}(\vec{u}_2)$ describe identical atoms and equal to 1 otherwise.

The atom speeds before and after the collision process are

$$
\vec{u}_1 = \vec{u}_{\mathrm{cm}} + \frac{m_2}{m_1 + m_2}\vec{u}_{\mathrm{r}}, \tag{4.17a}
$$

$$
\vec{u}_2 = \vec{u}_{\mathrm{cm}} - \frac{m_1}{m_1 + m_2}\vec{u}_{\mathrm{r}}, \tag{4.17b}
$$

where $\vec{u}_{cm}$ is the center-of-mass velocity:

$$\vec{u}_{cm} = \frac{m_1 \vec{u}_1 + m_2 \vec{u}_2}{m_1 + m_2}.$$ (4.18)

The differential collisional rate is evaluated by changing the variables from $\vec{u}_1$ and $\vec{u}_2$ to $\vec{u}_r$ and $\vec{u}_{cm}$. According to (4.17), the Jacobian of this transformation is

$$\frac{\partial(u_{1x}, u_{2x})}{\partial(u_{rx}, u_{cmx})} = \begin{vmatrix} \dfrac{\partial u_{1x}}{\partial u_{rx}} & \dfrac{\partial u_{1x}}{\partial u_{cmx}} \\ \dfrac{\partial u_{2x}}{\partial u_{rx}} & \dfrac{\partial u_{2x}}{\partial u_{cmx}} \end{vmatrix} = 1.$$ (4.19)

Hence,

$$\mathrm{d}j_{col} = \kappa_c N_1 N_2 \sigma_{c,col} \vec{u}_r f_{prob}(\vec{u}_r) f_{prob}(\vec{u}_{cm}) \mathrm{d}\vec{u}_r \, \mathrm{d}\vec{u}_{cm}.$$ (4.20)

On considering the velocity distributions of Equation (4.11) and integrating for $u_{cmz} > 0$, Equation (4.20) gives the collision rate per unit volume in an effusive flux:

$$j_{col} = \kappa_c N_1 N_2 \sigma_{c,col} \frac{4}{3} \left(\frac{2\pi k_B T}{m_1 + m_2} \right)^{\frac{1}{2}}.$$ (4.21)

For a one-component gas, $\kappa_c = 1/2$,

$$j_{col} = N^2 \sigma_{c,col} \frac{2}{3} \left(\frac{\pi k_B T}{m} \right)^{\frac{1}{2}}$$ (4.22)

and the collision rate per atom released from the surface is

$$j_{col,a} = P_v \sigma_{c,col} \frac{2}{3} \left(\frac{\pi}{m k_B T} \right)^{\frac{1}{2}},$$ (4.23)

where P_v is the vapor pressure at the condensed-phase temperature.

4.2   Time-of-flight instruments

4.2.1   Overview of mass/energy analyzers

Mass and energy characterization of laser-ablated species provide valuable information for the study of laser-induced desorption processes. Mass analyzers (a general schematic diagram is given in Figure 4.1) can be classified into scanning mass analyzers and time-of-flight (TOF) mass analyzers on the basis of their principles of operation.

Scanning mass analyzers

In scanning mass analyzers, electromagnetic fields separate ions according to their mass-to-charge ratios. In the magnetic mass spectrometer, ions pass through a magnetic sector after acceleration to high velocities. A magnetic field is applied in the direction normal to the ion motion. Assuming that the accelerating field has a potential V_f, the kinetic

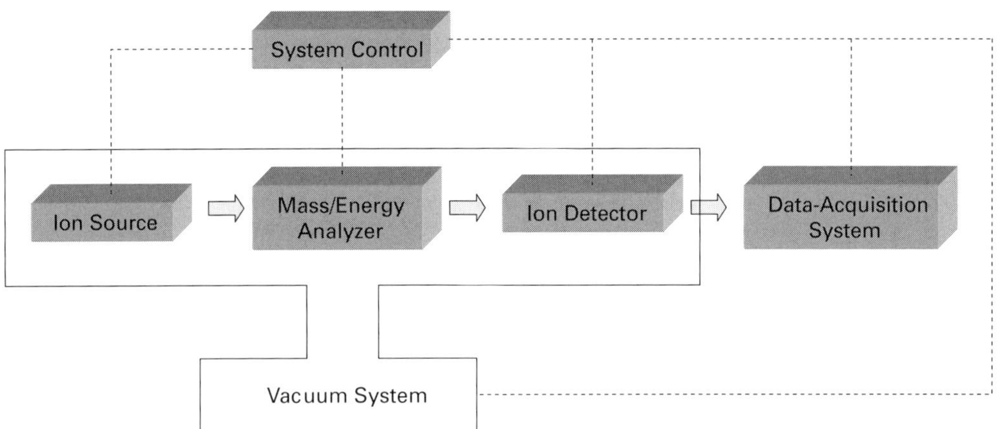

Figure 4.1. A schematic diagram of a general mass/energy analyzer system.

energy gained by the ions after passing through the electric field is

$$q_c V_f = mu^2/2, \tag{4.24}$$

where q_c, m, and u represent the charge, mass, and velocity of the ions.

According to the Lorentz force law,

$$q_c u B = mu^2/r. \tag{4.25}$$

Hence, the working equation for a magnetic-sector spectrometer is

$$m/q_c = B^2 r^2/(2V_f). \tag{4.26}$$

The magnetic-sector mass spectrometer operates either by holding the field constant at V_f and scanning the magnetic field, or by holding B constant and scanning V_f. The different mass-to-charge ratios are scanned across a slit placed ahead of the detector measuring the ion current. Double-focusing magnetic-sector mass analyzers offer the best quantitative performance, with high resolution, sensitivity, and dynamic range. However, they are not well suited for pulsed ionization methods and are usually larger and heavier than other mass analyzers.

The quadrupole mass spectrometer (QMS) is essentially a "mass filter" enforced by combined DC and RF potentials on quadrupole rods that allow passage only of ions having a selected mass-to-charge ratio. The QMS instruments have good reproducibility, and are compact and relatively less costly, but they offer limited resolution.

Time-of-flight mass spectrometers

A time-of-flight (TOF) mass spectrometer measures the mass-dependent time it takes for the ionized species to reach the detector. Ions of the same kinetic energy will move with different velocities. The kinetic energy gained by the ions from the electric field is

$$mu^2/2 = q_c V_f. \tag{4.27}$$

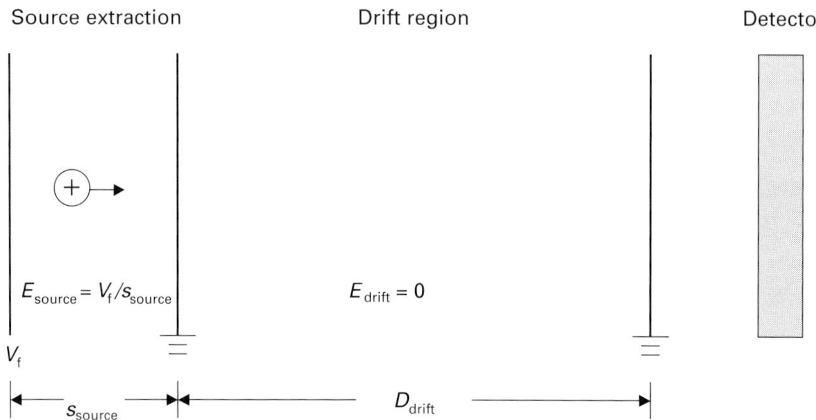

Figure 4.2. A schematic diagram of a TOF mass spectrometer.

The flight time it takes for an ion to traverse an electric field of length L is

$$t_{\text{f}} = L/u. \tag{4.28}$$

Consequently, the working equations for a TOF mass spectrometer are

$$t_{\text{f}} = L\sqrt{\frac{m}{2q_{\text{c}}V_{\text{f}}}}, \tag{4.29a}$$

$$m = \frac{2q_{\text{c}}V_{\text{f}}t_{\text{f}}^2}{L^2}. \tag{4.29b}$$

It is noted that the ions leaving the ion source have a distribution of kinetic energies and are generated over a time interval. The resolution of the mass spectrometer may be affected by these factors. The TOF instruments are the fastest mass analyzers, acquiring complete spectra instantaneously, and they have high mass range and ion-transmission ratio. Figure 4.2 shows a schematic diagram of a TOF mass spectrometer. A positive voltage V_{f} is applied across the source region, s_{source}, that is defined by the backing plate and the extraction grid. All ions of mass m and charge state Z are then accelerated to the same kinetic energy:

$$\frac{mu^2}{2} = ZeV_{\text{f}}s_{\text{source}}. \tag{4.30}$$

The exiting velocities, however, depend on the mass-to-charge ratio:

$$u = \sqrt{\frac{2ZeV_{\text{f}}s_{\text{source}}}{m}}. \tag{4.31}$$

The ions then pass through a much longer drift region (of length D_{drift}) with constant velocities. The time it takes for the ions to fly from the source to the detector is

$$t_{\text{f}} = D_{\text{drift}}\sqrt{\frac{m}{2ZeV_{\text{f}}s_{\text{source}}}}. \tag{4.32}$$

The time-of-flight data are then converted into a mass spectrum by using known values
of the accelerating electric field and the lengths of the source region and drift tube:

$$\frac{m}{Z} = 2e\,V_f s_{\text{source}} \left(\frac{t_f}{D_{\text{drift}}}\right)^2 . \tag{4.33}$$

In practice, the following empirical equation is used for calibrating the mass spectrum
and determining unknown masses:

$$\frac{m}{Z} = at_f^2 + b, \tag{4.34}$$

where the constants a and b are found by measuring the flight times of ions having at
least two known masses. The mass resolution of a mass spectrometer is determined by
the full width at half maximum (FWHM) of the ion peak. A resolution of 100 means
that the peak for an ion of $m/Z = 100$ is one mass unit wide. The resolution is related
to the flight-time spread Δt_f:

$$\frac{m}{\Delta m} = \frac{t_f}{2}\,\Delta t_f. \tag{4.35}$$

Consequently, in order to improve the mass resolution of a TOF instrument, the
spread in flight times for a given m/Z value must be minimized. As analyzed by Wiley
and McLaren (1955) and Cotter (1962), the magnitude of the spread is affected by
uncertainties in the time and location of ion formation, and the initial kinetic energy of
the ions (Figure 4.3).

Techniques aiming to improve the mass resolution include time-lag focusing, pulsed
ionization, double-field acceleration, and reflectron arrangements. The resolution of a
linear TOF mass spectrometer is typically in the range 100–300, whereas the resolution
of the mass reflectron can be greater than 1000. As shown in Figure 4.3(a), ions of the
same mass-to-charge ratio and endowed with the same kinetic energy tend to maintain
a constant difference in time and space across the drift tube. Pulsing electric fields
and laser-assisted ionization with lasers of pulse duration tens of nanoseconds are
utilized to minimize these effects. Figure 4.3(b) depicts the effect of uncertainty in ion-
formation location. When two ions of the same mass-to-charge ratio are formed with
the same initial velocity, the ion formed closer to the backing plate will be accelerated
to higher kinetic energy to attain higher velocity through the drift region. The effects
of uncertainties related to the initial spatial distribution of ions can be minimized by
adjusting the extraction field. A major factor limiting the mass resolution of linear mass
spectrometers is due to the initial kinetic-energy distribution (Figure 4.3(c)). Mamyrin
et al. (1973) introduced the mass-reflectron device, consisting of a series of retarding
plates at the end of the flight tube (Figure 4.4). The faster ions arrive earlier at the
reflectron and penetrate it deeper, achieving longer residence times than those of their
slower counterparts. Eventually the reflectron fields reverse the direction of the ions and
accelerate them out of the reflectron. By adjusting the reflectron field strength, all ions
can be made to arrive at the detector at the same time.

(a) $\Delta t_0 \neq 0$

(b) $\Delta S_0 \neq 0$

(c) $\Delta U_0 \neq 0$

(d) $V_1 \neq V_2$

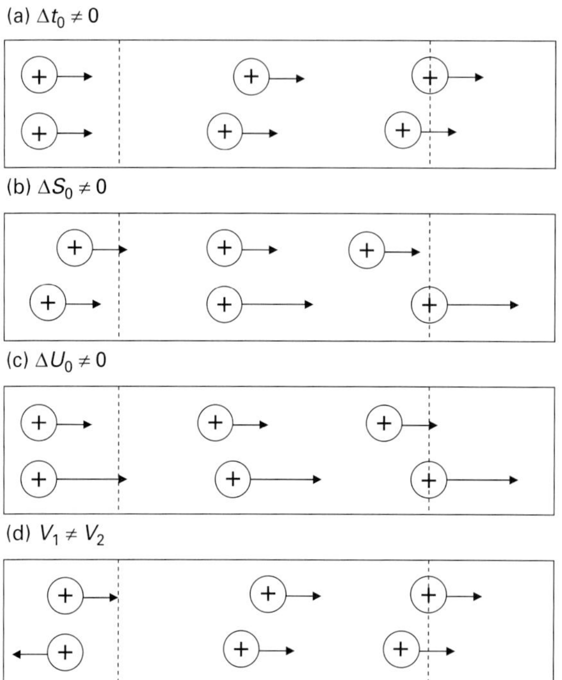

Figure 4.3. Effects of initial time, space, and kinetic-energy distributions on mass resolution: (a) two ions formed at different times; (b) two ions formed at different locations in the ion-extraction field; (c) two ions formed with different kinetic energies; and (d) two ions with the same initial kinetic energy that have initial velocities in opposite directions (Cotter, 1992).

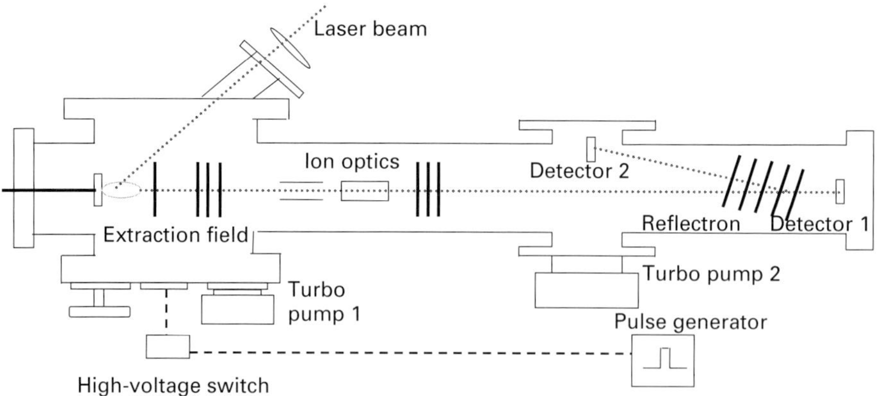

Figure 4.4. A schematic diagram of a TOF mass spectrometer, composed of a source chamber, acceleration electric fields, ion optics, a flight tube, a reflectron assembly, two detectors, a vacuum system, a laser optical system, and a data-acquisition and control system.

Laser mass spectrometry

In *laser-ablation/ionization mass spectrometry*, a pulsed laser is used to ablate and ionize the sample in a single-step process. The initiation of the plasma and its temporal evolution result from many complex physical processes, as shall be explained in Chapter 5. Intact neutral molecules are desorbed from the sample at low laser energy densities. In *laser-ablation/post-ionization spectrometry*, these neutral species are ionized via pulsed laser-induced multiphoton ionization (MPI), or by electron ionization. Photo-ionization occurs if the photon energy exceeds the ionization potential. Since typical ionization potentials for metals are in the range of 10 eV, this is not an efficient process. Multiphoton ionization is a higher-order process that can be achieved with high powers of the ionizing laser. Resonant MPI occurs when the frequency of the ionizing laser (typically a narrow-line-width dye laser) is tuned to an intermediate electronic state of the molecule. Extremely high selectivity and sensitivity can be achieved with resonant MPI.

Time-of-flight measurements characterize the translational kinetic energies of the ejected plume. In the absence of collisional and plasma effects or chemical reactions in the plume, it would appear that the TOF kinetic-energy distributions can be referred to the surface thermal condition with probability density given by Equation (4.12). If the surface flux is approximated by a point source, and the particle stream is collisionless with $u_z^2 \gg u_x^2, u_y^2$ and $u_z = L/t_f$, the detector signal is

$$I_d(t) \sim \frac{1}{t_f^4} \exp\left(-\frac{m(L/t_f)^2}{2k_B T}\right). \tag{4.36}$$

In the limit of a small surface vapor flux, the expansion of vapor into vacuum is collisionless, the energy distribution is Boltzmannian, and the mean translational energy should be indicative of the surface phase temperature, as indicated by Equation (4.13). With the development of a stream velocity $\tilde{u}$, the flight distance relative to coordinates moving at $\tilde{u}$ becomes $\tilde{L} = L - \tilde{u}t_f$, so Equation (4.36) becomes

$$I_d(t) \sim \frac{1}{t_f^4} \exp\left(-\frac{m[(L - \tilde{u}t_f)/t_f]^2}{2k_B T}\right). \tag{4.37}$$

The temporal behavior of the density signal can be converted into a flux-sensitive energy distribution:

$$\tilde{f}_{prob}(E_{tr}) \sim E_{tr} \exp\left(-\frac{E_{tr}}{k_B \tilde{T}} + 2\tilde{G}\tilde{u}\sqrt{\frac{E_{tr}}{k_B \tilde{T}}} - \tilde{G}^2 \tilde{u}^2\right), \tag{4.38}$$

where $\tilde{G}^2 = m/(2k_B \tilde{T})$. Kelly and Dreyfus (1988) have argued that $\tilde{u}$ is at least the sonic velocity at the outer edge of the Knudsen layer. If this is assumed, then, for a monatomic perfect gas, $\tilde{u} = \sqrt{5/6}\tilde{G}$, and Equation (4.38) becomes

$$\tilde{f}_{prob}(E_{tr}) \sim E_{tr} \exp\left(-\frac{E_{tr}}{k_B \tilde{T}} + 2\sqrt{\frac{5}{6}\frac{E_{tr}}{k_B \tilde{T}}} - \frac{5}{6}\right). \tag{4.39}$$

The mean kinetic energy of the above distribution is $\bar{E}_{tr} = 3.67 k_B \tilde{T}$.

4.3 Kinetic distributions of ejected particles

4.3.1 Thermal expectations

Stritzker *et al.* (1981), and Pospiesczyk *et al.* (1983) used a QMS in a TOF arrangement to determine the kinetic energies of evaporated Si and GaAs targets by 20-ns-long pulsed ruby-laser irradiation. From these energy distributions, they extracted the temperature of the evaporated atoms by fitting Maxwellian distributions from gas-kinetic theory. Furthermore, by assuming that this temperature represented the lattice temperature, they concluded that the process was thermal. Gibert *et al.* (1993) applied highly sensitive resonance ionization mass spectrometry (RIMS) to investigate the emission of neutral and ionized Fe atoms induced by N_2 laser irradiation ($\lambda = 337$ nm) in ultrahigh vacuum (10^{-10} torr). The experiment was conducted at low laser fluences (<275 mJ/cm^2), well below the plasma-formation regime. The kinetic energies of the neutral atoms derived from the TOF distributions at various heights from the irradiated surface yielded most probable velocities and temperatures in broad agreement with the thermal expectations (Figure 4.5). The thresholds for detection of 100 atoms or ions were 25 mJ/cm^2 and about 100 mJ/cm^2, correspondingly. On the other hand, the number of Fe$^+$ ions released per pulse from the surface at the highest laser fluence of 275 mJ/cm^2 was about three orders of magnitude less than the number of neutral atoms. The same method of investigation was used to probe soft laser sputtering of InP(100) surfaces by Dubreuil and Gibert (1994). For low fluence values (<190 mJ/cm^2) it was found that sputtering results mainly from absorption and excitation of defect sites. At higher laser fluences, the process assumes a thermal-like behavior. Measurements of the kinetic energies of the neutral ablated In and P atoms indicated that the transition between these regimes occurred at a temperature of $\sim$1400 K, which is close to the melting temperature of InP, namely 1333 K.

4.3.2 The role of electronic excitations

Experiments have provided evidence that electronic excitations may be involved in desorption processes from metals at low photon energies. Hoheisel *et al.* (1993) showed that metal atoms can be desorbed from small metal clusters by low-power CW laser radiation. The dependence on wavelength, power, and cluster size indicated that collective plasmon excitations play a role in stimulating desorption. Lee *et al.* (1993) observed bimodal energy distributions of gold atoms desorbed from a continuous film with pulsed laser irradiation at $\lambda = 532$ nm with duration $t_{\text{pulse}} = 7$ ns, and attributed the 0.3-eV high-energy peak to an electronic de-excitation process involving surface plasmons. It is noted that the experimental geometry used in this work was conducive to plasmon excitation, since the gold film was coated onto a glass prism and total internal reflection (TIR) was used to enhance plasmon excitation via evanescent wave. Kim and Helvajian (1994) observed energetic desorption of $\sim$0.7-eV energetic Ag$^+$ ions from continuous films utilizing both a TIR geometry and a direct surface-irradiation scheme.

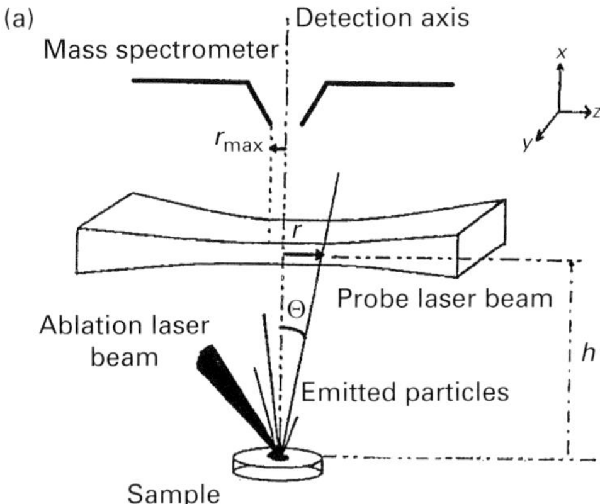

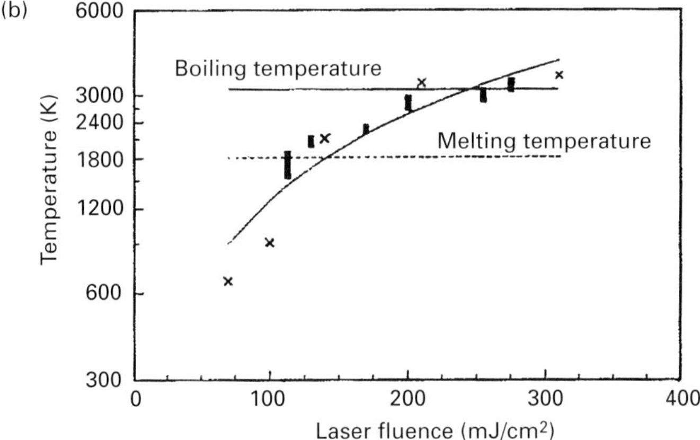

Figure 4.5. (a) A laser-sputtering experiment and RIMS detection geometry. (b) Temperature of the sputtered Fe atoms as a function of the laser fluence. ✗ Kinetic temperature deduced from the time-of-flight distributions. ■ Excitation temperature deduced from the sublevel population distributions. —— Fit by the solution of the one-dimensional heat-flow equation for laser-sputtering yields. (c) The time-of-flight distribution of sputtered Fe atoms for various heights above the target surface and two laser fluences (upper part 275 mJ/cm²; lower part 170 mJ/cm²). The relative Fe-atom density values are deduced from the RIMS signals. The curves are the best fits produced by half-range Maxwellian functions with respective temperatures 3130 and 2210 K (lower part). (d) Numbers of neutral and ionized Fe atoms sputtered per laser shot as a function of the laser fluence. The neutral-atom emission yield represents the yield for the $a\,^3D_4$ ground state only and does not take into account the contribution of higher excited states when the temperature increases. The dotted line gives an indication of the total atom yield. Note that, at the threshold, about 100 atoms are detected by RIMS. From Gibert *et al.* (1993), reproduced with permission from the American Institute of Physics.

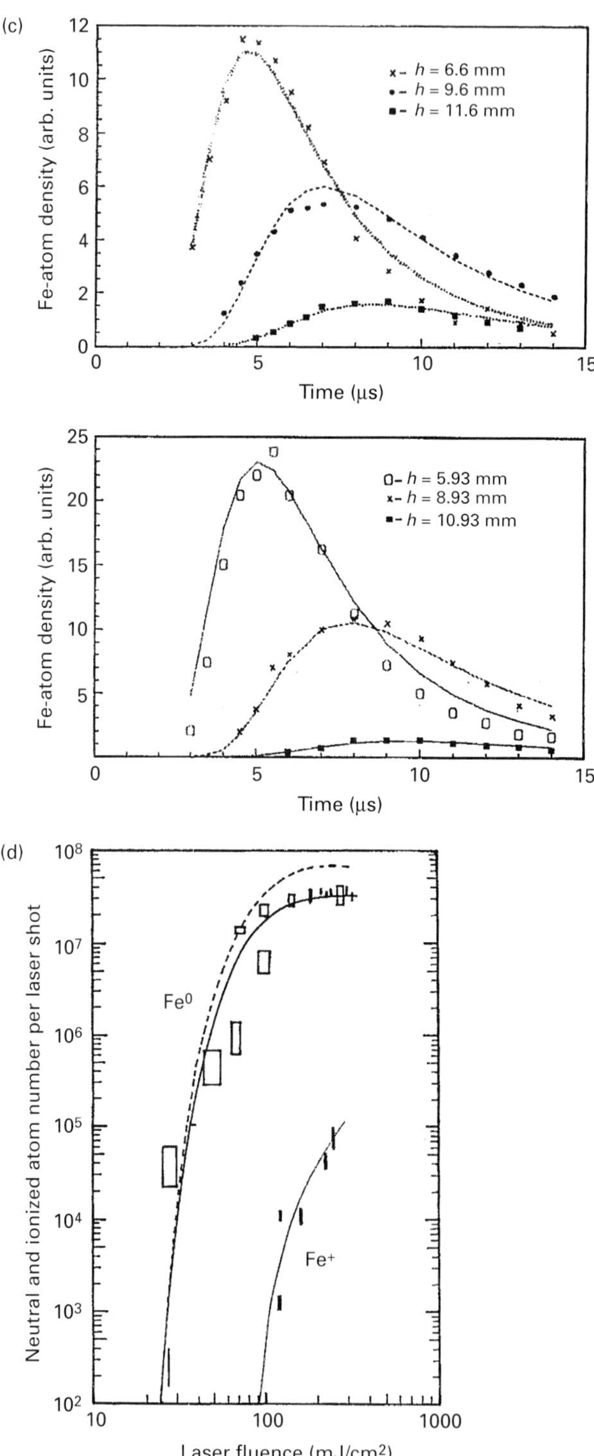

Figure 4.5. (*cont.*)

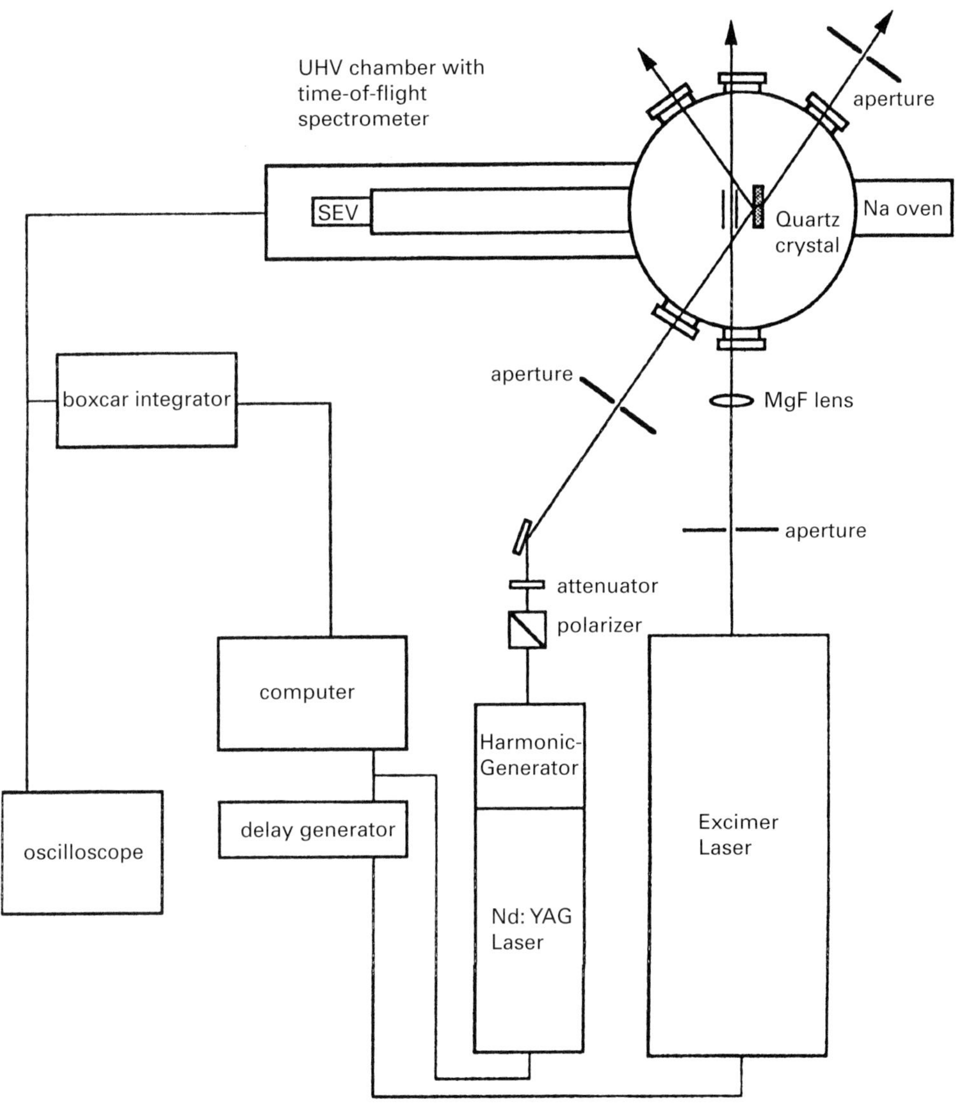

Figure 4.6. The experimental arrangement for laser-induced-desorption studies of metal atoms. Desorption is accomplished by the light of a pulsed (7 ns) Nd : YAG laser and its higher harmonics at $\lambda = 1064, 532, 355$, and 266 nm. The desorbed atoms are photoionized with the light of an excimer laser fired after a variable delay time and finally identified by a time-of-flight mass spectrometer. From Götz *et al.* (1996a), reproduced with permission from Springer-Verlag.

Laser-induced desorption of metal atoms has been studied by Götz *et al.* (1996a, 1996b). The experimental setup shown in Figure 4.6 basically consists of an ultrahigh-vacuum chamber, two lasers for stimulating desorption and ionizing the desorption products, and a TOF mass spectrometer. Sodium was deposited *in situ* by thermal evaporation from a source at 300 K onto a quartz crystal maintained at 80 K and characterized by optical transmission spectroscopy. The samples were irradiated with the fundamental and higher harmonics of a Nd : YAG laser of pulse width 7 ns at $\lambda = 1064, 532, 355$, and 266 nm at an angle of incidence of $50°$ with respect to the normal to the sample surface. At a distance of 20 mm above the target, the desorbed atoms were ionized with an excimer-laser beam at $\lambda = 193$ nm. The measured TOF distributions $I_d(t_f)$ were converted into energy distributions by applying the relation

$$f_{\text{prob}}(E_{\text{tr}}) = C I_d(t_f) t_f^2, \tag{4.40}$$

where C is a constant and t is the temporal delay between the desorbing and the ionizing pulses.

Figure 4.7 displays a series of three kinetic-energy distributions that were obtained by desorbing atoms at $\lambda = 355$ nm from the target kept at 80 K. At very low Na coverages (Figure 4.7(a)) two maxima appear in the kinetic-energy distribution, at 0.16 and 0.33 eV. These values remain fixed as the coverage increases in Figures 4.7(b) and (c). The effect of the laser fluence is shown in Figure 4.8.

At energy densities higher than $F = 32$ mJ/cm^2, the contributions of desorbed atoms possessing low kinetic energies mask the peak at 0.16 eV. Their distribution follows an exponential decay, but the peak at 0.33 eV is still detectable at $F = 52$ mJ/cm^2. Irradiation by the wavelengths 266 and 532 nm revealed no evidence of a bimodal kinetic-energy distribution, but only the 0.16-eV peak, independently of the applied laser fluence and the polarization of the laser light (Figure 4.9).

On the other hand, irradiation with $\lambda = 1064$ nm required energy densities exceeding 80 mJ/cm^2 and did not produce any distinct peak, but rather a broad exponential decay, similar in nature to the one depicted in Figure 4.8(d). Given that (a) the energy of the observed peaks is substantially independent of the laser fluence, substrate temperature, and surface coverage, and (b) the temperature experienced by the Na film is moderate (Hoheisel *et al.*, 1993), it is concluded that the desorption at $\lambda = 266, 355$, and 532 nm and low fluences occurs as a nonthermal process. The transition to the thermal process as the laser fluence increases can be traced by observing the broadened exponential decay of the kinetic-energy distribution that eventually overwhelms the energy peaks. Evidently, desorption brought about by irradiation at the wavelength $\lambda = 1064$ nm is observed at substantially higher energy densities, displaying the broad profile that is characteristic of a thermal process. The fact that bimodal distributions are observed for $\lambda = 355$ nm suggests the effect of the wavelength on the desorption process. Wavelengths of $\lambda = 266, 355$, and 532 nm carry photon energies of 4.64, 3.48, and 2.32 eV, respectively, that are much higher than the binding energy of the surface atoms, 0.7 eV, implying that the initial excitation decays by the release of heat to a lower-energy state before desorption occurs. Depending on the photon energy used to stimulate desorption, the process populates a state that is located more than 1 eV above the ground level.

These experimental results suggest a desorption mechanism along the lines of the Menzel–Redhead–Gomer mechanism (Menzel and Gomer, 1964). As depicted in

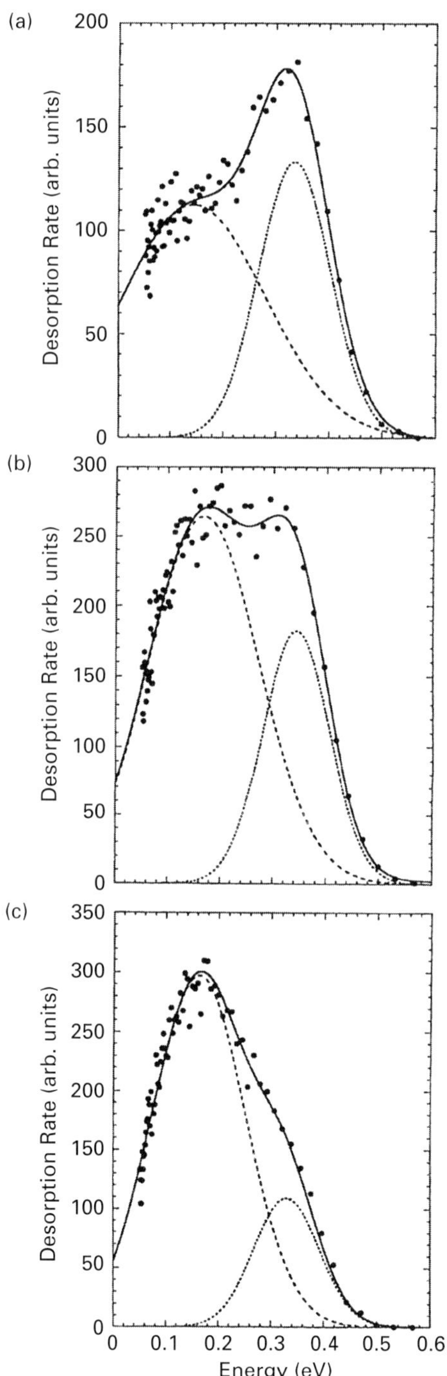

Figure 4.7. Kinetic-energy distributions of Na atoms desorbed with pulsed (7 ns) laser light of $\lambda = 355$ nm from the surface of small Na particles with average radii of (a) $R_{\mathrm{ave}} = 5$ nm, (b) $R_{\mathrm{ave}} = 7$ nm, and (c) $R_{\mathrm{ave}} = 10$ nm. The laser fluence was $F = 6.6$ mJ/cm^2. From Götz *et al.* (1996a), reproduced with permission from Springer-Verlag.

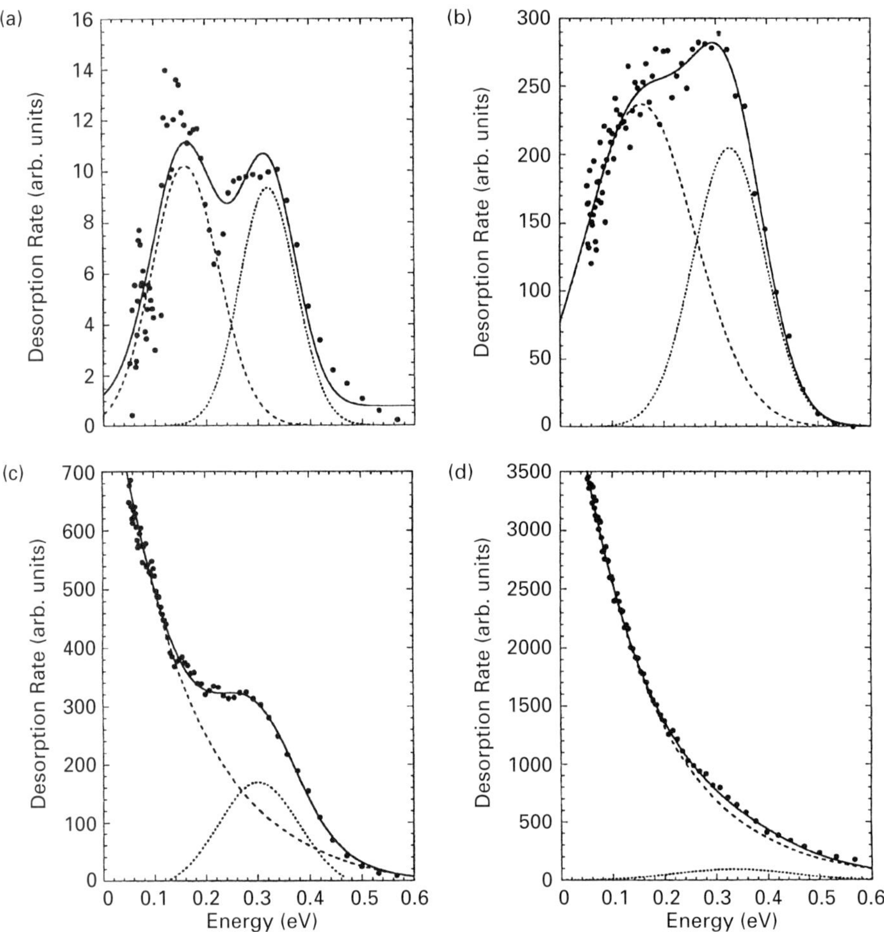

Figure 4.8. Kinetic-energy distributions of Na atoms desorbed with pulsed (7 ns) laser light of $\lambda = 355$ nm from the surface of small Na particles with fluences of (a) $F = 0.3$ mJ/cm^2, (b) $F = 6.6$ mJ/cm^2, (c) $F = 32$ mJ/cm^2, and (d) $F = 52$ mJ/cm^2. The average particle size was 7 nm. From Götz *et al.* (1996a), reproduced with permission from Springer-Verlag.

Figure 4.10, absorption of laser light at the wavelengths $\lambda = 266$, 355, and 532 nm initially populates different electronic levels denoted as IV, III, and II, respectively. Each of these excitations is followed by rapid relaxation into a lower-lying state that has to be antibonding and identical for the three wavelengths. The atoms are then accelerated away from the surface, gaining kinetic energy E_A. Its value depends upon the lifetime of the repulsive state that in turn determines the kinetic energy of the particles released at photon energy 0.16 eV. For $\lambda = 1064$ nm and photon energy of 1.16 eV (depicted by level I), the antibonding state cannot be reached and the photon energy absorbed can only be converted into heat, causing thermal desorption.

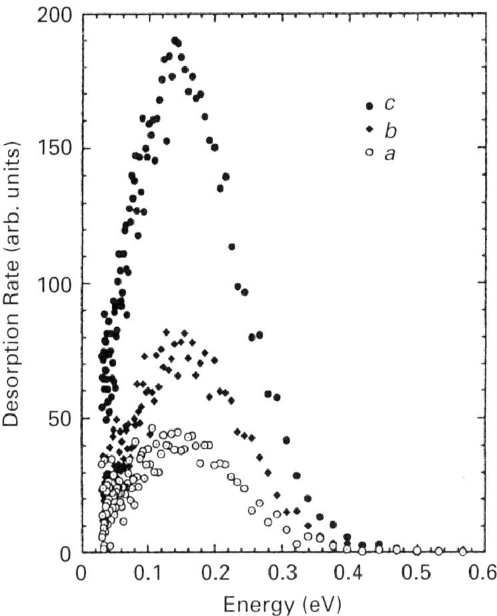

Figure 4.9. Kinetic-energy distributions of Na atoms desorbed with pulsed (7 ns) laser light of $\lambda = 532$ nm from the surface of small Na particles for three coverages of a, 5.6×10^{14} atoms/cm^2, b, 2.9×10^{15} atoms/cm^2, and c, 5.9×10^{15} atoms/cm^2. The incident laser fluence was $F = 6.6$ mJ/cm^2. From Götz *et al.* (1996a), reproduced with permission from Springer-Verlag.

4.3.3   The effect of topography development

It was previously remarked that the development of the temperature field in the heating of a bulk metal target with an excimer-laser beam of millimeter size is normally understood as a one-dimensional phenomenon. However, the development of surface topography has been observed in various systems. For metals under atmospheric conditions, Kelly and Rothenberg (1985) observed the development of surface topography under multiple-pulse irradiation. Bennett *et al.* (1995) examined the pulsed laser heating and melting of gold in a 10^{-6}-torr vacuum and demonstrated that a steady-state surface topography grows from the molten phase over several hundreds of pulses. Heat-transfer modeling of the near-surface thermal conditions and a hydrodynamic stability analysis ascribed the origin of the observed characteristic periodicity to the acceleration that the molten layer experiences due to the volumetric expansion upon melting. The characteristics of the distribution of translational energy in the ablation plume released from the surface showed that both the mean translational energy and the ablation yield achieve a steady state, together with the surface topography. In the aforementioned study, the material-removal rates were kept small (10 Å/pulse) and it was verified that the laser intensity was not sufficient to induce plasma effects in the vapor plume.

Figures 4.11(a)–(d) demonstrate the effect of laser fluence on the steady-state kinetic-energy distribution in the plume. The laser fluence was varied from a near-threshold

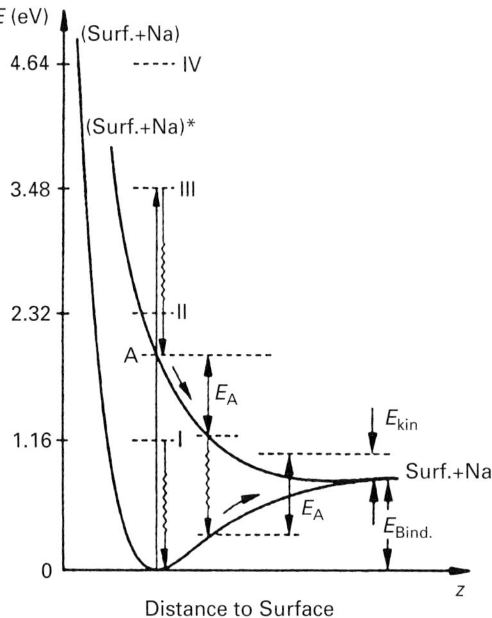

Figure 4.10. A schematic diagram of the potential-energy diagram of a surface atom illustrating non-thermal desorption of metal atoms within the framework of the Menzel–Gomer–Redhead mechanism. Laser light with $\lambda = 266$, 355, and 532 nm initially populates different electronic levels denoted by IV (4.64 eV), III (3.48 eV), and II (2.32 eV), respectively. For the sake of clarity only the excitation into state III is shown. Rapid relaxation consequently leads to a repulsive state A; the atom is accelerated and desorbs with $E_{kin} = 0.16$ eV. For near-infrared light with $\lambda = 1064$ nm (level I, 1.16 eV) the antibonding state cannot be reached, preventing non-thermal desorption. From Götz *et al.* (1996a), reproduced with permission from Springer-Verlag.

fluence (~ 0.68 J/cm^2) to about 1.0 J/cm^2 over the course of this experiment. For these QMS measurements, the detector was located normal to the surface and centered over the ablation area, and the laser beam was incident at $45°$ from the surface normal. Details of the experimental apparatus are given in Krajnovich (1995). Depicted in the figures are inverted TOF measurements in comparison with the theoretically predicted Boltzmann distribution. The first observation to be made is the apparent success of the Boltzmann distribution in describing the energy distribution in the plume. The second finding is that the measured mean kinetic energies (up to several electron-volts) correspond to temperatures far exceeding the thermal expectations (1 eV $\sim$ 11 620 K). The temperature in Equation (4.36) implicitly refers to the vapor temperature rather than the surface temperature. Nevertheless, it was shown that the Boltzmann distribution fits the experimental results better than does the "stream-velocity-corrected" distribution derived in Equation (4.39). The implication is that the 10 or 20 collisions per particle obtained by a simple estimate for the material-removal rates considered are not sufficient to impart a significant stream velocity.

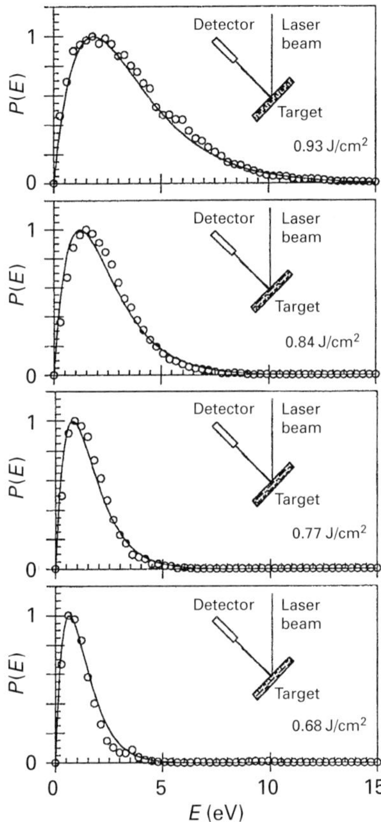

Figure 4.11. (a) The effect of laser fluence on the translational energy distribution of neutral Au atoms. The laser angle of incidence is $45°$ and the detector is normal to the sample. The dots correspond to the inverted TOF signal, while the solid line represents a Boltzmann distribution having the same mean energy as the inverted TOF signal. From Bennett *et al.* (1995), reproduced with permission from the American Institute of Physics.

The size of the surface features developed in the above study was comparable to the depth of the molten zone and the thermal diffusion depth in the material, thereby invalidating the assumption of one-dimensional field development and a homogeneous surface temperature. To retain this assumption, Bennett *et al.* (1996) devised an experiment in which, after cleaning and chemical etching, the target was preconditioned by surface melting effected by $N = 30$ pulses delivered onto the gold target, which was preheated statically at $T_0 = 1100$ K. This treatment produced a reasonably smooth surface (Figure 4.12), thus ensuring accuracy of peak-surface-temperature prediction within a 200-K confidence window. The average yields and mean translational energies exhibited little dependence on the partitioning ratio between the energy delivered by the laser and the steady target preheating while the calculated peak surface temperature was kept constant (Figure 4.13(a)). Similar behavior was observed for Ag, Cu, and a Cu–Au alloy, whereas Ni exhibited a thermal-like trend.

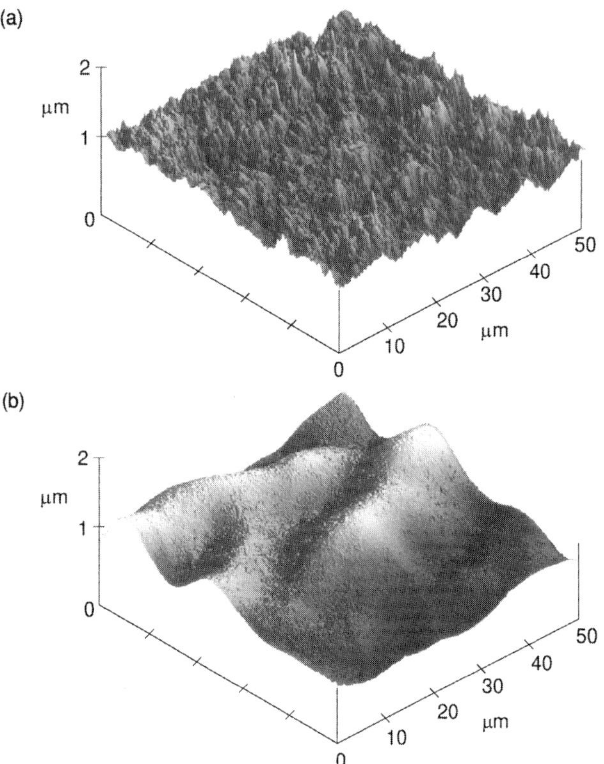

Figure 4.12. (a) An AFM image of a gold surface after etching in aqua regia. (b) The same surface after preconditioning at high temperature ($T_0 = 1100$ K, $F = 0.55$ J/cm^2, $N = 30$ pulses). Note that the height scale is exaggerated by more than $10\times$ relative to the scale of the x ad y axes in both images. The maximum slope on the smoothed surface is only 3–4°. From Bennett *et al.* (1995), reproduced with permission from the American Institute of Physics.

Following the initial thermal expectation for the mass efflux from the target, it can be argued that the integrated TOF signal, Y, can be related to the vapor pressure, P_v, and temperature:

$$Y \sim P_v \sim \exp\left(-\frac{L_v}{k_B T}\right). \tag{4.41}$$

Hence

$$\ln Y \sim -\frac{L_v}{k_B T} + \text{const.} \tag{4.42}$$

Figure 4.13(b) shows that this relation fits the experimental results for polycrystalline Au. With the exception of Ni, the experimental data appear to support the view that the surface vaporization is thermally mediated, in accordance with the observation on excimer-laser ablation of Cu by Dreyfus (1991). From the results of this study, it can be stated that the use of the mean translational energy to extract the peak surface temperature is questionable. On the other hand, the ablation yield obeys the behavior predicted from

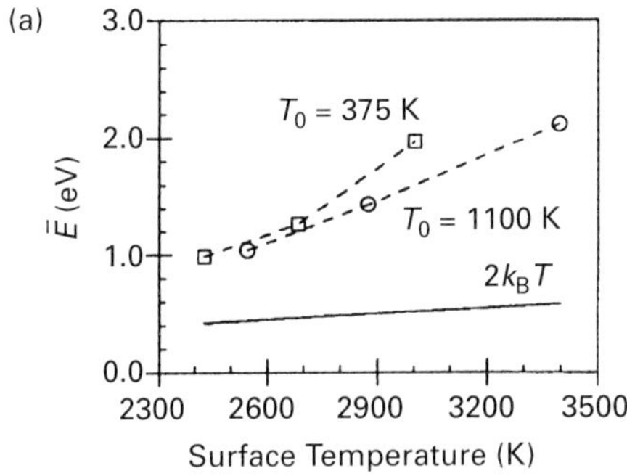

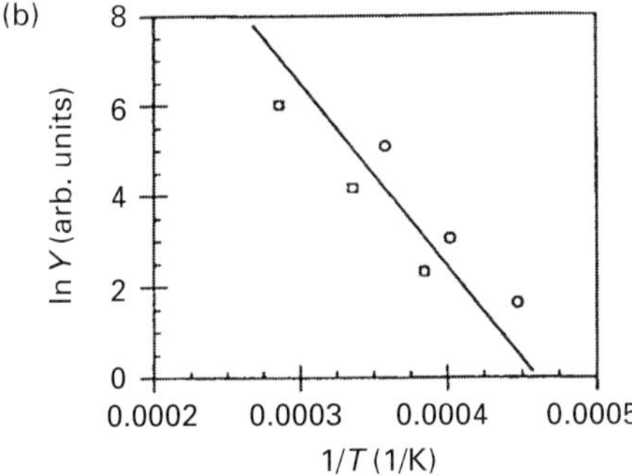

Figure 4.13. (a) Measured and thermally anticipated ($2k_\mathrm{B}T$) translational energies as a function of calculated peak surface temperature. The data pertain to the first ten laser pulses on the annealed target surface. (b) The natural logarithm of the TOF integrated signal from the polycrystalline Au target versus the reciprocal of the calculated peak surface temperature. The solid line corresponds to the anticipated slope $-L_v/(k_\mathrm{B}T)$ for a thermal process (Bennett *et al.*, 1996). From Bennett *et al.* (1996), reproduced with permission from the American Physical Society.

thermal considerations, but then again it is difficult to obtain precise measurement of the ablation yield in absolute terms.

References

Bennett, T. D., Grigoropoulos, C. P., and Krajnovich, D. J., 1995, "Near-Threshold Laser Sputtering of Gold," *J. Appl. Phys.*, **77**, 849–864.

Bennett, T. D., Krajnovich, D. J., and Grigoropoulos, C. P., 1996, "Separating Thermal, Electronic, and Topography Effects in Pulsed Laser Melting and Sputtering of Gold," *Phys. Rev. Lett.*, **76**, 1659–1562.

Carey, V. P., 1999, *Statistical Thermodynamics and Microscale Thermophysics*, Cambridge, Cambridge University Press.

Cotter, R. J., 1992, "Time-of-Flight Mass Spectrometry for the Structural Analysis of Biological Molecules," *Anal. Chem.*, **64**, 1027A–1039A.

Dreyfus, R. W., 1991, "Cu^0, C^+, and Cu_2 from Excimer-Ablated Copper," *J. Appl. Phys.*, **69**, 1721–1729.

Dubreuil, B., and Gibert, T., 1994, "Soft Laser Sputtering of InP(100) Surface," *J. Appl. Phys.*, **76**, 7545–7551.

Gibert, T., Dubreil, B., Barthe, M. F., and Debrun, J. L., 1993, "Investigation of Laser Sputtering of Iron at Low Fluence Using Resonance Ionization Mass Spectrometry," *J. Appl. Phys.*, **74**, 3506–3513.

Götz, T., Bergt, M., Hoheisel, W., Träger, F., and Stuke, M., 1996a, "Non-thermal Laser-Induced Desorption of Metal Atoms with Bimodal Kinetic Energy Distribution," *Appl. Phys. A*, **63**, 315–320.

1996b, "Laser Ablation of Metals: The Transition from Non-thermal Processes to Thermal Evaporation," *Appl. Surf. Sci.*, **96–98**, 280–286.

Hoheisel, W., Vollmer, M., and Träger, F., 1993, "Desorption of Metal Atoms with Laser Light: Mechanistic Studies," *Phys. Rev. B*, **48**, 17 463–17 476.

Kelly, R., and Dreyfus, R. W., 1988, "Reconsidering the Mechanisms of Laser Sputtering with Knudsen-Layer Formation Taken into Account," *Nucl. Instrum. Meth. Phys. Res. B*, **32**, 341–348.

Kelly, R., and Rothenberg, J. E., 1985, "Laser Sputtering: Part III. The Mechanism of the Sputtering of Metals at Low Energy Densities," *Nucl. Instrum. Meth. Phys. Res. B*, **7/8**, 755–763.

Kim, H.-S., and Helvajian, H., 1994, "Laser-Induced Ion Species Ejection from Thin Silver Films," in *Laser Ablation: Mechanisms and Applications – II. Second International Conference*, ed. J. C. Miller and D. B. Geohegan, New York, American Institute of Physics, pp. 38–43.

Krajnovich, D. J., 1995, "Laser Sputtering of Highly Oriented Pyrolytic Graphite at 248 nm," *J. Chem. Phys.*, **102**, 726–743.

Lee, I., Calcott, T. A., and Arakawa, E. T., 1993, "Desorption Studies of Metal Atoms Using Laser-Induced Surface-Plasmon Excitation," *Phys. Rev. B*, **47**, 6661–6666.

Mamyrin, B. A., Karataev, V. I., Shmikk, D. V., and Zagulin, V. A., 1973, "The Mass Reflectron, a New Nonmagnetic Time-of-Flight Mass Spectrometer with High Resolution," *Sov. Phys. JETP*, **37**, 45–48.

Menzel, D., and Gomer, R., 1964, "Desorption from Metal Surfaces by Low-Energy Electrons," *J. Chem. Phys.*, **41**, 3311–3328.

Pospieszczyk, A., Harith, M. A., and Stritzker, B., 1983, "Pulsed Laser Annealing of GaAs and Si: Combined Reflectivity and Time-of-Flight Measurements," *J. Appl. Phys.*, **54**, 3176–3182.

Stritzker, B., Pospieszcyck, A., and Tagle, J. A., 1981, "Measurement of Lattice Temperature of Silicon during Pulsed Laser Annealing", *Phys. Rev. Lett.*, **47**, 356–358.

Wiley, W. C., and McLaren, I. H., 1955, "Time-of-Flight Mass Spectrometer with Improved Resolution," *Rev. Sci. Instrum.*, **26**, 1150–1154.

5 Dynamics of laser ablation

C. Grigoropoulos and S. Mao

5.1 Introduction

Various theoretical models have been proposed to describe material removal from a solid surface heated by laser irradiation. The thermal models of Afanas'ev and Krokhin (1967), Anisimov (1968), and Olstad and Olander (1975) represent early theoretical contributions to this problem. Chan and Mazumder (1987) developed a one-dimensional steady-state model describing the damage caused by vaporization and liquid expulsion due to laser–material interaction. Much of the above work was driven by laser applications such as cutting and drilling, and was thus focused primarily on modification of the target's morphology, with no particular interest in the detailed description of the properties and dynamics of the evaporated and ablated species. Moreover, these models dealt with continuous-wave (CW) laser sources, or relatively long (millisecond) time scales.

During the first stage of interaction between the laser pulse and the solid material, part of the laser energy is reflected at the surface and part of the energy is absorbed within a short penetration depth in the material. The energy absorbed is subsequently transferred deeper into the interior of the target by heat conduction. At a later stage, if the amount of laser energy is large enough (depending upon the pulse length, intensity profile, wavelength, and thermal and radiative properties of the target material), melting occurs and vaporization follows. The vapor generated can be ionized, creating high-density plasma that further absorbs the incident laser light. Effects of this laser-plasma shielding have been shown via the simplified one-dimensional model of Lunney and Jordan (1998). The physical picture of laser-energy interaction with evaporating materials at high fluence then becomes quite complicated. Of interest are the descriptions of the vaporization and ionization processes, the associated fluid motions and gas-dynamics phenomena (including the vapor/plasma expansion and possible formation of shock waves against the ambient environment pressure) (Phipps and Dreyfus, 1993).

Phipps *et al.* (1988) developed a simple model to predict the ablation pressure and the impulse exerted on laser-irradiated targets for laser intensities exceeding the plasma-formation threshold. The model was shown by Phipps and Dreyfus (1993) to follow the experimental trends for the mass-loss rate and the ablation depth within a factor of two. In their computational studies, Aden *et al.* (1990, 1992) dealt with the laser-induced expansion of metal vapor against a background pressure. Vertes *et al.*

(1989, 1993) developed a one-component, one-dimensional model to describe the expansion of laser-generated plasmas. The model incorporated the conservation equations for mass, momentum, and energy. Singh and Narayan (1990) proposed a theoretical model for simulating laser-plasma–solid interactions, assuming that the plasma formed initially undergoes a three-dimensional isothermal expansion followed by an adiabatic expansion. This model yielded athermal, non-Maxwellian velocity distributions of the atomic and molecular species, as well as thickness and compositional variations of the deposited material as functions of the target–substrate distance and the irradiated-spot size.

Several studies have focused on the role of Knudsen-layer formation in laser vaporization, sputtering, and deposition. Kelly and Dreyfus (1988a) showed that, under the condition that the particle emission is driven by a thermal mechanism, the ejection velocities at the target surface are described by a half-range Maxwellian distribution. For as few as three collisions per particle, a Knudsen layer forms and is confined within a distance of a few mean free paths from the solid surface. Within this Knudsen layer, the density distribution function evolves to a full-range Maxwellian in a center-of-mass coordinate system. Moreover, Kelly and Dreyfus (1988b) showed that Knudsen-layer formation leads to forward-peaking of the kinetic-energy distributions of the ejected particles. Angular profiles following $\cos^4\theta$ to $\cos^{10}\theta$ functions were thus expected, in general agreement with experimental measurements. For relatively high evaporation yields exceeding, for instance, half of a monolayer per 10-ns laser pulse, Kelly (1990) showed that the resulting gas-phase interactions cause the Knudsen layer to evolve into an unsteady, adiabatic expansion, resembling the case of a gun firing a finite charge into an infinite, one-dimensional barrel. An explicit solution for the speed of sound and the gas velocity was obtained, emphasizing the importance of the Mach number and the specific heat ratio $\gamma_{\mathrm{ad}} = C_p/C_v$ in the interpretation of experimental time-of-flight (TOF) measurements. Comparison of the gas-dynamics model with numerical solutions of the flow equations and with direct simulations of the Boltzmann equation by Sibold and Urbassek (1991) showed reasonable agreement.

Finke *et al.* (1990a) and Finke and Simon (1990) treated the steady-state formation of the Knudsen layer numerically by solving the Boltzmann equation using an integral approach. Their solution yielded temperature, mass density, and velocity distributions in the Knudsen layer, as well as the decrease of pressure along the ejected vapor stream. Furthermore, Finke *et al.* (1990b) studied the energy supply of a collisional plasma with a low degree of ionization, coupled with kinetic considerations, with reference to steady-state laser-welding problems.

Chen *et al.* (1995, 1996) showed that dynamic source and partial-ionization effects dramatically accelerate the expansion of laser-ablated material in the direction perpendicular to the target, leading to highly forward-peaking plumes. Leboeuf *et al.* (1996) conducted dynamical computations of the plume propagation in vacuum and in an ambient gas using particle-in-cell hydrodynamics, continuum gas dynamics, and scattering models. The first two approaches were shown to reproduce the broad hydrodynamic features of plume expansion in vacuum and in ambient gas. On the other hand, the scattering model captured trends of plume splitting that had been observed by emission and absorption spectroscopy diagnostics once a critical background gas pressure had been reached. This result indicated that models adopting collisional scattering schemes may be more suitable for the long-mean-free-path flows encountered under typical pulsed-laser-deposition (PLD) conditions. In another study (Capewell and Goodwin, 1995), the Monte Carlo method for simulating rarefied gas dynamics was applied to simulate the dynamics of a hot Si plume expanding into background Ar gas. The results showed qualitative agreement with images of ablation plumes in PLD experiments.

5.2 Laser-induced plasma formation

5.2.1 Relaxation in plasmas

Plasmas can be visualized as mixtures of two gases, an ion gas and an electron gas, with masses m_i and m_e, respectively. Since $m_e \ll m_i$, collisional transfer of energy between these systems is difficult, as is manifested by the substantial differences between their average kinetic energies over relatively long time periods. The most frequently used expression for the electron–ion collision time is taken from Spitzer (1962):

$$\tau_{ei} = \frac{3}{4} \left(\frac{m_e}{2\pi} \right)^{\frac{1}{2}} \frac{k_B^{3/2} T^{3/2}}{Z^4 e^4 N_i \ln \Lambda}. \tag{5.1}$$

In the above, Z is the average charge state of the ions, e the proton charge, and $\ln \Lambda$ the Coulomb logarithm,

$$\Lambda = \frac{3}{2e^3} \left(\frac{k_B^3 T^3}{\pi N_e} \right)^{\frac{1}{2}}. \tag{5.2}$$

When the difference between the ion and electron temperatures is not too large, the change of temperatures can be described by a simple relaxation equation:

$$\frac{dT_e}{dt} = \frac{T_i - T_e}{\tau_{ei}}. \tag{5.3}$$

For weakly ionized gases, the above equation is still valid, provided that the exchange with neutral atoms is taken into account. In this case, the relaxation time constant is modified as follows:

$$\frac{1}{\tau_{ei+n}} = \frac{1}{\tau_{ei}} + \frac{1}{\tau_{en}}. \tag{5.3}$$

The relaxation time in collisions of electrons with neutral species is

$$\frac{1}{\tau_{en}} \approx N_e \bar{u}_e \sigma_{c,e,elast} \frac{2m_e}{m_i}, \tag{5.4}$$

where $\sigma_{c,e,elast}$ is the average cross section for elastic electron–atom collisions.

The plasma absorption coefficient, γ_{pl}, is

$$\gamma_{pl} = \frac{2\omega}{c} \, \mathrm{Im} \left[\left(1 - \frac{\omega_{pl}^2}{\omega^2 \left[1 + (i/\tau_{ei}) \right]} \right)^{\frac{1}{2}} \right], \tag{5.5}$$

where $\omega = 2\pi\nu$ is the angular frequency of the laser light. The term $\omega_{\mathrm{pl}} = 4\pi(N_e e^2/m_e)^{1/2}$ is the electron plasma frequency.

5.2.2   Plasma absorption

The equilibrium electron and ion densities, N_e and N_i, can be calculated using the Saha–Eggert equation (Spitzer, 1962):

$$\frac{N_e N_i}{N_n} = \frac{Q_e Q_i}{Q_n}\frac{m_e m_i}{m_e + m_i}\exp\left(-\frac{\Psi_i}{k_B T}\right). \tag{5.6}$$

In the above, Q_e, Q_i, and Q_n represent the internal partition functions of electrons, ions, and neutral species, respectively; Ψ_i is the ionization potential of a neutral atom; and m_e and m_i are the electron and ion masses.

Absorption due to free–free transitions

It is well known that radiation is emitted from a free electron subject to the Coulomb field exerted by a positively charged ionic core. In this Bremsstrahlung emission process, the electron loses kinetic energy and slows down. Laser radiation is therefore absorbed via *inverse Bremsstrahlung* by an electron that consequently accelerates past the ion. The radiation absorption due to inverse Bremsstrahlung via scattering of electrons with ions is quantified by the continuous spectral absorption cross section, $\sigma_{c,\mathrm{IB},\nu}$, and coefficient, $\gamma_{\mathrm{IB},\nu}$:

$$\sigma_{c,\mathrm{IB},\nu} = \frac{4}{3}\left(\frac{2\pi}{k_B T_e}\right)^{\frac{1}{2}}\frac{Z^2 e^6}{hcm_e^{3/2}\nu^3}N_i g_{\mathrm{ff}}\left[1 - \exp\left(-\frac{h\nu}{k_B T_e}\right)\right], \tag{5.7a}$$

$$\gamma_{\mathrm{IB},\nu} = \sigma_{c,\mathrm{IB},\nu}N_e, \tag{5.7b}$$

where g_{ff} is the Gaunt factor that is introduced in a quantum-mechanical theory, c the speed of light and h Planck's constant. The term within the last bracket represents the effective decrease in absorption caused by induced emission. Bremsstrahlung emission (and absorption) can occur via collisions with scattering centers that may be atoms, molecules, or ions. In typical laser-induced plasmas, $\sigma_{c,\mathrm{IB},\nu}$ is $O(10^{-20}\ \mathrm{cm}^2)$. While this is still small relative to the cross sections of atomic absorption, it is sufficient to produce intense interactions with laser fluences in the range of 10^{19} photons/cm^2 (Dreyfus, 1993). According to Zel'dovich *et al.* (1966), the relationship between the differential cross section for the emission of a photon $h\nu$ and the cross section for the elastic scattering of a photon is given by

$$\frac{d\sigma_{c,\nu}}{d\nu} = \frac{8}{3}\frac{e^2 u^2}{c^3 h\nu}\sigma_{c,\mathrm{tr}}, \tag{5.8}$$

where u is the absolute electron velocity and $\sigma_{c,\mathrm{tr}}$ is the so-called transport scattering cross section. The Bremsstrahlung absorption cross sections for ions and neutral species compare as

$$\frac{(d\sigma_{c,\nu})_{ion}}{(d\sigma_{c,\nu})_{neut}} \approx \frac{\pi r_0^2}{\sigma_{c,tr}} = \frac{\pi Z^2 e^4}{\sigma_{c,tr}(h\nu)^2} = \frac{\pi a_0^2}{\sigma_{c,tr}}\left(\frac{2\Psi_{i,H}}{h\nu}\right)^2 Z^2, \tag{5.9}$$

where $a_0 = h^2/(4\pi me^2)$ is the Bohr radius and $\Psi_{i,H} = 13.5$ eV is the ionization potential of hydrogen. Since $\sigma_{c,tr}/(\pi a_0^2) \sim 1\text{--}10$ and we are dealing with photon energies in the range of a few electron-volts, it is estimated that the effectiveness of neutral species for Bremsstrahlung absorption is one or two orders lower than that of ions. However, it should be mentioned that neutral species are ejected in much larger numbers than ions during the early phase of ablation. Hence, it is conceivable to assume that inverse Bremsstrahlung due to neutral species does contribute to absorption, leading to a temperature rise and thus further enhancing absorption (Amoruso, 2004).

Absorption due to bound–free transitions

Consider photo-ionization, i.e. the absorption of a photon accompanied by the transition of an electron past the ionization threshold into the continuous energy spectrum. The number density of excited atoms in the nth state is given by the Boltzmann distribution:

$$N_n = N_1 \frac{g_n}{g_1} e^{-\frac{E_n - E_1}{k_B T}} = N_1 \frac{g_n}{g_1} e^{-\frac{\Psi_i}{k_B T}\left(1 - \frac{1}{n^2}\right)}, \tag{5.10}$$

where $g_n = 2n^2$ is the statistical weight of the nth level and N_1 is the atom number density in the ground state. The cross section for absorption of a photon of energy $h\nu$ is

$$\sigma_{c,PI,\nu n} = \frac{64\pi^4}{3\sqrt{3}} \frac{e^{10} m_e Z^4}{h^6 c\nu^3 n^5}. \tag{5.11}$$

The total spectral coefficient of absorption

The absorption coefficient representing bound–free transitions is given by

$$\gamma_{PI,\nu} = \sum_{n^*}^{\infty} N_n \sigma_{c,PI,\nu}. \tag{5.12}$$

The lower bound of the summation is determined by the condition that the photon energy is greater than the binding energy of the electron in the atom. The composite coefficient of bound–free absorption is

$$\gamma_{PI,\nu}^T = \frac{64\pi^4}{3\sqrt{3}} \frac{e^{10} m_e Z^4 N}{h^6 c\nu^3} \sum_{n^*}^{\infty} \frac{1}{n^3} e^{-\frac{\Psi_i}{k_B T}\left(1 - \frac{1}{n^2}\right)}. \tag{5.13}$$

On adding the absorption contribution due to inverse Bremsstrahlung, the total absorption coefficient is

$$\gamma_\nu^T = \gamma_{PI,\nu}^T + \gamma_{IB,\nu} = \frac{64\pi^4}{3\sqrt{3}} \frac{e^{10} m_e Z^4 N}{h^6 c\nu^3} \left[\sum_{n^*}^{\infty} \frac{1}{n^3} e^{-\frac{\Psi_i}{k_B T}\left(1 - \frac{1}{n^2}\right)} + \frac{e^{-\frac{\Psi_i}{k_B T}}}{2\Psi_i/(k_B T)}\right]. \tag{5.14}$$

Provided that the photon energy is much lower than the ionization potential ($h\nu \ll \Psi_{\mathrm{i}}$), the spectral coefficient of absorption is given by the Kramers–Unsöld formula:

$$\gamma_\nu = \frac{16\pi^2}{3\sqrt{3}} \frac{e^6 Z^2 k_{\mathrm{B}} T N}{h^4 c \nu^3} \mathrm{e}^{-\frac{\Psi_{\mathrm{i}} - h\nu}{k_{\mathrm{B}} T}}. \tag{5.15}$$

For ultraviolet (UV) wavelengths, direct photo-ionization of the vapor atoms is expected to be a main ionization mechanism. For high plasma temperatures, the absorption is dominated by free–free inverse Bremsstrahlung. Phipps and Dreyfus (1993) note that, under most experimental conditions, electrons are probably present because of multiphoton ionization of atomic or molecular species in the plume, ionization from excited states in the plume, ejection of ions and electrons from the surface, and thermionic emission from the surface. Whatever the mechanism, the threshold for plasma initiation is significantly lower than what would be expected from equilibrium considerations. As previously noted, close to the plasma-ignition threshold, coupling of the laser energy with the plasma may occur via mechanisms other than inverse Bremsstrahlung. However, that mechanism regulates the energy transfer as the laser pulse progresses. Most of the laser absorption occurs in a vapor layer confined close to the target surface. In that layer, the electron–ion density is very high (10^{20}–10^{23} cm^{-3}) and the plasma pressure may exceed hundreds of kilobars. This higher-pressure region may launch a compressional wave toward the target surface (Ho *et al.*, 1996), leading to transverse flow of the trailing ejected particles or removal of material in the form of molten droplets. The penetration depth for light absorption in the hot, dense plasma region situated over the target surface may be in the micrometer range, producing optically thick conditions.

According to Zel'dovich, all elementary excitation and ionization processes can be divided into two categories: (a) excitation and ionization of atoms by collisions with other particles, and (b) photo-processes in which the role of the exciting agent is played by a photon instead of a particle. In the first category, a distinction must be made between electron-impact collisions and inelastic collisions with heavy particles. Accordingly, the following basic ionization reactions can be written:

$$A + e = A^+ + e + e, \tag{5.16a}$$

$$A + B = A^+ + B + e, \tag{5.16b}$$

$$A + h\nu = A^+ + e. \tag{5.16c}$$

The reverse reactions, proceeding from right to left, result in recombination of electrons with ions.

$$A + e = A^* + e, \tag{5.17a}$$

$$A + B = A^* + B, \tag{5.17b}$$

$$A + h\nu = A^*. \tag{5.17c}$$

The first two reverse processes represent de-excitation of excited atoms by collisions of the second kind, while the third reverse process depicts spontaneous emission of an excited atom. It is then possible to ionize not only neutral species, but also excited atoms:

$$A^* + e = A^+ + e + e, \tag{5.18a}$$

$$A^* + B = A^+ + B + e, \tag{5.18b}$$

$$A^* + h\nu = A^+ + e. \tag{5.18c}$$

Even though the number of neutral atoms is usually much higher than the number of excited atoms, the latter make a non-negligible contribution to ionization, since the ionization of excited atoms requires lower energy. Zel'dovich *et al.* (1966) derived the rate equations for (i) electron-impact ionization of ground-state atoms, (ii) electron-impact excitation of atoms from the ground state, and (iii) electron-impact ionization of excited states.

5.2.3　High-irradiance ablation

Phipps and Dreyfus (1993) and Phipps *et al.* (1988) developed a theory correlating experimental results on high-irradiance ($I \geq 10^{11}$ W/cm^2), short-pulse ($t_{\text{pulse}} \approx 10$ ns) ablation of materials. The theory is based on the following assumptions: (1) one-dimensional flow, (2) an inverse-Bremsstrahlung absorption mechanism, (3) Mach-1 flow (the Chapman–Jouguet condition, $M = 1$), (4) ideal-gas behavior, (5) macroscopic charge neutrality, and (6) self-regulating plasma opacity, giving an optical thickness of unity at the laser wavelength after a brief transition time, t_1. For surface absorbers in vacuum, an estimate of the onset of plasma dominance is given by the relationship

$$I\sqrt{t_{\text{pulse}}} \geq B. \tag{5.19}$$

The above holds over a five-orders-of-magnitude range in pulse duration from 5 ms to 10 ns and for laser wavelength in the range 0.25–10 mm for $B_{\text{max}} \approx 8 \times 10^8$ W s$^{1/2}$ m^{-2} with the plasma-formation threshold at $B \approx B_{\text{max}}/2$. The following expressions are derived.

Pressure:

$$P = 1.84 \times 10^{-4} \frac{\Sigma^{9/16}}{A^{1/8}} \frac{I^{3/4}}{(\lambda\sqrt{t_{\text{pulse}}})^{1/4}} \text{ (Pa).} \tag{5.20}$$

Electron temperature:

$$T_{\text{e}} = 2.98 \times 10^3 \frac{A^{1/8} Z^{3/4}}{(Z+1)^{5/8}} (I\lambda\sqrt{t_{\text{pulse}}})^{1/2} \text{ (K).} \tag{5.21}$$

Electron density:

$$N_{\text{e}} = 1.135 \times 10^{15} \frac{A^{5/16}}{Z^{1/8}(Z+1)^{9/16}} \frac{I^{1/4}}{(\lambda\sqrt{t_{\text{pulse}}})^{3/4}} \text{ (m}^{-3}\text{).} \tag{5.22}$$

Ablation rate:

$$\dot{m} = 26.6 \frac{\Sigma^{9/8} I^{1/2}}{A^{1/4}\lambda^{1/2}t_{\text{pulse}}^{1/4}} \text{ (}\mu\text{g m}^{-2}\text{ s}^{-1}\text{).} \tag{5.23}$$

The parameter

$$\Sigma = \frac{A}{2[Z^2(Z+1)]^{1/3}}.$$

(5.24)

A is the average ionic mass number, Z the average ionic charge.

It is recalled that the high-irradiance theory assumes a constant optical thickness and therefore plasma transmission. Second, there is no provision in the theory for the physical mechanism of the evaporation process (and hence of the actual surface temperature). Third, the high-irradiance theory does not take into account the optical properties of the target. This may be important when the large spectral variation of the surface reflectivity is recalled.

5.3   Modeling of ablation-plume propagation

5.3.1   The Euler equation for the dynamics of a compressible gas

The vaporized material ejected from the liquid surface is modeled as a compressible, inviscid ideal gas (Ho *et al.*, 1995). The dynamic state of the metal vapor phase is described by the compressible and nondissipative Euler conservation equations for mass, momentum, and energy.

Mass conservation:

$$\frac{\partial \rho}{\partial t} + \nabla \cdot (\rho \vec{V}) = 0.$$

(5.25)

Momentum conservation:

$$\frac{\partial \vec{V}}{\partial t} + \vec{V} \cdot \nabla \vec{V} = -\frac{1}{\rho} \nabla P.$$

(5.26)

When an intense laser pulse interacts with metal vapor, a significant amount of the laser radiation is absorbed by the ionized particles. Consequently, the radiation absorption and emission may couple strongly with the plume hydrodynamics. The energy-conservation equation in the absorbing and radiatively participating plume is

$$\rho\left(\frac{\partial \varepsilon}{\partial t} + \nabla \varepsilon \cdot \vec{V}\right) = -P\nabla \cdot \vec{V} + \gamma_{\mathrm{pl}} I_{\mathrm{las}} \exp(-\gamma_{\mathrm{pl}}|z|) + \nabla \cdot (-\vec{q}_{\mathrm{r}}),$$

(5.27)

where the exponentially decaying term accounts for the absorption of laser energy by the plume and $\vec{q}_{\mathrm{r}}$ is the radiative-flux vector. The radiation source term is defined via

$$\nabla \cdot \vec{q}_{\mathrm{r}} = \gamma_{\mathrm{av}}(4\pi I_{\mathrm{b}} - G).$$

(5.28)

In the above equation, $G = \int_{4\pi} I \, d\Omega$, where the integration over the solid angle Ω is carried out over a sphere, I_{b} is the blackbody radiation intensity, and γ_{av} is the Planck mean absorption coefficient defined as

$$\gamma_{\mathrm{av}} = \frac{\displaystyle\int_0^\infty I_{\mathrm{b}\nu} \gamma_\nu \, d\nu}{\displaystyle\int_0^\infty I_{\mathrm{b}\nu} \, d\nu}.$$

(5.29)

Utilizing Equation (5.7a), the above expression yields:

$$\gamma_{av} = \left(\frac{128}{27}k_B\right)^{\frac{1}{2}} \left(\frac{\pi}{m_e}\right)^{\frac{3}{2}} \frac{Z^2 e^6 g_{ff}}{hc^3} \frac{N_e N_i}{T^{7/2}}. \tag{5.30}$$

In order to calculate the radiation heat flux, the radiation transport equation (RTE) (Siegel and Howell, 1992) for an absorbing and emitting, but nonscattering, medium is solved for the total, directional radiative intensity, I:

$$\vec{s} \cdot \nabla I(\vec{r}, s) = -\gamma_{av} I(\vec{r}, s) + \gamma_{av} I_b. \tag{5.31}$$

Two more equations define the total energy per unit mass, ε_t,

$$\varepsilon_t = \varepsilon + \frac{1}{2}(u^2 + v^2), \tag{5.32}$$

and the equation of state for an ideal gas,

$$P = (\gamma_{ad} - 1)\rho\varepsilon, \tag{5.33}$$

where γ_{ad} is the ratio of the specific heats, $\gamma_{ad} = C_p/C_v$. A species-transport equation is used to specify the mass content of metal vapor per unit volume. Because of the ideal-gas assumption, and since the kinetic energy is much larger than the dissipation or interaction energies, homogeneous flow conditions are assumed. Thus, the species-transport equation is simplified to

$$\frac{\partial y}{\partial t} + \vec{V} \cdot \nabla y = 0, \tag{5.34}$$

where $y = \rho_v/\rho$ is the local mass fraction of the ejected vapor.

The boundary conditions across the discontinuity (Knudsen layer) at the vapor/liquid interface are as follows.

Mass conservation:

$$\rho_v(-V_v - V_{int}) = \rho_l(V_l - V_{int}). \tag{5.35a}$$

Momentum conservation:

$$P_v - P_l = \rho_v(-V_v - V_{int})(V_l + V_v). \tag{5.35b}$$

Energy conservation:

$$H_l + L_v = H_v + \frac{1}{2}\left(V_v^2 - V_l^2\right). \tag{5.35c}$$

Computations were done for an aluminum target irradiated by an excimer laser of fluence $F = 25$ J/cm^2 with an Ar background gas atmosphere of $P_\infty = 10^{-3}$ atm (Ho *et al.*, 1996). Figures 5.1 and 5.2(a)–(c) show the development of the pressure and velocity vector fields in the vapor phase at various times ($t = 40$, 100, and 500 ns). The shock front produced by the high-pressure plasma propagates with an

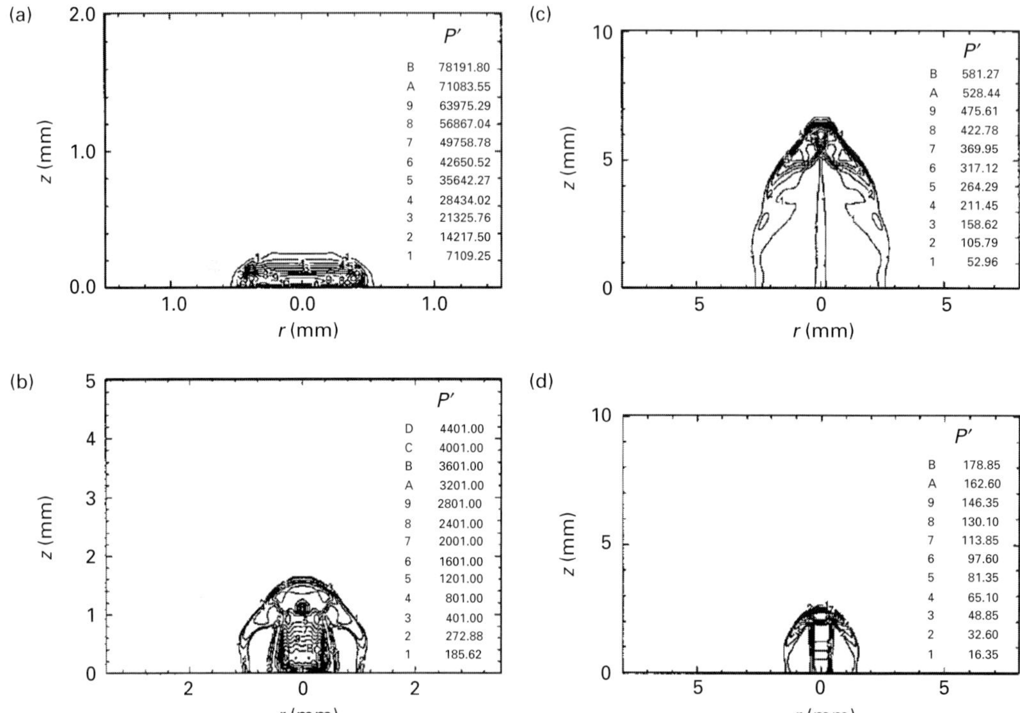

Figure 5.1. Dimensionless pressure, P', contours for an aluminum target subjected to excimer-laser pulses of duration 26 ns and fluence, $F = 25$ J/cm^2 at (a) $t = 40$ ns, (b) $t = 100$ ns, (c) $t = 500$ ns, and (d) $t = 500$ ns, assuming a transparent vapor plume. The ambient pressure is $P_\infty = 10^{-3}$ atm and the diameter of the laser spot is $r_{\text{las}} = 0.5$ mm. From Ho *et al.* (1996), reproduced with permission by the American Institute of Physics.

average speed of about 12 000 m/s into the ambient gas. An average shock-wave speed is predicted under the transparent-vapor-plume assumption depicted in Figures 5.1 and 5.2(d). A major assumption in this analysis stems from the use of the local-thermal-equilibrium framework, i.e. the existence of a one-component system whose energy is represented by a single temperature. This is unlikely to prevail, especially during the early stages of the ablation process. The inclusion of nonequilibrium absorption via photo-ionization and emission by recombination is a key element in further work, together with recondensation in the plume. The local fraction of metal vapor has been calculated by considering advective mixing and utilizing ideal-gas approximations for the internal-energy function. This appears reasonable, since the plume ejection is primarily driven by the high rise in pressure.

5.3.2 Molecular-dynamics models

Molecular-dynamics (MD) simulations offer the possibility of capturing the interplay of the different processes involved in ablation-plume evolution, including cluster

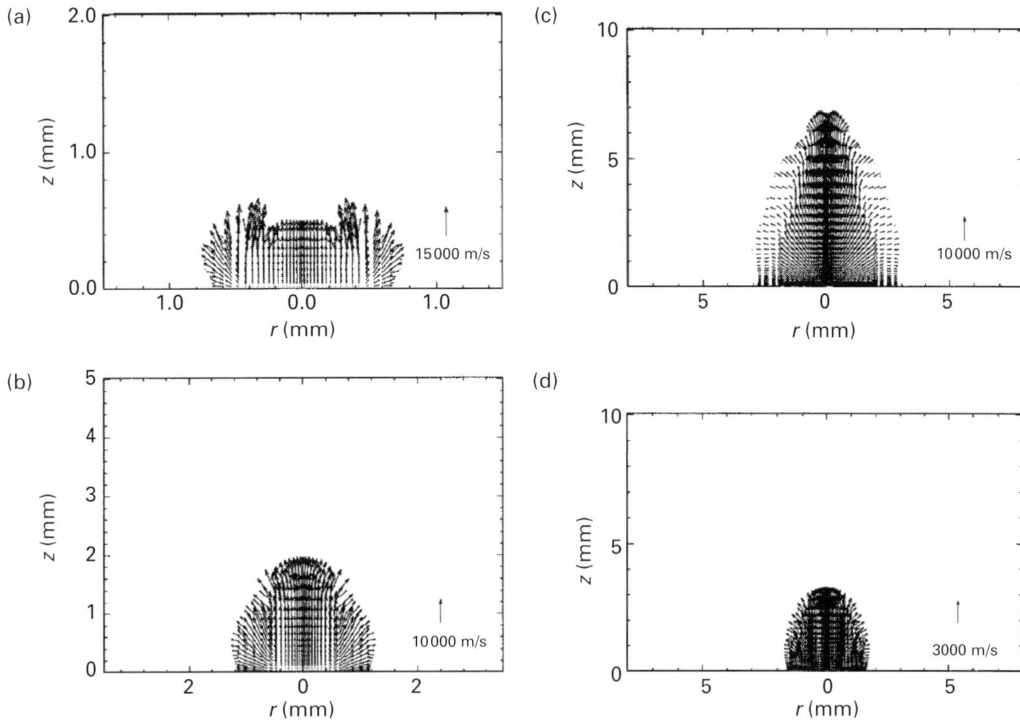

Figure 5.2. Velocity vectors for the conditions of Figure 5.1. From Ho *et al.* (1996), reproduced with permission by the American Institute of Physics.

formation. As discussed by Zhigilei (2003), the size of atomic-scale MD simulations of ablative events is simply too large for meaningful computations extending well into the picosecond and nanosecond regimes to be performed. In the breathing-sphere model (Zhigilei *et al.*, 1997) it is assumed that each molecule (or group of atoms) can be represented by a single particle that has the true translational degrees of freedom but approximate internal degrees of freedom. This model was applied by Zhigilei (2003) in the context of ablation of an idealized molecular solid, although the predictions represented general trends. It was found that the cluster composition of the ablation plume strongly depends on the irradiation conditions. At sufficiently high laser fluences, the phase explosion of the overheated material leads to the formation of a foamy transient structure of interconnected liquid regions that subsequently decompose into a mixture of liquid droplets, gas-phase molecules, and clusters. A typical simulation example for early plume propagation in this regime is given in Figure 5.3. As overheating becomes weaker with increasing depth, the fraction of the liquid phase increases and large droplets are formed in the tail of the plume. Simulations performed with shorter pulses show that, in the regime of stress confinement, thermoelastic stress triggers hydrodynamic motion leading to the ejection of larger and more numerous droplets (Figure 5.4) than in the regime of thermal confinement.

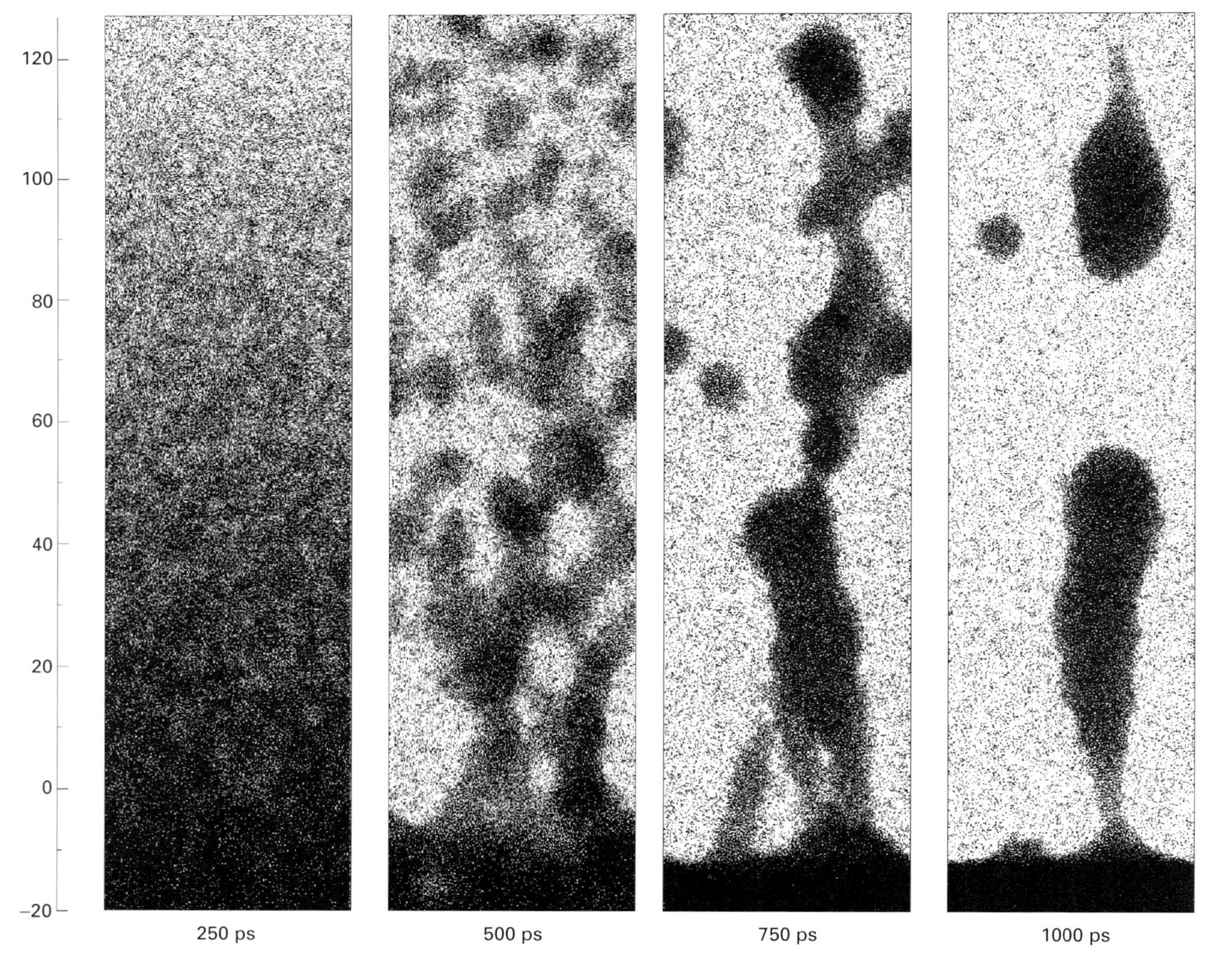

Figure 5.3. Snapshots from the simulation of laser ablation in the regime of thermal confinement. The laser pulse is of duration 150 ps and fluence 61 J/m^2 (1.75 times the ablation threshold fluence). From Zhigilei (2003), reproduced with permission by Springer-Verlag.

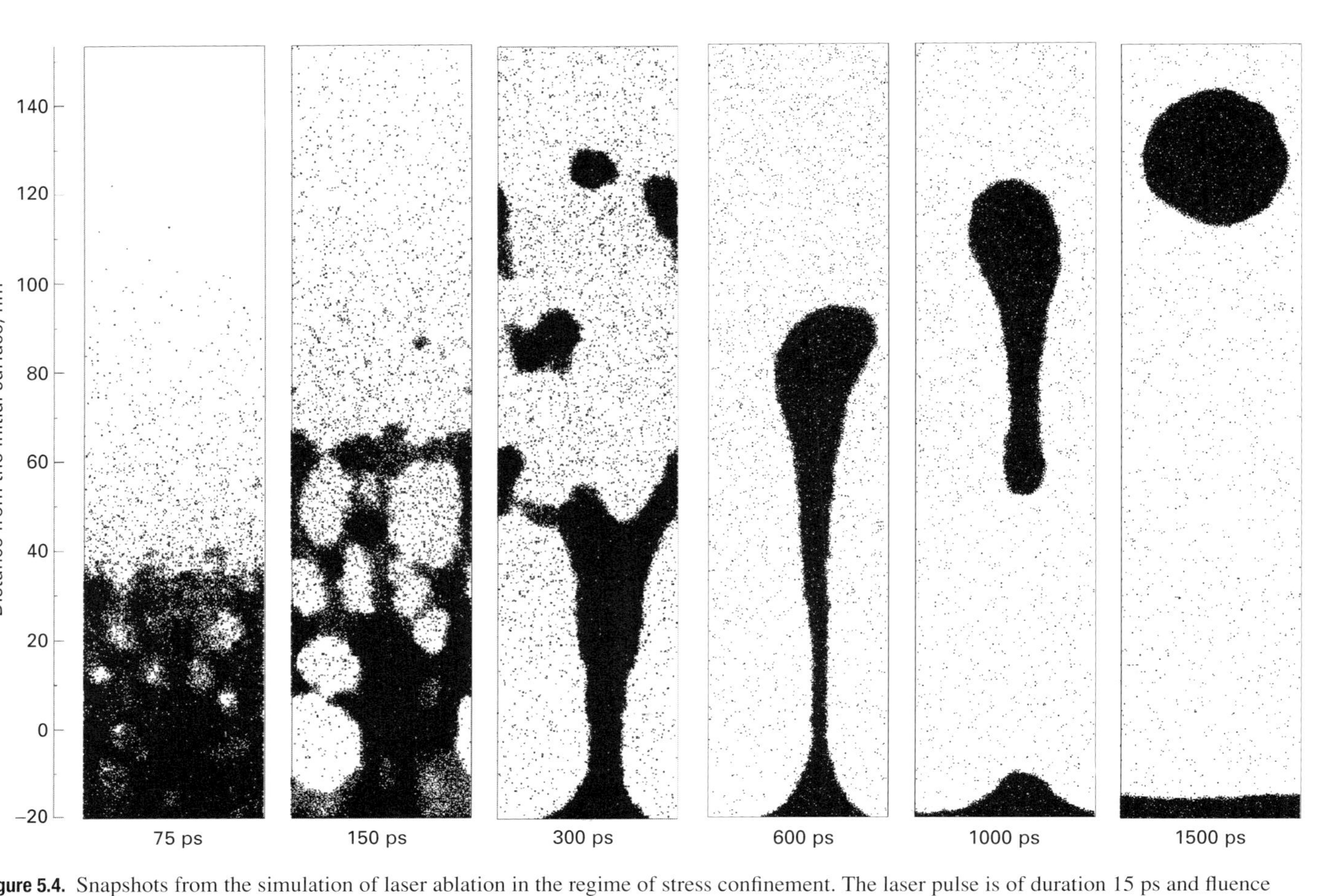

Figure 5.4. Snapshots from the simulation of laser ablation in the regime of stress confinement. The laser pulse is of duration 15 ps and fluence 40 J/m^2. From Zhigilei (2003), reproduced with permission by Springer-Verlag.

5.3.3   Direct Monte Carlo simulations

Shock-wave models describe well the pulsed-laser–material interaction in the presence of ambient background gas pressures above a few hundred millitorr. On the other hand, in the regime of a limited number of collisions per particle that corresponds typically to a pressure range of 0.1–200 mTorr, an appropriate treatment may be pursued via Monte Carlo simulations. Cools (1993) investigated the effect of elastic collisions between ablated particles and background-gas atoms on the kinetic-energy and spatial distributions of atoms arriving at the deposition substrate. Evolution from a forward distribution to a Maxwell–Boltzmann distribution at the gas temperature, so-called "thermalization," was observed in a pressure range that varies over two orders of magnitude, depending on the initial energy of the atoms and the ratio of the molecular masses of background gas and ablated atoms. A self-focusing phenomenon was found to occur during the first stages of thermalization. At low pressures, this effect tends to redistribute the ablated atom flux toward the surface normal. The effect of the background-gas pressure on the deposition stoichiometry was investigated by Itina *et al.* (1997), who considered laser ablation of a binary target into a diluted gas of pressure ≤ 100 mTorr. It was found that different pressures are required for thermalization of particles with different masses. Correspondingly, the deposition rate of light particles diminishes at lower ambient background-gas pressures than does that of heavy particles. An increase of gas pressure results in an increase in the uniformity of the stoichiometry distribution on the receiving substrate. Results indicated that compositionally uniform films can be obtained at appropriate pressures at which the intrinsic compositional nonuniformities are concealed due to species-dependent gas-scattering effects.

5.4   Diagnostics of laser-ablated plumes

5.4.1   Probe-beam-deflection diagnostics

Beam-deflection schemes are based on the detection of changes in the refractive index of the medium due to thermal, pressure, or concentration gradients. These probes have been demonstrated for applications including absorption spectroscopy (Boccara *et al.*, 1980; Sigrist, 1986; Tam, 1986) and thermal-diffusivity measurements (a survey is given in Park *et al.*, 1995). The deflection-angle transients are described by

$$\phi(t) = \left| \int_{\vec{s} \in \text{path}} \frac{1}{n} \nabla n(\vec{s}, t) \times \mathrm{d}\vec{s} \right| . \tag{5.36}$$

In general, the deflection angle has one component parallel and one component perpendicular to the sample surface. The deflection angle is detected by a position-sensitive detector such as a bi-cell or quadrant detector and a knife edge.

When a time-modulated or pulsed laser beam irradiates the target, the temperature field diffuses in the ambient backing medium and produces photothermal deflection

(PTD) of the probing beam on a time scale representative of the evolution of thermal transport in the system. Even in the simple-heating case, below the melting and ablation thresholds, the PTD response is sensitive to the boundary conditions on the temperature field, the material properties, the deflecting medium, the shape, temporal duration, or modulation frequency of the irradiating beam, and the location and size of the probing laser beam. If the profile of the laser beam is Gaussian with dependence on the radial coordinate, r, the temperature field in the backing medium depends on (r, z, t) and the calculation of the probe-beam deflection has to take into account the corresponding refractive-index changes, together with the finite size and profile of the probing laser beam.

Since the deflection signal carries information developed along the path of the beam, a simple ray-tracing approach might not be sufficient for precise calculation of the transient deflection angle, particularly as the cross-sectional dimensions of the probe beam approach the extent of the heat-affected zone. Temporal and spatial effects were examined by Shannon et $al.$ (1992), who showed that, in addition to the magnitude and phase of the deflection signal, the temporal shape of the deflection signal provides fundamental information about the transient heat flow in the target and the backing medium. In a subsequent study, Shannon et $al.$ (1994) showed that the PTD method could indicate the transition to the molten phase.

The thermoelastic expansion launches in the backing medium a pressure pulse that disturbs the probing laser beam, producing photoacoustic deflection (PAD). The compressive PAD pulse scales with the transient heating, whereas the cooling time is generally slower, producing a much weaker rarefaction wave. If the pressure dependence of the refractive index of the backing medium is known, then the deduction of absolute values of the pressure field can be accomplished, provided that the spatial shape of the acoustic wave is accounted for (Diaci, 1992). If the laser-beam energy exceeds the ablation threshold, the ablation products interfere with the probing laser beam. If ablation is performed in a backing atmosphere higher than 1 Torr, the ablation products push the background gas, creating a shock wave that travels with supersonic speed. The strength of the shock wave and the traveling speed decay as the distance from the target is increased. By varying the separation of the probe beam's axis from the target surface, deflection schemes can yield the shock speed (Sell et $al.$, 1991). The ablation products trail the leading shock front, but the deflection scheme is not sufficiently species-specific to provide clear information. Enloe et $al.$ (1987) and Chen and Yeung (1988) applied probe deflection to analyze density gradients in laser ablation in vacuum. Enloe (1987) used the probe-beam deflection to show that the change of the refractive index in plasmas is negative and proportional to the electron density, N_e:

$$\delta n = -\left(\frac{e^2}{2\pi m_e v^2} \right) N_e. \tag{5.37}$$

On the other hand, changes in concentrations of neutral species are positive. In general, probe-deflection diagnostics provide an appealing means for analyzing thermal conditions of the target and the generated ablation plume, but interpretation of the results requires isolation of the physical phenomenon.

5.4.2   Absorption and emission spectroscopy

Consider two nondegenerate levels of an atom or a molecule with energies E_1 and E_2. If an atom in the lower state $|1\rangle$ absorbs a photon of frequency $v = (E_2 - E_1)/h$, it is excited to the higher state $|2\rangle$. The probability per second that an atom will absorb a photon, $d\tilde{P}_{12}/dt$, is proportional to the number of photons of energy hv per unit volume, $N(v)$, and is expressed by

$$\frac{d\tilde{P}_{12}}{dt} = B_{12}N(v), \tag{5.38}$$

where the proportionality constant B_{12} is the Einstein coefficient of induced absorption. The radiation field can also induce stimulated emission via the transition $|2\rangle \rightarrow |1\rangle$ whose probability is

$$\frac{d\tilde{P}_{21}}{dt} = B_{21}N(v), \tag{5.39}$$

where B_{21} is the Einstein coefficient for stimulated emission. Whereas stimulated emission is a coherent, wavevector-maintaining process, spontaneous emission of a photon of energy hv through an incoherent process is also possible. The corresponding coefficient for spontaneous emission is

$$A_{21} = \frac{8\pi h v^3}{c^3}B_{21}. \tag{5.40}$$

Since A_{21} scales with v^3, spontaneous emission processes are important in the visible/UV parts of the spectrum. Since the atomic energy levels are discrete and distinct, each element has a set of characteristic "signature" emission lines. The detection of these lines can be used to identify the presence of a particular element.

The Maxwell–Boltzmann equation defines the ratio of the number of atoms in an excited state to the number of atoms in the ground state in thermal equilibrium;

$$\frac{N_1}{N_0} = \frac{g_0}{g_1}e^{-\frac{E_1}{k_B T}}. \tag{5.41}$$

The important assumption in this case is that the system is in local thermal equilibrium (LTE). Consequently, the number of atoms of a particular state is given by

$$N_1 = N\frac{g_1 e^{-\frac{E_1}{k_B T}}}{\sum\limits_{i=0}^{\infty} g_i e^{-\frac{E_i}{k_B T}}} = \frac{N g_1 e^{-\frac{E_1}{k_B T}}}{Z}. \tag{5.42}$$

The line intensity is given by the relationship

$$S_1 = A_{10}\frac{N_0 E_1}{\tau}\frac{g_1}{g_0}e^{-\frac{E_1}{k_B T}}. \tag{5.43}$$

The values for the degeneracies and the Einstein coefficients of each energy state can be obtained using standard references. By using Equation (5.43), the relative populations of each excited state can be found by comparing the respective intensities. Transmission and absorption imaging of excimer-laser-ablated plumes was performed by Schittenhelm *et al.* (1998) to measure the spatial plasma density distribution.

5.4.3   Laser-induced fluorescence spectroscopy

Fluorescence is the emission of radiation that results from absorption of radiation from another excitation source. An atom or a molecule is excited to a higher energy level as the result of the absorption of an excitation photon. A fluorescence photon is emitted when the excited atom or molecule relaxes down to a lower energy state, and the energy of the photon is equal to the energy difference between the two levels. The photon absorption and the subsequent energy transition occur very rapidly, typically on time scales of the order of picoseconds. Therefore, the emitted radiation persists only for as long as the duration of the excitation source to which it is subjected. When the excitation radiation and the emission are of the same wavelength the process is called resonance fluorescence. A more typical case is when the emitted photons have a longer wavelength and lower energy than the excitation photons. In these processes, the excited species undergo transitions involving intermediate energy levels that may, but need not, result in emission. Examples of non-photon-emitting processes include collisional quenching and conversion into vibrational energy.

The simplest case of fluorescence to analyze is the so-called two-level model. The principal assumption in this model is that only two energy levels are involved in population change during the excitation and stimulated emission. The rate equation for the excited-state (level-2) population, N_2, is

$$\frac{\mathrm{d}N_2}{\mathrm{d}t} = N_1(W_{12} + Q_{12}) - N_2(W_{21} + Q_{21} + A_{21}), \tag{5.44}$$

where N_1 is the population of the ground level 1, W_{12} and W_{21} are the absorption rate of the excitation radiation and the rate of stimulated emission, Q_{12} and Q_{21} are the collisional excitation and quenching rates, and A_{21} is the Einstein coefficient of spontaneous emission. Assuming that the total number density of the energy-state population is constant, i.e. $N_1 + N_2 = N_\mathrm{T} =$ constant, and that the system is at equilibrium, i.e. $\mathrm{d}N_1/\mathrm{d}t = \mathrm{d}N_2/\mathrm{d}t = 0$, it follows that

$$N_2 = \frac{N_\mathrm{T}(W_{12} + Q_{12})}{Q_{12} + Q_{21} + A_{21} + W_{21} + W_{12}}. \tag{5.45}$$

Simplifications of the above equation are often made. When the excitation power is low, $Q_{21} + A_{21} \gg W_{12} + W_{21}$, and the population density of the excited level is proportional to the absorption rate of the excitation radiation:

$$N_2 = \frac{W_{12}N_\mathrm{T}}{Q_{21} + A_{21}}. \tag{5.46}$$

The number of photons (N_p) incident on a fluorescence detector is proportional to the density of the excited-state population. It is given by

$$N_p = \frac{N_2 A_{21} V_c e_f \Omega_c}{4\pi}, \tag{5.47}$$

where V_c is the fluorescence-probe volume, e_f is the efficiency of the optical collection system, and Ω_c is the solid angle of the optical collection system. The fluorescence signal given by a detector is

$$S_f = N_p h c v_{21} \phi_p G_p, \tag{5.48}$$

where h is Planck's constant, c the speed of electromagnetic propagation, v_{21} the fluorescence-transition frequency, ϕ_p the quantum efficiency of the detector, and G_p the detector gain.

5.4.4   Plume interactions with background gas

The interaction of the laser-ablation plume with background gas is important for applications such as PLD. During PLD of YBCO films in ambient gas the background oxygen scatters, attenuates, and interacts with the ablated plume. As discussed by Geohegan (1992), during the initial stage of expansion, the plume experiences little slowing. At some point, a visible shock structure is formed, whereby the plume appears to split into a fast and a slower component. At longer times, the plume continues to slow down. Figure 5.5 depicts this sequence using time-resolved integrated imaging.

Figure 5.6 depicts experimental setups for imaging plume emission and capturing emission lines (Chu and Grigoropoulos, 2000). One can construct x–t diagrams by plotting the location versus the time corresponding to the peak emission signal. The most probable velocity is then found by taking the slope of this curve. Figure 5.7(a) shows the x–t curves for neutral titanium ablated by several laser fluences under vacuum. Since collisions occur in the laterally expanding plume, the centerline velocity can be expected to decrease. The decrease in centerline velocity is more significant for the lower fluence. The effects of the background argon pressure on the centerline velocity are shown in Figure 5.7(b). The background pressure suppresses plume expansion and consequently increases collisions within the plume. This is evident from the decrease in centerline velocity and, as will be seen later, the overall emission intensity.

Spectrally integrated and temporally resolved images of plumes ablated at 5 J/cm^2 under background argon pressures ranging from vacuum to 20 Torr have been acquired. Some selected images are shown in Figure 5.8. Since the images are spectrally integrated over the near-UV to visible range, they include emission from both neutral and ionized species. A spectrum taken of the plume revealed that only emission from Ti and Ti$^+$ is detectable, and no appreciable emission was detected from higher ionized states of Ti or the background Ar. At pressures above 50 mTorr, a separation of faster- and slower-moving components occurs in the plume. The intensity of the emission is also greater with higher background pressure as a result of there being a greater number of collisions. The location at which the separation of faster- and slower-moving components occurs is

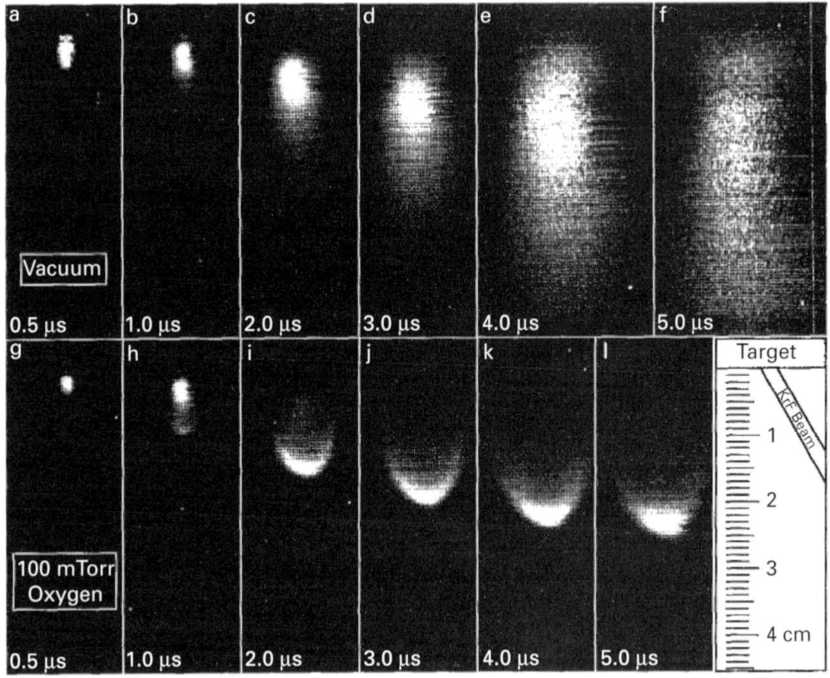

Figure 5.5. ICCD photographs of the visible plasma emission (exposure times 20 ns) following 1.0-J/cm^2 KrF/YBCO ablation into 1×10^{-6} Torr, (a)–(f), and 100 mTorr, (g)–(l), oxygen at the indicated delay times from the arrival of the laser pulse. The 0.2 cm $\times$ 0.2 cm 248-nm laser pulse irradiated the YBCO target at an angle of 30° as shown. From Geohegan (1992), reproduced with permission by the American Institute of Physics.

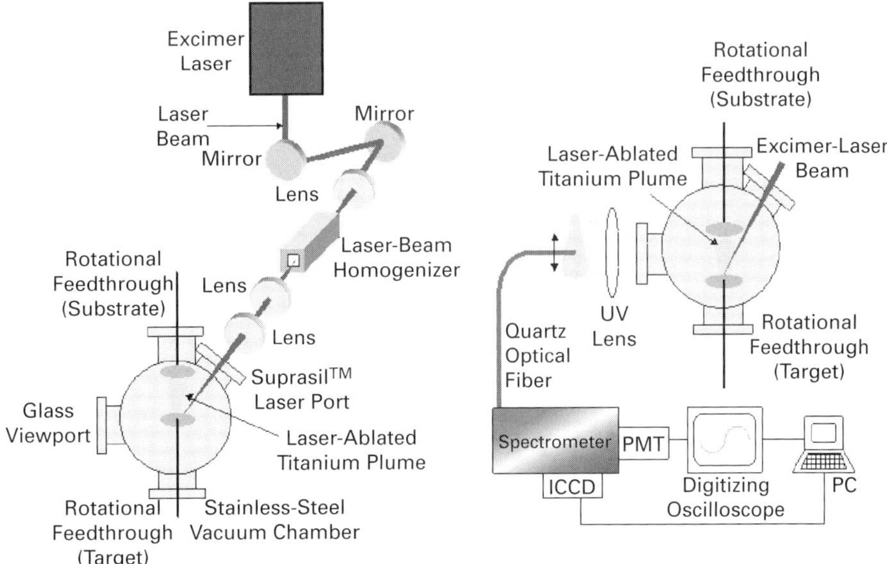

Figure 5.6. Experimental setups for imaging plume emission and capturing emission lines.

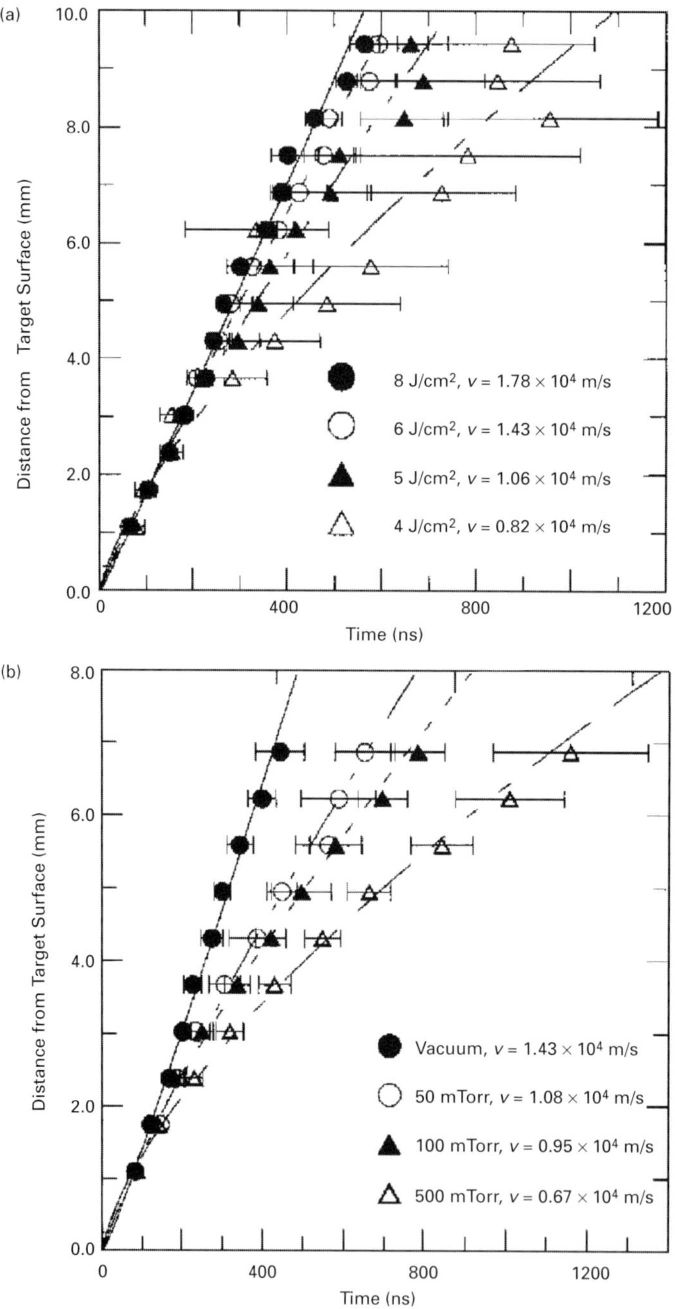

Figure 5.7. The left graph shows effects of laser fluence on the most probable centerline velocity of neutral titanium atoms in an expanding plume in vacuum. The velocities shown are average values over the measurement range. The right panel depicts effects of background Ar pressure on the most probable centerline velocity of neutral titanium atoms in the plume. The fluence is $F = 7$ J/cm^2. The velocities shown are average values over the measurement range. From Chu and Grigoropoulos (2000), reproduced with permission by the American Society of Mechanical Engineers.

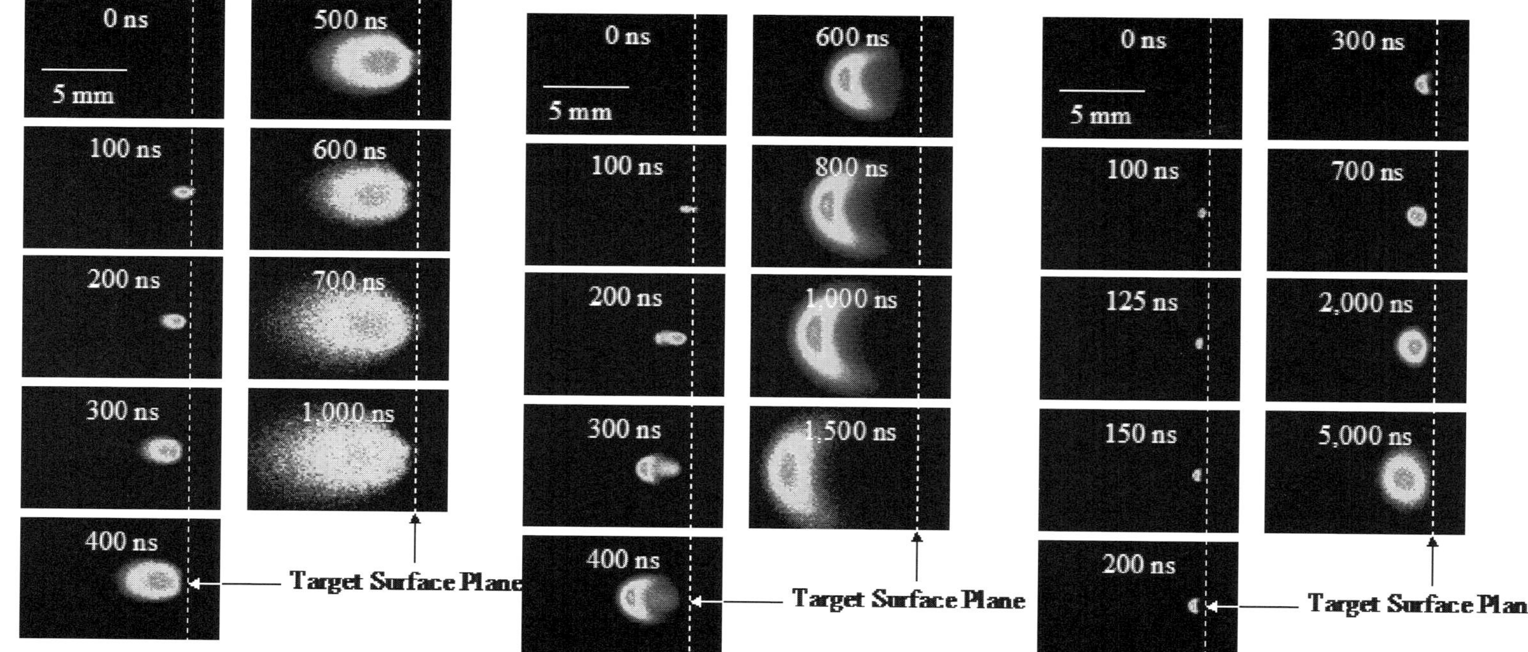

Figure 5.8. Spectrally integrated images with 10-ns gating in normalized pseudocolor. All were taken at $F = 5$ J/cm^2, the left image under vacuum, the center image at $P = 100$ mTorr, and the right image at $P = 1$ Torr. From Chu and Grigoropoulos (2000), reproduced with permission by the American Society of Mechanical Engineers.

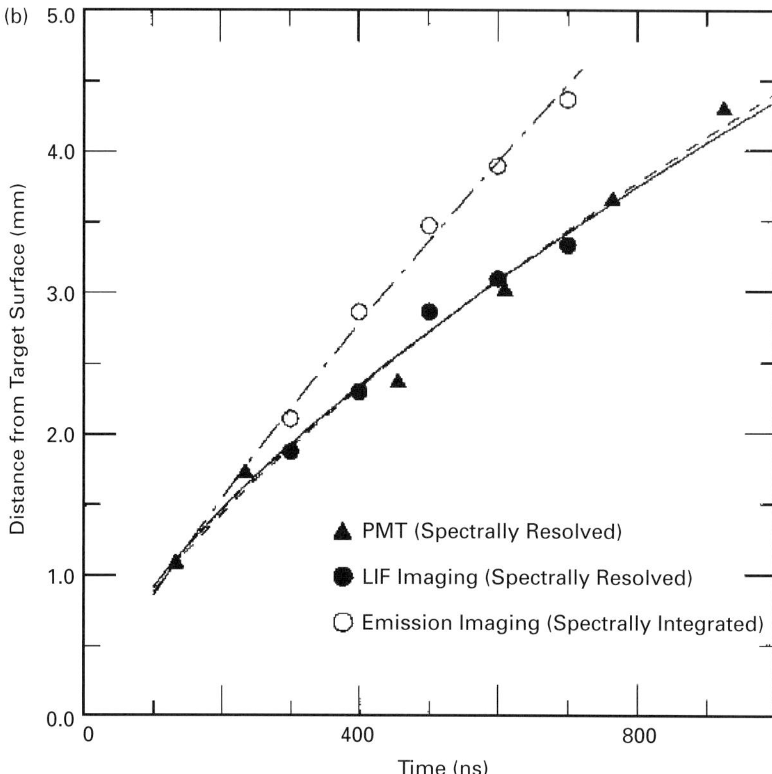

Figure 5.10. (*cont.*)

forward part of the plume is primarily composed of titanium ions. This observation is most probably due to the fact that the ions acquire greater kinetic energy through the redistribution of the absorbed incident laser energy by the plasma plume.

There have been theoretical studies of laser ablation under vacuum (or low-pressure) conditions (Bogaerts *et al.*, 2003), but few authors have focused on laser ablation under a background gas pressure of 1 atm. Investigators (Aden *et al.*, 1992; Ho *et al.*, 1995; Capitelli *et al.*, 2000) often apply the hydrodynamic equations only for the vapor species, whereas it is obvious that a binary gas mixture (e.g. metal vapor and background gas) as well as the interactions between vapor and gas need to be considered. A hydrodynamic model that describes the behavior of both vapor and background gas has been developed by Gusarov *et al.* (Gusarov *et al.*, 2000; Gusarov and Smurov, 2003) and applied to expansion in a background gas at 1 atm for both millisecond and nanosecond laser irradiance, but without taking into account plasma formation. A comprehensive numerical model for nanosecond laser ablation of metallic targets in background gas at 1 atm, with consideration of plasma formation and laser absorption in the plasma, was developed recently (Chen and Bogaerts, 2005; Bogaerts and Chen, 2005). It has been found that target heating, melting, and vaporization, as well as the vapor and background gas density, plume-expansion velocity and temperature, degree of ionization, and densities of ions and electrons in the plume, and hence also the plasma shielding, all increase with increasing laser irradiance.

The effect of laser-pulse duration shows two different trends, depending on whether the laser irradiance or the laser fluence is kept constant. At constant laser irradiance, the target heating, melting, and vaporization increase with laser-pulse duration, and this applies also to the densities of vapor and background-gas atoms and ions, and electrons in the plume, as well as to the plume-expansion velocity and temperature, because the laser fluence rises with increasing pulse duration. At fixed laser fluence, on the other hand, the target-heating and evaporation rates are greater for shorter laser pulses, because of the rise in laser irradiance. The total melt and evaporation depths are slightly greater for longer laser pulses, because the target is exposed to the laser for a longer time. Plasma shielding is less pronounced for longer pulses, because of the lower irradiance, so the net laser fluence reaching the target increases slightly with increasing pulse duration. The plume temperature and electron density, during or shortly after the laser pulse, become higher for shorter pulses, because of the higher laser irradiance at fixed laser fluence. However, after the laser pulse has finished (e.g. 100 ns), it is observed that the total laser fluence, not the pulse duration, determines the plume behavior, because the plume and plasma characteristics look very similar for different pulse durations, at constant laser fluence.

Target and plume characteristics (e.g. plume density, velocity, temperature, and degree of ionization in the plasma) at 532 nm were only slightly lower than those at 266 nm, because of the competing effects of target-surface reflectivity and laser plasma absorption. At 1064 nm, the target and plume characteristics were calculated to be significantly lower than at 266 and 532 nm, which was attributed mainly to the high surface reflectivity. The laser-irradiance threshold for target melting and vaporization and for plasma formation at 1064 nm was considerably higher than that at 266 and 532 nm.

5.5 Picosecond-laser plasmas

The formation and subsequent evolution of plasmas produced during interactions of high-power laser beams with solid materials (laser ablation) are topics of much practical interest. Applications include synthesis of novel materials (Lowndes *et al.*, 1996; Geohegan *et al.*, 1998), laser-ablation chemical analysis (Russo, 1995), ultrafast X-ray generation (Murnane *et al.*, 1991), and fast ignition schemes for inertial-confinement fusion, among others. In contrast to plasma initiation by direct photo-ionization of a free gas (Li *et al.*, 1992; Clark and Milchberg, 1997), a solid target at the focus of the incident laser beam generates an extra dimension of complexity of plasma generation during laser-ablation processes. While substantial progress in understanding nanosecond-laser ablation of solid materials has been achieved during the past few decades, there is little knowledge of the spatial and temporal development of plasmas induced by laser ablation of solids on picosecond and faster time scales.

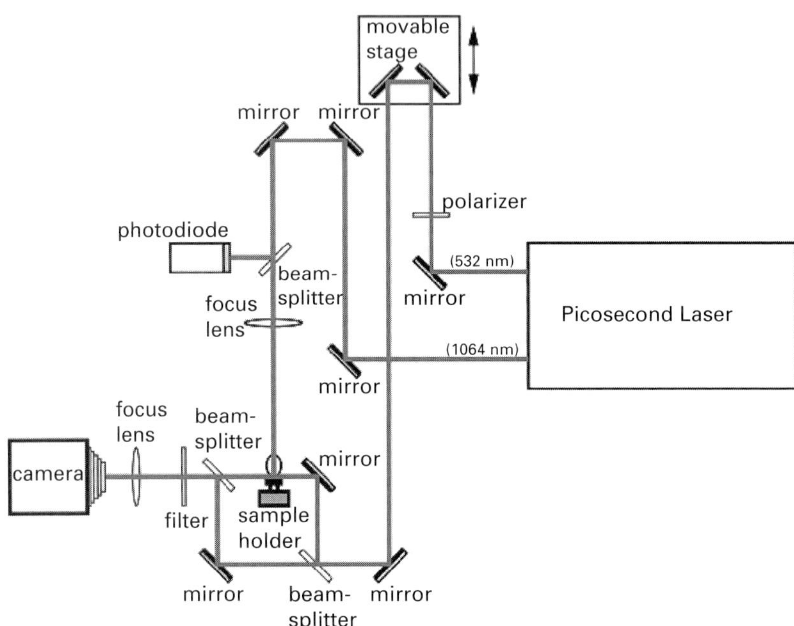

Figure 5.11. Three shadowgraph images of the laser-ablation plasma at 20 ps (a)–(c) and 1200 ps (d)–(f). The incident laser energy is given in each image. Laser input was from the above. From Mao *et al.* (2000a), reproduced with permission by the American Institute of Physics.

5.5.1   Fundamentals of picosecond-laser plasmas

Figure 5.11 diplays an experimental setup for time-resolved shadowgraph and interferometry imaging (Mao *et al.*, 2000). Figure 5.12(a) is a typical time-resolved interferogram of the picosecond plume. At very early times during the laser-ablation process ($t \sim 0$), the plasma plume was found to have an electron number density of the order of 10^{19} cm^{-3} near the target surface. Figure 5.12(b) shows the plume's electron-number-density profile along the incident laser axis at a delay time of 150 ps. The solid curve is a least-squares fit for an exponential decay. The plasma has an electron number density exceeding 10^{20} cm^{-3} from the target surface to a distance approximately 40 μm away from the target.

Figure 5.13(a) shows three shadowgraphs taken at a delay time of 20 ps with various incident laser energies. The picosecond plume, or the early-stage plasma, barely appears at 3.5 mJ, which corresponds to a laser irradiance of approximately 1.5×10^{12} W/cm^2. With increasing incident laser energy, both the longitudinal and the lateral extension of the picosecond plume increase. However, there is no other plume present at this early time. At a delay time of 1200 ps (Figure (5.13b)), a hemispherical plume dominates the shadowgraph for laser energies close to the picosecond plume or plasma-forming threshold ($\sim$3.5 mJ). This vapor plume of target mass is frequently observed during nanosecond and longer-pulsed laser ablation of solids.

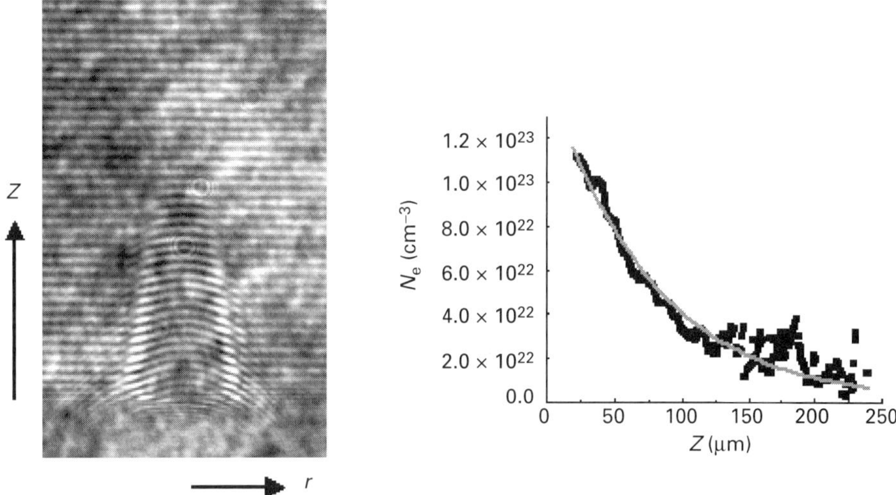

Figure 5.12. A schematic diagram showing the setup of the experiments. An interferometer is used to take shadowgraph and interferogram images of the picosecond-laser-ablation plasma.

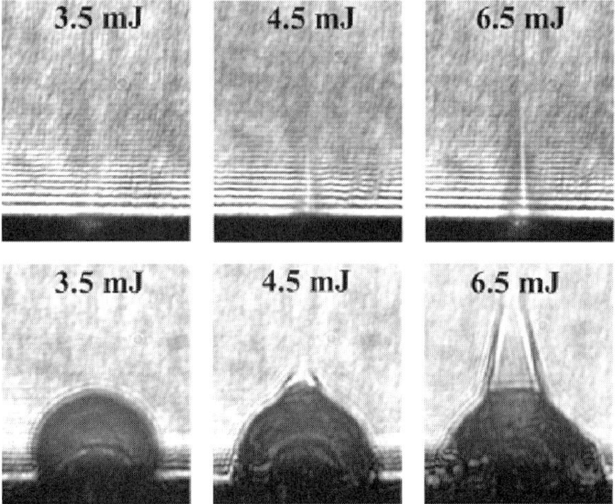

Figure 5.13. The electron-number-density profile along the incident laser axis. The solid curve is a least-squares fit of the experimental data showing exponential decay. The insert shows an interferogram of the picosecond-laser-ablation plasma. From Mao *et al.* (2000a), reproduced with permission by the American Institute of Physics.

Such a high number of electrons close to the target surface indicates that the electrons cannot originate from direct ionization of air. The ambient air has a density between 2×10^{19} and 3×10^{19} cm^{-3}. If the electrons inside the plasma come from direct ionization of air, each air molecule would have to provide about ten free electrons. This is impossible in the case at hand because the plasma temperature barely exceeds the second ionization potential of the atoms (e.g. nitrogen) in air. The high electron density close to the target (many times larger than the air density) suggests that electrons emitted from the target surface play a significant role in the initiation of this early-stage plasma. Energetic electrons have been detected (Petite *et al.*, 1992) from metal surfaces irradiated by picosecond laser pulses, although their importance for plasma initiation was not previously realized.

Let V represent the velocity of the particle species inside the plasma, then conservation of particle number density N, momentum p, and energy ε can be expressed as

$$\frac{\partial N}{\partial t} + \frac{\partial (N \cdot V)}{\partial z} = S_{\mathrm{n}},$$
$$\frac{\partial p}{\partial t} + \frac{\partial (p \cdot V)}{\partial z} = -\frac{\partial P}{\partial z} + NqE + Nf_{\mathrm{c}}, \qquad (5.49)$$
$$\frac{\partial \varepsilon}{\partial t} + \frac{\partial (\varepsilon \cdot V)}{\partial z} = -\frac{\partial (P \cdot V)}{\partial z} + NqE \cdot V + Nf_{\mathrm{c}} \cdot V + Q_{\mathrm{ab}},$$

with P the particle pressure, q the unit charge of an electron, E the electric field inside the plasma, f_{c} the collisional force between particles, and Q_{ab} the energy source due to absorption of laser light.

The particle momentum and energy can be written as

$$p = NmV,$$
$$\varepsilon = \frac{3}{2}Nk_{\mathrm{B}}T + \frac{1}{2}mV^{2}, \qquad (5.50)$$

where m is the particle mass, T is the particle temperature, and k_{B} is the Boltzmann constant.

In the density equation, S_{n} is the source term for electron and ion generation in the bulk of the plasma. Cascade ionization is the dominant ionization mechanism over multiphoton and recombination processes. The momentum equation accounts for the electric force (zero for atoms) and the force resulting from particle–particle collisions. The nonlinear (ponderomotive) force (Kruer, 1988), which is important for the propagation of high-intensity laser pulses in plasma (Young *et al.*, 1988), was found to be insignificant in the laser-energy range of the present study. The energy balance includes the work done by the forces exerted on particles, as well as the direct energy increment from absorbing laser light (for electrons only).

Electron emission takes place as the surface electrons absorb the energy of one or more incident photons, or acquire sufficient kinetic energy after heating to overcome a surface energy barrier. Electron emission is therefore allowed to provide initial seed electrons for plasma development above the laser-ablated target. While pulsed-laser-induced electron emission from solids has been studied for decades, its significance for the initiation of laser-ablation plasmas had not been quantified in previous theoretical work. In the simulation, surface electron emission due to thermionic and photoelectric effects was calculated using a coupled energy-transport model (for electron and lattice energies) for picosecond-laser–metal interactions. The electrons emitted from the laser-heated target collide with atoms in the ambient gas, which may result in cascade breakdown of the gas.

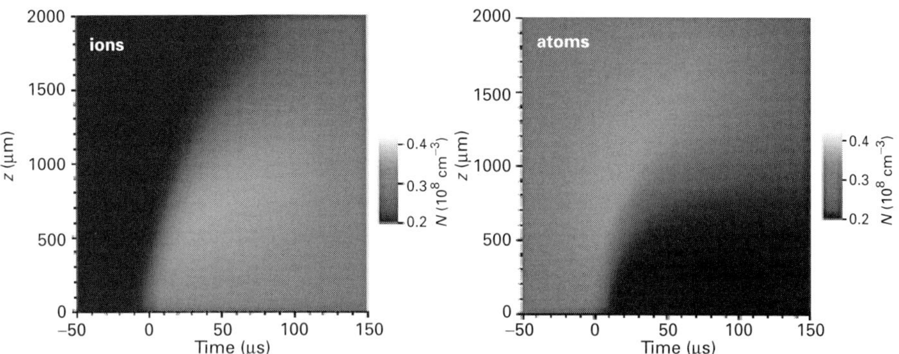

Figure 5.14. Spatial–temporal evolution of ion and atom density inside a plasma. Pseudocolors are used to represent the magnitudes of the densities. From Mao *et al.* (2000b), reproduced with permission by the American Institute of Physics.

Figure 5.14 shows the evolution of ion and atom densities above a Cu target. In the map for ions (atoms), the pale shade corresponds to a high density of ions (atoms). With N_i and N_a representing the ion and atom densities respectively, full ionization of the gas, $N_i \sim N_a$ (1 atm), occurs just when the ablation laser pulse passes its maximum energy. Significant ionization starts from the gas adjacent to the target surface where energetic electrons abound due to surface electron emission during the rising stage of the picosecond-laser pulse.

The electrons ejected from the target absorb laser energy principally by inverse-Bremsstrahlung processes during laser irradiation. They subsequently ionize the ambient gas by impact ionization, leading to cascade breakdown. The spatial–temporal movement of the ionization front is characterized by an approximate boundary separating the pale and black regions in the map for both ions and atoms. The ionization front advances approximately 750 μm during the first 35 ps (from approximately 0 ps to 35 ps), but it takes more than 100 ps to traverse a further 750 μm.

Figure 5.15 gives the electron-temperature profile in a spatial–temporal diagram. As is evident from the picture, electrons reach their maximum temperature just after the laser intensity passes its peak value ($t = 0$). The temperature of the plasma is of the order of 10^5 K, equivalent to about 10 eV. This value is approximately equal to the first ionization energy of atoms in ambient gas, but smaller than their second ionization energy. As a consequence, we expect that most of the ions in the plasma bear only one positive charge. This result suggests that each gas atom can provide at most one free electron to the plasma. The experimentally measured large number of electrons cannot come from laser-induced direct gas breakdown. Electron emission from the target surface, as discussed earlier, will supply a large quantity of electrons to the plasma.

Figure 5.16 shows shadowgraphs (a) and phase-shift maps (b) of the picosecond-laser ablation plasma at four delay times. The phase-shift value of the interference fringe is directly proportional to the value of the local electron number density (Jahoda and

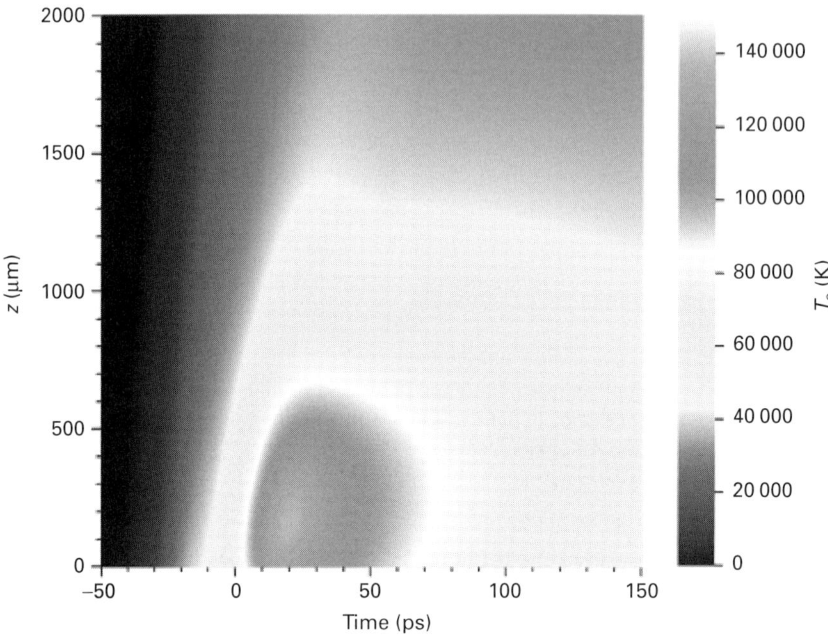

Figure 5.15. Spatial–temporal evolution of the electron temperature of the picosecond-laser-ablation plasma. From Mao *et al.* (2000b), reproduced with permission by the American Institute of Physics.

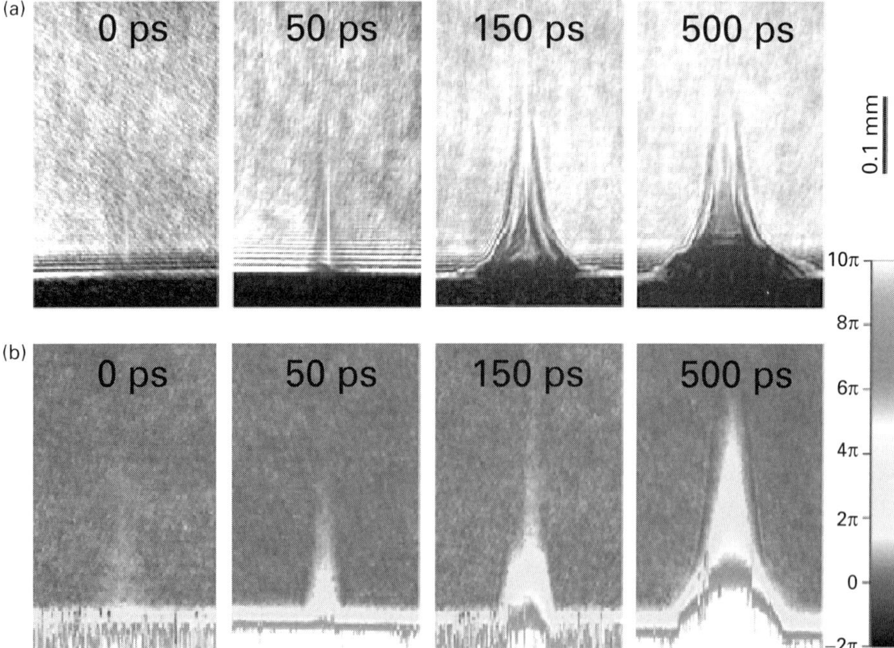

Figure 5.16. Shadowgrams (a) and phase-shift maps (b) of the plasma at four delay times. From Mao *et al.* (2000a), reproduced with permission by the American Institute of Physics.

Sawyer, 1971). The laser energy was approximately 9 mJ, corresponding to a laser irradiance of approximately 4×10^{12} W/cm^2. A weakly absorbing plasma is present, with an electron number density of the order of 10^{19} cm^{-3} near the target surface. The phase-shift maps (Figure 5.16(b)) indicate that the electron number density of this plasma increases substantially during the first 100 ps. After about 200 ps, atomic and ionic mass from the target appears in the shadowgraphs (the dark region close to the target). This material vapor plume corresponds to a growing white area in phase-shift maps, where the strong absorption of the probe beam weakens the resolution of the interferogram and no longer yields useful phase-shift values. As the material vapor plume expands away from the target, it gradually becomes hemispherical in shape. Large lateral expansion of the picosecond-laser ablation plasma is apparent, while the longitudinal expansion of the early-stage plasma is not very substantial after about 200 ps.

Figure 5.17 shows calculated electron-number-density profiles in the form of iso-density contours. These distance–time profiles provide a measure for the longitudinal expansion of the plasma. As shown in Figure 5.17(a), the electron number density near the target surface increases from 6×10^{19} cm^{-3} to 10^{20} cm^{-3} in about 15 ps (from -5 ps to 10 ps), in agreement with the experimental observations. With speeds of the order of 10^8–10^9 cm/s, the iso-density fronts expand rapidly at early times. However, after approximately 35 ps, when most of the laser-pulse energy has been delivered, the 10^{20}-cm^{-3} front starts to move backward, while the 6×10^{19}- and 8×10^{19}-cm^{-3} fronts expand at significantly reduced speeds. Such suppression of longitudinal plasma expansion is consistent with the experimental measurements. The experimental results that correspond to an electron number density of 7×10^{19} cm^{-3} are given in Figure 5.17(b). The insert is a series of plasma shadowgraphs taken at four delay times.

There are minor discrepancies between the calculated results and the measured ones. The calculated longitudinal expansion fronts are slightly greater than the measured ones; experimentally, the plasma also expands in the radial direction, whereas only a one-dimensional expansion was considered in the simulation. Despite such discrepancies, the general agreement between the simulation and the experimental data justifies the theoretical model.

The suppression of plasma expansion can be attributed to the development of an electric field, E, inside the plasma. One cause of the field is the ejection of energetic electrons from the target surface, which leaves positively charged ions on the target surface. In addition, because ionized gas atoms inside the plasma are much heavier than electrons, they can be considered immobile while electrons rapidly expand away from the target. As a consequence, there is a net positive charge close to the target surface and a negative charge away from the target. The resulting electric field that is directed against the incident laser beam acts to suppress further forward movement of the electrons inside the plasma. The calculated development in space and time of the electric field above the Cu target is shown in Figure 5.18. An electric field as strong as 4×10^6 V/m exists above the target.

For fundamental understanding and practical applications, it is instructive to ascertain to what extent the laser-ablation plasma absorbs the incident picosecond-laser energy

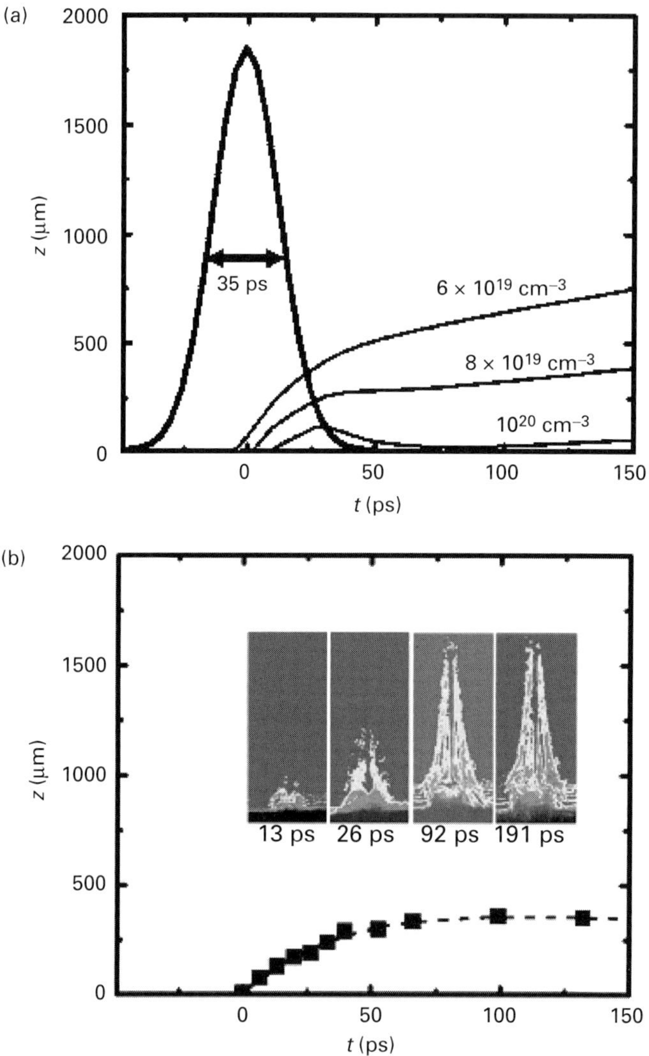

Figure 5.17. (a) Calculated space–time iso-density contours for the electrons inside the plasma. The 35-ps-FWHM ablation laser pulse is also shown for arbitrary units. (b) The experimental data for longitudinal plasma extension. The insert shows plasma pictures recorded at four different times. Laser input is from the above. From Mao *et al.* (2000b), reproduced with permission by the American Institute of Physics.

and thereby reduces the efficiency of energy deposition onto the target. The longitudinal extent of the air plasma remains approximately constant after about 100 ps, and the plasma expands principally in the lateral (radial) direction. A material plume starts to appear after a few hundred picoseconds; it overlaps with part of the air plasma as it expands. As determined from both interferometry and simulations, the electron number density of the plasma is of the order of 10^{20} cm^{-3}. At post-pulse times, the expanding

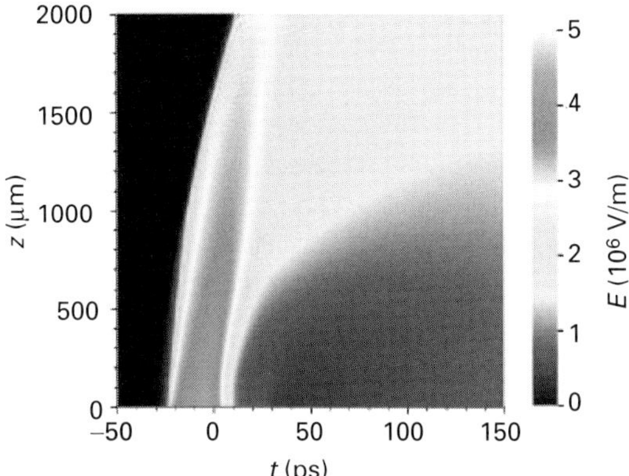

Figure 5.18. Spatial–temporal evolution of the electric field above the laser-ablated target. From Mao *et al.* (2000b), reproduced with permission by the American Institute of Physics.

plasma forms an approximate column with a high electron number density near the outer radii and a low-density region near the incident laser axis.

After about 100 ps, the lateral expansion follows the relation $r \sim t^{1/2}$. This dependence is consistent with the similarity solution for expansion resulting from an instantaneous line source of energy release (Sedov, 1959). As discussed by Li *et al.* (1992), the energy driving the radial expansion of the plasma can be estimated using the equation

$$r = C\left(\frac{E_\mathrm{p}}{\rho}\right)^{\frac{1}{4}} t^{1/2}. \tag{5.51}$$

Here C is a constant approximately equal to unity, E_p is the energy deposited per unit length which drives the expansion, ρ is the mass density of the air, and t is time. For the experimental conditions in this work, before the 35-ps laser pulse ends, the plasma is gradually heated by the absorption of light from collisional ionization (the inverse-Bremsstrahlung process) near the laser axis. The laser-spot radius on the target is about 50 μm; it takes some time (of the order of 100 ps) for the density and pressure profiles of the plasma to evolve toward the similarity solution, which assumes an instantaneous release of energy from a line source along the laser axis.

References

Aden, M., Beyer, E., and Herziger, G., 1990, "Laser-Induced Vaporization of Metal as a Riemann Problem," *J. Phys. D*, **23**, 655–661.

Aden, M., Beyer, E., Herziger, G., and Kunze, H., 1992, "Laser-Induced Vaporization of a Metal Surface," *J. Phys. D*, **25**, 57–65.

Afanas'ev, Y. V., and Krokhin, O. N., 1967, "Vaporization of Matter Exposed to Laser Emission," *Sov. Phys. JETP*, **25**, 639–645.

Amoruso, S., Toftman, B., and Schou, J., 2004, "Thermalization of a UV Laser Ablation Plume in a Background Gas: From a Directed to a Diffusionlike Flow," *Phys. Rev. B*, **69**, 056403.

Anisimov, S. I., 1968, "Vaporization of Metal Absorbing Laser Radiation," *Sov. Phys. JETP*, **27**, 182–183.

Boccara, A. C., Fournier, D., and Badoz, J., 1980, "Thermo-Optical Spectroscopy: Detection by the 'Mirage Effect,'" *Appl. Phys. Lett.*, **36**, 130–132.

Bogaerts, A., Chen, Z. Y., Gijbels, R., and Vertes, A., 2003, "Laser Ablation for Analytical Sampling: What Can We Learn from Modeling," *Spectrochim. Acta B: Atom. Spectrosc.*, **58** 1867–1893.

Bogaerts A., and Chen, Z. Y., 2005, "Effect of Laser Parameters on Laser Ablation and Laser-Induced Plasma Formation: A Numerical Modeling Investigation," *Spectrochim. Acta B: Atom. Spectrosc.*, **60**, 1280–1307.

Capewell, D. L., and Goodwin, D. G., 1995, "Monte Carlo Simulations of Reactive Pulsed Laser Deposition," *Proc. SPIE*, **2403**, 49–59.

Capitelli, M., Capitelli, F., and Eletskii, A., 2000, "Non-equilibrium and Equilibrium Problems in Laser-induced Plasmas," *Spectrochim. Acta B: Atom. Spectrosc.*, **55**, 559–574.

Chan, C. L., and Mazumder, J. E., 1987, "One-Dimensional Steady-State Model for Damage by Vaporization and Liquid Expulsion due to Laser–Material Ablation," *J. Appl. Phys.*, **62**, 4579–4586.

Chen, G., and Yeung, E. S., 1988, "A Spatial and Temporal Probe for Laser-Generated Plumes Based on Density Gradients," *Anal. Chem.*, **60**, 864–865.

Chen, K. R., Leboeuf, J. N., Wood, R. F. *et al.*, 1995, "Accelerated Expansion of Laser-Ablated Materials near a Solid Surface," *Phys. Rev. Lett.*, **75**, 4706–4709.

Chen, K. R., Leboeuf, J. N., Wood, R. F. *et al.*, 1996, "Mechanisms Affecting Kinetic Energies of Laser-Ablated Materials," *J. Vac. Sci. Technol. A*, **14**, 1111–1114.

Chen, Z. Y., and Bogaerts, A., 2005, "Laser Ablation of Cu and Plume Expansion into 1 atm Ambient Gas, *J. Appl. Phys.*, **97**, 063305.

Chu, S. C., and Grigoropoulos, C. P., 2000, "Determination of Kinetic Energy Distribution in a Laser-Ablated Titanium Plume by Emission and Laser-Induced Fluorescence Spectroscopy," *J. Heat Transfer*, **122**, 771–775.

Clark, T. R., and Milchberg, H. M., 1997, "Time- and Space-Resolved Density Evolution of the Plasma Waveguide," *Phys. Rev. Lett.*, **78**, 2373–2376.

Cools, J. C. S., 1993, "Monte Carlo Simulations of the Transport of Laser-Ablated Atoms of a Diluted Gas," *J. Appl. Phys.*, **74**, 6401–6406.

Diaci, J., 1992, "Response Functions of the Laser Beam Deflection Probe for Detection of Spherical Acoustic Waves," *Rev. Sci. Instrum.*, **63**, 5306–5310.

Dreyfus, R. W., 1993, "Interpreting Laser Ablation Using Cross-Sections," *Surface Sci.*, **23**, 177–181.

Enloe, C. L., Gilgenbach, R. M., and Meachum, J. S., 1987, "A Fast, Sensitive Laser Deflection System Suitable for Transient Plasma Analysis," *Rev. Sci. Instrum.*, **58**, 1597–1600.

Finke, B. R., and Simon, G., 1990, "On the Gas Kinetics of Laser-Induced Evaporation of Metals," *J. Phys. D*, **23**, 67–74.

Finke, B. R., Finke, M., Kapadia, P. D., Dowden, J. M., and Simon, G., 1990a, "Numerical Investigation of the Knudsen Layer, Appearing in the Laser-Induced Evaporation of Metals," *Proc. SPIE*, **1279**, 127–134.

Finke, B. R., Kapadia, P. D., and Dowden, J. M., 1990b, "A Fundamental Plasma Based Model for Energy Transfer in Laser Material Processing," *J. Phys. D*, **23**, 643–654.

Garrelie, E., Champeaux, C., and Catherinot, A., 1999, "Study by a Monte Carlo Simulation of the Influence of a Background Gas on the Expansion Dynamics of a Laser-Induced Plasma Plume," *Appl. Phys. A*, **69**, 45–50.

Geohegan, D. B., 1992, "Fast Intensified-CCD Photography of $YBa_2Cu_3O_{7-x}$ Laser Ablation in Vacuum and Ambient Oxygen," *Appl. Phys. Lett.*, **60**, 2732–2734.

Geohegan, D. B., Puretzky, A. A., Duscher, G., and Pennycook, S. J., 1998, "Time-Resolved Imaging of Gas Phase Nanoparticle Synthesis by Laser Ablation," *Appl. Phys. Lett.*, **72**, 2987–2989.

Girault, C., Damiani, D., Aubreton, J., and Catherinot, A., 1989, "Time-Resolved Spectroscopic Study of the KrF Laser-Induced Plasma Plume Created above an YBaCuO Superconducting Target," *Appl. Phys. Lett.*, **55**, 182–184.

Gusarov, A. V., Gnedovets, A. G., and Smurov, I., 2000, "Gas Dynamics of Laser Ablation: Influence of Ambient Atmosphere," *J. Appl. Phys.*, **88**, 4352–4364.

Gusarov, A. V., and Smurov, I., 2003, "Near-Surface Laser–Vapour Coupling in Nanosecond Pulsed Laser Ablation," *J. Phys. D*, **36**, 2962–2971.

Ho, J.-R., Grigoropoulos, C. P., and Humphrey, J. A. C., 1995, "Computational Model for the Heat Transfer and Gas Dynamics in the Pulsed Laser Evaporation of Metals," *J. Appl. Phys.*, **78**, 4696–4709.

Ho, J.-R., Grigoropoulos, C. P., and Humphrey, J. A. C., 1996, "Gas Dynamics and Radiation Heat Transfer in Excimer Laser Ablation of Aluminum," *J. Appl. Phys.*, **79**, 7205–7215.

Itina, T. E., Marine, W., and Autric, M., 1997, "Monte Carlo Simulation of Pulsed Laser Ablation from Two-Component Target into Diluted Ambient Gas," *J. Appl. Phys.*, **82**, 3536–3542.

Jahoda, F. C., and Sawyer, G. A., 1971, *Methods of Experimental Physics*, R. H. Lovberg and H. R. Greim, eds., Vol. 9B, New York, Academic Press.

Kelly, R., 1990, "On the Dual Role of the Knudsen Layer and Unsteady, Adiabatic Expansion in Pulse Sputtering Phenomena," *J. Chem. Phys.*, **92**, 5047–5056.

Kelly, R., and Dreyfus, R. W., 1988a, "Reconsidering the Mechanisms of Laser Sputtering with Knudsen-Layer Formation Taken into Account," *Nucl. Instrum. Meth. Phys. Res. B*, **32**, 341–345.

1988b, "On the Effect of Knudsen-Layer Formation on Studies of Vaporization, Sputtering and Desorption," *Surf. Sci.*, **198**, 263–276.

Kelly, R., and Rothenberg, J. E., 1985, "Laser Sputtering: Part III. The Mechanism of the Sputtering of Metals at Low Energy Densities," *Nucl. Instrum. Meth. Phys. Res. B*, **7/8**, 755–763.

Kruer, W. L., 1988, *The Physics of Laser Plasma Interactions*, Redwood City, CA, Addison-Wesley.

Leboeuf, J. N., Chen, K. R., Donato, J. M. *et al.*, 1996, "Modeling of Plume Dynamics in Laser Ablation Processes for Thin Film Deposition of Materials," *Phys. Plasmas*, **3**, 2203–2209.

Li, Y. M., Broughton, J. N., Fedosejevs, R., and Tomie, T., 1992, "Formation of Plasma Columns in Atmospheric Pressure Gases by Picosecond KrF Laser Pulses," *Opt. Commun.*, **93**, 366–377.

Lowndes, D. H., Geohegan, D. B., Puretzky, A. A., Norton, D. P., and Rouleau, C. M., 1996, "Synthesis of Novel Thin-Film Materials by Pulsed Laser Deposition," *Science*, **273**, 898–903.

Lunney, J. G., and Jordan, R., 1998, "Pulsed Laser Ablation of Metals," *Appl. Surf. Sci.*, **127–129**, 941–946.

Mao, S. S., Mao, X., Greif, R., and Russo, R. E., 2000a, "Dynamics of an Air Breakdown Plasma on a Solid Surface during Picosecond Laser Ablation," *Appl. Phys. Lett.*, **76**, 31–33.

2000b, "Simulation of a Picosecond Laser Ablation Plasma," *Appl. Phys. Lett.*, **76**, 3370–3372.

Murnane, H. M., Kapteyn, H. C., Rosen, M. D., and Falcone, R. W., 1991, "Ultrafast X-Ray Pulses from Laser-Produced Plasmas," *Science*, **251**, 531–536.

Olstad, R. A., and Olander, D. R., 1975, "Evaporation of Solids by Laser Pulses, I. Iron," *J. Appl. Phys.*, **46**, 1499–1505.

Park, H. K., Grigoropoulos, C. P., and Tam, A. C., 1995, "Optical Measurements of Thermal Diffusivity of a Material," *Int. J. Thermophys.*, **16**, 973–995.

Petite, G., Agostini, P., Trainham, R. P., Mevel, E., and Martin, P., 1992, "Origin of the High-Energy Electron Emission from Metals under Laser Irradiation," *Phys. Rev. B*, **45**, 12 210–12 217.

Phipps, C. R. Jr., and Dreyfus, R. W., 1993, "The High Laser Irradiance Regime," in *Laser Ionization Mass Analysis*, edited by A. Vertes, R. Gijbels, and F. Adams, New York, John Wiley and Sons, pp. 369–431.

Phipps, C. R. Jr., Turner, T. P., Harrison, R. F. *et al.*, 1988, "Impulse Coupling to Targets in Vacuum by KrF, HF, and CO_2 Single-Pulse Lasers," *J. Appl. Phys.*, **64**, 1083–1096.

Russo, R. E., 1995, "Laser Ablation," *Appl. Spectrosc.*, **49**, A14–A25.

Schittenhelm, H., Gallies, G., Straub, A., Berger, P., and Hügel, H., 1998, "Measurements of Wavelength-Dependent Transmission in Excimer Laser-Induced Plasma Plumes and their Interpretation," *J. Phys. D*, **31**, 418–427.

Sedov, L. I., 1959, *Similarity and Dimensional Methods in Mechanics*, New York, Academic Press.

Sell, J. A., Heffelfinger, D. M., Ventzek, P. L. G., and Gilgenbach, R. M, 1991, "Photoacoustic and Photothermal Beam Deflection as a Probe of Laser Ablation of Materials," *J. Appl. Phys.*, **69**, 1330–1336.

Shannon, M. A., Rostami, A. A., and Russo, R. E., 1992, "Photothermal Deflection Measurements for Monitoring Heat Transfer During Modulated Laser Heating of Solids," *J. Appl. Phys.*, **71**, 53–63.

Shannon, M. A., Rubinsky, B., and Russo, R. E., 1994, "Detecting Laser-Induced Phase Change at the Surface of Solids via Latent Heat of Melting with a Photothermal Deflection Technique," *J. Appl. Phys.*, **75**, 1473–1485.

Sibold, D., and Urbassek, H. M., 1991, "Kinetic Study of Pulsed Desorption Flows into Vacuum," *Phys. Rev. A*, **43**, 6722–6734.

Siegel, R., and Howell, J. R., 1992, *Thermal Radiation Heat Transfer*, 3rd edn, Bristol, PA, Hemisphere.

Sigrist, M. W., 1986, "Laser Generation of Acoustic Waves in Liquids and Gases," *J. Appl. Phys.*, **60**, R83–R121.

Singh, R. K., and Narayan, J., 1990, "Pulsed-Laser Evaporation Technique for Deposition of Thin Films: Physics and Theoretical Model," *Phys. Rev. B*, **41**, 8843–8859.

Spitzer, L. Jr., 1962, *Physics of Fully Ionized Gases*, New York, John Wiley Interscience.

Tam, A. C., 1986, "Applications of Photoacoustic Sensing Techniques," *Rev. Mod. Phys.*, **58**, 381–431.

Vertes, A., Gijbels, R., and Adams, F. eds., 1993, *Laser Ionization Mass Analysis*, New York, John Wiley.

Vertes, A., Juhasz, P., De Wolf, M., and Gijbels, R., 1989, "Hydrodynamic Modeling of Laser Plasma Ionization Processes," *Int. J. Mass Spectrosc. Ion Processes*, **94**, 63–85.

Young, P. E., Baldis, H. A., Drake, R. P., Cambell, E. M., and Estabrook, K. G., 1998, "Evidence of Ponderomotive Filamentation in Laser Produced Plasma," *Phys. Rev. Lett.*, **61**, 2336–2339.

Zel'dovich, Ya. B., and Raizer, Yu. B., 1966, *Physics of Shock Waves and High-Temperature Hydrodynamic Phenomena*, New York, Academic Press.

Zhigilei, L. V., 2003, "Dynamics of the Plume Formation and Parameters of the Ejected Clusters in Short-Pulse Laser Ablation," *Appl. Phys. A*, **76**, 339–350.

Zhigilei, L. V., Kodali, P. B. S., and Garrison, B. J., 1997, "Molecular Dynamics Model for Laser Ablation and Desorption of Organic Solids," *J. Phys. Chem. B*, **101**, 2028–2037.

6 Ultrafast-laser interactions with materials

C. Grigoropoulos and S. Mao

6.1 Introduction

Lasers that can produce coherent photon pulses with durations in the femtosecond regime have opened up new frontiers in materials research with extremely short temporal resolution and high photon intensity. The ultrafast nature of femtosecond lasers has been used to observe, in real time, phenomena including chemical reactions in gases (Zewail, 1994) and electron–lattice energy transfer in solids (Shah, 1996). On the other hand, ultra-short laser pulses impart extremely high intensities and provide precise laser-ablation thresholds at substantially reduced laser energy densities. The increasing availability of intense femtosecond lasers has sparked a growing interest in high-precision materials processing. In contrast to material modification using nanosecond or longer laser pulses, for which standard modes of thermal processes dominate, there is no heat exchange between the pulse and the material during femtosecond-laser–material interactions. As a consequence, femtosecond laser pulses can induce nonthermal structural changes driven directly by electronic excitation and associated nonlinear processes, before the material lattice has equilibrated with the excited carriers. This fast mode of material modification can result in vanishing thermal stress and minimal collateral damage for processing practically any solid-state material. Additionally, damage produced by femtosecond laser pulses is far more regular from shot to shot. These breakdown characteristics make femtosecond lasers ideal tools for precision material processing.

Thorough knowledge of the short-pulse-laser interaction with the target material is essential for controlling the resulting modification of the target's topography. The use of ultra-short pulses with correspondingly high laser intensities reduces the extent of heat diffusion into the target, facilitating instantaneous material expulsion. This enables high-aspect-ratio cuts and features, free of debris and lateral damage (e.g. Momma *et al.*, 1998; Pronko *et al.*, 1995; Liu *et al.*, 1997; Wu, 1997). Therefore, the ablation process is stable and reproducible. As a result, the produced structure size is not limited by thermal or mechanical damage, i.e. melting, formation of burr and cracks, etc. Thus, the minimal achievable structure size is limited mainly by diffraction to the order of a wavelength (Korte *et al.*, 1999). It has also to be recognized that ultrafast-laser pulses enforce high intensities that trigger nonlinear absorption effects that may dominate the interaction process. One of the most important repercussions is the efficient processing of transparent dielectrics, which has a number of applications, enabling for example three-dimensional binary-data storage (Glezer *et al.*, 1996).

6.2 Femtosecond-laser interaction with metals

6.2.1 The relaxation-time approximation and two-step models

The relaxation-time approximation to the Boltzmann transport equation is given according to the treatment by Qiu and Tien (1993). The electron density distribution function, $f(\vec{r}, \vec{V}, t)$, evolves according to the Boltzmann transport equation:

$$\frac{\partial f}{\partial t} + V_x \frac{\partial f}{\partial x} + \frac{\mathcal{F}_x}{m} \frac{\partial f}{\partial V_x} = \left[\frac{\partial f}{\partial t}\right]_{\mathrm{s}}, \tag{6.1}$$

where $\mathcal{F}_x$ is the external force and m the electron mass. When the electron and lattice temperatures T_e and T_l are higher than the Debye temperature T_D, the following approximations can be made for the scattering term on the right-hand side:

$$\left[\frac{\partial f}{\partial t}\right]_{\mathrm{s}} = \left[\frac{\partial f}{\partial t}\right]_{\mathrm{s1}} + \left[\frac{\partial f}{\partial t}\right]_{\mathrm{s2}}, \tag{6.2a}$$

$$\left[\frac{\partial f}{\partial t}\right]_{\mathrm{s1}} = \frac{1}{2}\Lambda(1 - f_{\mathrm{eq}})\left(f_{\mathrm{eq}} - \frac{1}{2}\right)\left(1 - \frac{T_\mathrm{l}}{T_\mathrm{e}}\right)\frac{T_\mathrm{D}}{T_\mathrm{e}} E^{-\frac{1}{2}}, \tag{6.2b}$$

$$\left[\frac{\partial f}{\partial t}\right]_{\mathrm{s2}} = \frac{1}{2^{4/3}}\Lambda(f_{\mathrm{eq}} - f)\frac{T_\mathrm{l}}{T_\mathrm{D}} E_{\mathrm{F0}} E^{-\frac{3}{2}}, \tag{6.2c}$$

where

$$\Lambda = \frac{3\pi^2 \mathcal{P}^2 (m/2)^{1/2}}{m_\mathrm{a} k_\mathrm{B} T_\mathrm{D}}\left(\frac{3}{4\pi\Delta}\right)^{\frac{1}{3}},$$

$\mathcal{P}$ is the transient matrix element, Δ is the unit-cell volume, m_a is the atomic mass, f_{eq} is the equilibrium electron Fermi–Dirac distribution, and E_{F0} is the Fermi energy at $T_\mathrm{e} = 0$ K. The scattering term given by (6.2b) becomes zero when the electron temperature becomes equal to the phonon temperature. Since $0.5 < f_{\mathrm{eq}} < 1$ when $E < E_\mathrm{F}$, $0 < f_{\mathrm{eq}} < 0.5$ when $E > E_\mathrm{F}$, and $f_{\mathrm{eq}} = 0$ when $E = E_\mathrm{F}$, this term changes sign as E goes from $E > E_\mathrm{F}$ to $E < E_\mathrm{F}$. For temperatures $T_\mathrm{e} > T_\mathrm{l}$, this term is positive and negative for energies above and below the Fermi surface, respectively, implying an inelastic process transporting a net energy from the electrons to the lattice. The second term given by (6.2c) becomes zero when electrons are in thermal equilibrium (i.e. $f = f_{\mathrm{eq}}$) and is elastic, i.e. it does not cause net energy transfer between electrons and lattice. A relaxation time, τ, is identified in relation to this process:

$$\left[\frac{\partial f}{\partial t}\right]_{\mathrm{s2}} = \frac{f_{\mathrm{eq}} - f}{\tau}, \tag{6.3a}$$

$$\tau = 2^{4/3}\Lambda^{-1}\frac{T_\mathrm{D}}{T_\mathrm{l}} E_{\mathrm{F0}}^{-1} E^{3/2} = \tau_{\mathrm{eq}}(T_\mathrm{l}). \tag{6.3b}$$

Equation (6.1) is written as

$$\frac{\partial f}{\partial t} + V_x \frac{\partial f}{\partial x} + \frac{\mathcal{F}_x}{m} \frac{\partial f}{\partial V_x} = \frac{f_{\text{eq}} - f}{\tau} + \left[\frac{\partial f}{\partial t} \right]_{\text{sl}}. \tag{6.4}$$

Linearization of this equation implies

$$V_x \frac{\partial f}{\partial x} \approx V_x \frac{\partial f_{\text{eq}}}{\partial x}, \tag{6.5a}$$

$$\frac{\mathcal{F}_x}{m} \frac{\partial f}{\partial V_x} \approx \frac{\mathcal{F}_x}{m} \frac{\partial f_{\text{eq}}}{\partial V_x}. \tag{6.5b}$$

The following relations are recalled:

$$\frac{\partial f_{\text{eq}}}{\partial x} = \frac{\partial f_{\text{eq}}}{\partial T_{\text{e}}} \frac{\partial T_{\text{e}}}{\partial x} = -\left[\frac{E}{T_{\text{e}}} + T_{\text{e}} \frac{\mathrm{d}}{\mathrm{d}T_{\text{e}}} \left(\frac{E_{\text{F}}}{T_{\text{e}}} \right) \right] \frac{\partial f_{\text{eq}}}{\partial E} \frac{\partial T_{\text{e}}}{\partial x}, \tag{6.6a}$$

$$\frac{\partial f_{\text{eq}}}{\partial V_x} = \frac{\partial f_{\text{eq}}}{\partial E} \frac{\partial E}{\partial V_x} = m V_x \frac{\partial f_{\text{eq}}}{\partial E}, \tag{6.6b}$$

$$\mathcal{F}_x = -e\mathcal{E}_x, \tag{6.6c}$$

where $\mathcal{E}_x$ is the electric field in the x-direction. Utilizing (6.5) and (6.6), the following relation is obtained for the density function:

$$f(E) = f_{\text{eq}}(E) + \tau V_x \left[\frac{E}{T_{\text{e}}} + T_{\text{e}} \frac{\mathrm{d}}{\mathrm{d}T_{\text{e}}} \left(\frac{E_{\text{F}}}{T_{\text{e}}} \right) \right] \frac{\partial f_{\text{eq}}}{\partial E} \frac{\partial T_{\text{e}}}{\partial x}$$
$$+ \tau e \mathcal{E}_x V_x - \tau \frac{\partial f}{\partial t} + \tau \left[\frac{\partial f}{\partial t} \right]_{\text{sl}}. \tag{6.7}$$

The electrical current, J, and energy flux, Q_x, are given by

$$J = -\int_{-\infty}^{+\infty} \int_{-\infty}^{+\infty} \int_{-\infty}^{+\infty} e V_x D_{\text{den}}(\vec{V}) f(\vec{V}) \mathrm{d}V_x \, \mathrm{d}V_y \, \mathrm{d}V_z, \tag{6.8a}$$

$$Q_x = \int_{-\infty}^{+\infty} \int_{-\infty}^{+\infty} \int_{-\infty}^{+\infty} E V_x D_{\text{den}}(\vec{V}) f(\vec{V}) \mathrm{d}V_x \, \mathrm{d}V_y \, \mathrm{d}V_z, \tag{6.8b}$$

where $D_{\text{den}}(\vec{V})$ is the density of states. The prediction of (6.7) can then be utilized to estimate the current and energy flux. It is noted that f_{eq} and $[\partial f/\partial t]_{\text{sl}}$ are symmetrical about the surface $V_x = 0$, while the scattering-, diffusion-, and electrical-force-driven term is nonzero only in a narrow region around the Fermi surface. Therefore, the contributions of the $\tau(\partial f/\partial t)$ term to the electric and heat current are approximated as

$$J = -\tau_{\text{F}} \frac{\partial J}{\partial t} + e K_1 \left[e E_x + T_{\text{e}} \frac{\mathrm{d}}{\mathrm{d}T_{\text{e}}} \left(\frac{E_{\text{F}}}{T_{\text{e}}} \right) \frac{\partial T_{\text{e}}}{\partial x} \right] + \frac{e}{T_{\text{e}}} K_2 \frac{\partial T_{\text{e}}}{\partial x}, \tag{6.9a}$$

$$Q_x = -\tau_{\text{F}} \frac{\partial Q}{\partial t} - e K_2 E_x - \left[\frac{K_3}{T_{\text{e}}} + K_2 T_{\text{e}} \frac{\mathrm{d}}{\mathrm{d}T_{\text{e}}} \left(\frac{E_{\text{F}}}{T_{\text{e}}} \right) \right] \frac{\partial T_{\text{e}}}{\partial x}, \tag{6.9b}$$

$$K_n = -\frac{2}{3m} \int_0^\infty \tau(E) D_{\text{den}}(E) E^n \frac{\partial f_{\text{eq}}}{\partial E} \, \mathrm{d}E . \tag{6.9c}$$

On noting that $J = 0$ and $\partial J/\partial t = 0$, Equations (6.9a) and (6.9b) can be combined:

$$Q_x = -k_e \frac{\partial T_e}{\partial x} - \tau_F \frac{\partial Q}{\partial t}, \tag{6.10a}$$

$$k_e = \left(K_1 K_3 - K_2^2\right)/(K_1 T_e). \tag{6.10b}$$

The above one-dimensional heat-flux relation can be generalized:

$$Q = -k_e \nabla(T_e) - \tau_F \frac{\partial Q_x}{\partial t}. \tag{6.11}$$

The electron heat capacity can be obtained from

$$C_e = \begin{cases} \gamma_C T_e, & T_e < T_F, \\ 3/(2N_e k_B), & T_e > T_F, \end{cases} \tag{6.12}$$

where γ_C is a material constant and T_F is the Fermi temperature.

From a kinetic standpoint, the electron thermal conductivity, k_e, can be written as:

$$k_e = \frac{1}{3} \frac{C_e V^2}{\nu}, \tag{6.13}$$

where V is the electron mean-square velocity and the frequency $\nu = \nu_{e-e} + \nu_{e-ph}$. For $T_e \ll T_F$, Equation (6.13) is reduced to

$$k_e = \frac{\pi^2 N_e k_B^2 T_e}{3m_e \nu}, \tag{6.14}$$

where the frequency $\nu \propto T_l$. Consequently, the electron thermal conductivity is related to the conventionally measured thermal conductivity, $k_{eq}(T_l)$, by

$$k_e = \frac{T_e}{T_l} k_{eq}(T_l). \tag{6.15}$$

At high temperatures, the electron thermal conductivity can be modeled (Anisimov and Rethfeld, 1997) by

$$\kappa_e = C_{ke}\theta_e \frac{\left(\theta_e^2 + 0.16\right)^{5/4}\left(\theta_e^2 + 0.44\right)}{\left(\theta_e^2 + 0.092\right)^{1/2}\left(\theta_e^2 + b\theta_l\right)}, \tag{6.16}$$

where $\theta_e = T_e/T_F$ and $\theta_l = T_l/T_F$, while C_{ke} and b are material constants that can be determined from experimental data. For gold, $C_{ke} = 353$ W K^{-1} m^{-1} and $b = 0.16$. Under equilibrium conditions, the electron thermal conductivity k_e follows a $T_e^{5/2}$ law near the Fermi temperature. Sub-picosecond-laser excitation experiments (Milchberg et $al.$, 1988) show that the electrical conductivity significantly decreases for electron energies from 5 to 40 eV to reach a resistivity-saturation regime.

Table 6.1. Thermalization properties of metals

Metal[a]	Electron–phonon coupling factor G (10^{16} W m^{-3} K^{-1})		Thermalization time $t_{\rm e} = C_{\rm e}/G$ (ps)
	Predicted	Measured	
Copper (Cu)	14	4.8 ± 0.7^b	0.6
Silver (Ag)	3.1	2.8^d	0.6
Gold (Au)	2.6	2.8 ± 0.5^b	0.8
Chromium (Cr)	45	42 ± 5^b	0.1
Tungsten (W)	27	26 ± 3^b	0.2
Vanadium (V)	648	$523 \pm 37^b, 170^c$	0.06
Niobium (Nb)	138^e	387 ± 36^b	0.05
Titanium (Ti)	202	185 ± 16^b	0.05
Lead (Pb)	62	12.4 ± 1.4^b	0.4

[a] Assumed numbers of free electrons per atom are 0.5 for Cr, 1.0 for Cu, Ag, Au, W, Ti, and Pb, and 2.0 for V and Nb.
[b] Brorson *et al.* (1990).
[c] Elsayed-Ali *et al.* (1987).
[d] Groenveld *et al.* (1992).
[e] Yoo *et al.* (1990).

6.2.2   Two-step models

Energy-conservation equations for both electrons and phonons and the heat-flux equation (6.11) generate the hyperbolic two-step (HTS) radiation-heating model:

$$C_{\rm e}(T_{\rm e})\frac{\partial T_{\rm e}}{\partial t} = -\nabla \cdot \vec{Q} - G(T_{\rm e} - T_{\rm l}) + Q_{\rm ab}, \qquad (6.17{\rm a})$$

$$C_{\rm l}(T_{\rm l})\frac{\partial T_{\rm l}}{\partial t} = G(T_{\rm e} - T_{\rm l}), \qquad (6.17{\rm b})$$

$$\tau_{\rm F}\frac{\partial \vec{Q}}{\partial t} + \kappa\,\nabla(T_{\rm e}) + \vec{Q} = 0, \qquad (6.17{\rm c})$$

$$C_{\rm e}(T_{\rm e}) = \gamma_{\rm C} T_{\rm e}, \qquad k_{\rm e} = k_{\rm eq}\frac{T_{\rm e}}{T_{\rm l}}, \qquad (6.17{\rm d})$$

where $Q_{\rm ab}$ is the volumetric radiation-absorption term and the electron–phonon coupling constant is derived (Qiu and Tien, 1993) as

$$G = \frac{9}{16}\frac{Nk_{\rm B}^2 T_{\rm D}^2 V_{\rm F}}{\Lambda_{\rm F}(T_{\rm l})T_{\rm l}E_{\rm F}}. \qquad (6.18)$$

In the above, $V_{\rm F}$ is the speed of electrons possessing the Fermi energy and $\Lambda_{\rm F}$ is the electron mean free path. Table 6.1 gives values of the electron–phonon coupling constant, G. The main physical picture is that the energy deposition is accomplished first via radiation absorption by the electrons and then by the subsequent exchange between electrons and lattice.

If the heating time is much longer than the thermalization time, the effect of the hyperbolic transport is small and the parabolic two-step (PTS) model (Anisimov *et al.*,

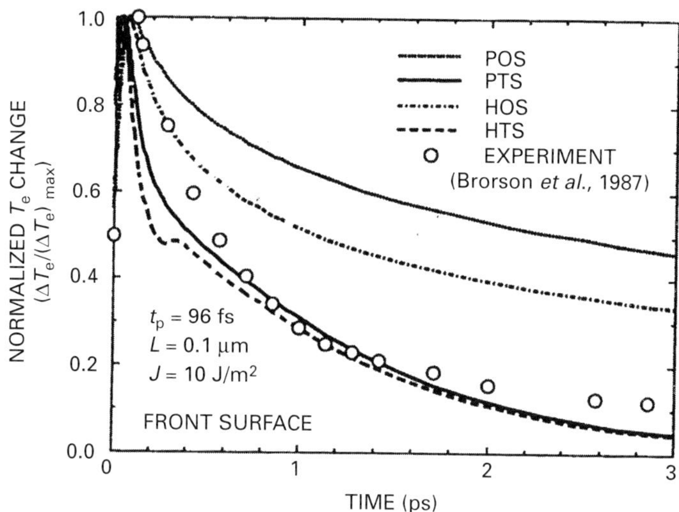

Figure 6.1. Comparison of predicted electron-temperature changes with experimental data at the front surface. The pulse duration is 96 fs, the film thickness is 0.1 μm, and the fluence is $F = 10$ J/m². From Qiu and Tien (1993), reproduced with permission by the American Society of Mechanical Engineers.

1974) is

$$C_e(T_e)\frac{\partial T_e}{\partial t} = -\nabla \cdot \vec{Q} - G(T_e - T_l) + Q_{ab},\tag{6.19a}$$

$$C_l(T_l)\frac{\partial T_l}{\partial t} = G(T_e - T_l),\tag{6.19b}$$

$$k_e\,\nabla(T_e) + \vec{Q} = 0.\tag{6.19c}$$

If the heating time is much longer than the thermalization time, the temperature difference between electrons and lattice is negligible and, as a result, the classical diffusion approximation (i.e. the parabolic one-step (POS) model) holds, meaning that the lattice and electron systems are at equilibrium. On the other hand, if the electron–phonon coupling is neglected, the HTS model reduces to the hyperbolic one-step (HOS) model that has been proposed to account for finite "thermal-speed" effects:

$$C\frac{\partial T}{\partial t} = -\nabla \cdot \vec{Q} + Q_{ab},\tag{6.20a}$$

$$\tau_F\frac{\partial \vec{Q}}{\partial t} + k_{eq}\,\nabla(T) + \vec{Q} = 0.\tag{6.20b}$$

Figure 6.1 shows the comparison of predicted electron-temperature changes with experimental data at the front surface of a 0.1-μm-thick gold film subjected to femtosecond-laser excitation. After the exposure to the laser pulse, the absorbed radiation energy is removed from the electron system through the interaction with phonons and the electron temperature drops quickly. Both the POS model and the HOS model neglect

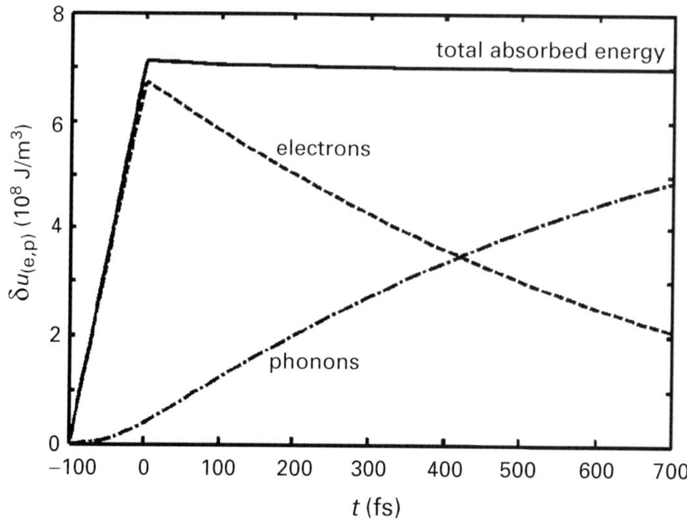

Figure 6.11. The transient increases in energy of the electron gas, δU_e, and phonon gas, δU_{phon}, and total absorbed energy, δU. The laser pulse was assumed to have the same parameters as in Figures 6.11 and 6.12. From Rethfeld *et al.* (2002), reproduced with permission by the American Physical Society.

6.3   Femtosecond-laser interaction with semiconductor materials

Depending on how the incident photon energy, $h\nu$, compares with the band gap, E_{bg}, absorption of ultra-short-pulse radiation takes place via single-photon and/or multiphoton interband absorption or free-carrier, intraband absorption. Interband absorption creates electron–hole pairs with an initial kinetic energy of $h\nu - E_{bg}$, whereas free-carrier absorption events endow the free carriers with an additional kinetic energy, $h\nu$. On a time scale that is less than 100 fs, the carriers thermalize to a Fermi–Dirac distribution via carrier–carrier collisions. Recombination and impact-ionization processes allow this thermalization to equilibrium number-density distributions. For the relevant carrier densities, $N > 10^{18}$ cm^{-3}, the dominant recombination process is Auger recombination, a three-body interaction process, whereby two carriers interact with a third carrier, increasing the electron temperature, T_e, but reducing the carrier number density, N. The impact-ionization process occurs when an energetic carrier creates an electron–hole pair while losing energy. The carriers then attempt to reach thermal equilibrium with the lattice, initially by emitting longitudinal-optical (LO) phonons. The carrier-LO relaxation time, τ_e, depends both on carrier temperature and on lattice temperature. However, for silicon it is approximated as constant at 0.5 ps (Van Driel, 1987). The LO phonon population attempts to thermalize with other lattice modes through phonon–phonon interaction. Even though some phonon modes might not attain thermal equilibrium until longer times, most of the deposited laser energy will be converted into a near-thermal equilibrium lattice state within times of order tens of picoseconds.

The dominant interband absorption mechanism under short-pulse excitation is two-photon absorption as opposed to long-pulse irradiation since the one-photon absorption length is nearly $10\,\mu$m in bulk, crystalline silicon and the probability of nonlinear processes increases with increasing intensity. The two-photon absorption coefficient, β (cm/GW), is given by Van Stryland *et al.* (1985):

$$\beta = (3.1 \pm 0.5) \times 10^3 \frac{\sqrt{E_{\mathrm{p}}} F_2(2h\nu/E_{\mathrm{bg}})}{n^2 E_{\mathrm{bg}}^3}, \tag{6.21}$$

where E_{p} (eV) is a nearly material-independent constant, E_{bg} (eV) is the semiconductor bandgap, n is the real part of the complex refractive index, and F_2 is a function defined in Van Stryland *et al.* (1985) that is determined utilizing band-structure considerations.

The assumption of local quasi-equilibrium implies that the electron distribution follows the Fermi–Dirac function. The particle number density, electron energy, and lattice energy are then calculated via the relaxation time approximation to Boltzmann's transport equation. Given the short pulse duration (0.1 ps) relative to the electron–lattice relaxation time ($\sim$0.5 ps), the relaxation-time approximation has limitations. Nevertheless, a qualitative description of the thermal and nonthermal heating processes can be established. Assuming that the quasi-Fermi level remains in the middle of the band gap, ignoring band-gap shrinkage according to temperature variation, and considering both linear and two-photon absorption, the balance equation for energy carriers can be written as follows:

$$\frac{\partial N}{\partial t} + \nabla \cdot (-D_0 \nabla N) = \frac{(1 - R)\gamma I(x, t)}{h\nu} + \frac{(1 - R)^2 \beta I^2(x, t)}{2h\nu}$$
$$- \gamma_{\mathrm{Aug}} N^3 + \delta(T_{\mathrm{e}})N, \tag{6.22}$$

where R is the reflectivity, γ is the linear absorption coefficient, β is the two-photon absorption coefficient, D_0 is the ambipolar diffusivity, $h\nu$ is the photon energy quantum, γ_{Aug} is the Auger-recombination coefficient, and δ is the impact-ionization coefficient. The gradient of the laser-beam intensity is then written as follows:

$$\frac{\mathrm{d}I}{\mathrm{d}x} = -\gamma I - \beta I^2 - \Theta N I, \tag{6.23}$$

where Θ is the free-carrier absorption coefficient.

The total energy-balance equations for the electron and lattice systems using the relaxation-time approximation can be written as follows:

$$\frac{\partial U_{\mathrm{e}}}{\partial t} + \nabla \cdot (-k_{\mathrm{e}} \nabla T_{\mathrm{e}}) = (1 - R)(\alpha + \Theta N)I(x, t)$$
$$+ (1 - R)^2 \beta^2 I^2(x, t) - G_{\mathrm{e}}(T_{\mathrm{e}} - T_{\mathrm{l}}), \tag{6.24a}$$
$$\frac{\partial U_{\mathrm{l}}}{\partial t} + \nabla \cdot (-k_{\mathrm{l}} \nabla T_{\mathrm{l}}) = G_{\mathrm{e}}(T_{\mathrm{e}} - T_{\mathrm{l}}), \tag{6.24b}$$

$m^*_{\text{opt}} = 0.18$, the relaxation time $\tau_{\text{Drude}} = 1.1$ fs, and the two-photon-absorption coefficient at 625 nm $\beta = 50 \pm 10$ cm/GW. Furthermore, the critical density for transition to ultrafast phase transition was about 10^{22} cm^{-3}. Figure 6.16 gives the electron–hole density as a function of the laser fluence derived from experimental data, together with the numerical predictions.

Theoretical studies have also been performed to assess the mechanisms of the ultrafast "plasma-annealing" and phase-transformation processes. Theoretical studies by Stampfli and Bennemann (1990, 1992) showed that the transverse-acoustic (TA)-phonon system becomes unstable if more than 8% of the valence electrons are excited to the conduction band. Atomic displacements were found to increase to around 1 Å within 100–200 fs if 15% of the electrons are excited. On the basis of these results, the authors concluded that the resulting TA-phonon instability would lead to a rapid exchange of energy between the electrons and the atomic lattice in the form of mechanical work. A different approach was adopted by Silvestrelli *et al.* (1996), who performed *ab initio* molecular-dynamics calculations on the basis of finite-temperature density-functional theory to simulate ultrafast laser heating of silicon. Since the electron relaxation time (Agassi, 1984) is much shorter than the electron–lattice relaxation time, the electron system remains in internal equilibrium at the initial laser-induced temperature and the ions are allowed to evolve freely. The simulation results showed that high concentrations of excited electrons can change the effective ion–ion interactions, thereby dramatically weakening the covalent bond and leading to a melting transition to a metallic state, which, in contrast to ordinary liquid silicon, is characterized by a high coordination number. The calculated ionic temperature after an elapsed time of about 100 fs was $\sim$1700 K, i.e. close to the melting temperature of c-Si (Figure 6.17). These results are in sharp departure from the hypothesis that the phase transformation occurs with the lattice remaining relatively cold through a mechanical instability due to phonon softening.

Femtosecond X-ray pulses have been used to study lattice dynamics (Rose-Petruck *et al.*, 1999) and ultrafast melting (Chin *et al.*, 1999; Larsson *et al.*, 1998; Lindenberg *et al.*, 2000, Siders *et al.*, 1999; Cavalleri *et al.*, 2000) associated with the fundamental phase-transition process. Visible probe light is absorbed within a short penetration depth and cannot accurately resolve nanometer-sized lattice distortion. Pulsed hard-X-ray sources are therefore advantageous for measuring atomic rearrangement and structural dynamics inside the target material. In addition to these observations, lattice disordering was detected by depletion of diffracted X-ray images at the irradiation spot, indicating the occurrence of a nonthermal solid-to-liquid phase transition. Figure 6.18 gives the schematic setup for the X-ray probe experiments (von der Linde *et al.*, 2001). The 30-fs, 800-nm laser pulse output generated at 20 Hz by a Ti : Al$_2$O$_3$ laser was used for both sample excitation and X-ray generation. A split portion of these pulses was focused onto a moving Cu wire in vacuum, resulting in a point source of Cu Kα photons. The radiation emitted into two closely spaced lines, Kα_1 and Kα_2, was diffracted by the sample that was excited by the optical pump pulse and detected by a sensitive X-ray CCD detector. As shown in Figure 6.19, the diffraction signal dropped significantly within a few hundred femtoseconds in the region subjected to intensity sufficient for imparting homogeneous

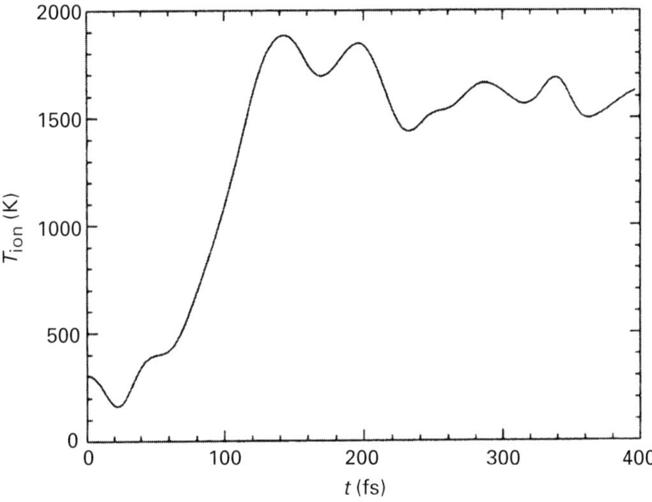

Figure 6.17. The time dependence of the instantaneous ionic temperature, defined as

$$T_{\rm ion}(t) = \frac{M}{(3N-3)k_{\rm B}} \sum_{l=1}^{N} v_l^2(t),$$

where $k_{\rm B}$ is the Boltzmann constant, M is the Si ion mass, $v_i(t)$ is the ionic velocity at time t, and $N = 64$ is the number of atoms in the MD supercell representing the Brillouin zone. From Silvestrelli *et al.* (1996), reproduced with permission by the American Physical Society.

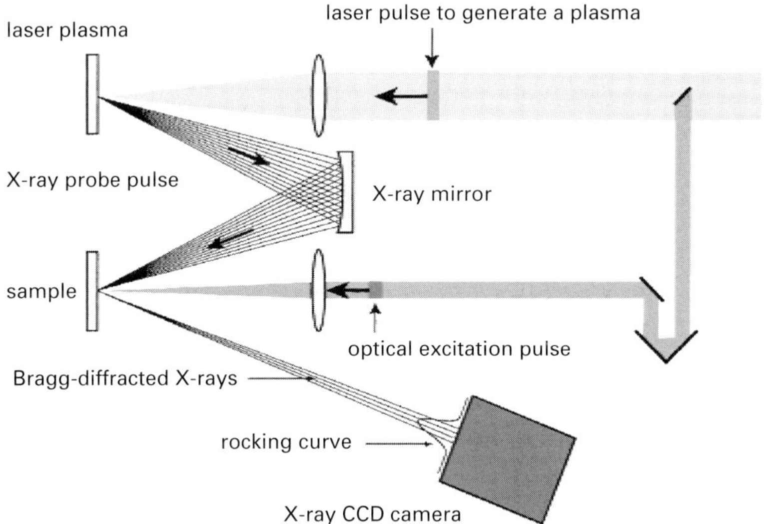

Figure 6.18. The setup for the visible-pump, X-ray-probe experiments. From von der Linde *et al.* (2001), reproduced with permission from Cambridge University Press.

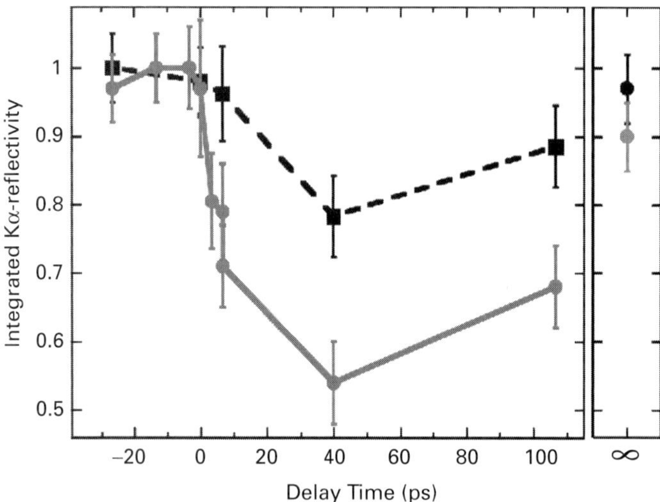

Figure 6.19. Time-resolved Kα X-ray reflectivity from a 160-nm Ge(111) film, integrated over the central pumped region (solid red line) and over a region vertically displaced by $\sim$0.2 mm from the center (dashed black line). Measurement of non-thermal melting in germanium. From Siders *et al.* (1999), reproduced with permission by the American Association for the Advancement of Science.

ultrafast melting. The significant drop in integrated diffracted intensity signified loss of crystalline order. At infinite time delays, the diffraction signal recovered to $\sim$90% of the initial value (Siders *et al.*, 1999), showing restoration of crystalline order, with the departure due either to amorphization or to ablative material loss. Since the integrated X-ray reflectivity scales with the material thickness, it was concluded that approximately 30–50 nm of the film undergoes ultrafast disordering. As argued previously, it has to be appreciated that the corresponding melting speed would have to be of the order of 10^4 m/s, which exceeds the speed of sound. In contrast, at near-threshold intensities the experiments show inhomogeneous melting evolving over a longer temporal scale that is consistent with a thermal phase-transition pathway.

6.4.2   Femtosecond-laser ablation of crystalline silicon

Ultrashort pulsed laser processing of crystalline silicon has been studied experimentally (Choi and Grigoropoulos, 2002). The energy density provided by this laser beam focused to a 100-μm spot was sufficiently strong to ablate silicon. A pump-and-probe experiment was implemented, utilizing a time-delayed frequency-doubled ($\lambda =$ 400 nm) beam for *in situ* reflectance measurements and observation by ultrafast microscopy. The deposition of the femtosecond-laser radiation generated a high-density electron–hole plasma that subsequently triggered ablation at about 10 ps (Figure 6.20).

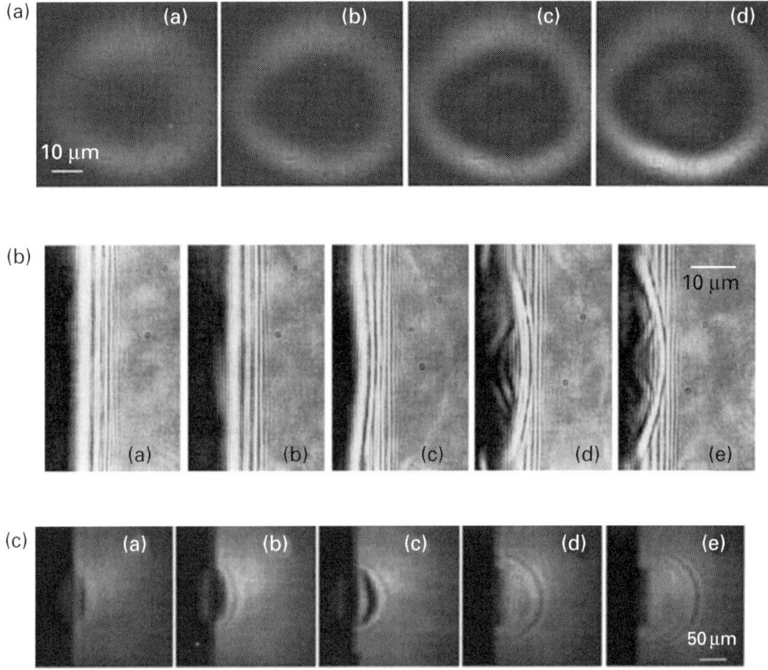

Figure 6.20. (a) Reflection images of a crystalline silicon surface subjected to femtosecond laser irradiation with $F = 1.5$ J/cm^2. The darkening of the core region at 5–10 ps indicates initiation of the ablation process. (b) Shadowgraph images at early times. Material expulsion is visible at $t = 100$ ps. (c) Shadowgraph images of the shock wave at long times. The ablation plume emerges from the surface. (d) The envelope of the shock wave at various elapsed times. The initial speed exceeds 1600 m/s. From Choi *et al.* (2002), reproduced with permission by the American Institute of Physics.

Most of the ablated material was expelled on the nanosecond temporal scale. A shock wave was launched into the atmospheric-pressure air background. The position of this shock wave was monitored and analyzed by applying blast theory for an instantaneous point-source explosion.

Single shots at various laser fluences were used to fabricate micro-sized features as shown in Figure 6.21. Atomic-force microscopy (AFM) was used to measure detailed profiles of the modified features. In the lower-fluence regime, the density of hot electrons is relatively low and laser energy is mainly deposited in the shallow region defined by the optical penetration depth. Two-photon absorption and the electron-conduction term can be approximately taken into account by the introduction of an effective "laser + electron heat" penetration depth. At higher fluences, the contribution of electron heat conduction becomes important, and the heat-affected region is defined by the electron-heat penetration depth. The ablation efficiency increased up to the fluence level around 10 J/cm^2, but, at higher fluences than 100 J/cm^2, it decreased significantly. In order to study the coupling mechanisms of the high-fluence femtosecond laser pulse with the

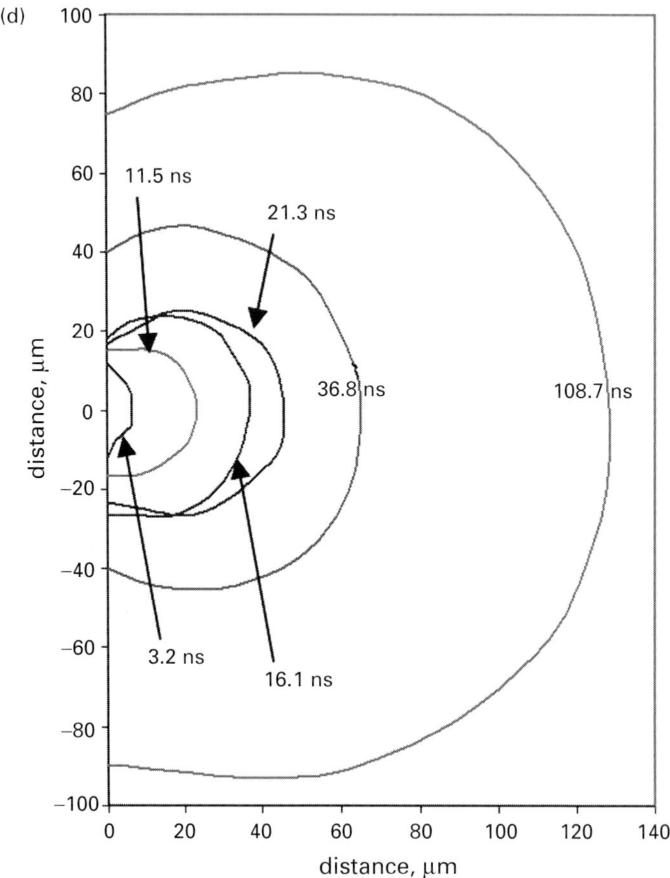

Figure 6.20. (*cont.*)

target, time-resolved-pump-and-probe imaging data were presented as in Figure 6.22 for two fluence levels (10 and 1400 J/cm^2). Strong resistance during the expansion of the ablated plume leads to the generation of higher recoil pressure. This causes an increase in the redeposition and resolidification of the ablated materials, corresponding to a reduction of the ablation efficiency (Hwang *et al.*, 2006).

6.5   Generation of highly energetic particles

Basic models involving collisional absorption, transport, hydrodynamics, and fast-electron and hard-X-ray generation were reviewed in Gibbon and Forster (1996), focusing mainly on extremely high laser intensities (above 10^{16} W/cm^2). Processing with femtosecond laser pulses offers unique characteristics of minimal thermal damage with low and well-defined ablation thresholds. These are deemed advantageous for micro-machining fine structures and sensitive device components. Nanosecond laser pulses

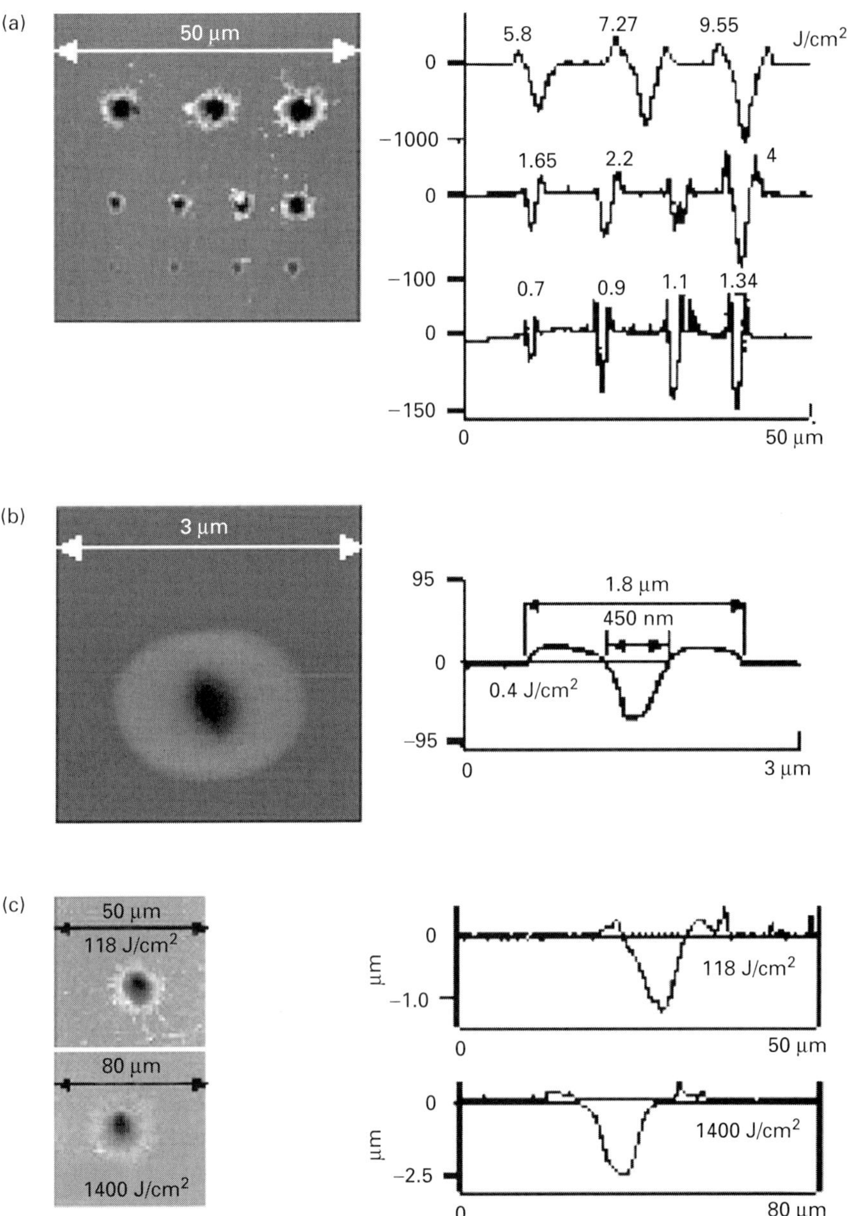

Figure 6.21. An AFM image and cross-sectional profile of a sub-micrometer hole being made to open in a c-Si wafer by single femtosecond laser pulses of $\lambda = 800$ nm at medium energy densities (a), at near-threshold energy density (b), and at high energy densities (c); and the ablation efficiency as a function of the energy density (d). From Hwang *et al.* (2006), reproduced with permission by the American Institute of Physics.

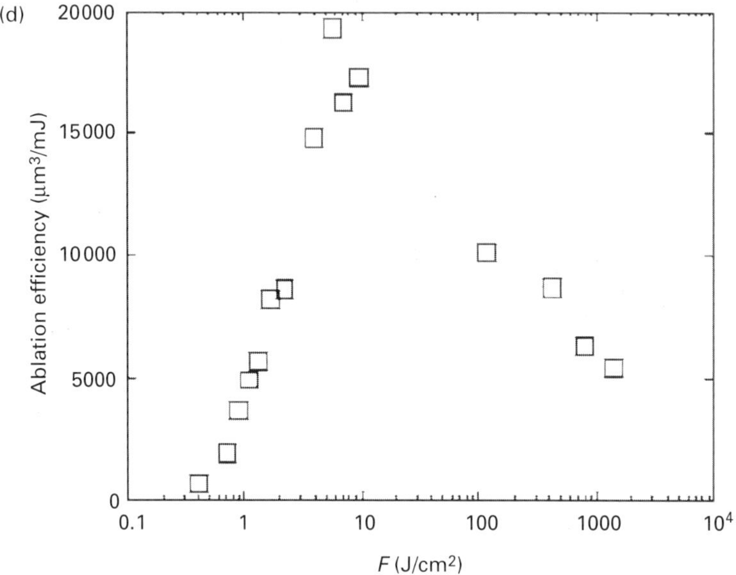

Figure 6.21. (*cont.*)

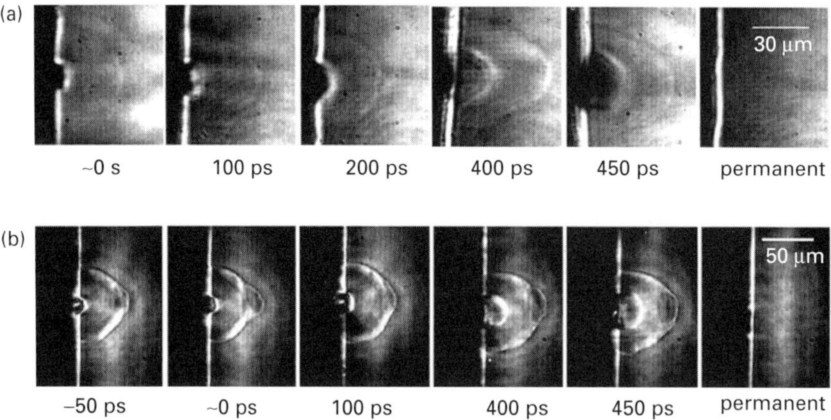

Figure 6.22. Time-resolved shadowgraphs (side view) of the silicon-ablation process: (a) $F = 11.2$ J/cm² (285 nJ) (b); $F = 1400$ J/cm² (36.7 µJ). From Hwang *et al.* (2006), reproduced with permission by the American Institute of Physics.

tend to generate explosive phase change, strong shock waves, and formation of particle clusters. Ultra-short laser pulses impart weaker mechanical effects, since plasmas form after the expiration of the laser pulse and hence do not absorb part of the incident laser irradiation. The high intensities of femtosecond laser pulses are likely to induce strong particle kinetics and ionization. These effects can be utilized to synthesize new materials. For example, energetic ions and neutral species present in the laser-ablated plume have been found to enhance initial nucleation on the substrate surface and allow epitaxial growth at lower substrate temperature. On the other hand, they can cause resputtering

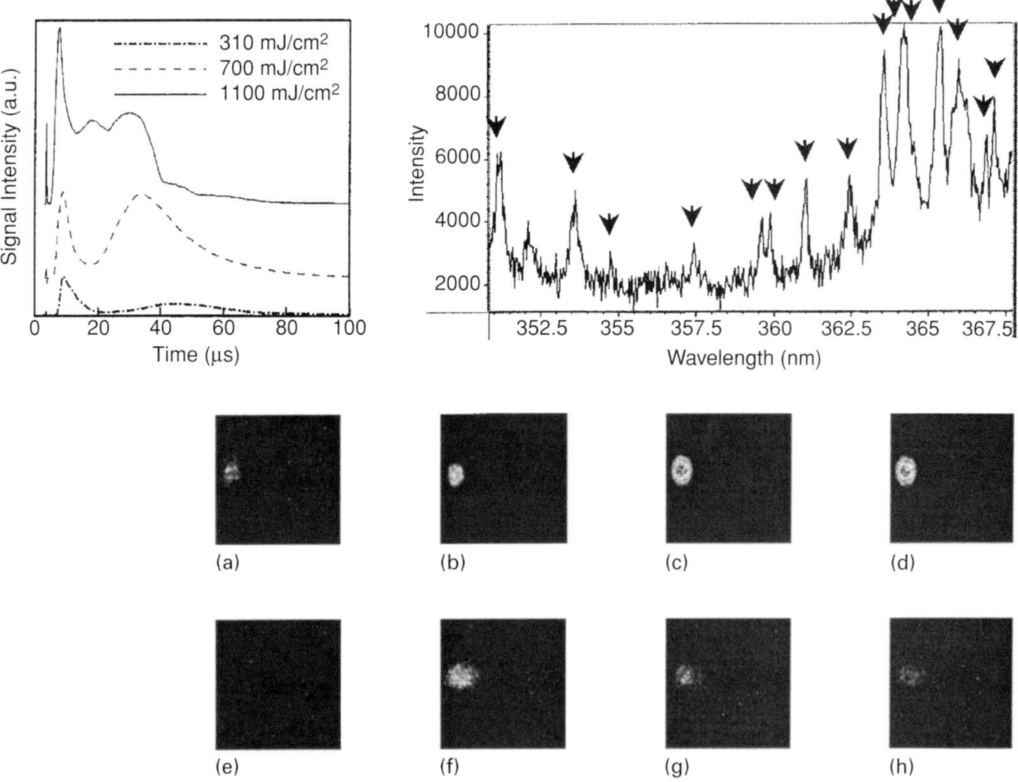

Figure 6.23. Left: time-of-flight spectra of laser-ablated titanium ions at various laser fluences; the most probable velocity of fast titanium ions as a function of laser fluence; upper right: typical titanium plume-emission spectrum induced by femtosecond-laser pulses at $F = 6$ J/cm^2 (both neutral-atom and ion emission lines were identified); and bottom right: time-resolved titanium plume emission images captured by a gated ICCD camera. The applied laser fluence was 5 J/cm^2, and the working pressure 100 μTorr. The first five images ((a)–(e)) were obtained with a gate width of 10 ns. A longer gate width of 500 ns was employed for the last three images ((f)–(h)). The gate delay time was 20 ns, 80 ns, 120 ns, 140 ns, 180 ns, 1.42 μs, 2.42 μs, and 2.92 μs, respectively. From Ye and Grigoropoulos (2001), reproduced with permission by the American Institute of Physics.

in the process of film growth. The ablation mechanisms are strongly dependent both on the laser-pulse parameters and on the material properties.

Femtosecond-laser ablation of metals has been studied via time-of-flight (TOF) and emission-spectroscopy measurement (Ye and Grigoropoulos, 2001). Laser pulses of 80 fs (FWHM) at $\lambda = 800$ nm were delivered by a Ti : sapphire femtosecond-laser system. These ion TOF spectra were utilized to determine the velocity distribution of the ejected ions. While nanosecond-laser ablation typically generates ions of energy a few tens of electron-volts, femtosecond-laser irradiation even at moderate energy densities can produce energetic ions with energies in the range of a few keV. The most probable energy of these fast ions is proportional to the laser fluence. Figure 6.23 shows typical titanium-ion TOF spectra taken at various laser fluences. No extraction field was used in these measurements. Each spectrum was averaged over 300 shots. All spectra exhibited a spike at an elapsed time of about 3.6 μs. This was believed due to the soft X-rays emitted by the plasma shortly after the laser pulse struck the sample surface. Three ion peaks can be distinguished in the spectrum at the highest laser fluence (1100 mJ/cm^2). The first of these peaks is narrower than the other two and includes the more energetic ions. As the laser fluence drops, only two peaks are discernible, and they shift to longer TOF values while becoming broader in temporal distribution. The cause of the appearance of the third peak at the highest laser fluence may be due to ion clusters of a different mass or charge state and needs further investigation. The titanium-ion velocity can be as high as 2.0×10^7 cm/s even at the moderate laser energy densities applied in this experiment.

6.6   Ultrafast phase explosion

The mechanisms of decomposition of a metal (nickel) during femtosecond-laser ablation have been studied using molecular-dynamics (MD) simulations (Cheng and Xu, 2005). It was found that phase explosion is responsible for gas-bubble generation and the subsequent material removal at lower laser fluences. The phase explosion occurs as a combined result of heating, thermal expansion, and the propagation of the tensile-stress wave induced by the laser pulse. When the laser fluence is higher, it was revealed that critical-point phase separation plays an important role in material removal.

As discussed in Chapter 3, various mechanisms, such as phase explosion and critical-point phase separation, have been proposed to explain laser ablation. Phase explosion is homogeneous bubble nucleation at close to the spinodal temperature (slightly below the critical temperature), during which gas-bubble nucleation occurs simultaneously in a superheated, metastable liquid. The temperature–density $(T-\rho)$ and pressure–temperature $(p-T)$ diagrams of the phase-explosion process are illustrated in Figure 6.24 (Kelly and Miotello, 1996). During rapid laser heating, the liquid can be raised to a temperature above the normal boiling temperature (point A), which corresponds to a state of superheating in the region between the binodal line and the spinodal line on the phase diagram, the metastable zone. When the material approaches the spinode (point B), intense fluctuation could overcome the activation barrier for the vapor embryos to grow into nuclei. This activation barrier decreases in height as the material gets closer to the spinode, causing a drastic increase of the nucleation rate, which turns the material into a mixture of vapor and liquid droplets. Therefore, the spinodal line is the limit of superheating in the metastable liquid, and no homogeneous structure will exist beyond it when the liquid is heated. Experimental work has shown that phase explosion occurred during nanosecond-laser ablation of a metal (Song and Xu, 1998; Xu, 2001).

During femtosecond-laser ablation, an important factor that needs to be considered is the extraordinary heating rate. Heating above the critical temperature directly from the solid phase becomes possible (point A in Figure 6.25), followed by expansion leading to formation of the thermodynamically unstable region (B), causing material

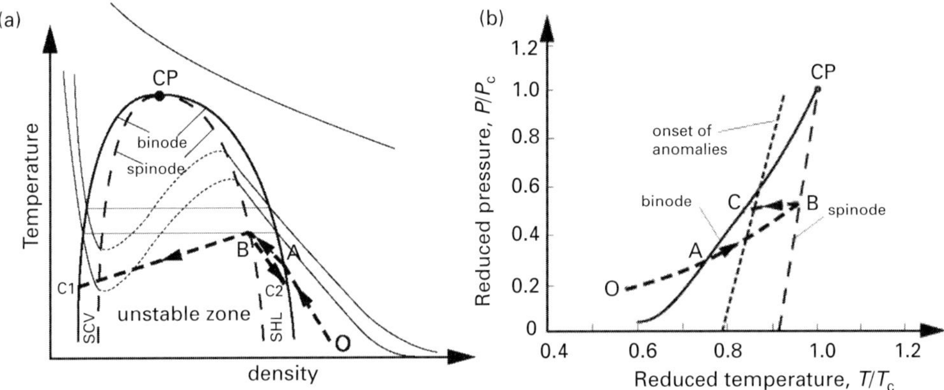

Figure 6.24. (a) T–ρ and (b) p–T diagrams of phase explosion. The dome presented as a solid line is the binode. The dome presented as a dashed line is the spinode. SHL, super-heated liquid. SCV, super-cooled vapor. CP, critical point. From Cheng and Xu (2005), reproduced with permission by the American Physical Society.

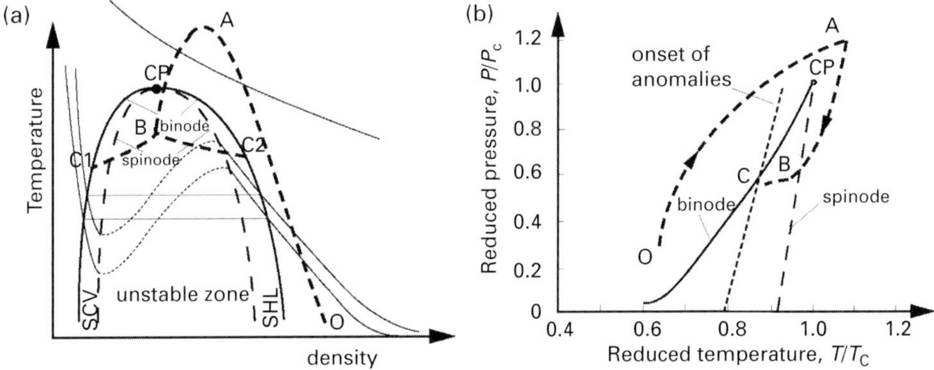

Figure 6.25. (a) T–ρ and (b) p–T diagrams of critical-point phase separation. The dome presented as a solid line is the binode. The dome presented as a dashed line is the spinode. SHL, super-heated liquid. SCV, super-cooled vapor. CP, critical point. From Cheng and Xu (2005), reproduced with permission by the American Physical Society.

decomposition (Skripov and Skripov, 1979). This material-decomposition process, from solid to supercritical fluid to the unstable region, is termed critical-point phase separation.

The mechanism leading to ablation was studied by analyzing the thermodynamic trajectories of groups of atoms that undergo phase separation. Figure 6.26 shows the groups of atoms analyzed for the laser fluence of 0.3 J/cm^2 at 120 ps. According to Figure 6.26, groups 2 and 4 have turned into gas at 120 ps, while groups 1, 3, and 5 are in the liquid phase (and will remain as liquid). Their thermodynamic trajectories of densities and temperatures during the ablation process are shown in Figure 6.27. The arrows indicate the temporal progress, while the numbers along the trajectories mark the time in picoseconds. From Figure 6.27, it is seen that groups 2, 3, and 4, which experience material separation, cross both the binodal line and the spinodal line. These

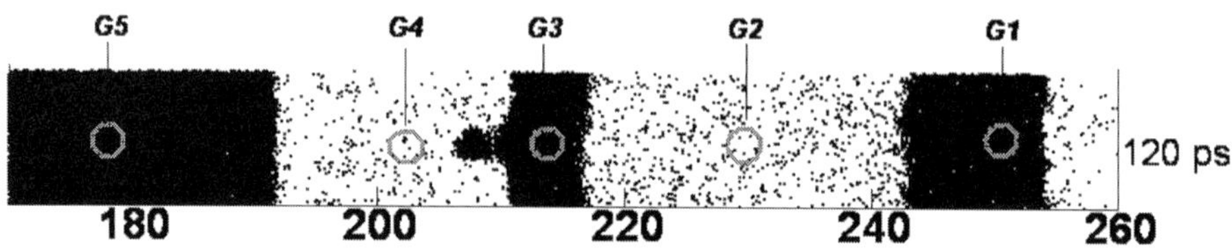

Figure 6.26. Positions of groups of atoms at a laser fluence of 0.3 J/cm^2. From Cheng and Xu (2005), reproduced with permission by the American Physical Society.

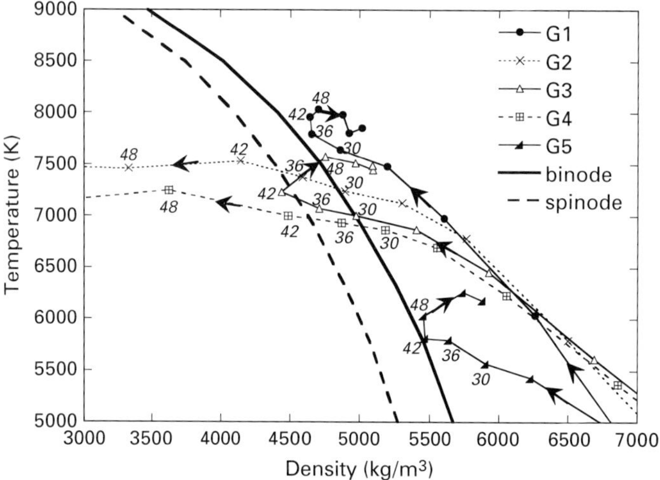

Figure 6.27. Thermodynamic trajectories of groups of atoms at a laser fluence of 0.3 J/cm^2. From Cheng and Xu (2005), reproduced with permission by the American Physical Society.

three groups undergo a phase-separation process, with group 2 and 4 turning into vapor. On the other hand, groups 1 and 5, which do not touch the spinodal, do not undergo phase change. This indicates that the phase change of the material is directly related to whether it reaches the spinodal line or not. Recall that, when liquid enters the metastable region and approaches the spinode, it will undergo the phase-explosion process and turn into a mixture of liquid and vapor. Therefore, the thermodynamic trajectories of the groups suggest that phase explosion occurs at this laser fluence.

Figure 6.28 shows the thermodynamic trajectories of several groups of atoms at a higher laser fluence, 0.65 J/cm^2. The locations of these groups of atoms at 90 ps are marked in Figure 6.29. From Figure 6.28, it is seen that all three groups are first raised to temperatures higher than the critical temperature and become a super-critical fluid. After expansion, their temperature decreases, and they enter the unstable zone below the critical point as the phase separation occurs at about 30 ps. Groups 1 and 2 evolve into gas, while group 3 becomes liquid. No phase separation occurs during the initial heating period (from 1 ps to about 30 ps), although the density decreases continuously. The material remains homogeneous until it enters the unstable zone after expansion, and the liquid (group 3) precipitates out from the homogeneous phase.

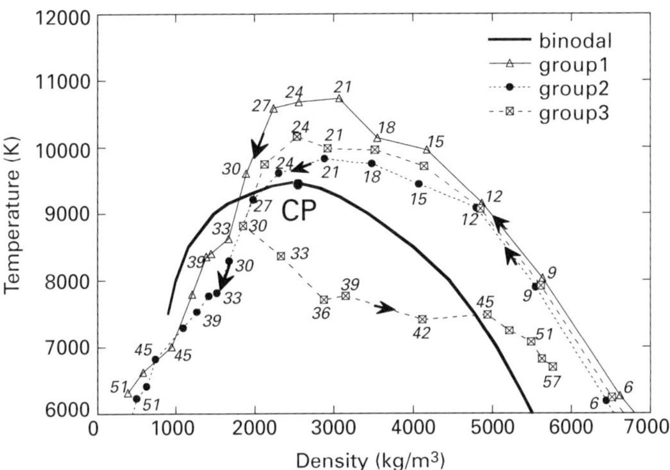

Figure 6.28. Thermodynamic trajectories of groups of atoms. The fluence is 0.65 J/cm^2. From Cheng and Xu (2005), reproduced with permission by the American Physical Society.

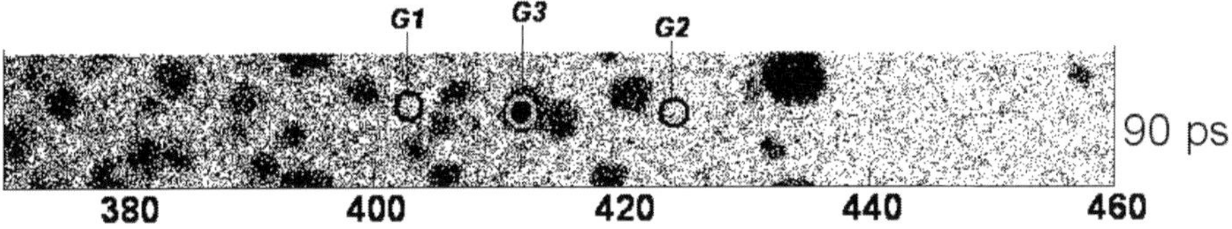

Figure 6.29. Positions of groups of atoms on the T–ρ diagram at 90 ps at a laser fluence of 0.65 J/cm^2. From Cheng and Xu (2005), reproduced with permission by the American Physical Society.

The thermodynamic trajectories of the groups of atoms described above are clearly different from those at lower laser fluences, but follow that of critical-point phase separation shown in Figure 6.25(a). Heating above the critical point, followed by the expansion into the unstable zone that causes phase separation, has clearly been illustrated, which agrees with the theoretical description of critical-point phase separation. Similar thermodynamic trajectories are found for laser fluences of 1.0 and 1.5 J/cm^2. Therefore, it is concluded that critical-point phase separation plays an important role in material decomposition.

In summary, the mechanisms of femtosecond-laser ablation of a nickel target were studied using MD simulations in a laser fluence range commonly used for materials processing. Two distinct laser-fluence regimes were identified, and attributed to different ablation mechanisms. At lower laser fluences, the peak temperature reached is below the critical temperature, and material decomposition occurs through phase explosion. Bubble nucleation occurs inside the metastable liquid at temperatures attained as the spinode is approached, and is assisted by the tensile stress developed during laser heating. At higher laser fluences critical-point phase separation occurs. The initial peak temperature reached exceeds the critical temperature. The super-critical fluid enters the unstable zone after relaxation and loses its homogeneity, causing phase separation.

6.7   Nonlinear absorption and breakdown in dielectric materials

The conventional view of pulsed-laser–material interactions, with wavelength between the near infrared (IR) and near ultraviolet (UV), includes the transfer of electromagnetic energy to electronic excitation, followed by electron–lattice interactions that convert energy into heat. However, the processes of material response following intense femtosecond-laser irradiation are far more complex, particularly for wide-band-gap dielectrics. When a dielectric material is subject to intense femtosecond-laser irradiation, the refractive index of the material may become intensity-dependent, and a large amount of excited electrons may be generated by IR pulses in "transparent" dielectrics. Relaxation channels of electronic excitation in wide-band-gap materials may produce intrinsic defects, leading to photo-induced damage in the otherwise "defect-free" medium. These fundamentally nonlinear processes have stimulated substantial research efforts regarding both the understanding of the complexity of femtosecond-laser interactions with dielectrics and the applications of the underlying microscopic mechanisms to innovative materials fabrication. An overview of advances toward understanding the fundamental physics of femtosecond-laser interactions with dielectrics that are important for materials-processing applications has been given by Mao *et al.* (2004).

Although laser-induced breakdown (LIB) in optically transparent materials had been studied extensively since the advent of lasers, progress in femtosecond-laser technology facilitated studies over a regime of high intensities, allowing the decoupling of thermal effects that are invariably present under long-pulse irradiation. Du *et al.* (1994) monitored the threshold of LIB in fused silica by transmission and scattered-plasma-emission measurements over a wide range of laser pulse widths. The fluence threshold versus the pulse width was shown to attain a $\sqrt{t_{\mathrm{pulse}}}$ dependence for $t_{\mathrm{pulse}} > 10\,\mathrm{ps}$ that is representative of thermal-diffusion effects. The deviation from this trend below 10 ps was studied by Stuart *et al.* (1995, 1996) experimentally and theoretically. Figure 6.30 shows the dependence of the damage-threshold fluence in fused silica and CaF_2 on the pulse width.

For femtosecond-laser interactions with dielectrics, in addition to their classical value in elucidating the origin of LIB in optical materials, structural modifications of dielectrics are of particular significance to bulk micro-structuring that creates sub-wavelength "voxels." As an example, femtosecond laser pulses can be focused inside transparent dielectric materials in a layer-by-laser fashion. High-density, three-dimensional optical storage was achieved as the result of femtosecond-laser-induced sub-micrometer structural transition that locally alters the refractive index at the laser pulse's focus (Day *et al.*, 1999; Kawata *et al.*, 1995; Glezer *et al.*, 1996). Similarly, three-dimensional photonic band-gap lattices were realized by spatially organized micro-patterning of transparent dielectrics using femtosecond laser pulses (Sun *et al.*, 2001).

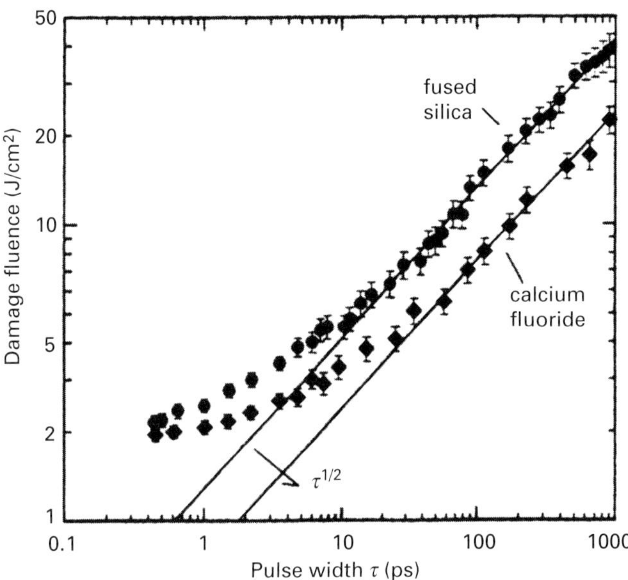

Figure 6.30. Observed values of the damage threshold at 1053 nm for fused silica (full circles) and CaF$_2$ (full rhombi). Solid lines are $\sqrt{t_{\mathrm{pulse}}}$ fits to long-pulse results. The estimated uncertainty in the absolute fluence is $\pm15\%$. From Stuart *et al.* (1995), reproduced with permission by the American Physical Society.

6.7.1 Carrier excitation, photo-ionization, and avalanche ionization

The problem of *carrier excitation* and ionization in the case of wide-band-gap materials subjected to a laser electromagnetic field has been addressed extensively in the literature. The balance between different ionization channels during femtosecond-laser interactions with dielectric materials is still under discussion. In the simplest case, the laser can deposit energy into a material by creating an electron–hole plasma through single-photon absorption. However, for wide-band-gap dielectrics, the cross section of such linear absorption is extremely small. Instead, under intense femtosecond-laser irradiation, nonlinear processes such as multiphoton ionization, tunnel ionization, or avalanche ionization become the dominant mechanisms that create free carriers inside the materials.

For irradiation of wide-band-gap materials using femtosecond laser pulses with wavelength near the visible (from near IR to near UV), a single laser photon does not have sufficient energy to excite an electron from the valence band to the conduction band. Simultaneous absorption of multiple photons must be involved to excite a valence-band electron, with the resulting *photo-ionization* rate strongly depending on the laser intensity (Figure 6.31(a)). The rate of multiphoton absorption can be expressed as $\sigma^{(m)} I^m$, where I is the laser intensity and $\sigma^{(m)}$ is the cross section of m-photon absorption for excitation of a valence-band electron to the conduction band. The number of photons required is determined by the smallest m that satisfies the relation $m(\hbar\omega) > E_{\mathrm{bg}}$, where E_{bg} is the band-gap energy of the dielectric material, $\hbar\omega$ is the photon energy, and $\hbar$

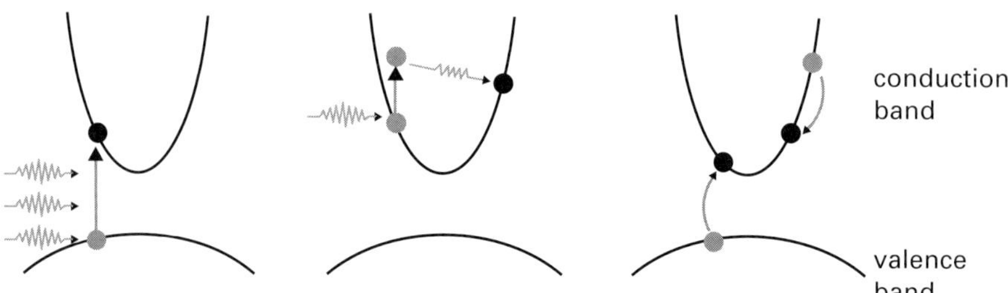

Figure 6.31. Schematic illustrations of (a) multiphoton ionization, (b) free-carrier absorption, and (c) impact ionization. From Sundaram and Mazur (2002), reproduced with permission by Nature Publishing Group.

is the Planck constant. Let us emphasize here one characteristic of femtosecond-laser irradiation: since it allows the use of higher peak intensities than is possible with conventional pulses, it reinforces intrinsic high-order interband transitions by comparison with the ever-present defect-related processes (of a lower order).

A second photo-ionization process, *tunneling ionization*, may come into play during femtosecond-laser interactions with dielectrics under an extremely strong laser field, for example, when the laser pulse is very short (e.g. <10 fs). In this regime, the laser produces a strong periodic band-bending, which allows valence electrons to tunnel directly to the conduction band in a time shorter than the laser period. The crossover between multiphoton and tunneling ionization can be characterized by the Keldysh parameter (Keldysh, 1965), $\gamma_{\text{Ke}} = \omega(2m^*E_{\text{g}})^{1/2}/(eE)$, where m^* and e are the effective mass and charge of the electron, and E is the intensity of a laser electric field oscillating at frequency ω. When γ_{Ke} is much larger than unity, as is the case for most material-related investigations of laser interactions with dielectrics, multiphoton ionization dominates the excitation process. As a numerical example, $\gamma_{\text{Ke}} = 1$ corresponds to an intensity of approximately 4×10^{13} W/cm^2 for irradiation of fused silica with a laser pulse at 800 nm.

An electron being excited to the conduction band of a wide-band-gap dielectric material can absorb several laser photons sequentially, moving itself to higher energy states where free-carrier absorption is efficient (Figure 6.31(b)). The complex refractive index, $n^c = n - ik$, is related to the complex dielectric function ε^c, which, according to the Drude model, can be expressed by

$$\varepsilon^c = 1 - \omega_{\text{p}}^2 \left[\frac{\tau^2}{1 + \omega^2\tau^2} - i\frac{\tau^2}{\omega\tau(1 + \omega^2\tau^2)} \right], \tag{6.27}$$

with the scattering time τ being typically a fraction of a femtosecond. For wide-band-gap dielectrics under intense laser irradiation, strong electron interactions with the lattice are characterized by both polar and nonpolar phonon scattering (Fischetti *et al.*, 1985; Arnold and Cartier, 1992). In the expression for ε^c, ω_{p} is the plasma frequency defined by

$$\omega_{\rm p} = \sqrt{\frac{e^2 N}{\varepsilon_0 m^*}}, \tag{6.28}$$

where N is the carrier density and ε_0 is the electric permittivity. When $\omega_{\rm p} \ll \omega$, i.e., the carrier density is well below 10^{21} cm^{-3}, the absorption coefficient can be derived as

$$\gamma = \frac{\tau}{nc} \frac{\omega_{\rm p}^2}{1 + \omega^2 \tau^2}, \tag{6.29}$$

where c is the speed of light.

When the electron density generated by photo-ionization reaches a high level (e.g. $\omega_{\rm p} \sim \omega$), a large fraction of the remaining femtosecond laser pulse can be absorbed. It is interesting to note that high-energy (e.g. three times the band-gap energy) carriers can also be created in materials in which the electron–phonon scattering rate is low, such that multiple electron–phonon collisions could not occur in one laser pulse. Carrier heating could be produced through direct interbranch single-photon or multiphoton absorption, in a way quite similar to the valence-to-conduction interband absorption discussed above. In all materials, both processes should certainly be taken into account, which one dominates depending essentially on the strength of the electron–phonon coupling.

Avalanche ionization involves free-carrier absorption followed by impact ionization (Figure 6.31(c)). Since the energy of an electron in the high energy states exceeds the conduction-band minimum by more than the band-gap energy, it can ionize another electron from the valence band, resulting in two excited electrons at the conduction-band minimum (Bloembergen, 1974). These electrons can again be heated by the laser field through free-carrier absorption, and, once they have enough energy, impact more valence-band electrons. This process can repeat itself as long as the laser electromagnetic field is present and intense enough, leading to a so-called electronic avalanche. The growth of the conduction-band population by this avalanche process has the form $\beta_{\rm av}N$, where $\beta_{\rm av}$ is the avalanche-ionization rate, a phenomenological parameter that accounts for the fact that only high-energy carriers can produce impact ionization.

Avalanche ionization requires seed electrons to be present in the conduction band, which can for instance be excited by photo-ionization. The following rate equation has been proposed to describe the injection of electrons into the conduction band of dielectrics by femtosecond-to-picosecond laser pulses, under the combined action of multiphoton excitation and avalanche ionization (Stuart *et al.*, 1996):

$$\frac{{\rm d}N}{{\rm d}t} = \alpha I N + \sigma^{(m)} N I^m, \tag{6.30}$$

where α is a constant. For dielectric materials in which free-carrier losses (e.g. self-trapping and recombination) occur on a time scale comparable to the femtosecond-laser pulse duration (e.g. quartz and fused silica), this population equation should be modified

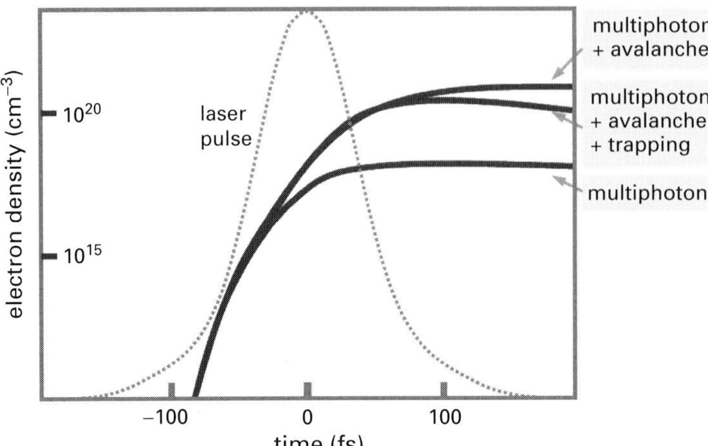

Figure 6.32. Schematic illustrations of 100-fs-laser-induced electron-density evolution under three different excitation–relaxation conditions: multiphoton ionization only, and multiphoton plus avalanche ionization, and multiphoton plus avalanche ionization with carrier trapping. Multiphoton ionization provides seed electrons for avalanche ionization, whereas trapping offers a channel for electron-density reduction. The Gaussian laser pulse is also illustrated. From Mao *et al.* (2004), reproduced with permission by Springer-Verlag.

as follows:

$$\frac{dN}{dt} = \alpha I N + \sigma^{(m)} N I^m + \sigma_x N_{STE} I^{m_x} - \frac{N}{\tau_x},$$
$$\frac{dN_{STE}}{dt} = -\sigma_x N_{STE} I^{m_x} + \frac{N}{\tau_x}. \tag{6.31}$$

In the above expressions, the contribution from self-trapped excitons of density N_{STE} that builds up during the pulse duration is included (which may in some cases be bimolecular recombination, depending on the carrier density) (Li *et al.*, 1999; Petite *et al.*, 1999). σ_x is the multiphoton cross section (of order m_x) for self-trapped excitons and τ_x is the characteristic trapping time. A schematic illustration of the effect of the self-trapping on a femtosecond-laser-excited electron population is shown in Figure 6.32.

More recently, variations of the classical avalanche process that may play a role for sufficiently short laser pulses (e.g. <40 fs) have been investigated theoretically. One such mechanism is collision-assisted multiphoton avalanche, in which some valence electrons are excited to the conduction band by conduction electrons with energy smaller than the threshold for impact ionization, by absorbing several laser photons during inelastic electron–electron collisions. Another mechanism is hole-assisted multiphoton absorption, which is similar to the so-called enhanced ionization of molecules in strong laser fields (Seideman *et al.*, 1995). Through its Coulomb field, a hole exponentially enhances the multiphoton absorption rate of atoms at adjacent lattice sites. As soon as new holes are created, they continue the same trend that could lead to a collision-free electronic avalanche.

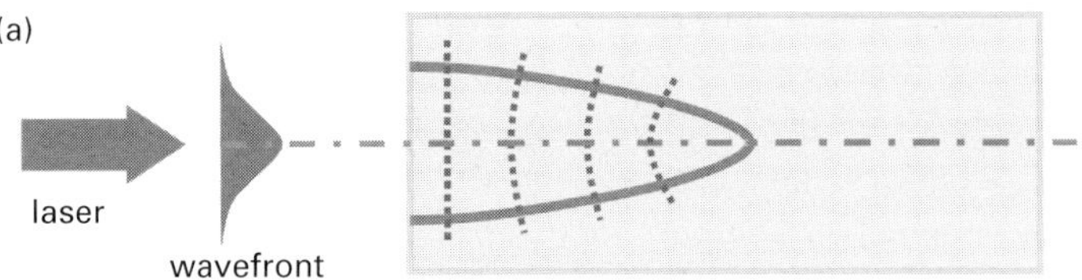

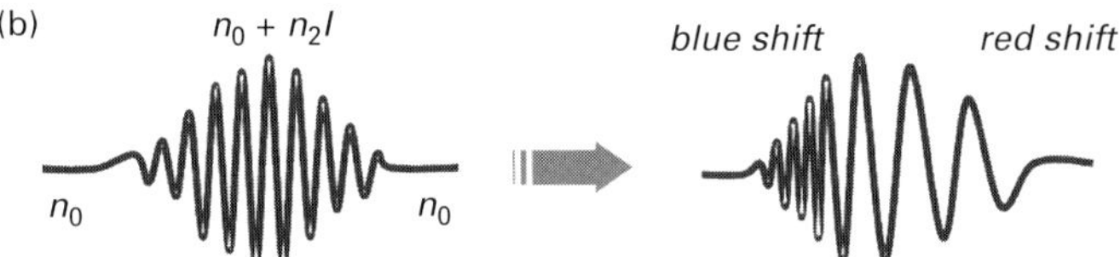

Figure 6.33. Schematic illustrations of (a) self-focusing and (b) self-phase modulation resulting from a nonlinear refractive index. From Mao *et al.* (2004), reproduced with permission by Springer-Verlag.

6.7.2 Nonlinear propagation

When a laser pulse propagates through a dielectric material, it induces microscopic displacement of the bound charges, forming oscillating electric dipoles that add up to the macroscopic polarization. For isotropic dielectric materials such as fused silica, the resulting index of refraction (real part) can be derived as (Kelley, 1965)

$$n = \sqrt{1 + \chi^{(1)} + \frac{3}{4}\chi^{(3)}E^2},\tag{6.32}$$

where $\chi^{(1)}$ and $\chi^{(3)}$ are the linear and nonlinear susceptibility, respectively. In a more convenient form,

$$n = n_0 + n_2 I,\tag{6.33}$$

where $I = \frac{1}{2}\varepsilon_0 c n_0 E^2$ is the laser intensity, and $n_0 = \sqrt{1 + \chi^{(1)}}$ and $n_2 = 3\chi^{(3)}/(4\varepsilon_0 c n_0^2)$ are the linear and nonlinear part of the refractive index, respectively. A nonzero nonlinear refractive index n_2 gives rise to many nonlinear optical effects as an intense femtosecond laser pulse propagates through dielectric materials.

Self-focusing and self-phase modulation

The spatial variation of the laser intensity $I(r)$ can create a spatially varying refractive index in dielectrics. Owing to the typical Gaussian spatial profile of a femtosecond laser pulse, the index of refraction is larger toward the center of the pulse. The spatial variation of n causes a lens-like effect that tends to focus the laser beam inside the dielectrics (Figure 6.33(a)). If the peak intensity of the femtosecond laser pulse exceeds a critical power for self-focusing (Shen, 1984), $P_{\mathrm{cr}} = 3.77\lambda^2/(8\pi n_0 n_2)$, the collapse of the pulse to a singularity is predicted. Nevertheless, other mechanisms such as defocusing due to nonlinear ionization may balance self-focusing and prevent pulse collapse inside dielectric materials.

As the result of spatial self-focusing, the on-axis intensity of femtosecond laser pulses inside dielectrics, especially at its temporal peak, can be significantly larger than its original value. Consequently, the pulse may be sharpened temporally with a steeper rise and decay of the temporal profile (pulse sharpening). Since the intensity $I(t)$ of femtosecond laser pulses is highly time-dependent, the refractive index depends on time. Analogously to self-focusing, the phase of the propagating pulse can be modulated by the time-domain envelope of the pulse itself (self-phase modulation). With a nonzero nonlinear refractive index n_2, the derivative of the phase $\Phi_{\mathrm{ph}}(z, t)$ of the pulse with respect to time becomes (Shen, 1984)

$$\frac{\mathrm{d}\Phi_{\mathrm{ph}}}{\mathrm{d}t} = \omega - \frac{n_2 z}{c}\frac{\mathrm{d}I(t)}{\mathrm{d}t}. \tag{6.34}$$

The time-varying term of the phase produces frequency shifts that broaden the pulse spectrum as illustrated in Figure 6.33(b). Spectral broadening depends on the nonlinear index of refraction n_2 and the time derivative of the laser pulse intensity.

Plasma defocusing

Since various nonlinear ionization mechanisms generate an electron–hole plasma inside wide-band-gap dielectric materials, this plasma has a defocusing effect for femtosecond-laser pulse propagation. The electron density is highest in the center of the pulse and decreases outward in the radial direction with a typical Gaussian spatial intensity profile. The real part of the refractive index is modified by the generation of the electron–hole plasma (for $\omega_{\mathrm{p}}/\omega \ll n_0$) (Shen, 1984),

$$n = n_0 - \frac{N}{2n_0 N_{\mathrm{c}}}, \tag{6.35}$$

where $N_{\mathrm{c}} = \omega^2 \varepsilon_0 m^*/e^2$ is the characteristic plasma density when the plasma frequency is equal to the laser frequency. It is clear that the presence of electron–hole plasmas results in a decrease in the refractive index, in contrast to the Kerr effect. As a result, the refractive index is smallest on the beam axis and the beam is defocused by the plasma, which acts as a diverging lens, balancing self-focusing.

6.7.3   Defect generation

In general, energy from intense femtosecond laser pulses absorbed by a solid material can be converted into elementary electronic excitations – electrons and holes, which relax and reduce their energy inside the solid through both delocalized and localized carrier–lattice interaction channels (Song and Williams, 1993; Haglund and Itoh, 1994). For some wide-band-gap dielectric materials, the most important relaxation mechanism is the localization of the energy stored in the electron–hole pair that creates self-trapped carriers, especially self-trapped excitons (STEs), which provide the energy necessary for localized lattice re-arrangement and thus defect accumulation.

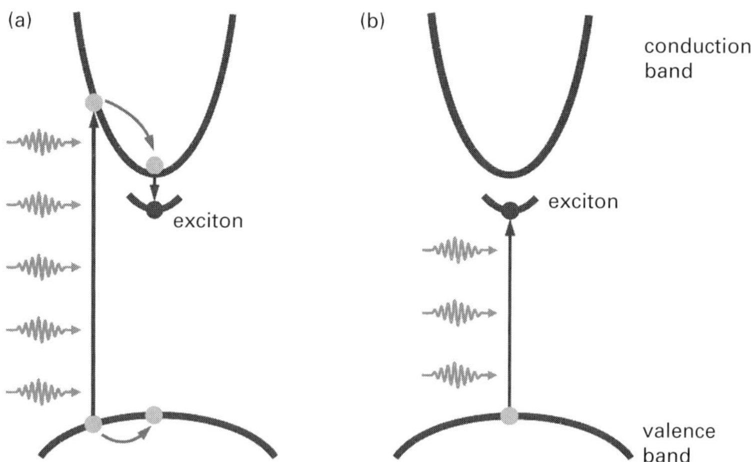

Figure 6.34. Schematic illustrations of the exciton level and two basic routes for exciton generation: (a) inelastic scattering of the multiphoton-excited electrons and (b) direct resonance absorption of multiple photons. From Mao *et al.* (2004), reproduced with permission by Springer-Verlag.

Excitons

Through nonlinear ionization, the interaction of an intense femtosecond laser pulse with wide-band-gap dielectrics causes electronic excitations that promote an electron from the valence band to the conduction band, leaving a hole in the valence band. An electron and a hole may be bound together by Coulomb attraction, constituting what is collectively referred to as an exciton, a concept of electrically neutral electronic excitation (Ueta *et al.*, 1986). Figure 6.34 shows a schematic diagram of exciton energy levels in relation to the conduction-band edge. While excitons can be either weakly or tightly bound, in wide-band-gap materials with a typically small dielectric constant, they are strongly bound and localized near a single atom. Excitons may be promoted by inelastic scattering (Vasil'ev *et al.*, 1999) of the excited electrons that slows the electrons in the conduction band (Figure 6.34(a)), or by direct resonant absorption of multiple photons (Figure 6.34(b)). The binding of electron–hole pairs into excitons is a very fast process, which often takes less than 1 ps in wide-band-gap materials (Haglund and Itoh, 1994).

Excitons are unstable with respect to their recombination process; they can relax through delocalized and localized channels. For wide-band-gap dielectrics that are strong-coupling solids, a localized trapping mechanism rather than scattering is more probable for excitons. Consequently, the electronic excitation energy in these materials is localized by the creation of STEs, which are formed as the result of free-exciton relaxation, or when a self-trapped hole traps an electron (Toyozawa, 1980).

Exciton self-trapping

The major interest in STEs in dielectrics comes from the fact that they are a means of converting electronic excitation into energetic atomic processes such as defect formation. Self-trapping generally describes carriers localized on a lattice site initially free of lattice defects (e.g. vacancies, interstitials, or impurities). A charged carrier in a deformable lattice creates an attractive potential well; the trapping results from a small atomic displacement that deepens the potential well in which the carrier resides. In general, localized lattice deformation may result from short-range covalent molecular bonding or long-range electrostatic polarization associated with ion displacements. Thermal fluctuations can provide the energy for at least one particular lattice site with enough instantaneous deformation for the self-trapping to begin.

Excitons can be trapped by their interactions with lattice distortion to form STEs. Holes may also be trapped at the distortion of the lattice which, after the trapping of an electron, creates a STE. Materials that display self-trapping are predominantly insulators with wide band gaps, such as alkali halides and SiO_2. In alkali-halide crystals, which have an energy band gap ranging from 5.9 eV (NaI) to 13.7 eV (LiF), a self-trapped exciton consists of an electron bound by the Coulomb field of the surrounding alkali ions and a hole that occupies an orbital of a halogen molecular ion (X_2^-). Similarly, in SiO_2, which is constructed from SiO_4 tetrahedra with silicon at the center and an oxygen atom at each of the four corners (Trukhin, 1992), the self-trapping process is accompanied by a strong distortion of the SiO_2 lattice. Weakening of the Si—O—Si bond yields an oxygen atom leaving its equilibrium position in the tetrahedron, forming silicon and oxygen dangling bonds. The hole of the self-trapped exciton stays on the oxygen dangling bond and the electron is on the silicon dangling bond.

Energy transport of STEs is by means of hopping diffusion rather than by band-like mode. As STEs recombine, they produce a characteristic luminescence that can be studied by time-resolved spectroscopy (Thoma *et al.*, 1997; Guizard *et al.*, 1996). For example, high-purity quartz emits a blue luminescence ($\sim$2.8 eV) under irradiation, which corresponds to a large Stokes shift relative to the band gap. For wide-band-gap dielectric materials, the localized relaxation channel that leads to the production of STEs is correlated with the formation and accumulation of transient and permanent lattice defects.

Origins of intrinsic defects

Optical excitation can be sufficient to generate vacancies and interstitials in perfect dielectric lattices. Defect formation may be classified as of extrinsic or intrinsic type depending on whether the defect is derived from a precursor. Advances (Song and Williams, 1993) in the study of STE structures have provided the basis for a new level of understanding of the mechanisms of intrinsic defect formation. In the absence of exciton self-trapping, electronic excitation would remain completely delocalized in a perfect dielectric material. Exciton self-trapping can provide the energy required (of the order of electron-volts) to initiate intrinsic defects, including vacancy–interstitial pairs where an atom is displaced in the course of the decay of electronic excitations.

Excitonic mechanisms of defect formation are well established in laser-irradiated halides and SiO_2, among many other wide-band-gap materials with strong electron–lattice couplings. F-centers and H-centers are the primary defects that are the immediate products of self-trapped-exciton decay in alkali halides (Figure 6.35). After initial non-linear ionization that generates electrons and holes, the process of defect formation

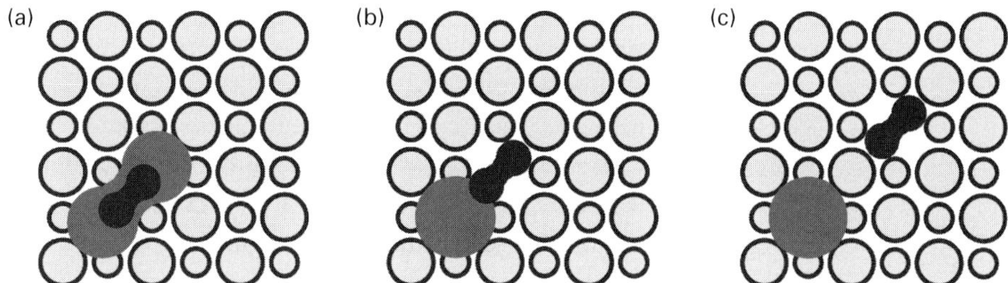

Figure 6.35. Schematic illustrations of defect formation from self-trapped excitons: (a) an on-center self-trapped exciton, (b) an off-center self-trapped exciton, and (c) an F–H pair in alkali halides. Small and large circles represent alkali and halogen ions, respectively. From Mao *et al.* (2004), reproduced with permission by Springer-Verlag.

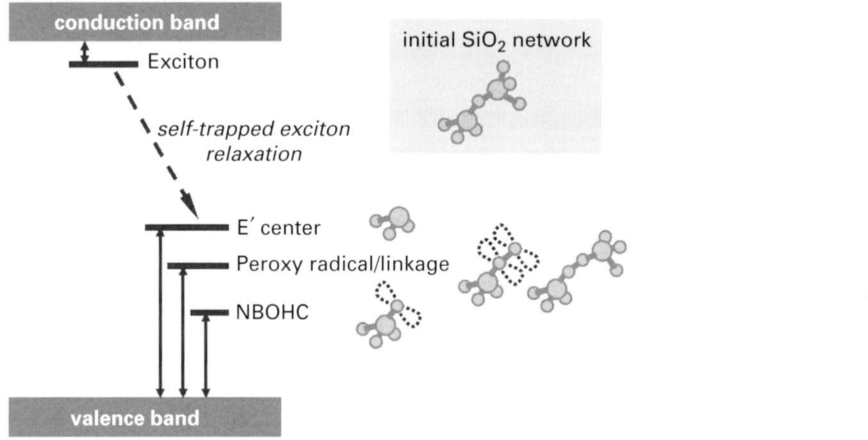

Figure 6.36. A schematic illustration of exciton and intrinsic-defect energy levels in SiO_2. From Mao *et al.* (2004), reproduced with permission by Springer-Verlag.

starts from exciton creation, followed by self-trapping of the exciton. An isomeric transformation occurs from a self-trapped exciton to a Frenkel-defect pair comprising an F-center, a halogen vacancy with a bound electron, and an H-center, an interstitial halogen ion bound to a lattice halogen ion by a hole. Off-center relaxation is the crucial step toward decomposition of the self-trapped exciton, since a self-trapped exciton is gradually changed to a stable vacancy–interstitial defect pair by displacing the H-center away from its point of creation, out of the range for recombination with the electron wave function bound to the F-center.

In SiO_2, the E′ (oxygen vacancy) and nonbridging oxygen-hole centers are the analogs of the F and H centers in alkali halides. The oxygen vacancy in SiO_2 is essentially a dangling silicon bond. The displaced oxygen atom goes into the nonbridging oxygen-hole center state (Si—O·), which may end up in a peroxy linkage (Si—O—O—Si) or radical (Si—O—O·), an isomer of a self-trapped exciton after covalently coupling to another oxygen atom at an interstitial site. Figure 6.36 gives a simplified schematic diagram of energy levels in SiO_2. The point defects resulting from decay of self-trapped excitons add more energy levels, analogously to the effect of impurities.

As the result of intense femtosecond-laser irradiation, during which the electronic defects resulting from the decay of self-trapped excitons grow in number, defect clusters may form, which can yield macroscopic structural damage in the material. In addition, the significant transient increase in volume associated with exciton self-trapping could create a shock-wave-like perturbation that eventually damages the otherwise perfect lattice.

6.7.4   Damage of dielectrics

Laser-induced damage in wide-band-gap dielectric materials is known to be an extremely nonlinear process. There is no doubt that damage in pure wide-band-gap materials is associated with rapid buildup of conduction electrons. Many experimental and theoretical studies have been performed to determine the damage mechanisms, with the majority of these efforts focused on the damage or breakdown threshold as a function of the laser-pulse duration. It is well established that for pulse durations of 10 ps or longer, for which thermal diffusion comes to play, the threshold laser fluence (energy density) for material damage depends on the laser-pulse duration (Stuart *et al.*, 1996; Lenzner, 1999; Tien *et al.*, 1999) according to a $\sqrt{t_{\mathrm{pulse}}}$ scaling. Nevertheless, for femtosecond-laser interactions with wide-band-gap dielectrics, when the pulse duration is much shorter than the characteristic time for thermal diffusion, the damage threshold deviates from such a square-root scaling.

There have been many theoretical attempts (Stuart *et al.*, 1996; Tien *et al.*, 1999) aimed at determining the mechanism of femtosecond-laser-induced dielectric breakdown. It is important to understand the relative roles of various ionization and relaxation channels in femtosecond-laser-induced dielectric breakdown, in particular near the damage threshold. Once a dense electron–hole plasma has been generated in the bulk of a defect-free wide-band-gap material as the result of femtosecond-laser excitation, several mechanisms that may lead to damage or optical breakdown can be foreseen. Damage could be caused by melting or vaporization of the solid, following the strong phonon emission by the laser-generated conduction electrons. It may also be due to the outcome of generation and accumulation of intrinsic defects such as vacancy–interstitial pairs. As discussed before, creation and decay of STEs in dielectrics are at the origin of the intrinsic defects. Consequently, theoretical models that describe femtosecond-laser-induced optical breakdown in perfect dielectrics should implement the evolution of excitons (their formation, self-trapping, and relaxation), in addition to various ionization and delocalized recombination mechanisms.

6.7.5   Femtosecond-laser-induced carrier dynamics

Imaging experiments

One simple method to study femtosecond-laser-induced carrier excitation in (transparent) dielectrics is ultrafast imaging. Experiments (Mao *et al.*, 2003) were performed using a femtosecond time-resolved pump–probe setup to image the electron–hole plasma. A

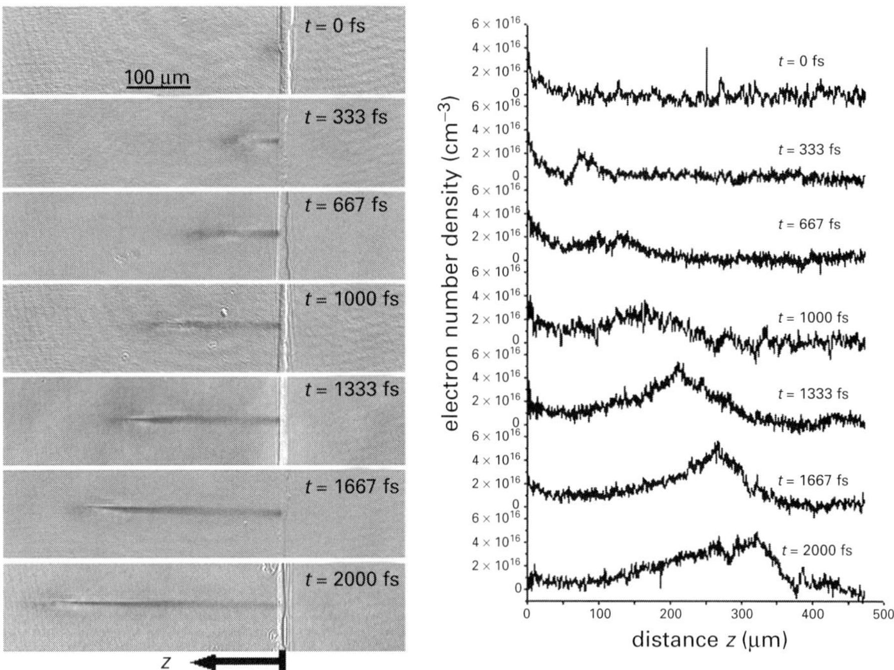

Figure 6.37. (a) Time-resolved images of femtosecond-laser-induced electronic excitation inside a silica glass; (b) evolution of the electron-number-density profile inside a femtosecond-laser-irradiated silica glass. From Mao *et al.* (2003), reproduced with permission by the American Institute of Physics.

high-power femtosecond laser at its fundamental wavelength (800 nm) was used as the pump beam, which has a duration of approximately 100 fs (FWHM). The 800-nm laser beam was focused to a spot size of diameter 50 μm onto a silica-glass sample using a lens with focal length $f = 15$ cm. After a beamsplitter, one arm of the 800-nm output passed an optical delay stage and a KDP crystal, forming a probe beam at 400 nm perpendicular to the excitation-laser pulse. By moving the delay stage, the optical path of the probe beam could be varied, changing the time difference between the pump beam and the probe beam. Time zero was set when the peaks of the ablation laser beam and the probe beam overlapped in time at the sample surface. The resulting shadowgraph images represent the spatial transmittance of the probe pulse during laser irradiation of the sample, corrected for background intensity measured without laser excitation. The electron number density of the laser-induced plasma inside the silica can be estimated from the transmittance at various delay times.

Figure 6.37(a) shows a series of time-resolved images of the electron–hole plasma at the same laser irradiation, $I = 1.3 \times 10^{13}$ W/cm^2. At $t = 0$, only a small dark area appears close to the glass surface, that results from electron excitation by the leading edge of the femtosecond laser pulse. At longer delay times, the plasma filament grows longer, with the darkest section (strongest absorption) moving away from the glass surface into the bulk. From measuring the probe-pulse transmittance I_{p0}/I_{pd} of the time-resolved images, one can estimate the electron number density of the femtosecond-laser-excited plasma inside the silica glass at various delay times.

The electron number density of the laser-induced plasma shown in Figure 6.37(a) is plotted in Figure 6.37(b). At $t = 333$ fs, there is an electron-number-density maximum ($\sim 2 \times 10^{19}$ cm^{-3}) at $z = 80$ μm, which moves into the silica at later times. While the peak value of the electron number density increases with time, it reaches a maximum of approximately 5×10^{19} cm^{-3} at $t = 1333$ fs. This observation is consistent with the fact that a femtosecond laser pulse experiences initial self-focusing inside a dielectric material, followed by defocusing when the laser-induced electron excitation is strong enough to compensate for the laser-induced change in refractive index.

The density of the laser-induced electron–hole plasma as obtained from ultrafast imaging provides only an order-of-magnitude estimate or semi-quantitative information. Frequency-domain interferometry proves to be a powerful technique to elucidate the fundamental processes of femtosecond-laser-induced carrier dynamics in dielectric materials.

Interferometry experiments

While imaging experiments give access to the change of the imaginary part of the refractive index induced by a pump laser pulse, the change in the real part of the refractive index also provides essential information on the dynamics of excited carriers in dielectrics. Interferometry is the natural way to measure this quantity. A very powerful interferometric technique when dealing with broadband light sources is spectral interferometry, which has increasingly been implemented with ultra-short laser pulses for a wide variety of experiments, e.g. for the full temporal characterization of these pulses by the SPIDER technique (Iaconis and Walmsley, 1998; Quéré *et al.*, 2003), and for time-resolved experiments, especially in laser-generated plasmas (Geindre *et al.*, 2001).

The spectral or frequency-domain interferometry technique uses two pulses, separated in time by a delay t_d large compared with their duration and sent in a spectrometer. Provided that the spectral resolution of the spectrometer is much larger than the inverse of the delay, the measured spectrum is, therefore,

$$S(\omega) = 2S_0(\omega)[1 + \cos(\omega t_d + \Phi_{ph})]. \tag{6.36}$$

For the sake of simplicity, the two delayed pulses were assumed to be identical ("twin pulses") to derive this expression, a condition that is actually not required for this technique to apply. $S_0(\omega)$ is the spectral intensity of these pulses, and Φ_{ph} their relative phase. Since t_d is large compared with the spectral width of $S_0(\omega)$, $S(\omega)$ presents fringes, spaced by $2\pi/t_d$. The position of these fringes is determined by the relative phase Φ_{ph} of the two pulses.

Spectral interferometry can be used to probe the temporal dynamics of a system perturbed by a pump pulse. In this case, the first pulse probes the system before the pump pulse, and is thus used as a reference pulse. The second pulse probes the system at a delay t after the pump pulse. The perturbation induced by the pump pulse leads to a change $\Delta\Phi_{ph}(t)$ of the relative phase of the twin pulses. This phase shift $\Delta\Phi_{ph}$ results in a shift of the fringes in the spectrum of the twin pulse. Spectral interferometry

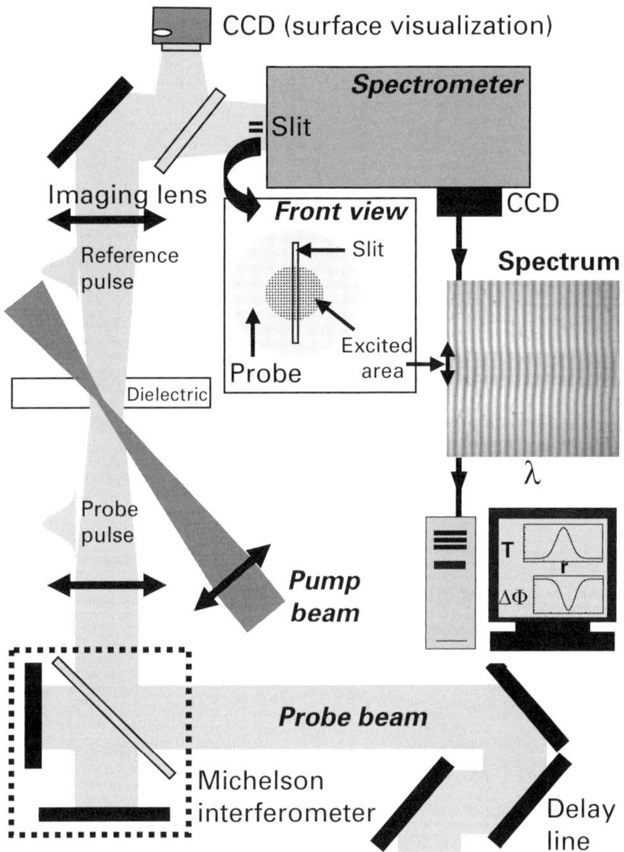

Figure 6.38. The experimental setup for spectral interferometry. The probe-beam waist is larger than the pump-beam waist (typically 10 μm), and the interaction area is imaged on the entrance slit of the spectrometer with a large magnification, so that the spatial profile of phase shift along the slit direction (r) is obtained. From Mao *et al.* (2004), reproduced with permission by Springer-Verlag.

uses this shift to measure $\Delta\Phi$. This technique has been used to probe laser-excited dielectrics (Figure 6.38) (Martin *et al.*, 1997). The twin pulses are transmitted through the dielectric sample. One interferogram is acquired without any pump pulse, as the reference. A second one is measured with an intense pump pulse exciting the dielectric between the reference and the probe pulse. In this configuration, the phase shift $\Delta\Phi_{ph}(t)$ is given by

$$\Delta\Phi_{ph}(t) = (2\pi L/\lambda)\Delta n(t), \tag{6.37}$$

where λ is the probe-beam wavelength, L the length of the probed medium (assumed to be homogeneously excited for simplicity), and $\Delta n(t)$ the instantaneous change in the real part of the refractive index that results from the pump-induced excitation. Note that, by using the contrast of the fringes, spectral interferometry also gives access to the

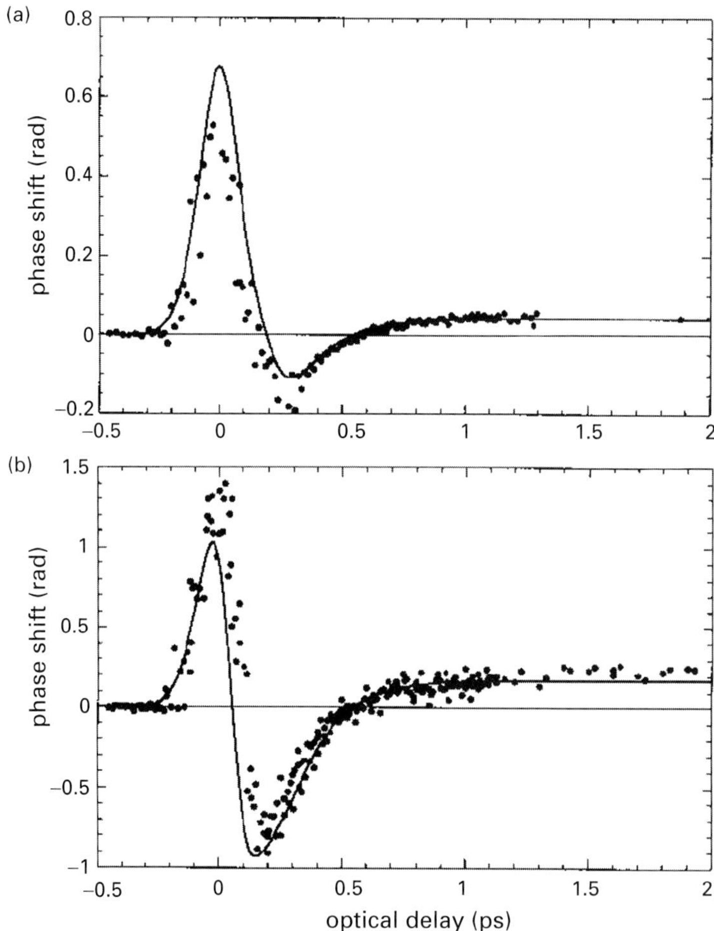

Figure 6.39. Temporal evolution of the phase shift in SiO_2 for two pump-laser intensities. The probe wavelength is 618 nm and the sample temperature 300 K. From Martin *et al.* (1997), reproduced with permission by the American Physical Society.

change in absorption coefficient, i.e. to the change in the imaginary part of the refractive index.

Two types of temporal behavior of $\Delta\Phi_{ph}(t)$ have been observed. In all cases, $\Delta\Phi_{ph}(t)$ is positive for short delays, when the pump and the probe temporally overlap in the dielectric, because of the pump-induced optical Kerr effect. $\Delta\Phi_{ph}(t)$ then becomes rapidly negative; according to the Drude model, this is due to the injection of electrons into the conduction band (see above). In some solids, $\Delta\Phi_{ph}(t)$ remains negative for several tens of picoseconds, whereas in others (e.g. SiO_2 in Figure 6.39), it becomes positive again. In SiO_2, this relaxation occurs with a time constant of 150 fs. It has been demonstrated that this second type of evolution is due to the trapping of most of the excited carriers as self-trapped excitons (Martin *et al.*, 1997). Since STEs correspond to localized states, the change in refractive index is given by the Lorentz model,

$$\Delta n = \frac{N_{\mathrm{STE}}e^2}{2n_0 m \varepsilon_0} \frac{1}{\omega_{\mathrm{tr}}^2 - \omega^2} \tag{6.38}$$

where N_{STE} is the STE density, n_0 the refractive index of the unperturbed solid, ω_{tr} the resonance frequency of the STE's first excited level (~ 6.2 eV in SiO_2), and ω the probe laser's central wavelength. If $\omega < \omega_{\mathrm{tr}}$, as is the case for the SiO_2 data in Figure 6.39, the presence of STEs leads to a positive phase shift.

These measurements have provided important information on the ultrafast dynamics of excited carriers in dielectrics. In diamond, MgO, and Al_2O_3, the negative phase shift was observed to persist for tens of picoseconds. This suggests that no trapping occurs on this time scale, or that the electrons form very shallow traps. Fast formation of STEs has been observed in NaCl, KBr, and SiO_2 (both amorphous and crystalline), leading to carrier lifetimes two orders of magnitude smaller. The difference in carrier dynamics can be qualitatively explained by general considerations about the STE's formation, in terms of lattice elasticity and deformation potential. A fundamental difference was also observed between the trapping kinetics in NaCl and SiO_2: the trapping time was independent of the excitation density in SiO_2, whereas carriers in NaCl trap faster when the excitation density is higher. This can be interpreted as evidence of direct exciton trapping in SiO_2, and of hole trapping followed by electron trapping in NaCl. For intense ultra-short laser interactions with dielectrics (Quéré *et al.*, 2001), the phase shift $\Delta\Phi_\infty$ measured at a sufficiently large delay after the laser pulse gives access to the excitation density N in the solid at the end of the laser pulse. If this density is not too high, $\Delta\Phi_\infty$ is directly proportional to N. At low intensities, $\Delta\Phi_\infty$ is observed to vary as I^6 in SiO_2 and as I^5 in MgO. The exponents of these power laws correspond in both cases to the minimum number of photons that the valence electrons have to absorb to be injected into the conduction band ($6\hbar\omega = 9.42$ eV $> E_{\mathrm{bg}}(SiO_2) \approx 9$ eV and $5\hbar\omega = 7.85$ eV $> E_{\mathrm{bg}}(\mathrm{MgO}) \approx 7.7$ eV). This proves that the dominant excitation process in this intensity range is perturbative multiphoton absorption by valence electrons. The optical breakdown threshold of SiO_2 measured at 800 nm and 60 fs by Stuart *et al.* (1996) falls within this range, suggesting that optical breakdown is not associated with an electronic avalanche. However, this result is in contradiction with the conclusions drawn from breakdown-threshold measurements (Stuart *et al.*, 1996), which suggest that the electronic avalanche should dominate multiphoton absorption even in the femtosecond regime. A model of optical breakdown reconciling the results from these two studies remains to be developed.

At higher intensities, a saturation of $\Delta\Phi_\infty$ is observed relative to these power laws. This saturation occurs because, at high intensity, the pump beam is strongly absorbed due to free-carrier absorption by conduction electrons, and even reflected by the target when the electron density becomes higher than the critical density at the pump frequency. Thus, in this regime, it is only in a thin layer (~ 200 nm) of material that the excitation density keeps increasing with increasing intensity. In this range, the occurrence of an electronic avalanche, due to the strong heating of the conduction electrons, cannot be excluded, although the data can be fitted with a purely multiphoton-injection law. Since the density increases with rising intensity only in a very thin layer, the sensitivity of the technique to these variations might be too low to distinguish between different excitation processes.

Time-resolved absorption

The STE creation by self-trapping of an electron–hole pair can be monitored using transient absorption spectroscopy, which consists of measuring at various delay times after electron–hole injection the sample absorption, and monitoring the appearance of selected absorption bands. In some materials, such as SiO_2, for which the absorption lies in the UV, it is possible only to perform single-wavelength measurements.

In alkali halides, where the absorption bands lie in the visible, it is possible to use as a probing pulse a white-light continuum generated by focusing an intense sub-picosecond laser pulse in, e.g., a water cell. Owing to various nonlinear effects, the spectrum of the laser pulse is broadened and can essentially cover the whole visible range, so a full absorption spectrum can be recorded simultaneously for each laser shot. This method has now been applied to a large number of alkali halides (Williams *et al.*, 1984; Shibata *et al.*, 1994) and has considerably helped, together with the above-mentioned interferometric measurements, to unravel the difficult issue of the so-called "excitonic" mechanisms of point-defect creation in irradiated wide-band-gap dielectrics.

6.7.6   Femtosecond-laser-pulse propagation

There has been a growing interest in femtosecond laser propagation in wide-band-gap dielectrics because the laser intensity can be much higher than the threshold for self-focusing. At such high intensities, the dynamics of femtosecond pulse propagation is considerably more complex, since it may be accompanied by nonlinear phenomena such as pulse splitting in both the space and the time domain. Spatial or temporal splitting of femtosecond laser pulses offers a mechanism for intense femtosecond-laser propagation inside dielectrics without encountering the catastrophic damage caused by self-focusing.

Self-focusing and defocusing

Femtosecond-laser-induced nonlinear self-focusing as well as the related filamentation phenomenon have been investigated for decades, for example, in air (Braun *et al.*, 1995; Brodeur *et al.*, 1997). However, there have been few femtosecond time-resolved studies of laser self-focusing and filamentation inside wide-band-gap dielectrics. Although self-focusing occurs during femtosecond-laser propagation, the density of free electrons at the focus does not increase indefinitely, but reaches a saturation value after about a couple of picoseconds. This phenomenon happens to be the consequence of a self-defocusing process caused by the generation of an electron–hole plasma. The balance between self-focusing due to the nonlinear Kerr effect and defocusing due to plasma formation can lead to self-channeling of the femtosecond laser pulse inside dielectrics (Tzortzakis *et al.*, 2001), if the beam radius is such that ionization comes into play before spatial splitting and material damage occur.

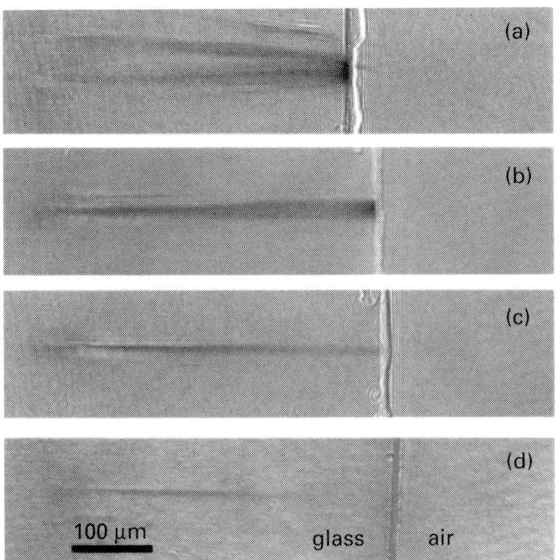

Figure 6.40. The intensity dependence of femtosecond-laser-induced electronic excitation inside a silica glass. From Mao *et al.* (2003), reproduced with permission by the American Institute of Physics.

Spatial splitting

When self-focusing is strong, a single input pulse can break up into several narrow filaments of light. Because the Gaussian spatial profile of the pulse is destroyed by self-focusing, the pulse cannot be focused to a diffraction-limited size. Figure 6.40 shows a series of shadowgraph images inside a silica sample taken at the same delay time (2000 fs) but at different laser intensities I. At $I = 5 \times 10^{12}$ W/cm^2, there is only one filament, a thin, dark stripe that results from the absorption of the probe beam by laser-excited electrons inside the silica glass. At $I = 2.5 \times 10^{13}$ W/cm^2, the primary filament splits into two at a location about 200 μm inside the silica glass. At even higher irradiance (e.g. 10^{14} W/cm^2), filament splitting, as a persistent phenomenon, starts right after the femtosecond-laser pulse has entered the glass sample.

The nonlinear Schrödinger equation, which is the leading-order approximation to the Maxwell equations, has been successful in describing the propagation of intense laser pulses such as self-focusing in nonlinear Kerr media with nonlinear refractive index n_2. Assuming that the laser pulse propagates in the z-direction, the basic nonlinear Schrödinger equation has the form

$$2i\kappa \, \frac{\partial A}{\partial z} + \nabla_\perp^2 A + \frac{2\kappa^2 n_2}{n_0}|A|^2 A = 0, \tag{6.39}$$

where A is the envelope amplitude of the propagating laser electric field, κ is the wavevector, $\kappa = \omega n/c$, and $\nabla_\perp^2$ is the transverse Laplacian operator. The second term represents diffraction, whereas the third term accounts for the contribution due to an intensity-dependent refractive index. If the input laser pulse is cylindrically symmetrical, then, according to the nonlinear Schrödinger equation, the pulse remains cylindrically symmetrical during propagation. Since the interpretation of multiple filamentation should include a mechanism that breaks the cylindrical symmetry, the standard explanation (Bespalov and Talanov, 1966) of multiple filamentation is that breakup of cylindrical symmetry is initiated by small or random inhomogeneity in the input laser pulse, leading to instabilities in the refractive-index profile.

An alternative deterministic explanation of filament splitting was proposed recently (Fibich and Ilan, 2001). It is based on the nonlinear Schrödinger equation with the inclusion of nonlinear perturbation that describes self-focusing in the presence of vectorial (polarization) as well as nonparaxial effects. Assuming that the input laser pulse is linearly polarized in the x-direction, the general form of the modified nonlinear Schrödinger equation can be derived as

$$2i\kappa \frac{\partial A}{\partial z} + \nabla_{\perp}^2 A + \frac{2\kappa^2 n_2}{n_0}|A|^2 A = \xi^2 \operatorname{Im}\left(A, \frac{\partial A}{\partial x}, \frac{\partial^2 A}{\partial x^2} \right) + O(\xi^4), \qquad (6.40)$$

where ξ ($\ll 1$) is the dimensionless parameter defined as the ratio of the laser wavelength to the pulse-spot parameter, $\xi = \lambda/(2\pi r_0)$. The asymmetry in the x and y derivatives of the vectorial perturbation terms in the new equation suggests that the symmetry-breaking mechanism can arise from the vectorial effect for a linearly polarized laser pulse. This vectorially induced symmetry breaking leads to multiple filamentation even when the linearly polarized input laser pulse is cylindrically symmetrical. Predictions made via the above modified nonlinear Schrödinger equation (Fibich and Ilan, 2001) have shown the emergence of two filaments, propagating forward in the z-direction, while moving away from each other along the x-direction. Multiple filamentation resulting from noise in the input beam should vary between experiments and be independent of the direction of initial polarization, while multiple filamentation resulting from vectorial effects should persist with experiments and depend on polarization.

Temporal splitting

In addition to spatial splitting, an intense femtosecond laser pulse may undergo temporal splitting as it propagates inside a dielectric material (Diddams *et al.*, 1998). Associated with temporal splitting is the fact that the spectrum of the femtosecond laser pulse can broaden significantly and eventually evolve into a supercontinuum or white-light generation at high intensities. While temporal splitting was predicted theoretically more than ten years ago, the experimental verification of temporal splitting came about only within the last decade. In one such experiment (Ranka and Gaeta, 1998), a near-Gaussian 78-fs, 795-nm laser pulse was focused to a spot size of 75 μm on the front face of a 3-cm-long silica-glass sample. The temporal behavior of the pulse was characterized by measuring the intensity cross correlation of the transmitted pulse with the initial input pulse. The spectrum of the transmitted beam was taken concurrently using a fiber-coupled spectrometer with resolution 0.3 nm.

The nonlinear Schrödinger equation with the inclusion of material dispersion was applied to predict temporal splitting of femtosecond-laser pulses inside dielectrics (Diddams *et al.*, 1998; Zozulya *et al.*, 1999).

Diddams *et al.* (1998) considered a femtosecond pulse initially focused in both space and time as the result of strong self-focusing and the associated pulse sharpening. As the peak intensity increases, the process of self-phase modulation also increases, which leads to the generation of new frequency components that are red-shifted near the leading edge of the pulse and blue-shifted near the trailing edge. Because of the positive group-velocity dispersion in most dielectrics, the wave trains of the laser pulse at different frequencies were shown to propagate at different speeds, with the red component faster than the blue, hence initiating pulse splitting.

While the nonlinear Schrödinger equation including normal group-velocity dispersion predicts temporal splitting with symmetry, the asymmetrical feature of the splitting pulses was examined using an equation beyond the slowly-varying-envelope approximation (Ranka and Gaeta, 1998; Zozulya *et al.*, 1999). Self-steepening and space–time focusing were found to shift the beam energy into one of the two split pulses formed by group-velocity dispersion. Similarly, multiphoton ionization and plasma formation were incorporated into the modified nonlinear Schrödinger equation. The resulting defocusing and nonlinear absorption of the trailing edge of the pulse tend to push the peak intensity to the leading edge (Gaeta, 2000).

6.8   Application in the micromachining of glass

Much attention has recently been paid to microfabrication of transparent materials by ultra-short laser pulses. When a highly intense ultra-short laser beam is focused inside the bulk of a transparent material, only the localized region in the neighborhood of the focal volume absorbs laser energy by nonlinear optical breakdown, leaving the rest of the target specimen unaffected (Schaffer *et al.*, 2001). Many applications taking advantage of this volumetric absorption have been investigated, including three-dimensional optical storage (Glezer *et al.*, 1996), fabrication of optical waveguides (Davis *et al.*, 1996; Will *et al.*, 2002), three-dimensional structuring by photo-polymerization (Kawata and Sun, 2003; Marcinkevicius *et al.*, 2001), and three-dimensional drilling (Li *et al.*, 2001).

Work on the fabrication of three-dimensional fluidic microchannels in optical glass using a liquid-assisted, femtosecond-laser ablation process was reported by Hwang *et al.* (2004). As shown in Figure 6.41(a), drilling was initiated from the rear surface to preserve a consistent absorption profile of the laser pulse energy and was followed by continuous scanning of the laser beam toward the front surface. Efficient removal of machined debris from fabricated microchannels is a critical issue for achieving holes of uniform diameters and high aspect ratios. Machining in the presence of a liquid assisted the ejection of the debris. The scanning speed of the laser pulses and waiting periods between series of the laser shots were varied to define optimal conditions for removing

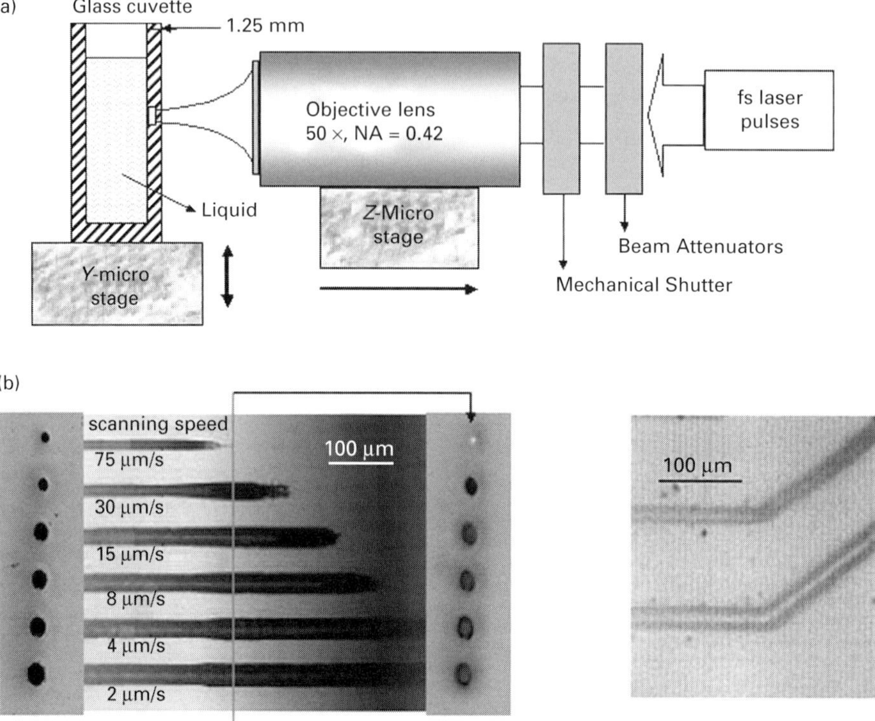

Figure 6.41. (a) A schematic diagram of the liquid-assisted glass-drilling process. (b) Examples of straight and bent channels in glass. From Hwang *et al.* (2004), reproduced with permission from Springer-Verlag.

the ablated debris from the channel. Figure 6.41(b) shows examples of fabrication of straight and bent channels in glass.

References

Agassi, D., 1984, "Phenomenological Model for Picosecond-Pulse Laser Annealing of Semiconductors," *J. Appl. Phys.*, **55**, 4376–4383.

Anisimov, S. I., Kapeliovich, B. L., and Perel'man, T. L., 1974, "Electron Emission from Metal Surfaces Exposed to Ultrashort Laser Pulses," *Sov. Phys. JETP*, **39**, 375–377.

Anisimov, S. I., and Rethfeld, B., 1997, "On the Theory of Ultrashort Laser Pulse Interaction with a Metal," *Proc. SPIE*, **3093**, 192–203.

Arnold, D., and Cartier, E., 1992, "Theory of Laser-Induced Free-Electron Heating and Impact Ionization in Wide-Band-Gap Solids," *Phys. Rev. B*, **46**, 15102–15115.

Bespalov, V. I., and Talanov, V. I., 1966, "Filamentary Structure of Light Beams in Nonlinear Liquids," *JETP Lett.*, **3**, 307.

Bloembergen, N., 1974, "Laser-Induced Electric Breakdown in Solids," *IEEE J. Quant. Electron.*, **QE-10**, 375–386.

Braun, A., Korn, G., Liu, X. *et al.*, 1995, "Self-Channeling of High-Peak-Power Femtosecond Laser Pulses in Air," *Opt. Lett.*, **20**, 73–75.

Brodeur, A., Chien, C. Y., Ilkov, F. A. *et al.*, 1997, "Moving Focus in the Propagation of Ultrashort Laser Pulses in Air," *Opt. Lett.*, **22**, 304–306.

Brorson, S. D., Fujimoto, J. G., and Ippen, E. P., 1987, "Femtosecond Electronic Heat-Transport Dynamics in Thin Gold Films," *Phys. Rev. Lett.*, **59**, 1962–1965.

Brorson, S. D., Kazeroonian, A., Moodera, J. S. *et al.*, 1990, "Femtosecond Room-Temperature Measurement of the Electron–Phonon Coupling Constant l in Metallic Superconductors," *Phys. Rev. Lett.*, **64**, 2172–2175.

Cavalleri, A., Siders, C. W., Brown, F. L. H. *et al.*, 2000, "Anharmonic Lattice Dynamics in Germanium Measured with Ultrafast X-Ray Diffraction," *Phys. Rev. Lett.*, **85**, 586–589.

Cheng, C., and Xu, X., 2005, "Material Decomposition near Critical Temperature during Femtosecond Laser Ablation," *Phys. Rev. B*, **72**, 165415-1–15.

Chin, A. H., Schoenlein, R. W., Glover, T. E. *et al.*, 1999, "Ultrafast Structural Dynamics in InSb Probed by Time-Resolved X-Ray Diffraction," *Phys. Rev. Lett.*, **83**, 336–339.

Choi, T.-Y., and Grigoropoulos, C. P., 2002, "Plasma and Ablation Dynamics in Ultrafast Laser Processing of Crystalline Silicon," *J. Appl. Phys.*, **92**, 4918–4925.

Davis K. M., Miura K., Sugimoto N., and Hirao K., 1996, "Writing Waveguides in Glass with a Femtosecond Laser", *Opt. Lett.*, **21**, 1729–1731.

Day, D., Min Gu, and Smallridge, A., 1999, "Use of Two-photon Excitation for Erasable–Rewritable Three-Dimensional Bit Optical Data Storage in a Photorefractive Polymer," *Opt. Lett.*, **24**, 948–950.

Diddams, S. A., Eaton, H. K., Zozulya, A. A., and Clement, T. S., 1998, "Characterizing the Nonlinear Propagation of Femtosecond Pulses in Bulk Media," *IEEE J. Sel. Topics Quant. Electron.*, **4**, 306–316.

Downer, M., Fork, R., and Shank, C., 1985, "Femtosecond Imaging of Melting and Evaporation at a Photoexcited Silicon Surface," *J. Opt. Soc. Am. B*, **2**, 595–599.

Du, D., Liu, X., Korn, G., Squier, J., and Mourou, G., 1994, "Laser-Induced Breakdown by Impact Ionization in SiO_2 with Pulse Widths from 7 ns to 150 fs," *Appl. Phys. Lett.*, **64**, 3071–3073.

Elsayed-Ali, H. E., Norris, T. B., Pessot, M. A., and Mourou, G. A., 1987, "Time-Resolved Observation of Electron–Phonon Relaxation in Copper," *Phys. Rev. Lett.*, **58**, 1212–1215.

Fann, W. S., Storz, R., Tom, H. W. K., and Bokor, J., 1992a, "Direct Measurement of Nonequilibrium Electron-Energy Distributions in Subpicosecond Laser-Heated Gold Films," *Phys. Rev. Lett.*, **68**, 2834–2837.

1992b, "Electron Thermalization in Gold," *Phys. Rev. B*, **46**, 13 592–13 595.

Fibich, G., and Ilan, B., 2001, "Vectorial and Random Effects in Self-Focusing and in Multiple Filamentation," *Physica D*, **157**, 112–146.

Fischetti, M. V., DiMaria, D. J., Brorson, S. D., Theis, T. N., and Kirtley, J. R., 1985, "Theory of High-Field Electron Transport in Silicon Dioxide," *Phys. Rev. B*, **31**, 8124–8142.

Gaeta, A. L., 2000, "Catastrophic Collapse of Ultrashort Pulses," *Phys. Rev. Lett.*, **84**, 3582–3585.

Geindre, J.-P., Audebert, P., Rebibo, S., and Gauthier, J.-C., 2001, "Single-Shot Spectral Interferometry with Chirped Pulses," *Opt. Lett.*, **26**, 1612–1616.

Gibbon, P., and Forster, E., 1996, "Short-Pulse Laser–Plasma Interactions," *Plasma Phys. Contr. Fusion*, **38**, 769–793.

Glezer, E. N., Milosavljevic, M., Huang, L. *et al.*, 1996, "Three-Dimensional Optical Storage inside Transparent Materials," *Opt. Lett.*, **21**, 2023–2025.

Groeneveld, R. H. M., Sprik, R., Wittebrood, M., and Lagendijk, A., 1992, "Effect of a Nonthermal Electron Distribution on the Electron–Phonon Energy Relaxation Process in Noble Metals," *Phys. Rev. B*, **45**, 5079–5082.

Guizard, S., Martin, P., Petite, G., D'Oliveira, P., and Meynadier, P., 1996, "Time-Resolved Study of Laser-Induced Colour Centres in SiO_2," *J. Phys. – Condens. Matter*, **8**, 1281–1290.

Guo, C., Rodriguez, G., and Taylor, A. J., 2001, "Ultrafast Dynamics of Electron Thermalization in Gold," *Phys. Rev. Lett.*, **86**, 1638–1641.

Haglund, R. F. Jr., and Itoh, N., 1994, "Electronic Processes in Laser Ablation of Semiconductors and Insulators," in *Laser Ablation. Principles and Applications,* edited by J. C. Miller, Berlin, Springer-Verlag, pp. 11–52.

Hwang, D., Choi, T., and Grigoropoulos, C. P., 2004, "Liquid-Assisted Femtosecond Laser Drilling of Straight and Three-Dimensional Microchannels in Glass," *Appl. Phys. A*, **79**, 605–612.

Hwang, D. J., Choi, T.-Y., and Grigoropoulos, C. P., 2006, "Efficiency of Silicon Micromachining by Femtosecond Laser Pulses in Ambient Air," *J. Appl. Phys.*, **99**, 083101–083106.

Iaconis, C., and Walmsley, I. A., 1998, "Spectral Phase Interferometry for Direct Electric-Field Reconstruction of Ultrashort Optical Pulses, *Opt. Lett.*, **23**, 792–796.

Kawata, S., and Sun, H. B., 2003, "Two-Photon Photopolymerization as a Tool for Making Micro-devices", *Appl. Surf. Sci.*, **208**, 153–158.

Kawata, Y., Ueki, H., Hastimoto, Y., and Kawata, S., 1995, Three-Dimensional Optical Memory with a Photorefractive Crystal," *Appl. Opt.*, **34**, 4105–4110.

Kelley, P. L., 1965, "Self-Focusing of Optical Beams," *Phys. Rev. Lett.*, **15**, 1005–1008.

Keldysh, L. V., 1965, "Ionization in Field of a Strong Electromagnetic Wave," *Sov. Phys. JETP*, **20**, 1307.

Kelly, R., and Miotello, A., 1996, "Comments on Explosive Mechanisms of Laser Sputtering," *Appl. Surf. Sci.*, **96–98**, 205–215.

Korte, F., Nolte, S., Chichkov, B. N. *et al.*, 1999, "Far-Field and Near-Field Material Processing with Femtosecond Laser Pulses," *Appl. Phys. A*, **69**, S7–S11.

Larsson, J., Heimann, P. A., Lindenberg, A. M. *et al.*, 1998, "Ultrafast Structural Changes Measured by Time-Resolved X-Ray Diffraction," *Appl. Phys. A*, **66**, 587–591.

Lenzner, M., 1999, "Femtosecond Laser-Induced Damage of Dielectrics," *Int. J. Mod. Phys. B*, **13**, 1559–1578.

Lenzner, M., Krüger, J., Santania, S. *et al.*, 1998, "Femtosecond Optical Breakdown in Dielectrics," *Phys. Rev. Lett.*, **80**, 4076–4079.

Li, M., Menon, S., Nibarger, J. P., and Gibson, G. N., 1999, "Ultrafast Electron Dynamics in Femtosecond Optical Breakdown of Dielectrics," *Phys. Rev. Lett.*, **82**, 2394–2397.

Li, Y., Itoh K., Watanabe, W. *et al.*, 2001, "Three-Dimensional Hole Drilling of Silica Glass from the Rear Surface with Femtosecond Laser Pulses," *Opt. Lett.*, **26**, 1912–1914.

Lindenberg, A. M., Kang, J., Johnson, S. L., 2000, "Time-Resolved X-Ray Diffraction from Coherent Phonons during a Laser-Induced Phase Transition," *Phys. Rev. Lett.*, **84**, 111–116.

Liu, X., Du, D., and Mourou, G., 1997, "Laser Ablation and Micromachining with Ultrashort Laser Pulses," *IEEE J. Quant. Electron.*, **33**, 1706–1716.

Mao, X. L., Mao, S. S., and Russo, R. E., 2003, "Imaging Femtosecond Laser-Induced Electronic Excitation in Glass," *Appl. Phys. Lett.*, **82**, 697–699.

Mao, S.S., Quéré, F., Guizard, S. *et al.*, 2004, "Dynamics of Femtosecond Laser Interactions with Dielectrics," *Appl. Phys. A*, **79**, 1695–1709.

Marcinkevicius, A., Juodkazis, S., Watanabe, M. *et al.*, 2001, "Femtosecond Laser-Assisted Three-Dimensional Microfabrication in Silica," *Opt. Lett.*, **26**, 277–279.

Martin, P., Guizard, S., Daguzan, Ph. *et al.*, 1997, "Subpicosecond Study of Carrier Trapping Dynamics in Wide-Band-Gap Crystals," *Phys. Rev. B*, **55**, 5799–5810.

Milchberg, H. M., Freeman, R. R., and Davey, S. C., 1988, "Behavior of a Simple Metal under Utrashort Pulse High Intensity Laser Illumination," *Proc. SPIE*, **913**, 159–163.

Momma, C., Nolte, S., Kamlage, G., von Alvensleben, F., and Tunnermann, A., 1998, "Beam Delivery of Femtosecond Laser Radiation by Diffractive Optical Elements," *Appl. Phys. A*, **67**, 517–520.

Perry, M. D., Stuart, B. C., Banks, P. S. *et al.*, 1999, "Ultrashort-Pulse Laser Machining of Dielectric Materials," *J. Appl. Phys.*, **85**, 6803–6810.

Petite, G., Guizard, S., Martin, P., and Quéré F., 1999, "Comment on 'Ultrafast Electron Dynamics in Femtosecond Optical Breakdown of Dielectrics,'" *Phys. Rev. Lett.*, **83**, 5182.

Pronko, P., Dutta, S., Squier, J. *et al.*, 1995, "Machining of Sub-micron Holes using a Femtosecond Laser at 800 nm," *Opt. Commun.*, **114**, 106–110.

Pronko, P. P., VanRompay, P.A., Horvath, C. *et al.*, 1998, "Avalanche Ionization and Dielectric Breakdown in Silicon with Ultrafast Laser Pulses," *Phys. Rev. B*, **58**, 2387–239.

Qiu, T. Q., and Tien, C.-L., 1993, "Heat Transfer Mechanisms During Short-Pulse Laser Heating of Metals," *J. Heat Transfer*, **115**, 835–841.

Quéré, F., Guizard, S., and Martin, P., 2001, "Time-Resolved Study of Laser-Breakdown in Dielectrics," *Europhys. Lett.*, **56**, 138–144.

Quéré, F., Itatani, J., Yudin, G. L., and Corkum, R. B., 2003, "Attosecond Spectral Shearing Interferometry, *Phys. Rev. Lett.*, **90**, 073902-1–6.

Ranka, J. K., and Gaeta, A. L., 1998, "Breakdown of the Slowly Varying Envelope Approximation in the Self-Focusing of Ultrashort Pulses," *Opt. Lett.*, **23**, 534–536.

Rethfeld, B., Kaiser, A., Vicanek, M., and Simon, G., 1999, "Femtosecond Laser-Induced Heating of Electron Gas in Aluminum," *Appl. Phys. A*, **69**, 109–112.

2002, "Ultrafast Dynamics of Nonequilibrium Electrons in Metals under Femtosecond Laser Irradiation," *Phys. Rev. B*, **65**, 214303-11.

Rose-Petruck, C., Jimenez, R., Guo, T. *et al.*, 1999, "Picosecond–Milliångström Lattice Dynamics Measured by Ultrafast X-Ray Diffraction," *Nature*, **398**, 310–312.

Schaffer, C. B., Brodeur, A., Garcia, J. F., and Mazur, E., 2001, "Micromachining Bulk Glass by Use of Femtosecond Laser Pulses with Nanojoule Energy," *Opt. Lett.*, **26**, 93–95.

Seideman, T., Ivanov M. Yu, and Corkum, P. B., 1995, "Role of Electron Localization in Intense-Field Molecular Ionization," *Phys. Rev. Lett.*, **75**, 2819–2822.

Shah, J., 1996, *Ultrafast Spectroscopy of Semiconductors and Semiconductor Nanostructures*, Berlin, Springer-Verlag.

Shank, C. V., Yen, R., and Hirlimann, C., 1983, "Time-Resolved Reflectivity Measurements of Femtosecond-Optical-Pulse-Induced Phase Transitions in Silicon," *Phys. Rev. Lett.*, **50**, 454–457.

Shen, Y. R., 1984, *The Principles of Nonlinear Optics*, New York, Wiley.

Shibata, T., Iwai, S., Tokisaki, T. *et al.*, 1994, "Femtosecond Spectroscopic Studies of the Lattice-Relaxation Initiated by Interacting Electron–Hole Pairs under Relaxation in Alkali-Halides," *Phys. Rev. B*, **49**, 13 255–13 258.

Siders, C. W., Cavalleri, A., Sokolowski-Tinten, K. *et al.*, 1999, "Detection of Nonthermal Melting by Ultrafast X-Ray Diffraction," *Science*, **286**, 1340–1342.

Silvestrelli, P. L., Alavi, A., Parrinello, M., and Frenkel, D., 1996, "*Ab Initio* Molecular Dynamics Simulation of Laser Melting of Silicon," *Phys. Rev. Lett.*, **77**, 3149–3152.

Skripov, V. P., and Skripov, A. V., 1979, "Spinodal Decomposition (Phase-Transition via Unstable States)," *Usp. Fiz. Nauk*, **128**, 193–231.

Sokolowski-Tinten, K., Bialkowski, J., and von der Linde, D., 1995, "Ultrafast Laser-Induced Order–Disorder Transitions in Semiconductors," *Phys. Rev. B*, **51**, 14 186–14 198.

Sokolowski-Tinten, K., Cavalleri, A., and von der Linde, D., 1999, "Single-Pulse Time- and Fluence-Resolved Optical Measurements at Femtosecond Excited Surfaces," *Appl. Phys. A*, **69**, 577–579.

Sokolowski-Tinten, K., and von der Linde, D., 2000, "Generation of Dense Electron–Hole Plasmas in Silicon," *Phys. Rev. B*, **61**, 2643–2650.

Song, K. H., and Xu, X., 1998, "Explosive Phase Transformation in Excimer Laser Ablation," *Appl. Surf. Sci.*, **127**, 111–116.

Song, K. S., and Williams, R. T., 1993, *Self-Trapped Excitons*, Berlin, Springer-Verlag.

Stampfli, P., and Bennemann, K. H., 1990, "Theory for the Instability of the Diamond Structure of Si, Ge, and C Induced by a Dense Electron–Hole Plasma," *Phys. Rev. B*, **42**, 7163–7173.

1992, "Dynamical Theory of the Laser-Induced Lattice Instability of Silicon," *Phys. Rev. B*, **46**, 10 686–10 692.

Stuart, B. C., Feit, M. D., Herman, S. *et al.*, 1995, "Laser-Induced Damage in Dielectrics with Nanosecond to Subpicosecond Pulses," *Phys. Rev. Lett.*, **74**, 2248–2251.

1996, "Nanosecond-to-Femtosecond Laser-Induced Breakdown in Dielectrics," *Phys. Rev. B*, **53**, 1749–1761.

Sun, H.-B., Xu, Y., Juodkazis, S. *et al.*, 2001, "Arbitrary-Lattice Photonic Crystals Created by Multiphoton Microfabrication," *Opt. Lett.*, **26**, 325–327.

Sundaram, S. K., and Mazur, E., 2002, "Inducing and Probing Non-thermal Transitions in Semiconductors using Femtosecond Laser Pulses," *Nature Mater.*, **1**, 217–224.

Thoma, E. D., Yochum, H. M., and William, R. T., 1997, "Subpicosecond Spectroscopy of Hole and Exciton Self-Trapping in Alkali-Halide Crystals," *Phys. Rev. B*, **56**, 8001–8011.

Tien, An-Chun, Backus, S., Kapteyn, H., Murnane M., and Mourou, G., 1999, "Short-Pulse Laser Damage in Transparent Materials as a Function of Pulse Duration," *Phys. Rev. Lett.*, **82**, 3883–3886.

Toyozawa, Y., 1980, *Relaxation of Elementary Excitations*, edited by R. Kubo and E. Hanamura, Berlin, Springer-Verlag, pp. 3–18.

Trukhin, A. N., 1992, "Excitons in SiO_2: A Review," *J. Non-Cryst. Solids*, **149**, 32–45.

Tzortzakis, S., Sudrie, L., Franco, M. *et al.*, 2001, Self-Guided Propagation of Ultrashort IR Laser Pulses in Fused Silica," *Phys. Rev. Lett.*, **87**, 213902/1–4

Ueta, M., Kanzaki, H., Kobayashi, K., Toyozawa, Y., and Hanamura, E., eds., 1986, *Excitonic Processes in Solids*, Berlin, Springer-Verlag.

Van Driel, H. M., 1987, "Kinetics of High-Density Plasmas Generated in Si by 1.06- and 0.53-μm Picosecond Laser Pulses," *Phys. Rev. B*, **35**, 8166–8176.

Van Stryland, E. W., Vanherzeele, H., Woodall, M. A. *et al.*, 1985, "Two Photon Absorption, Nonlinear Refraction, and Optical Limiting in Semiconductors," *Opt. Eng.*, **24**, 613–623.

Van Vechten, J., Tsu, R., and Saris, F., 1979, "Nonthermal Pulsed Laser Annealing of Si, Plasma Annealing," *Phys. Lett. A*, **74**, 422–426.

Vasil'ev, A. N., Fang, Y., and Mikhailin, V. V. 1999, "Impact Production of Secondary Electronic Excitations in Insulators: Multiple-Parabolic-Branch Band Model," *Phys. Rev. B*, **60**, 5340–5347.

von der Linde, D., Sokolowski-Tinten, K., Blome, C. *et al.*, 2001, "Generation and Application of Ultrashort X-Ray Pulses," *Laser Part. Beams*, **19**, 15–22.

Will, M., Nolte, S., Chichkov B. N., and Tünnermann A., 2002, "Optical Properties of Waveguides Fabricated in Fused Silica by Femtosecond Laser Pulses," *Appl. Opt.*, **41**, 4360–4364.

Williams, R. T., Craig, B. B., and Faust, W. L., 1984, "F-Center Formation in NaCl – Picosecond Spectroscopic Evidence for Halogen Diffusion on the Lowest Excitonic Potential Surface," *Phys. Rev. Lett.*, **52**, 1709–1712.

Wu, M., 1997, "Micromachining for Optical and Optoelectronic Systems," *Proc. IEEE*, **85**, 1833–1856.

Xu, X., 2001, "Heat Transfer and Phase Change during High Power Pulsed Laser Ablation of Metal," in *Annual Review of Heat Transfer*, edited by C.-L. Tien, V. Prasad, and F. P. Incropera, **12**, New York, Begell House, 79.

Ye, M., and Grigoropoulos, C. P., 2001, "Time-of-Flight and Emission Spectroscopy Study of Femtosecond Laser Ablation of Titanium," *J. Appl. Phys.*, **89**, 5183–5190.

Yoo, K. M. Zhao, X. M., Siddique, M. *et al.*, 1990, "Femtosecond Thermal Modulation Measurements of Electron–Phonon Relaxation in Niobium," *Appl. Phys. Lett.*, **56**, 1908–1910.

Zewail, A. H., 1994, *Femtochemistry: Ultrafast Dynamics of the Chemical Bond*, Singapore, World Scientific.

Ziman, J. M., 1964, *Principles of the Theory of Solids*, Cambridge, Cambridge University Press.

Zozulya, A. A., Diddams, S. A., Van Engen, A. G., and Clement, T. S., 1999, "Propagation Dynamics of Intense Femtosecond Pulses: Multiple Splittings, Coalescence, and Continuum Generation," *Phys. Rev. Lett.*, **82**, 1430–1433.

7 Laser processing of thin semiconductor films

7.1 Modeling of energy absorption and heat transfer in pulsed-laser irradiation of thin semitransparent films

During transient heating of semitransparent materials at the nanosecond scale, the thermal gradients across the heat-affected zone are accompanied by changes in the complex refractive index of the material. These changes, coupled with wave interference, modify the energy absorption, and thus the temperature field, in the target material. These effects are taken into account in a rigorous manner using thin-film optics theory as outlined in Chapter 1. Consider for example a silicon layer illuminated by a laser beam of nanosecond pulse duration. The energy absorption in the semiconductor material depends upon the temperature in the film. For the time scales examined, the temperature penetration in the structure is small, so the bottom substrate surface remains at the ambient temperature, T_∞. On the other hand, losses to the ambient from the irradiated surface are negligible compared with the incident laser energy density. Initially the structure is isothermal, at the ambient temperature. The temperature field in the semiconductor film induces changes in the refractive index of the material. The semiconductor film is thus treated as a stratified multilayer structure, composed of N layers of varying complex refractive index.

The absorption in the mth layer is evaluated thus:

$$Q_{\text{ab},m} = Q_{\text{las}}(t)\frac{\mathrm{d}S_m}{\mathrm{d}z}, \tag{7.1}$$

where S is the local magnitude of the Poynting vector.

The heat transfer in the jth layer is

$$\rho_j(T_j)C_{p,j}(T_j)\frac{\partial T_j}{\partial t} = \frac{\partial}{\partial x}\left(k_j(T_j)\frac{\partial T_j}{\partial x}\right) + \frac{\partial}{\partial y}\left(k_j(T_j)\frac{\partial T_j}{\partial y}\right)$$

$$+ \frac{\partial}{\partial z}\left(k_j(T_j)\frac{\partial T_j}{\partial z}\right) + Q_{\text{ab}}(x, y, z, t, T_j). \tag{7.2}$$

The temperature field distribution can therefore be determined. It is mentioned that a direct method for the computation of the magnitude of the Poynting vector was presented by Mansuripur *et al.* (1982).

Transient optical transmission and reflection measurements have been reported for the investigation of the irradiation of c-Si on sapphire structures on picosecond (Lompré *et al.*, 1983) and nanosecond (Jellison *et al.*, 1986) time scales. The development of the transient temperature field during heating of a 0.2-μm-thick amorphous silicon (a-Si) film deposited on a fused quartz substrate, by pulsed KrF excimer laser ($\lambda = 0.248$ μm) irradiation was studied (Park *et al.*, 1992). Static reflectivity and transmissivity measurements were used to obtain the thin-film optical properties at the probing diode-laser wavelength ($\lambda = 0.752$ μm) at elevated temperatures. Experimental *in situ*, transient, optical transmission data are compared with conductive-heat-transfer modeling results. The a-Si layer is a strong absorber for the excimer laser light. The energy absorption is thus given by the simple exponential decay

$$Q_{ab}(z) = (1 - R)Q_{pk}e^{-\gamma z}. \tag{7.3}$$

In the above expression R is the normal-incidence reflectivity and γ is the absorption coefficient for the excimer laser light. These properties are virtually temperature-independent, but the variation with temperature of the complex refractive index of the a-Si film across the thin film thickness for the probing laser light is taken into account. The a-Si film is considered stratified in the z-direction in the manner described previously. The comparison between experiment and model for the laser-beam fluences $F = $ 19.62 and 31.6 mJ/cm^2 is shown in Figures 7.1(a) and (b). The calculated peak temperature for the fluence, $F = 31.6$ mJ/cm^2, is approximately 650 K. It is recalled that the thin-film optical properties have been measured up to this temperature range. At higher fluences, the agreement is not as close (Figures 7.1(c) and (d)). The effects of the film's thermal diffusivity and of the laser pulse shape were found to be relatively unimportant, but the transmissivity probe is quite sensitive to variations of the film thickness.

7.2   Continuous-wave (CW) laser annealing

Crystalline semiconductor films on insulating substrates can be advantageous for the fabrication of high-speed electronic devices. The recrystallization can be effected by radiative sources such as lasers (Celler, 1983), graphite-strip heaters (Fan *et al.*, 1983), incoherent lamps (Knapp and Picreaux, 1983), and electron beams. The basic process is zone recrystallization at the microscopic level, resulting in an increase of the crystal grain size and potential improvement of the electrical-transport properties.

Single-crystalline, thin-film material has been produced in insulated, lithographically patterned areas, usually of size 20 μm by 100 μm. Electronic devices fabricated using the material recrystallized by this method have good performance and reliability. However, consistent control of the crystal-growth orientation has been difficult and the strong tendency toward deep supercooling and rapid solidification may cause numerous defects. Stultz and Gibbons (1981), and Kawamura *et al.* (1982) reported production of single-crystalline thin silicon strips up to 50 μm wide and of length several centimeters by shaping the heating source profile to obtain concave solidification fronts. It is therefore evident that controlled thermal-gradient distribution across the solidification boundary is essential for oriented crystal growth.

Few experimental temperature measurements have been reported for processes involving radiative phenomena at high temperatures. In welding, for example, temperature

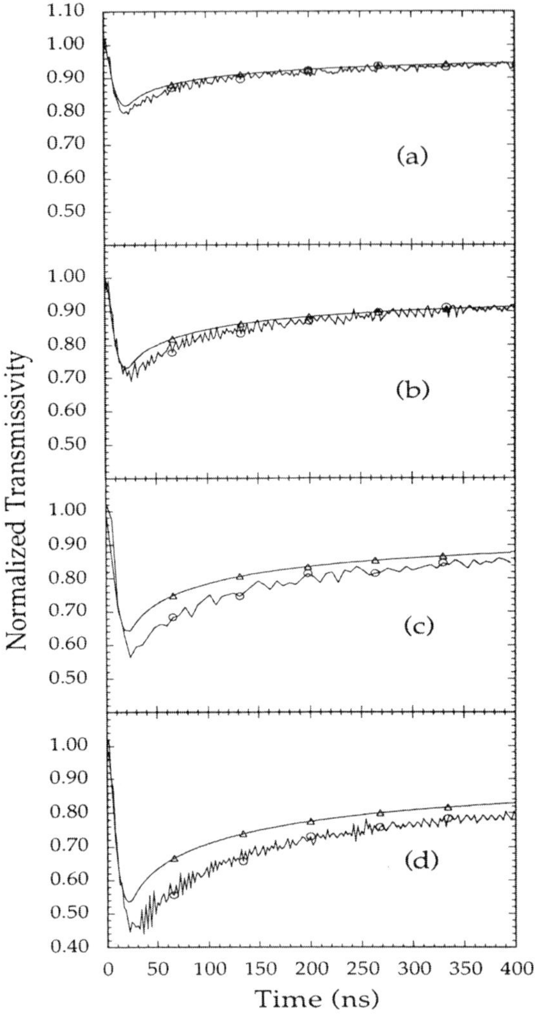

Figure 7.1. A comparison between the numerical prediction (smooth solid line) and the experimental transient transmissivity measurement (noisy signal) for a 0.2-μm-thick amorphous silicon layer, irradiated with a KrF excimer laser ($\lambda = 0.248$ μm). Results are shown for laser-beam fluences F of (a) 19.62 mJ/cm^2, (b) 31.6 mJ/cm^2, (c) 46.6 mJ/cm^2, and (d) 67.4 mJ/cm^2. The pulse length is $t_1 = 26$ ns. From Park *et al.* (1992), reproduced with permission from the American Institute of Physics.

measurements have been given by Kraus (1987). As pointed out by Dewitt and Rondeau (1989), the development of non-invasive temperature-measurement techniques depends on accurate knowledge of the radiative properties of the material. This is certainly true for thin-film laser annealing, where the length scales involved are typically microscopic. Sedwick (1981) measured the maximum temperature during CW-laser annealing of silicon film by using a modified optical pyrometer. The range of these measurements was 1300–1680 K, with an estimated accuracy of a few tens of degrees. Lemons and Bosch

(1982) measured the power spectra of the light emitted from molten silicon spots. These emissive power spectra were then correlated to temperature. This type of analysis is limited by the measurement resolution and by the insufficiency of our knowledge of the spectral dependence of the emissivity of silicon at high temperatures. Kodas *et al.* (1987) used micrometer-sized thin-film thermocouples made of intersecting nickel and gold lines embedded in a $Si–SiO_2$ structure to obtain surface temperatures during heating with a focused laser beam. It was found that the high thermal conductivity of the thermocouples provides heat-flow paths that greatly alter the induced temperature field.

Time-resolved reflectivity measurements provide a useful non-invasive diagnostic tool in laser materials processing. A mechanically scanned probing laser beam was employed (Grigoropoulos *et al.*, 1991b) to obtain spatially resolved reflectivity measurements in CW laser annealing of thin silicon films. Transient normal-incidence reflectivity measurements were also made (Grigoropoulos *et al.*, 1993). The important parameters in laser annealing are (1) the shape of the laser-beam irradiance distribution, (2) the laser beam's total power, and (3) the material translation speed. Several analytical models based on Green-function techniques have been used to predict the temperature distribution in CW-laser-annealed silicon layers at temperatures below the melting threshold. In an early study, the temperature field induced by a heat source moving over a semi-infinite silicon layer was given by Nissim *et al.* (1980). This analysis was extended by Burgener and Reedy (1982), who constructed an analytical heat-transfer model for thin-film CW annealing. Temperature distributions in multilayered structures subjected to pulsed irradiation by scanning laser sources have been calculated numerically by Mansuripur and Connell (1982), and using a Fourier-transform method by Anderson (1988). Kant and Deckert (1991) constructed an analytical model to calculate transient temperature distributions in multilayered optical disks used in magneto-optical recording. In these devices, the magnetic medium is heated by the laser source beyond the Curie point, a temperature at which the magnetization of the medium is lost. The Laplace transform in the time domain and the Fourier transform in the spatial domain were employed to reduce the transient heat-transfer equation to an ordinary differential equation. A straightforward numerical treatment entails the enthalpy formulation (e.g. Miaoulis and Mikic, 1986; Grigoropoulos *et al.*, 1986). The computational predictions of the temperature field can be converted to the surface reflectivity response to a He–Ne probing laser beam (Grigoropoulos *et al.*, 1993). As a general remark, all computational models developed to date assume thermal equilibrium conditions, and employ bulk thermal and radiative properties. Whereas exact knowledge of the thin-film diffusivity is relatively unimportant for the CW laser annealing process, which has a characteristic time scale in the millisecond range, the heat transfer is very sensitive to the thin-film radiative properties and the corresponding energy absorption.

7.3   Inhomogeneous semiconductor-film melting

An interesting phenomenon in laser melting of thin silicon films is the formation of microscopic patterns of coexisting solid and liquid regions. This phase coexistence is due to the optical response of the semiconductor material to the incident laser-light irradiation. Experimental evidence has shown that the partial-melting degree of order, shape and structure, depend on the intensity distribution, wavelength, and polarization of the laser beam. The abrupt increase in the reflectivity of silicon upon melting may drive a thermal instability that will be examined first. When the wavelength of the incident laser light is of the order of the size of the structures formed, the radiant energy absorption is dominated by wave-optics effects, as will be discussed next.

Lemons and Bosch (1982) were the first to report the coexistence of solid and liquid phases in thin silicon films irradiated with CW CO_2 and Ar^+ laser light. Crystallographic studies by Biegelsen *et al.* (1984) showed that the initial solid crystallite inclusions in the liquid phase have (100) texture, thus acting as seeds for the subsequent epitaxial growth of the semiconductor film on the amorphous substrate. The origin of partial melting using Ar^+-laser light sources ($\lambda = 0.5145$ μm) is traced in Figures (7.2(a)–(c)) (Grigoropoulos *et al.*, 1991a). As the incident laser-beam intensity is decreased by expanding the laser beam, a slight perturbation of the solid–liquid phase boundary is observed (Figure 7.2(a)). This perturbation develops into capillarity-limited dendritic penetration of solid crystallites into the molten zone (Figure 7.2(b)) and thereafter to a partially molten region (Figure 7.2(c)). The observed crystallite size and orientation did not seem to correlate with the wavelength of the incident laser light in these experiments.

Impurity segregation could in principle cause constitutional supercooling, leading to melting instabilities. Hawkins and Biegelsen (1983) attributed the film breakup to the differential increase in reflectivity upon melting. The radiant energy absorbed by a silicon layer when melted with a light source operating in the visible wavelength drops by a factor of two, compared with the energy that is absorbed by the solid layer just below the melting temperature. This significant reduction generates phase-change instabilities whose origin can be predicted by linear stability analysis as Jackson and Kurtze (1985) showed in the case of infinitely extensive semiconductor layers melted with sources of uniform irradiance distribution. Assuming that the partially melted silicon layer forms an array of periodic, aligned, and alternating solid and liquid stripes, steady-state profiles in the silicon layer were obtained. These temperature profiles indicate supercooling in the liquid and superheating in the solid material. Following the linear stability analysis of Mullins and Sekerka (1964), a small-amplitude perturbation was allowed at the phase-change interface. It was found that the stabilizing effect of the crystal–melt surface tension tends to suppress instabilities. The basic physical mechanism implies that, if a region of instability is encountered, the growth of the perturbations creates further subdivision until the stripe spacing is small enough. In addition, it was found that the spacing is relatively insensitive to the laser-beam power. As the power input into the sample is increased, the planar solid–liquid interfaces within the two-phase mixture may become morphologically unstable. At a critical value of the heat input, the periodic lamellar configuration breaks down and a complex array of cellular or periodic patterns appears. Grigoropoulos *et al.* (1987a, 1987b) investigated the stability of silicon phase boundaries melted by scanning laser light sources. It was found that the system stability is the combined result of competing effects. The differential absorptivity of the solid and liquid phases in the vicinity of the phase boundaries gives a stabilizing effect

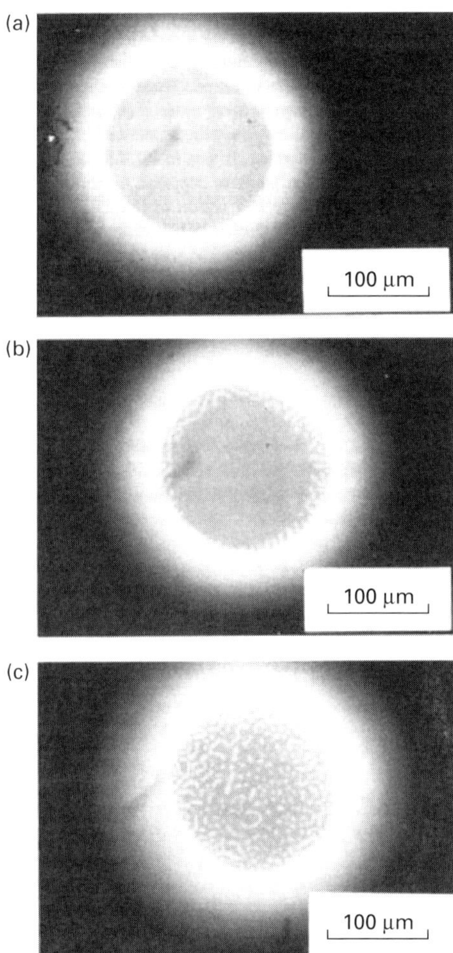

Figure 7.2. (a) A micrograph of a thin silicon layer during melting, with the laser-beam power $P_T = 1.8$ W, the silicon layer not moving, and the $1/e$ irradiance radius $w_{\lambda b} = 78$ μm. (b) A micrograph of a thin silicon layer during melting, with the laser-beam power $P_T = 1.8$ W, the silicon layer not moving, and the $1/e$ irradiance radius $w_{\lambda b} = 84$ μm. (c) A micrograph of a thin silicon layer during melting, with the laser-beam power $P_T = 1.8$ W, the silicon layer not moving, and the $1/e$ irradiance radius $w_{\lambda b} = 90$ μm. From Grigoropoulos *et al.* (1991a), reproduced with permission from the American Society of Mechanical Engineers.

by obstructing the growth of long-wavelength disturbances. The increase in surface reflectivity upon melting generates unstable basic-state temperature gradients that yield instability for a region of intermediate wavelengths. Finally, the stabilizing effect of surface tension dominates the short-wavelength disturbance behavior.

A purely thermal analysis is not capable of explaining the preferential orientation of laser-induced periodic structures in a direction perpendicular to the polarization of the incident laser light, having a pattern spacing in scale with the laser wavelength. Indeed, Ehlrich *et al.* (1983) observed ripple formation in IR ($\lambda = 1.064$ μm) and visible, frequency-doubled ($\lambda = 0.532$ μm) Nd : YAG laser-irradiated Si, Ge, and GaAs. The transient ripple formation was shown to arise from amplified surface scattering. Surface periodic structures were also observed by Fauchet and Siegman (1982) in Nd : YAG ($\lambda = 0.532$ μm, pulse duration of 80 ps) laser annealing of crystalline and ion-implanted Si and GaAs. Nemanich *et al.* (1983) demonstrated that phase coexistence in silicon films irradiated with CW CO_2 laser light can appear in the form of alternating solid and liquid stripes. These organized patterns were aligned perpendicularly to the polarization of the incident laser light, and had a constant periodicity, equal – within the margin of experimental error – to the laser light wavelength, $\lambda = 10.6$ μm. Using a linear stability analysis, Sipe *et al.* (1983) showed that shallow surface melt layers cannot sustain electromagnetic-field perturbations, and breakup has to occur through the thickness of the silicon layer. Preston *et al.* (1986a) studied the inhomogeneous phase patterns in silicon films irradiated with CW CO_2 laser light. They argued that the formation of aligned and alternating patterns could be explained by the interference shielding effects caused by the difference between the optical properties of the solid and liquid silicon phases. In a related study, Young *et al.* (1983) reported detailed experimental investigations on the periodic damage structures that were produced on solid, bulk Ge, Si, Al, and brass surfaces when they were irradiated by pulsed laser beams of wavelength $\lambda = 1.06$ and 0.532 μm. These results were compared with the theory developed by Sipe *et al.* (1983). It was verified that both the laser-induced and the initial sample roughness play a major role in the inhomogeneous energy deposition and subsequent evolution of the laser-melting formations.

Preston *et al.* (1987) showed that the response of the surface to normally incident radiation can be approximated by a uniform polarization within the solid regions and current distributions at the vacuum–melt and melt–substrate interfaces. The energy deposition in the molten regions was found to be higher than that in the solid regions. The details of this theoretical model, which is applicable to both periodic and nonperiodic microstructures with length scales comparable to, or less than, the wavelength of the incident laser light, were described by Preston *et al.* (1989). It was further shown by Preston *et al.* (1986b) that the ordered and disordered molten structures reflect different levels of balance between the absorption of laser energy by the semiconductor material and the heat flow from the illuminated region. Dworshack *et al.* (1990) examined effects of the angle of incidence of the laser beam, reporting observations and analytical studies of partial-melting morphologies that are formed in thin silicon films melted by TE- and TM-polarized CO_2 laser light.

Radiatively induced partial melting involves dendritic-growth phenomena that are driven and controlled by the coupling of electromagnetic radiation with coexisting phases. Little known, yet important, physical quantities, such as surface free energies and contact angles in bounded thin-film systems, can be studied at the microscopic level. From a theoretical point of view, it is a challenge to explain the observed selection of melting pattern and crystalline growth using the tools of nonlinear stability analysis. The study of scattering and diffraction of electromagnetic radiation caused by rough film surfaces and composite film structures is also relevant to this research.

7.4 Nanosecond-laser-induced temperature fields in melting and resolidification of silicon thin films

Excimer-laser crystallization is an efficient technology for obtaining high-performance poly-Si TFTs for advanced flat-panel-display applications. In order to improve both the device performance and uniformity, high-quality poly-Si films with controlled grain size and location are required. To accomplish this objective, several methods (Im and Sposili, 1996; Sposili and Im, 1996; Kim and Im, 1996b; Aichmayr *et al.*, 1999; Ishihara and Matsumura, 1997) utilizing spatially selective melting and lateral temperature modulation have been devised. A melt-mediated transformation scenario (Im *et al.*, 1993; Wood and Geist, 1986a, 1986b; Wood *et al.*, 1996) suggesting that the recrystallized Si morphology is determined by several complex phase transformations has been proposed. Optical diagnostic methods are appropriate for non-intrusively monitoring the melting and recrystallization phenomena. Since the optical properties depend on temperature and phase state, the reflectivity and transmissivity are good probing indicators of the laser annealing process. Analysis of time-resolved reflectivity data during the laser heating of silicon and germanium was used to determine the onset of melting and the melting duration (Stiffler and Thompson, 1988; Jellison *et al.*, 1986). For understanding the solidification mechanism, it is crucial to quantify the transient temperature field.

7.4.1 Explosive recrystallization

A variety of interesting recrystallization phenomena develops during the pulsed-laser annealing of a-Si samples. At low energy densities, fine-grained polycrystalline silicon is observed. Thompson *et al.* (1984) subjected a-Si films on sapphire substrates to ruby-laser pulses ($\lambda = 694$ nm) and measured simultaneously electrical-conductance transients that are indicators of the melt depth and the reflectance response to Ar^+-laser ($\lambda = 488$ nm) probing. The reflectivity traces provided evidence of melt development inside the sample, while cross-sectional transmission electron microscopy (TEM) revealed formation of a coarse-grained poly-Si layer over a fine-grained layer. Thompson *et al.* (1984) explained the experimental results by suggesting the occurrence of explosive crystallization (EC) inside the sample. At low energy densities, the laser energy melts only a thin primary liquid layer near the surface. As this liquid solidifies as poly-Si, the latent heat released from the liquid raises the temperature of the resolidified poly-Si above the melting point for amorphous silicon (presumed to be hundreds of kelvins below the equilibrium melting point for crystalline silicon) and the underlying a-Si begins to melt. This thin liquid layer is severely undercooled and can only resolidify to fine-grained material. The velocity of the explosive melt front was estimated to lie between 10 and 20 m/s.

Wood and Geist (1986a, 1986b) constructed a numerical model including undercooling, interface kinetics, and nucleation in heat-flow calculations to interpret the experimental results outlined above. The origin of the EC process was examined by Murakami *et al.* (1987). Their samples were 600-nm c-Si layers on sapphire, amorphized by implantation with high-energy Zn^+ ions and subjected to ruby-laser pulses.

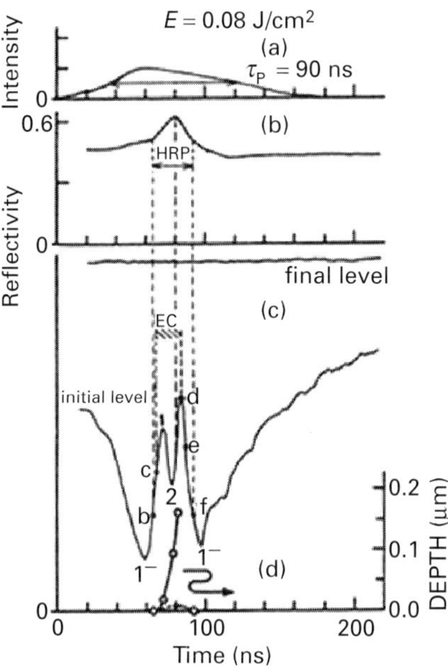

Figure 7.3. (a) The laser pulse shape, (b) the front-side time-resolved optical-reflectance (TROR) signal, (c) the backside TROR signal, and (d) the obtained depth of the interface between the self-propagating buried l-Si layer and inner a-Si. The duration of explosive crystallization (EC) is shown in (c). From Murakami *et al.* (1987), reproduced with permission from the American Physical Society.

Transient optical-reflectance measurements were performed simultaneously from the front and back sides, at the He–Ne laser wavelengths 633 and 1152 nm. Figure 7.3 shows the reflectance signals for the laser fluence $F = 0.08$ J/cm^2 that were interpreted using thin-film-optics calculations. While the front-side reflectance corresponds to a maximum surface melt depth of only 7 nm, the back-side reflectance trace clearly shows interference features. Such features would have to be generated by a l-Si–a-Si interface propagating to a maximum melt depth of 80–100 nm. Since this depth is much larger than the thickness of 7 nm measured by determining the front-side reflectance, it is reasonable to infer that the fringes are created by the interface a buried self-propagating l-Si layer forms with the underlying a-Si. Furthermore, the onset of EC is estimated to occur at point c in Figure 7.3, when the undercooled surface liquid layer is only 3 nm thick. At the time instant d of the maximum buried interface depth, the front-side reflectance still showed evidence of surface melting. After the end of the EC process, the reflectivity drops to e and f when the thin layer solidifies. A velocity of about 14 m/s could be estimated for the propagation of the buried interface.

7.4.2   Interface kinetics

The nonequilibrium phase transformations of semiconductors upon rapid laser irradiation of nanosecond and shorter duration are determined by the departure of the interface temperature from the equilibrium melting temperature. Upon melting, the interface should in principle exhibit overheating, $T_i > T_m$, whereas undercooling, $T_i < T_m$, drives the solidification process. The fundamental nature of the phase-transformation process is in essence specified by the interface response function, $V_i = V(T_i)$. Various numerical and experimental studies have been conducted to deduce the interface response function, but a definitive description has yet to be established. In the molecular-dynamics study by Kluge and Ray (1989), the Stilliger–Weber potential was utilized to analyze the (001) solidification of crystalline silicon. At not too high a deviation from the equilibrium melting point, the slope of the response function was found to be –9.8 K m^{-1} s.

As detailed by Stolk *et al.* (1993), the interface response function can be analyzed via two alternative approaches. In the transition-state theory (TST), the transitions between the solid and liquid phases are assumed to occur via an intermediate state, introducing a barrier to $U(T)$. In the diffusion-limited theory (DLT), the interface velocity is assumed to be related to the diffusivity of atoms in the liquid phase. Both descriptions result in a kinetic relation for the interface velocity of the form

$$U(T_{\text{int}}) = c \exp\left(-\frac{Q_{\text{act}}}{k_B T_{\text{int}}}\right)\left[1 - \exp\left(-\frac{\Delta g_{B,\text{ls}}}{k_B T_{\text{int}}}\right)\right], \tag{7.4}$$

where $\Delta g_{B,\text{ls}}$ is the difference in Gibbs free energy per atom between the liquid and the solid, and Q_{act} is an activation energy. The kinetic prefactor, $c = f \omega_0 d$, where ω_0 is an attempt frequency, d a distance over which the interface moves for a successful jump, and f the fraction of active sites at the interface. In the collision-limited model of TST the maximum freezing velocity is fundamentally limited by the sound velocity c_s in the solid. For c-Si, $c_s = 8433$ m/s. In the DLT, the collision frequencies at the interface are presumed limited by the diffusion of atoms in the liquid. In this case, Q_{act} represents the activation energy for self-diffusion of atoms in the liquid near the interface, and the attempt frequency $\omega_0 = D_0/\lambda^2$, where λ is a characteristic distance for diffusion and D_0 is the prefactor in the equation for the diffusion constant, $D(T) = D_0 \exp[-Q_{\text{act}}/(k_B T)]$.

An experiment was devised by Stolk *et al.* (1993) to determine the interface velocities for crystalline- and amorphous-solid–liquid-silicon transformations, $U_c(T)$ and $U_a(T)$, respectively. Samples consisting of c-Si/a-Si/c-Si layer structure were produced by ion implantation, with a 420-nm-thick amorphous layer buried underneath a 130-nm-thick single-crystalline layer. The structure was subjected to irradiation by a single ruby-laser ($\lambda = 694$ nm) pulse of FWHM pulse duration 32 ns and the transient optical-reflectivity response to probing by an AlGaAs laser operating at the near-IR wavelength $\lambda = 825$ nm was monitored *in situ*. Since the melting temperature of amorphous silicon is lower than the melting temperature of crystalline silicon, melting is initiated in the buried a-Si layer. The undercooled liquid phase will crystallize at the top c-Si layer. Because the latent heat of amorphous silicon is lower than the latent heat of crystalline silicon, the net transformation is exothermic, thereby promoting deeper melting of the buried a-Si via the self-sustained EC mechanism. The rapid motion of the advancing melting front creates interference of the reflected probing beam. By observing the fringe pattern generated, the EC velocity was estimated to be about 17.8 m/s. Figure 7.4 depicts the interface response functions of c-Si and a-Si that are consistent with the constraints

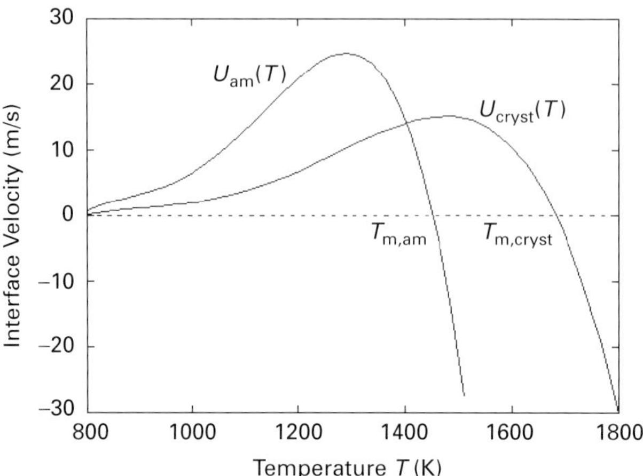

Figure 7.4. The interface response function of c-Si and a-Si derived from the experimental constraints (1)–(4) defined in the text. The solid points are the experimental data used in the analysis.

derived from the experimental observations. These constraints are as follows. (1) Both U_{cryst} and U_{am} are equal to zero at the equilibrium melting temperatures T_{mcryst} and T_{mam}. T_{mam} is assumed to be about 200 K lower than T_{mcryst}. (2) By further assuming that during epitaxial EC the freezing c-Si/l-Si and the melting l-Si/a-Si interfaces propagate at the same velocity, it is deduced that $U_{\mathrm{cryst}}(T_{\mathrm{int}}) = 17.8$ m/s and $U_{\mathrm{am}}(T_{\mathrm{int}}) = -17.8$ m/s. (3) Since the freezing of l-Si on Si(100) transforms from crystallization into amorphous growth if the interface velocity exceeds 15 m/s, it is inferred that the curves depicting the interface response functions must cross at 15 m/s and at a temperature below T_{mam}. (4) Because the amorphization velocity in picosecond-laser irradiation of crystalline silicon saturates at 25 m/s (Bucksbaum and Bokor, 1984), it is inferred that the a-Si interface response function attains a maximum at 25 m/s. Aided by numerical simulations, Stolk *et al.* (1993) showed that the TST model is likely to be invalid, and determined that the experimental data were consistent with the DLT model. The activation energy Q_{act} in the DLT model would be in the range 0.7–1.1 eV. This is high compared with the activation energy for self-diffusion in metals, implying that l-Si does not exhibit a purely metallic behavior. This hypothesis is supported by molecular-dynamics simulations (Stich *et al.*, 1989), which suggested that l-Si has a lower average coordination number than do most liquid metals, due to persistence of covalent bonding in the liquid phase. It is also argued that the solid in the vicinity of the interface would tend to cause local ordering in the liquid, hence increasing the barrier to self-diffusion.

7.4.3   Experimental diagnostics

In situ experiments combining measurements of time-resolved ($\sim$1 ns) electrical conductance and optical reflectance and transmittance at visible and near-IR wavelengths with thermal-emission measurements have been conducted to analyze the temperature history and the dynamics of melting and resolidification in thin amorphous and polycrystalline Si films (Hatano *et al.*, 2000). Time-resolved electrical-conductance measurement is utilized for obtaining the melting duration, the melt depth, and the solid–liquid-interface velocity. Spectrally resolved pyrometry based on Planck's blackbody-radiation intensity distribution enabled measurement of the transient temperature during the phase-transition process. Hence, the origin of the recrystallized material morphologies that critically depend on the applied laser energy density could be identified.

A schematic diagram of the experimental system indicating the various diagnostic probes is shown in Figure 7.5(a). The sample consists of a 50-nm-thick a-Si film deposited onto a fused-quartz substrate by LPCVD. A pulsed KrF excimer laser (wavelength $\lambda = 248$ nm, FWHM 25 ns) is utilized for heating the sample. Figure 7.5(b) shows the experimental setup for temperature measurement by detecting the transient thermal emission (Chen and Grigoropoulos, 1997). Emitted radiation is focused by two short-focal-length lenses onto a fast InGaAs photodetector with a rise time of 3 ns. The temperature history of the liquid–solid phase-change process is obtained by measuring the thermal-emission signals on the basis of Planck's blackbody-radiation intensity distribution law:

$$e_{\lambda b} = \frac{2\pi C_1}{\lambda^5} \frac{1}{\exp[C_2/(\lambda T)] - 1}, \tag{7.5}$$

where $e_{\lambda b}$ is the blackbody emissive power, $C_1 = 7.9555 \times 10^7$ W μm^4/m^2, and $C_2 = 1.4388 \times 10^4$ K μm. The thermal-emission signal collected by the detector can be expressed by

$$I_{\mathrm{d}}(T) = \frac{R_\Omega A}{\pi} \int_{\lambda_1}^{\lambda_2} \int_{\phi_1}^{\phi_1} \int_{\theta_1}^{\theta_2} \varepsilon_\lambda'(\lambda, \theta, \phi, T) \tau(\lambda) G(\lambda) e_{\lambda b}(\lambda, T) \mathrm{d}\theta \, \mathrm{d}\phi \, \mathrm{d}\lambda, \tag{7.6}$$

where T is the temperature, θ and ϕ are the polar and azimuthal angles, λ is the wavelength, R_Ω is the impedance of the oscilloscope (50 Ω), A is the area on the sample which is sensed by the detector, $\varepsilon_\lambda'(\lambda, \theta, \phi, T)$ is the spectral directional emissivity, $\tau(\lambda)$ is the transmissivity of the two lenses, and $G(\lambda)$ is the wavelength-dependent responsivity (A/W) of the InGaAs detector. To obtain the emissivity, the reflectivity and transmissivity data are used to calculate the absorptivity. Invoking Kirchhoff's law (Siegel and Howell, 1992), the spectral directional absorptivity is set equal to the emissivity. The specular front reflectivity, R_λ', and transmissivity, τ_λ', were measured to obtain the emissivity ($\varepsilon_\lambda' = 1 - R_\lambda' - \tau_\lambda'$) at the wavelength $\lambda = 1.52$ μm of the IR He–Ne laser. The transient reflectivity and transmissivity (Jellison *et al.*, 1986) were also measured to determine the melt duration. A CW He–Ne laser ($\lambda = 633$ nm) is used as the probing light source; the signals are focused onto the fast silicon p–n photodiode with nanosecond response time. The reflectivity increases and the transmissivity decreases during melting, since liquid Si has a higher reflectivity than does solid Si. The back-side reflectivity at the wavelength $\lambda = 633$ nm is measured at the incidence angle of 45° as an additional probe of the solidification process. Time-resolved electrical-conductance measurement

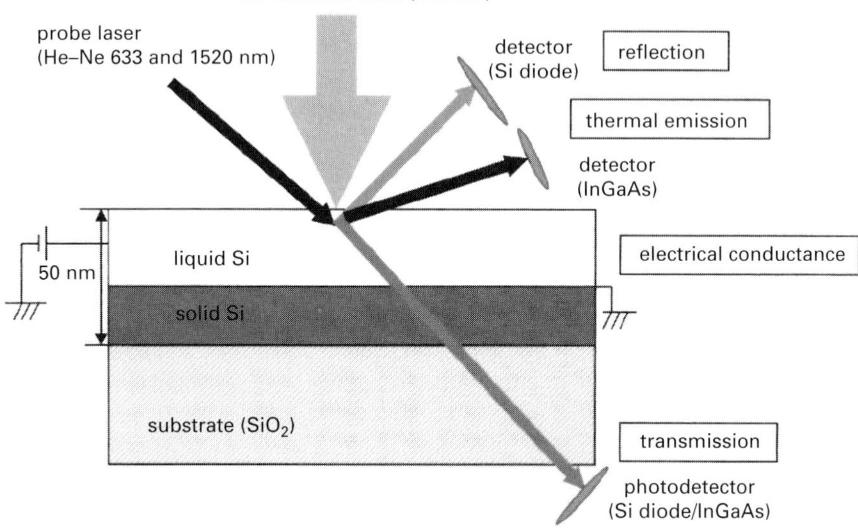

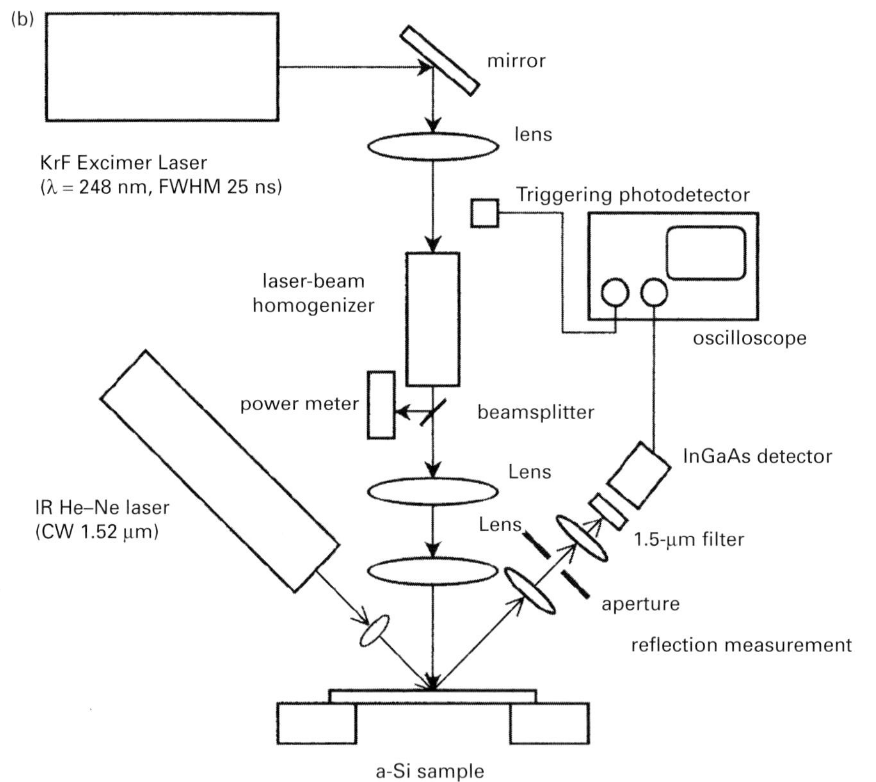

Figure 7.5. (a) A schematic diagram of the *in situ* diagnostic probes; (b) the experimental setup for measuring the thermal emission. The IR He–Ne laser is used for measuring the front-side reflectivity (as shown in the figure), transmissivity, and emissivity. From Hatano *et al.* (2000), reproduced with permission from the American Institute of Physics.

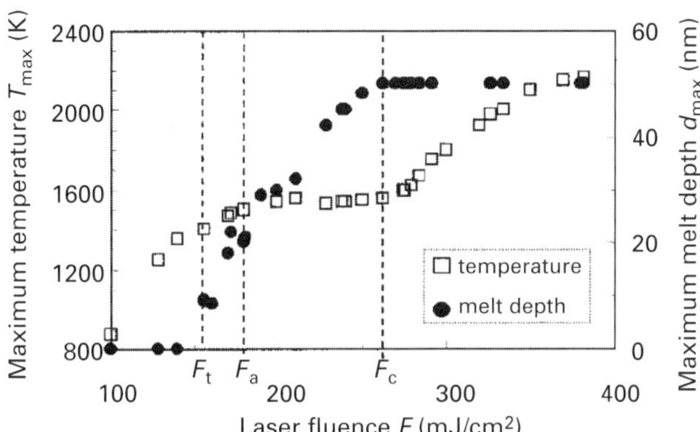

Figure 7.6. Dependences of the maximum temperature and melt depth on the laser fluence. From Hatano *et al.* (2000), reproduced with permission from the American Institute of Physics.

(Galvin *et al.*, 1982) is applied in order to obtain the melt duration, melt depth, and solid–liquid-interface velocity. The molten Si produces an abrupt rise in the conduction electron density such that the electrical conductivity reaches values typical of liquid metals. The electrical conductivity of molten Si (75 Ω^{-1} cm^{-1}) is much higher than that of solid Si (0.3 Ω^{-1} cm^{-1}) (Glazov *et al.*, 1969). Consequently, the total conductance of the Si is drastically increased due to the presence of a molten layer.

The dependence of the maximum melt depth on the laser fluence and the variation of the maximum temperature are shown in Figure 7.6. The conductance signal is essentially representative of the volume fraction of the liquid phase. For convenience, the melt depth is used as an indicator of the volume of liquid Si. The temperature is integrated over the absorption depth at the near-IR wavelength. The threshold fluence for surface melting is at 155 mJ/cm^2, while complete melting occurs at 262 mJ/cm^2. In the partial-melting regime that lies between these values, the melt depth increases with laser fluence. Since the absorbed laser energy in excess of the level needed for surface melting is consumed by the latent heat of phase change from solid a-Si to liquid, the maximum temperature remains nearly constant. The constancy of the measured value is subject to the condition that the melt depth exceeds the absorption depth in liquid silicon, which in the near-IR wavelength range is about 20 nm. This is the reason why the measured temperature rises at the fluence of 179 mJ/cm^2. In the complete-melting regime, the temperature increases with fluence, since the excess laser energy density beyond the threshold for complete melting is used to heat the liquid Si, consequently raising the peak temperature.

Figure 7.7 shows the relationship between the average grain size of recrystallized Si and the maximum temperature attained as a function of the laser fluence. The grain size strongly depends on fluence and therefore on the temperature history and the solid–liquid-interface velocity. Accordingly, the grain-size variation follows closely the regimes defined via both the conductance and temperature measurements. In the low-fluence range that corresponds to the partial melting regime, SEM shows a small but gradual increase in grain size with fluence. In the high-fluence range, which corresponds

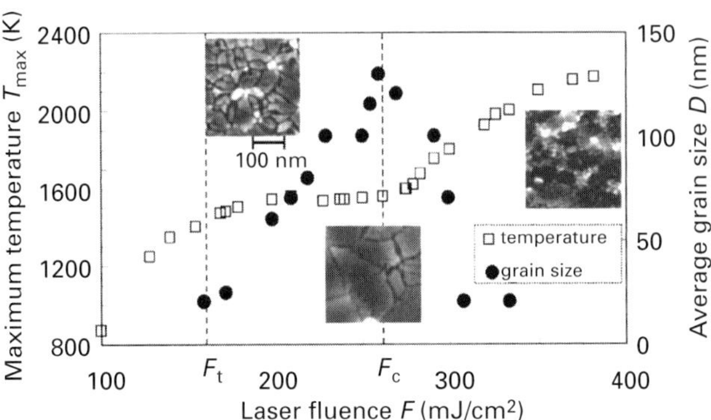

Figure 7.7. Dependences of the maximum temperature and average grain size on the laser fluence. From Hatano *et al.* (2000), reproduced with permission from the American Institute of Physics.

to the complete-melting regime, a dramatic reversal of the recrystallized microstructure is observed, even with only a slight increase in fluence. In the "near-complete"-melting regime, i.e. in the transition zone from partial melting to the complete-melting regime, a substantially enlarged grain size is obtained (Im and Sposili, 1996). As the radiant laser energy increases, the silicon layer becomes completely molten and the melting duration is elongated. This is clearly observed in the reflectivity and transmissivity traces. Two bumps, aligned with the melting and crystallization transitions and separated by a flat region, are shown in the emissivity curve displayed in Figure 7.8(a). The existence of the flat region indicates that the optical properties of liquid silicon do not depend strongly on the temperature variation. The constant emissivity value of about 0.15 matches well the value of 0.156 obtained from thin-film-optics calculations using the measured data. Substantial supercooling, followed by nucleation, is inferred by examining the temperature and melt-depth transients shown in Figure 7.8(b). The molten Si cools very rapidly, as shown in the transient temperature signal. Owing to the lack of heterogeneous nucleation sites, the melt is forced to experience substantial supercooling (Figure 7.8(b)). Accordingly, the transient temperature exhibits a dip in the neighborhood of 60–70 ns, which exactly coincides with the end of the full melting. The temperature dip is therefore interpreted as preceding the onset of homogeneous nucleation in supercooled liquid Si. The supercooling prior to nucleation is by about 230 K lower than the equilibrium melting point, hence triggering nucleation. Upon the inception of solidification, latent heat is released, raising the temperature of the film up to the melting point. Following this recalescence, growth of the solid continues as heat is being conducted into the substrate. The rate of nucleation increases in the deeply supercooled liquid and the homogeneous nucleation process results in a fine-grain recrystallized structure.

7.4.4 Recrystallization of poly-silicon versus amorphous silicon

The melting behavior for a-Si versus poly-Si as initial material was determined by examining the melting duration and melt depth extracted from transient conductance

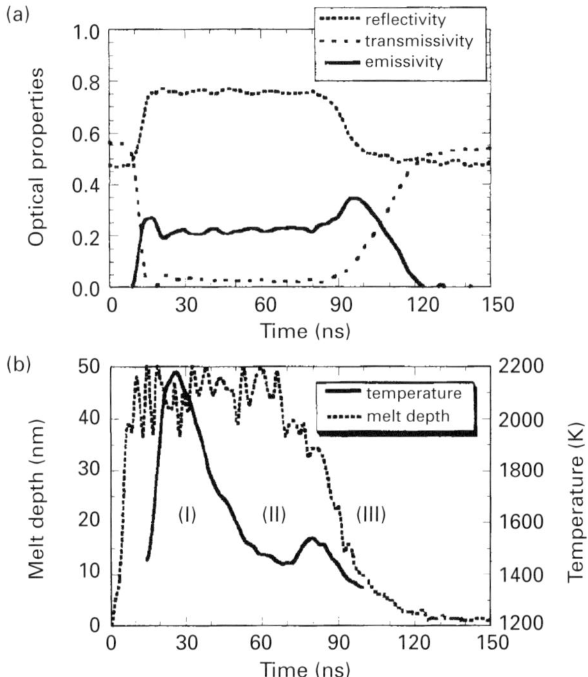

Figure 7.8. (a) The transient front-side reflectivity, transmissivity, emissivity ($\lambda = 1.52$ μm), and emission signal ($\lambda = 1.5$ μm) at the angle of $45°$; (b) temperature and melt-depth histories. The laser fluence $F = 365$ mJ/cm^2 generates complete melting. From Hatano *et al.* (2000), reproduced with permission from the American Institute of Physics.

measurements and is shown in Figure 7.9. The threshold fluences for surface melting, F_t, and complete melting, F_c, are lower for the a-Si film than for the poly-Si film. Both the melting duration and the melt depth in the poly-Si film are smaller than their counterparts for the a-Si film at the same fluence. These effects are mainly caused by the difference in melting point and thermal conductivity between a-Si and poly-Si material. Figure 7.10 compares the measured peak temperature values for a-Si and poly-Si. The melting temperature for poly-Si exhibits a plateau slightly below 1700 K. It is noted that the equilibrium melting temperature of crystalline silicon is 1685 K. It is verified that nanosecond-laser-heated a-Si melts at a temperature about 100–150 K lower than does crystalline Si.

7.5   Nucleation in the supercooled liquid

The rapid quenching of the liquid-silicon pool is inevitable due to the nanosecond time scale of a laser pulse. Owing to the thinness of the a-Si film ($\sim$50 nm) and the short laser pulse, most of the heat is conducted to the substrate, with minimal convection and radiation heat losses. During the time scale of a few tens of nanoseconds for

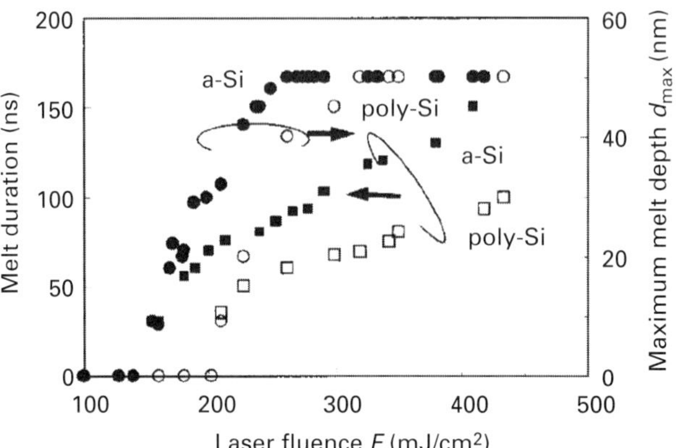

Figure 7.9. A comparison of the dependences of the melt duration and melt depth on the KrF excimer-laser fluence for a-Si versus poly-Si. All specimens are 50 nm thick. From Hatano *et al.* (2000), reproduced with permission from the American Institute of Physics.

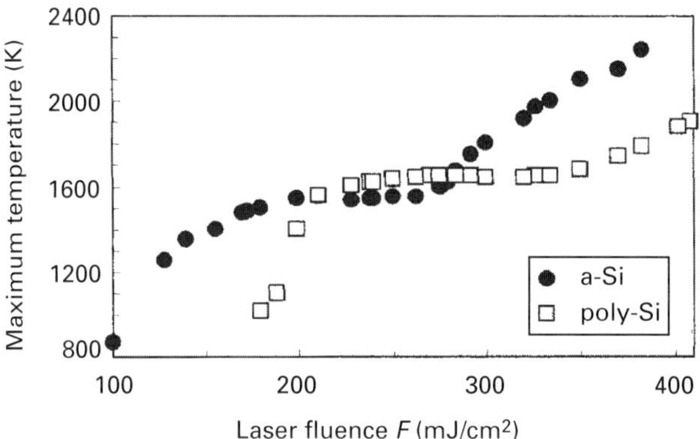

Figure 7.10. A comparison of the measured peak temperatures for a-Si and poly-Si as functions of the KrF excimer-laser fluence. From Hatano *et al.* (2000), reproduced with permission from the American Institute of Physics.

a laser pulse, the thermally affected zone in the poorly conducting substrate is not established. The liquid-silicon pool tends to be supercooled as revealed in the emission-temperature measurement. When the melt is supercooled below its equilibrium melting temperature, the driving force for crystallization is dramatically enhanced due to the Gibbs free-energy difference, which increases with the degree of supercooling (Herlach, 1994).

The concept of nucleation has satisfactorily explained the growth of materials in many metastable phase transformations. Particles of a new phase are assumed to form and change in size by statistical fluctuations. Particles reaching the critical size required for continuous growth can attain appreciable dimensions at the expense of the parent matrix and are called nuclei. Particles of sub-critical size will be called embryos in order

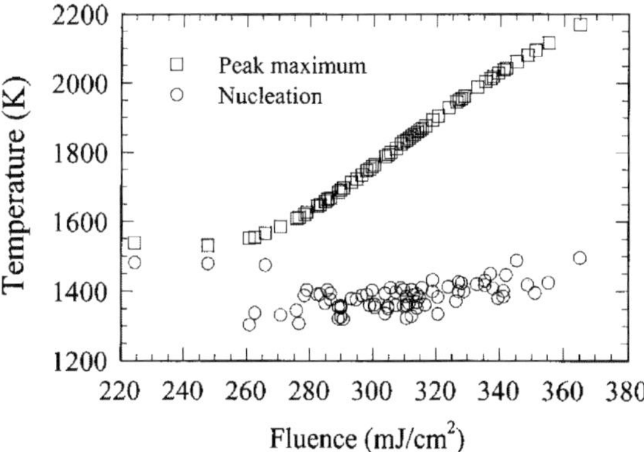

Figure 7.11. Measured dependences of the nucleation temperature and the peak temperature on the laser fluence. The 50-nm-thick a-Si film is melted by a single KrF excimer-laser pulse. From Moon *et al.* (2002), reproduced with permission from the American Society of Mechanical Engineers.

to differentiate them from nuclei. The interface between a solid embryo and surrounding supercooled liquid implies an activation barrier. With increasing supercooling, the Gibbs free-energy difference increases and exceeds the energy barrier so that an embryo can be transformed into a solid nucleus. The growth of embryos over the free-energy barrier is called *thermal nucleation.*

If the temperature change is rapid during phase transformations, the distribution of embryos cannot change significantly, because there is insufficient time for them to shrink or grow to a steady-state concentration. The term *athermal nucleation* refers to the process whereby an embryo becomes a nucleus as a consequence of a shrinking critical size. According to classical nucleation theory, the athermal and thermal mechanisms constitute two distinct nucleation paths through which sub-critical clusters can become supercritical (Fisher *et al.*, 1948). It is apparent that, when a liquid is quenched, the temporal reduction in critical cluster size that accompanies the cooling of the liquid can lead to athermal nucleation of solid matter. The specific details of solid nucleation in a supercooled liquid can influence the rates and conditions of the transformation itself, as well as determine the phase and the microstructure of the resulting material.

As shown in Figure 7.8(b), the occurrence of nucleation in the supercooled liquid after rapid quenching is manifested by a pronounced temperature dip. Figure 7.11 shows that the nucleation temperature extracted from the transient traces is almost constant with respect to the laser fluence. It is noted that at the fluence of 224 mJ/cm^2 the melt depth is 40 nm. Since the absorption depth in the liquid silicon over the IR wavelength range is around 20 nm, the emission measurement of the temperature of liquid silicon is valid only when the melt depth exceeds the absorption depth. Moreover, the thermal conductivity of liquid silicon is known to be close to that of a metal. Thus, the temperature profile in the liquid-silicon film is almost uniform over the entire liquid film. The measured data exhibit initiation of nucleation at a constant temperature level.

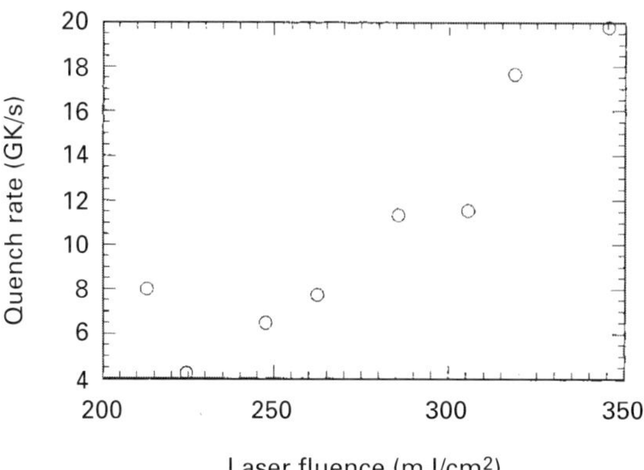

Figure 7.12. The measured dependence of the quench rate on the laser fluence. The 50-nm-thick a-Si film is melted by a single KrF excimer-laser pulse. From Moon *et al.* (2002), reproduced with permission from the American Society of Mechanical Engineers.

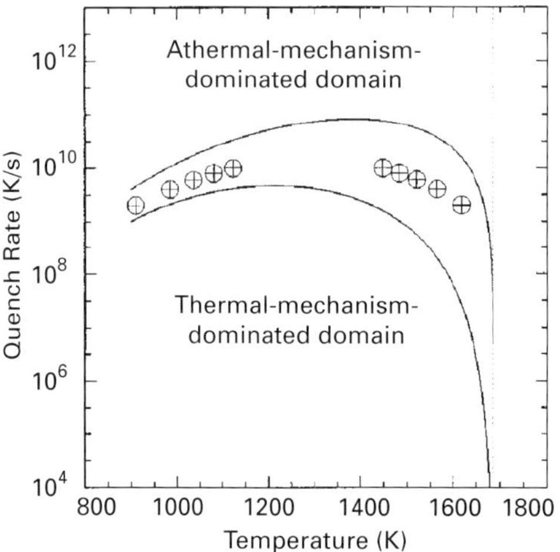

Figure 7.13. A mechanism diagram constructed for liquid Si. In addition to the upper and lower isomechanism lines, included in it are the points obtained form a set of numerical simulations at constant quench rates. The dotted line shows the melting point of Si at 1685 K. From Im *et al.* (1998), reproduced with permission from the American Institute of Physics.

Figure 7.12 depicts the quenching rate during the cooling process. The quenching rate is taken as the temporal slope of an imaginary line connecting the peak temperature with the corresponding nucleation temperature. The quenching rate increases in the full-melting regime, i.e. for laser fluences greater than 262 mJ/cm^2. Im *et al.* (1998) modeled the relationship between the quenching rate and the degree of supercooling (Figure 7.13). This diagram distinguishes three regimes: (i) the domain dominated by athermal nucleation, (ii) the region dominated by the thermal mechanism, and (iii) the domain where both mechanisms operate. The measured quenching rate in the full-melting regime belongs to the athermal-nucleation domain. However, in the case of

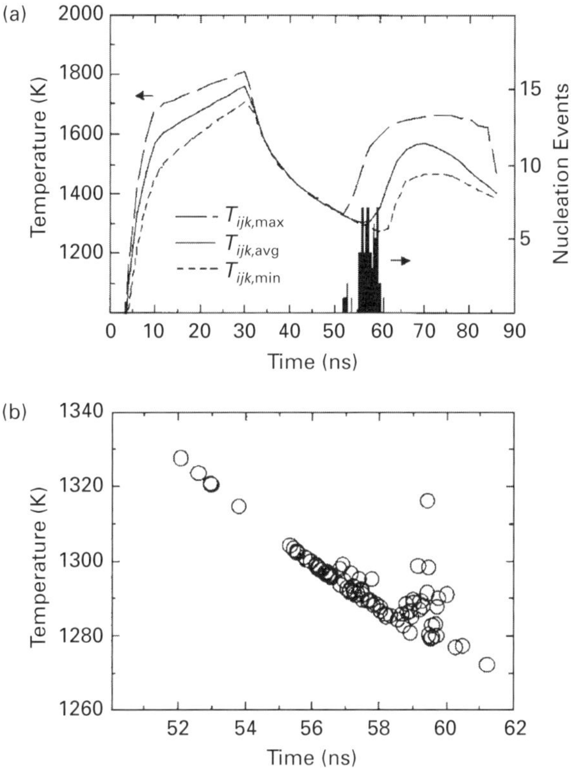

Figure 7.14. Node-temperature evolution in a single simulation showing maximum, minimum, and average temperatures in the film. The histogram corresponds to nucleation events that occurred during the transformation. (b) Some nucleation events, plotted as individual points at the time and temperature of occurrence. From Leonard and Im (2001), reproduced with permission from the American Institute of Physics.

partial melting, the main nucleation mechanism is thermal. In athermal nucleation, the shrinking of the critical size of a nucleus lowers the energy barrier. When this happens through a rapid quenching rate, quasi-crystalline structure tends to be formed. This argument could explain the formation of microcrystalline silicon in the full-melting regime in the laser-annealing process. In the partial-melting regime, where the initial melt depth is shallow, explosive crystallization is responsible for the formation of poly-silicon grains. In contrast, a larger crystal size is produced under near-complete-melting conditions via thermal nucleation.

A stochastic model for simulating nucleation of solids in supercooled liquid was constructed by Leonard and Im (2001). Heterogeneous nucleation is possible in nodes containing catalytic interfaces. The simulation considered only homogeneous nucleation, catalyzed at the l-Si–SiO$_2$ interface. Figure 7.14(a) shows the maximum, minimum, and average temperatures within the Si film together with a histogram of the nucleation events. As Figure 7.14(b) shows in more detail, the aforementioned nucleation mechanism takes effect over a considerable time period and temperature range. As the transformation proceeds, the film temperature increases, though it remains below the equilibrium melting temperature. Even though the assumed mechanism of nucleation at the l-Si–SiO$_2$ interface is not precisely defined and supported by experimental evidence or physical arguments, the essence of the model could be expanded to accommodate homogeneous nucleation and eventually treat complex problems of practical relevance, including lateral crystal growth.

7.6 Lateral crystal growth induced by spatially modified irradiation

Conventional excimer-laser crystallization (ELC) can produce grains of diameter hundreds of nanometers depending on the thickness of a-Si film irradiated. However, the processing window for the conventional technique is narrow because large grains can be obtained only in the so-called superlateral-growth (SLG) regime wherein the film is nearly fully melted so that the remaining solid silicon particles on the substrate surface act as seeds for grain growth. In addition, the grain size produced by conventional ELC is highly nonuniform, with randomly oriented grain boundaries that result in nonuniform device characteristics. Therefore recent research efforts on ELC have been focused on developing spatially controlled crystallization methods. Several methods have been shown to give laterally oriented grain growth. These methods include the use of a beam mask (Im and Sposili, 1996), diffraction mask (Ishikawa *et al.*, 1998), anti-reflective coating (Kim and Im, 1996b), phase-shift mask (Oh *et al.*, 1998), and the interference effect induced by a frequency-doubled Nd:YAG laser (Aichmayr *et al.*, 1999). The working principle in all these techniques relies on shaping the laser-energy profile that is irradiated onto an a-Si sample.

Figure 7.15(a) displays the experimental concept utilized for shaping the beam profile via a mask possessing a step-wise phase-shift (Oh *et al.*, 1998). The calculated normalized intensity distribution ($I/I_{\rm pk}$) as a function of the distance from the phase-shift step is displayed in Figure 7.15(b) for various phase-shift angles. Figure 7.16 depicts the SEM image obtained by using a phase-shift mask with $\theta = 53°$ and mask separation of 0.4 mm. For the incoming KrF ($\lambda = 248$ nm) laser intensity of $I_{\rm pk} = 900$ mJ/cm^2 and substrate temperature of 500 °C, grains as long as 7 μm were produced in 200-nm-thick a-Si films that had been deposited onto 850-nm-thick SiO$_2$ on silicon.

In the laser interference experiment performed by Aichmayr *et al.* (1999) single Nd:YAG laser pulses at $\lambda = 532$ nm were utilized to generate interference patterns on the surface of 300-nm-thick a-Si films. The grating period of the sinusoidal patterns is related to the laser wavelength λ and the angle of incidence θ via $\lambda/(2\sin\theta)$. For appropriate laser intensities, the a-Si film melts and crystallizes only around the interference maxima. At low laser intensities the grain-size distribution produced exhibits the trends typically displayed in large-area annealing. At higher laser intensity, namely 400 mJ/cm^2, with a grating period of 5 μm, lateral growth was obtained (Figure 7.17). A 1-μm-wide strip of amorphous material remains between the recrystallized stripes (region A). A narrow stripe of small-grain material, of width 280 nm, can be distinguished in region B. The

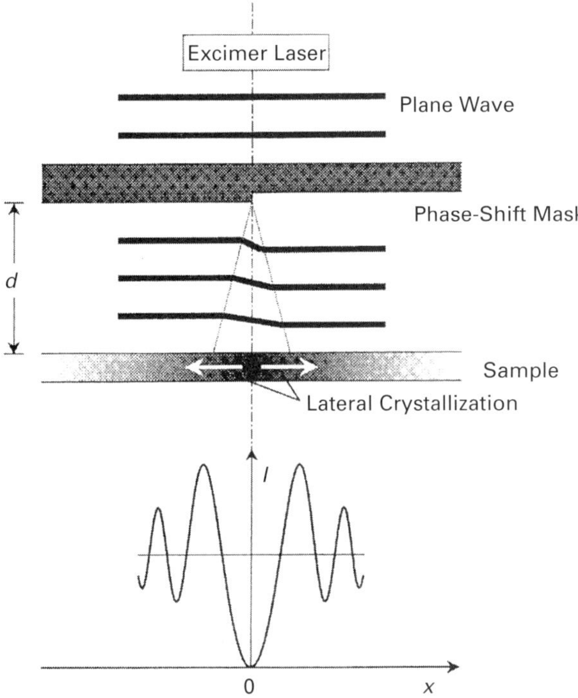

Figure 7.15. A schematic diagram of the phase-mask excimer-laser crystallization (PMELC) method. From Oh *et al.* (1998), reproduced with permission from the Institute of Pure and Applied Physics.

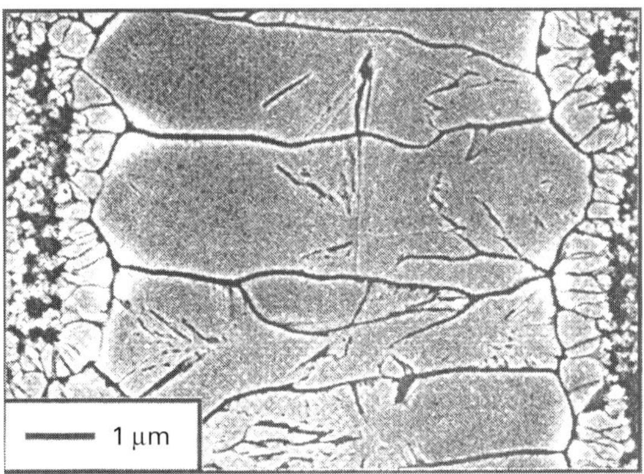

Figure 7.16. The top view of the crystallized Si film for the case of a mask–sample separation of 0.4 mm, $\theta = 53°$, and $I_{pk} = 900$ mJ/cm^2. From Oh *et al.* (1998), reproduced with permission from the Institute of Pure and Applied Physics.

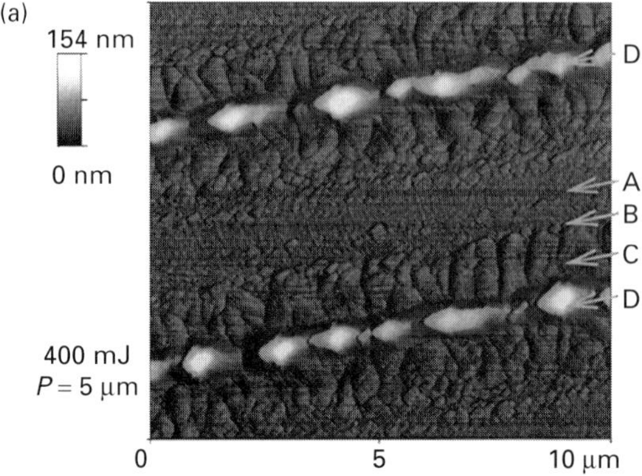

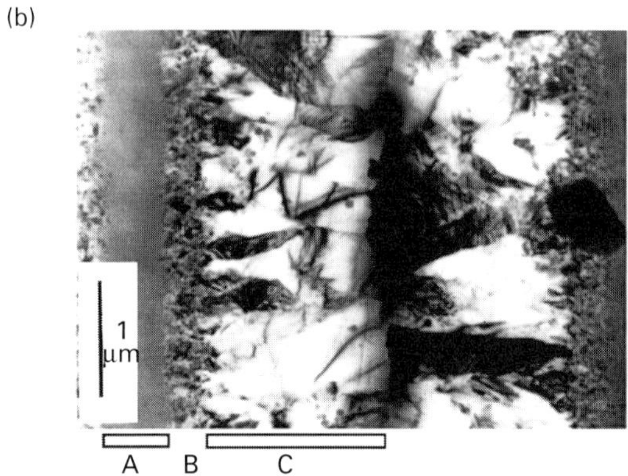

Figure 7.17. (a) Atomic-force microscopy (AFM) and (b) TEM micrographs of a sample fabricated by laser-interference crystallization with a pulse energy of 200 mJ/cm^2 per beam and a period of 5 μm. The micrograph in (b) is an enlargement of the picture in the upper panel. Fine-grained material on the edges of the line (labeled B) is now flanked by long grains (labeled C), oriented toward the center of the lines and reaching lateral dimensions of almost 2 μm. The protrusions in the middle of the line appear flat due to the saturation of the AFM signal. From Aichmayr *et al.* (1999), reproduced with permission from the American Institute of Physics.

most striking feature is the growth of long grains, of length up to 1.7 μm, extending from the boundary of region B toward the middle of the line.

To extend the grain size, Im *et al.* (1998) applied sequential lateral solidification by translating the sample along the lateral growth direction. Chevron-shaped beamlets tend to produce a single-grain central region (containing sub-boundaries) since the grain nucleated at the apex of the pattern tends to expand in all directions (Figure 7.18). In contrast, straight slits tend to produce high-angle boundaries normal to the advancing solidification front. This expectation is verified in the results displayed in Figure 7.19. Furthermore, the central portion of the chevron-processed single-crystal region is made free of sub-boundaries and other defects.

In order to induce lateral grain growth, a fluence gradient must be enforced such that the a-Si film is completely melted in the area exposed to higher laser fluence

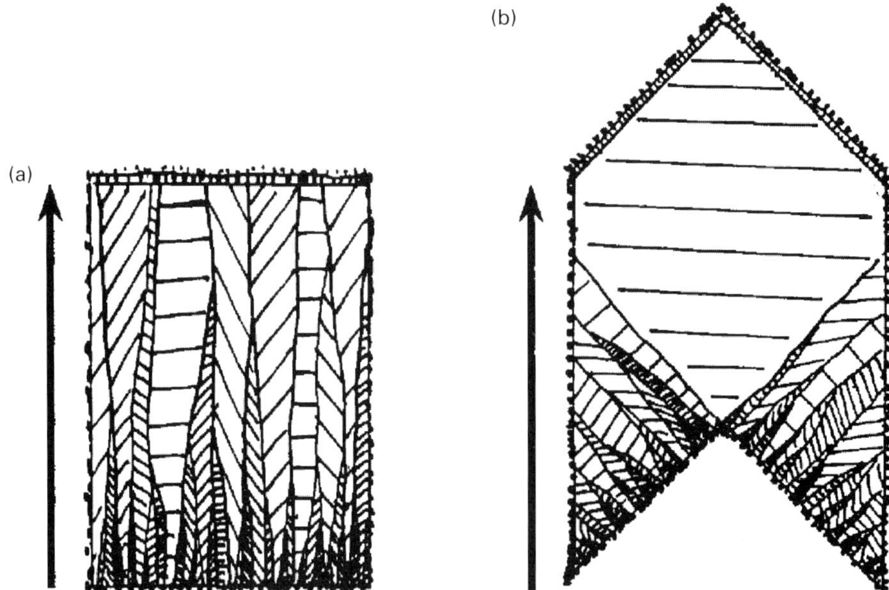

Figure 7.18. A schematic diagram showing the sequential-lateral-solidification microstructure for (a) a straight-slit and (b) a chevron-shaped beamlet. The arrow shows the solidification direction. From Im *et al.* (1999), reproduced with permission from Wiley-Interscience.

and partially melted in the adjacent area exposed to lower laser fluence. Under this condition, grains grow laterally toward the completely molten region. The lateral grain growth will eventually be arrested either by collision with lateral grains grown from the other side or by spontaneous nucleation triggered in the severely supercooled pool of molten silicon. Evidently, higher fluence gradients drive steeper temperature gradients. Since it takes a longer time for the hotter molten-silicon region to cool down to the temperature for spontaneous nucleation, the lateral grain growth can continue out to a longer distance. The relationship between the fluence gradient and the lateral growth length was demonstrated in Lee *et al.* (2000). The dependence of the lateral growth length on the fluence gradient is shown in Figure 7.20. The error bar associated with the lateral growth length stemming from statistical variations and the uncertainty of the measurements is about ± 50 nm. The lateral growth length is almost constant at about 500 nm for fluence gradients below 80 mJ/cm^2 μm but increases rapidly as the laser fluence gradient increases further. The directionality of the lateral grains is also improved by increasing the fluence gradient. Lateral grains of length about 1.5 μm can be obtained in a 50-nm-thick a-Si film by a single excimer-laser pulse without any substrate heating under a high fluence gradient (Figure 7.21).

A qualitative solidification model as depicted by the simplified one-dimensional temperature profiles shown in Figure 7.22 was proposed. Figure 7.22(a) represents the temperature distribution at times when the phase boundary advances toward the liquid region. The interface is at the highest temperature "T_{int}" because of the release of latent heat due to lateral solidification at the interface. The temperature of the bulk liquid

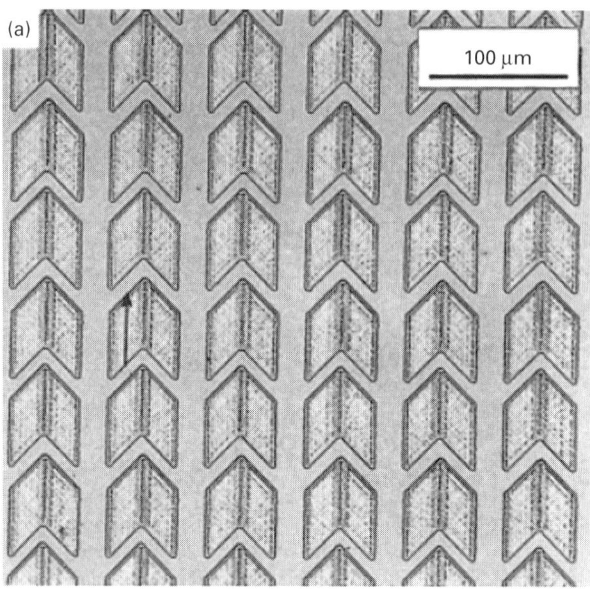

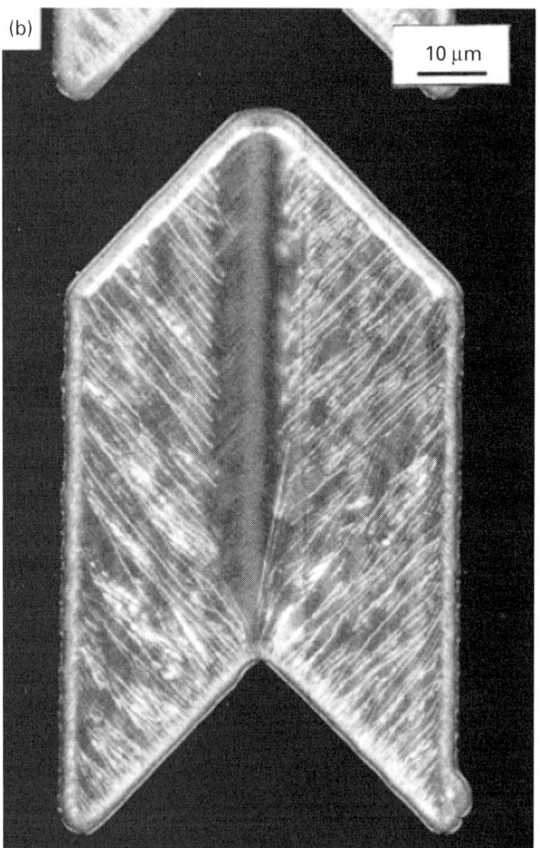

Figure 7.19. Optical micrographs of defect-etched sequential-lateral-solidification-processed film. (a) Low magnification. The arrow shows the solidification direction and the magnitude of the translation. (b) High magnification, dark-field image, with dotted lines demarkating the boundaries between the single-crystalline and columnar-crystal regions. From Im *et al.* (1999), reproduced with permission from Wiley-Interscience.

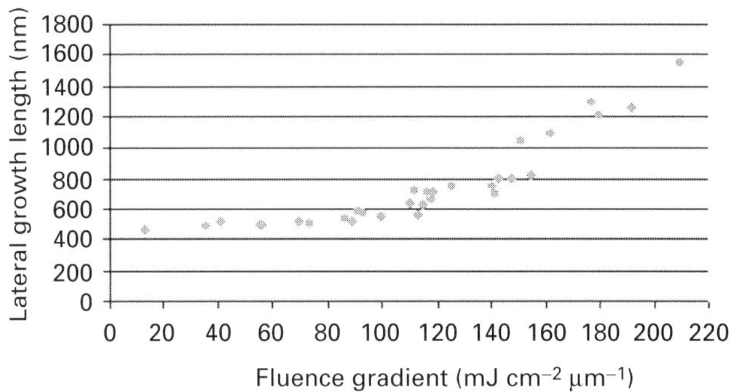

Figure 7.20. The dependence of the lateral growth length on the fluence gradient for a 50-nm-thick a-Si film. From Lee *et al.* (2000), reproduced with permission from the American Institute of Physics.

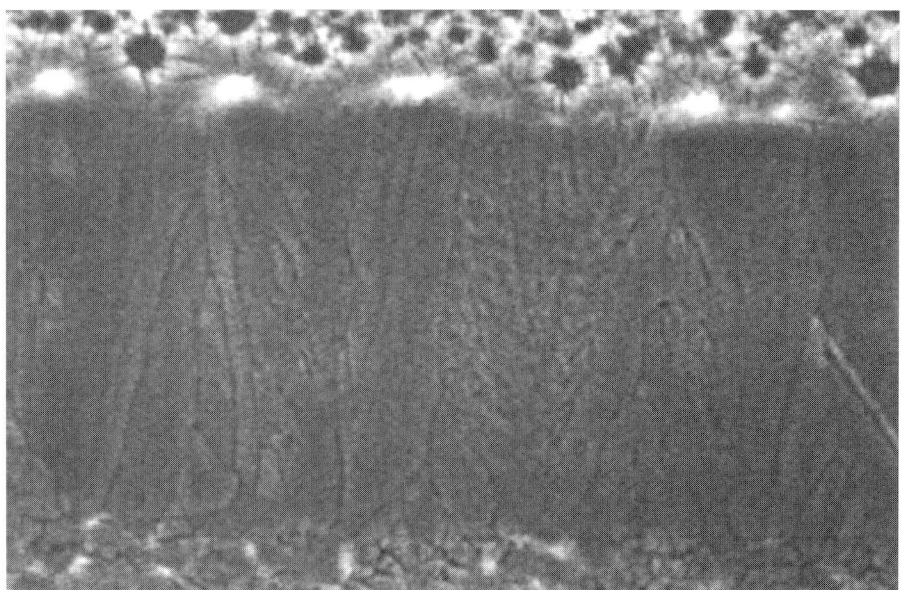

Figure 7.21. A lateral growth length of about 1.5 μm is obtained under a high fluence gradient. From Lee *et al.* (2000), reproduced with permission from the American Institute of Physics.

is at a supercooling temperature "T_{sc}" that is below the melting point but above the spontaneous-nucleation temperature "T_n." As time progresses, the bulk-liquid super-cooling temperature "T_{sc}" drops below the spontaneous-nucleation temperature "T_n" (Figure 7.22(b)). At this point, spontaneous nucleation begins in the bulk liquid, triggering solidification. However, the interface is still at the highest temperature "T_{int}" due to continuous release of latent heat at the interface by lateral solidification. As time proceeds beyond point "b," a second solid–liquid interface is produced because of spontaneous nucleation and solidification in the bulk liquid (Figure 7.22(c)). Owing to release of latent heat at the second interface, the temperature is raised to "T_{int2}."

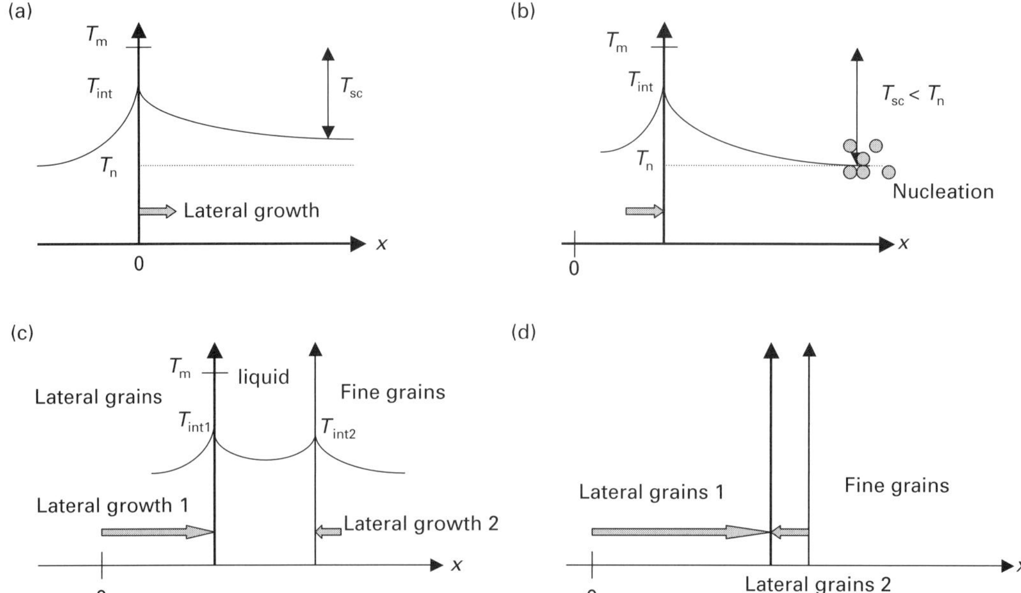

Figure 7.22. A simplified representation of the one-dimensional temperature distribution indicating the evolution of the crystal-growth process. From Lee *et al.* (2000), reproduced with permission from the American Institute of Physics.

The resulting temperature distribution is shown in Figure 7.22(c), with lateral grains on the left due to lateral solidification and fine grains on the right due to spontaneous nucleation, confining the liquid-silicon pool. Finally, the two interfaces meet when the solidification ends (Figure 7.22(d)). Utilizing transient-conductance measurements, a lateral-solidification velocity of about 7 m/s was obtained, this being consistent with this phase-transformation sequence.

It is evident that lateral growth is limited by spontaneous nucleation in the bulk liquid. If spontaneous nucleation could be suppressed or delayed, the lateral growth would continue out to a longer distance, hence producing longer lateral growth. In the case of a high fluence gradient, the temperature increases rapidly from the partially molten region toward the completely molten region. The higher local temperature in the completely molten region implies a correspondingly longer time being taken to reach the deep supercooling required for spontaneous nucleation. Therefore the increase in nucleation time, i.e. the time elapsed from the beginning of the lateral growth until the inception of spontaneous nucleation, is a crucial parameter for lateral growth. The success at producing long lateral grains for the 50-nm-thick a-Si films by imposing a high fluence gradient is mainly attributed to this effect.

In order to visualize the dynamics of the liquid–solid interface during lateral grain growth, high-spatial- and temporal-resolution laser flash photography was employed (Lee *et al.*, 2001a; Xu *et al.*, 2006). A schematic diagram of the experimental setup for flash laser photography is shown in Figure 7.23. Since the reflectivities of a-Si (~50% at

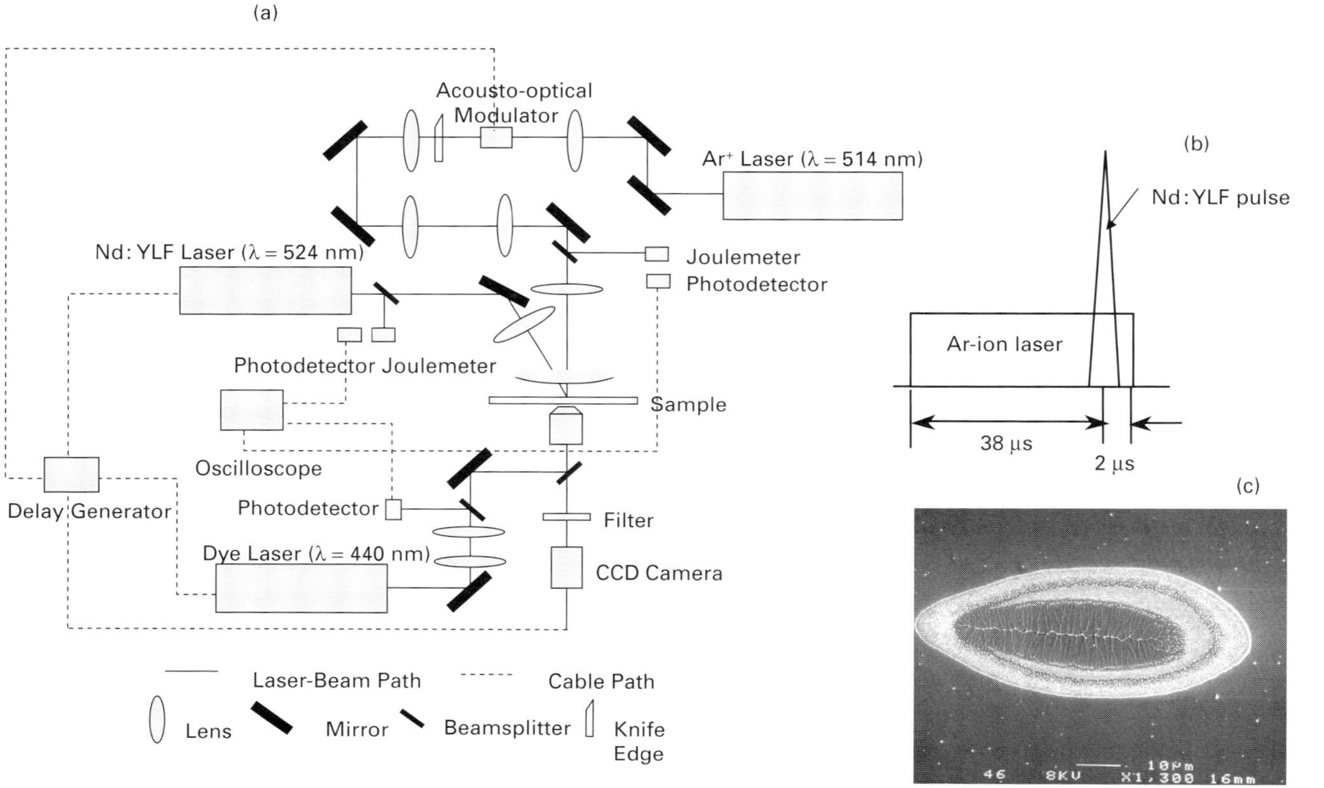

Figure 7.23. (a) A schematic diagram of the experimental setup of laser flash photography for probing the double-laser recrystallization process; (b) a graph indicating the timing of the superposed pulses; and (c) a SEM micrograph exhibiting large polycrystals. From Xu *et al.* (2006), reproduced with permission from the American Institute of Physics.

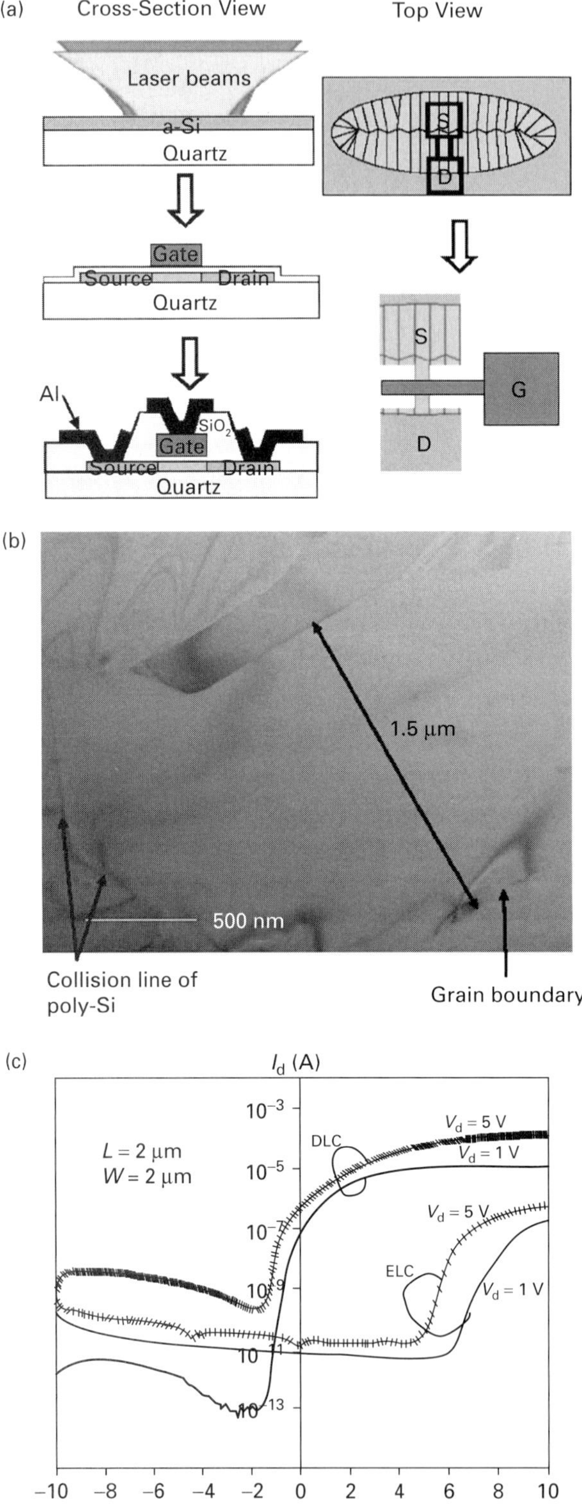

Figure 7.24 Images showing the sequence of the resolidification process. From Xu *et al.* (2006), reproduced with permission from the American Institute of Physics.

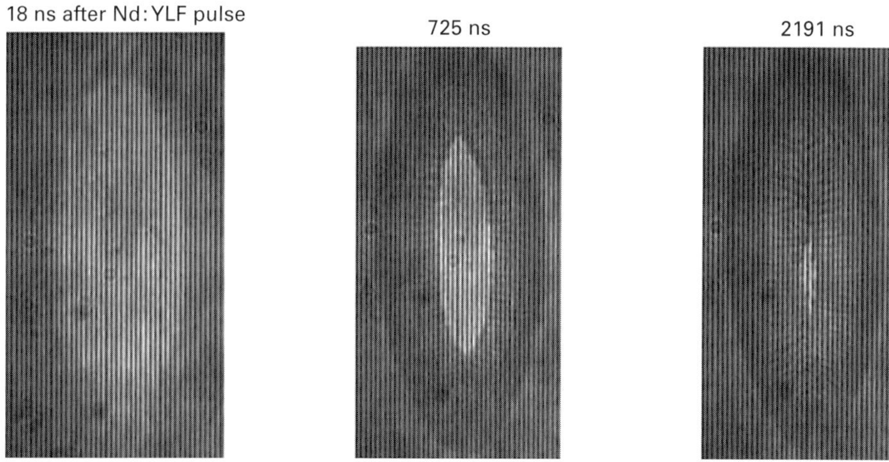

Figure 7.25. (a) A summary of the steps in the TFT fabrication process; (b) a SEM image indicating the width of the polycrystals; and (c) performance characteristics of the fabricated TFTs. From Xu *et al.* (2006), reproduced with permission from the American Institute of Physics.

room temperature), liquid silicon ($\sim$70%), and poly-Si ($\sim$30% at room temperature) are different at the illumination wavelength ($\lambda = 445$ nm), the melting and resolidification sequence can be identified (Moon *et al.*, 2000).

The typical grain microstructure induced by the double-laser recrystallization technique shows that lateral grains of width 1.5 μm grown inward from the edge of the inner ellipse collide along the center of the beam axis. The quality of the recrystallized material enables fabrication of high-peformace TFTs (Figure 7.24). The sequence of the resolidification process revealed by laser flash photography is shown in Figure 7.25. A few nanoseconds following the Nd:YLF laser pulse, a bright liquid-silicon region appears on the image, indicating that the a-Si film is then fully melted. The resolidification process at the initial 300 ns is carried out at a high speed, resulting in microcrystalline material in the outer region melted by the Nd:YLF laser alone. After about 300 ns, the molten-silicon region attains nearly the size of the final lateral-growth region. Beyond this time, the lateral-solidification velocity is estimated to be about 10 m/s. The entire resolidification process takes only a few microseconds to complete, as compared with a much shorter melt duration in single-beam laser recrystallization without substrate heating.

7.7 Mass transfer and shallow doping

With the rapid development in sub-micrometer electronics, it is becoming increasingly urgent to fabricate high-concentration ultra-shallow p$^+$ junctions for the source and drain of PMOS (p-channel metal–oxide–semiconductor) transistors (Ng and Lynch, 1987). This need presents a challenge to current semiconductor-doping technologies such as furnace diffusion and ion implantation. It is difficult to achieve both high concentration and ultra-shallow junctions by conventional furnace doping due to the nature of the solid-phase diffusion and the lack of controllability, even though many efforts have been made toward developing rapid-thermal-process techniques (Usami *et al.*, 1992). On the other hand, by the ion-implantation technique that has been studied extensively and implemented in the current semiconductor industry, it is still difficult to form high-concentration ultra-shallow p^+ junctions. The obstacles arise because the low ion-implantation energies required reach the limits of the capabilities of implantation equipment and from the greater inherent channeling effect in crystalline silicon for light ions such as boron (Liu and Oldham, 1983; Fan *et al.*, 1987). A modified ion-implantation method has been developed to amorphize the surface of the silicon wafer first, in order to reduce the channeling of boron during the implantation process. However, several post-anneals are needed and defects are often generated within the junctions, degrading device performance (Sands *et al.*, 1984).

To meet the requirements of both high doping concentration and ultra-shallow junction depth for PMOS transistors, it is necessary to investigate new techniques. It would be interesting, then, to look for new techniques that can generate the transformation to the liquid-silicon phase with extremely high boron diffusivity and would facilitate control of the diffusion time. Use of a pulsed UV excimer-laser as an intensive heating source can provide a new avenue. With the peak power as high as 10^8 W, the excimer-laser light can be strongly absorbed by the silicon surface within a depth of 10 nm, resulting in a thin layer of molten silicon near the surface. Since the mass diffusivity of boron in liquid silicon is greater than that in solid silicon by six orders of magnitude, excimer-laser irradiation can confine the diffusion within the thin layer of molten silicon at the nanosecond time scale. By controlling the laser pulse energy, the thickness of the thin layer of molten silicon can be incrementally changed from a few nanometers to micrometers. Therefore, it is possible to precisely control the maximum diffusion depth. Also the short pulse width of the excimer laser, which is of the order of 10 ns, provides a complementary way of adjusting the diffusion zone within the molten-silicon layer by varying the number of laser pulses, thereby essentially changing the mass-diffusion time. Meanwhile, the high boron diffusivity in the thin layer of liquid silicon results in a very high doping concentration in the region. A gas-immersion laser doping (GILD) process has been developed (Deutsch *et al.*, 1981), in which dopant gas molecules such as halogen compounds (BCl_3, PCl_3), hydrogenated compounds (AsH_3, B_2H_6, PH_3), and organic compounds ($B(CH_3)_3$, $Al(CH_3)_3$) are dissociated upon UV laser irradiation and incorporated into the thin layer of molten silicon. It has been demonstrated that sub-micrometer shallow junctions can be formed by this gas-phase laser doping technique and good device characteristics have been achieved (Weiner *et al.*, 1993; Slaoui *et al.*, 1990). Figure 7.26 shows the key process steps involved in the fabrication of GILD TFTs (Giust and Sigmon, 1997). Figure 7.27 plots sheet resistance in 90-nm-thick films versus the number and fluence of doping pulses from the XeCl excimer laser ($\lambda = 308$ nm). Prior to doping, all initially a-Si samples were laser-recrystallized using 100 laser pulses at the full-melting-threshold (FMT) fluence of 420 mJ/cm^2. At laser fluences below the FMT fluence, the sheet resistance

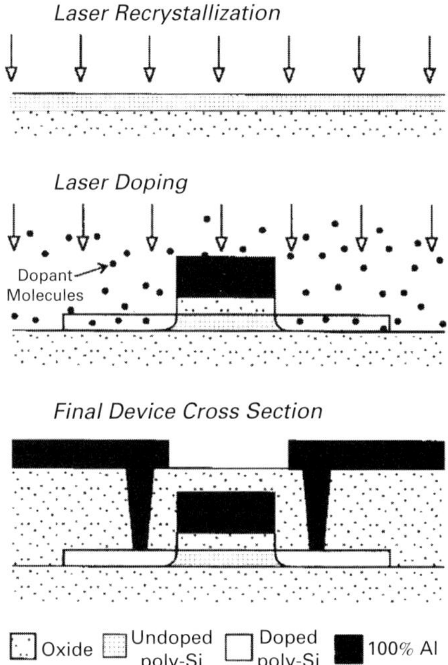

Figure 7.26. Key process steps in the fabrication of gass-immersion laser-doped (GILD) TFTs, showing (a) laser recrystallization of the polysilicon film to produce low-defect-density channel regions; (b) the GILD process, in which dopant atoms adsorbed onto the surface of the source and drain are incorporated into the melt; and (c) the cross section of the finished device, ready for electrical testing. From Giust and Sigmon (1997), reproduced with permission from the IEEE.

decreases with the number of pulses and increasing laser fluence. However, after 100 pulses at the FMT fluence, the poly-Si layer cracks. The GILD technique requires a vacuum chamber and mass-flow-control systems. Solid-state dopants have traditionally been used in the furnace diffusion process for fabrication of semiconductor devices, for the sake of simplicity. A technique was developed for fabrication of the ultra-shallow p^+ junction by excimer-laser doping of crystalline silicon with a solid dopant spin-on-glass (SOG) film (Zhang *et al.*, 1996).

The heat transfer in the silicon can be approximated as a one-dimensional conduction problem. As a first-order approximation, the thermal properties were considered independent of the concentration and the mass diffusivity independent of the temperature. Therefore, the mass diffusion is decoupled from the thermal transport. The mass diffusion can be solved numerically at each time step after the temperature field and the melt–solid interface have been found. Transient mass diffusion is typically modeled by the one-dimensional Fick equation:

$$\frac{\partial C}{\partial t} = \frac{\partial}{\partial z}\left(D\,\frac{\partial C}{\partial z}\right). \tag{7.7}$$

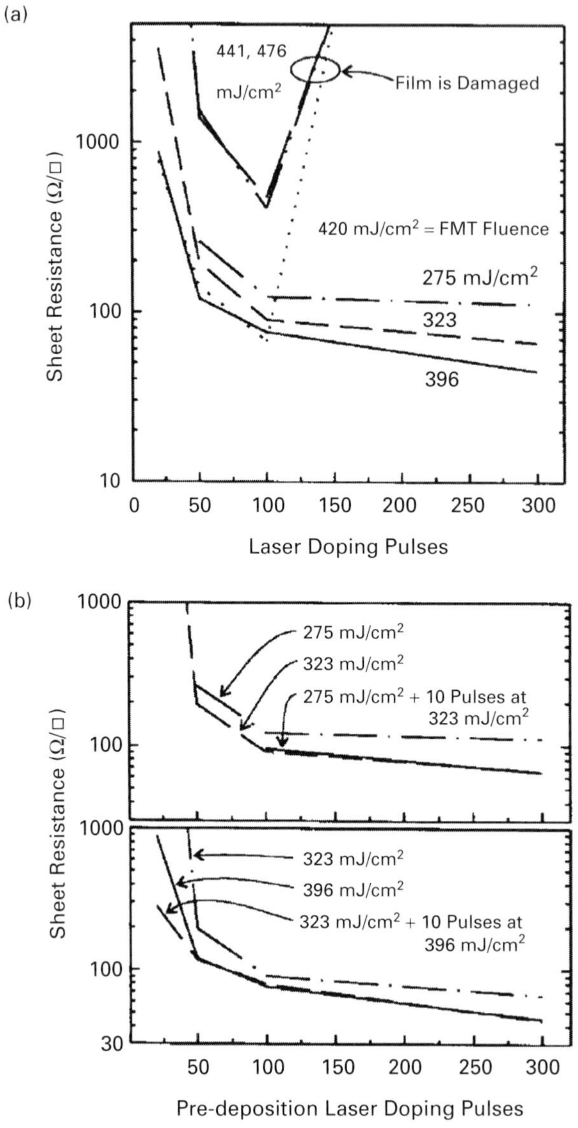

Figure 7.27. The measured Van der Pauw sheet resistance versus the number of laser-doping pulses for several laser fluences. The laser fluence needed to just fully melt the polysilicon film or "full-melt" threshold (FMT) is 420 mJ/cm². Note that irradiation above the FMT fluence damages the polysilicon film. From Giust and Sigmon (1997), reproduced with permission from the IEEE.

Since the mass diffusivity of dopant in the thin layer of liquid silicon induced by the pulsed laser irradiation is higher by several orders of magnitude than that in the solid phase (Kodera, 1963), the shape of the dopant profile is more "box-like," rather than the gradual decrease observed in most diffusion cases. Abrupt or "box-like" dopant profiles provide ideal p^+-junction properties (Weiner, 1993). The art in the formation of this "box-like" shape lies in the fact that, as the number of pulses increases, the diffusion of boron is limited by the maximum melting depth determined by the pulsed laser energy and pulse width. After the dopant diffusion reaches the maximum melting depth, more dopant atoms are piled up and accumulated in the molten layer instead of crossing the melt–solid interface, because of the very low mass diffusivity in the solid silicon. The diffusivity of boron may depend on its concentration if the region is heavily doped, compared with the silicon intrinsic carrier with concentration of the order of $10^{19}/cm^3$ at the melting temperature (Wolf and Tauber, 1986). It is also noted that the thermal conductivity of the heavily doped solid-silicon layer can be decreased significantly, by a factor of 2–5, in the temperature range from room temperature to 600 K (Tai *et al.*, 1988; Slack, 1964). At higher temperature, the thermal conductivity becomes less affected by the high dopant concentration. The decrease in thermal conductivity can lead to deeper and longer melting in silicon.

References

Aichmayr, G., Toet, D., Mulato, M. *et al.*, 1999, "Dynamics of Lateral Grain Growth during the Laser Interference Crystallization of a-Si," *J. Appl. Phys.*, **85**, 4010–4023.

Anderson, R. J., 1988, "A Method to Calculate the Laser Heating of Layered Structures," *J. Appl. Phys.*, **64**, 6639–6647.

Biegelsen, D. K., Fennell, L. E., and Zesch, J. C., 1984, "Origin of Oriented Crystal Growth of Radiantly Melted Silicon on SiO_2," *Appl. Phys. Lett.*, **45**, 546–548.

Bucksbaum, P. H., and Bokor, J., 1984, "Rapid Melting and Regrowth Velocities in Silicon Heated by Ultraviolet Picosecond Laser Pulses," *Phys. Rev. Lett.*, **53**, 182–187.

Burgener, M. L., and Reedy, R. E., 1982, "Temperature Distributions Produced in a Two Layer Structure by a Scanning CW Laser or Electron Beam," *J. Appl. Phys.*, **53**, 4357–4363.

Celler, G. K., 1983, "Laser Crystallization of Thin Si Films on Amorphous Insulating Substrates," *J. Cryst. Growth*, **63**, 429–447.

Chen, S. C., and Grigoropoulos, C. P., 1997, "Non-contact Nanosecond-Time-Resolution Temperature Measurement and Evaluation of Thermal Properties in Excimer Laser Heating of Ni–P Disk Substrates," *Appl. Phys. Lett.*, **71**, 3191–3193.

Deutsch, T. F., Ehrlich, D. J., Rathman, D. D., Silversmith D. J., and Osgood, R. M. Jr., 1981, "Electrical Properties of Laser Chemically Doped Silicon," *Appl. Phys. Lett.*, **39**, 825–827.

DeWitt, D. P., and Rondeau, R. E., 1989, "Measurement of Surface Temperatures and Spectral Emissivities During Laser Irradiance," *J. Thermophys.*, **3**, 153–159.

Dworshack, K., Sipe, J. E., and Van Driel, H. M., 1990, "Solid-Melt Patterns Induced on Silicon by a Continuous Laser Beam at Nonnormal Incidence," *J. Opt. Soc. Am. B*, **7**, 981–989.

Ehrlich, D. J., Brueck, S. R. J., and Tsao, J. Y., 1983, "Surface Electromagnetic Waves in Laser Materials Interactions," *Proc. Mater. Res. Soc.*, **13**, 191–196.

Fauchet, P. M., and Siegman, A. E., 1982, "Surface Ripples on Silicon and Gallium Arsenide under Picosecond Laser Illumination," *Appl. Phys. Lett.*, **40**, 824–826.

Fan, D., Huang, J., and Jaccodine, R. J., 1987, "Enhanced Tail Diffusion of Ion Implanted Boron in Silicon," *Appl. Phys. Lett.*, **50**, 1745–1747.

Fan, J. C. C., Tsaur, B.-Y., and Geis, M. W., 1983, "Graphite-Strip Heater Zone-Melting Recrystallization of Si Films," *J. Cryst. Growth*, **63**, 453–483.

Fisher, J. C., Hollomon, J. H., and Turnbull, D., 1948, "Nucleation," *J. Appl. Phys.*, **19**, 775–784.

Galvin, G. J., Thompson, M. O., Mayer, J. W. *et al.*, 1982, "Measurement of the Velocity of the Crystal–Liquid Interface in Pulsed Laser Annealing of Si," *Phys. Rev. Lett.*, **48**, 33–36.

Giust, G. K., and Sigmon, T. W., 1997, "Self-Aligned Aluminum Top-Gate Polysilicon Thin-Film Transistors Fabricated Using Laser Recrystallization and Gas-Immersion Laser Doping," *IEEE Electron Device Lett.*, **18**, 394–396.

Glazov, V. M., Chizhevskays, S. N., and Glagoleva, N. N., 1969, *Liquid Semiconductors*, New York, Plenum.

Grigoropoulos, C. P., Buckholz, R. H., and Domoto, G. A., 1986, "A Heat Transfer Algorithm for the Laser-Induced Melting and Recrystallization of Thin Silicon Layers," *J. Appl. Phys.*, **60**, 2304–2309.

1987a, "Stability of Silicon Phase Boundaries," *J. Appl. Phys.*, **62**, 474–479.

1987b, "A Thermal Instability in the Laser Driven Melting and Recrystallization of Thin Silicon Films on Glass Substrates," *J. Heat Transfer*, **109**, 841–847.

1988, "An Experimental Study on Laser Annealing of Thin Silicon Layers," *J. Heat Transfer*, **110**, 416–423.

Grigoropoulos, C. P., and Dutcher, W. E., 1992, "Moving Front Fixing in Thin Film Laser Annealing," *J. Heat Transfer*, **114**, 271–277.

Grigoropoulos, C. P., Dutcher, W. E., and Emery, A. F., 1991a, "Experimental and Computational Analysis of CW Argon Laser Melting of Thin Silicon Films" *J. Heat Transfer*, **113**, 21–29.

Grigoropoulos, C. P., Dutcher, W. E., and Barclay, K. E, 1991b, "Radiative Phenomena in CW Laser Annealing," *J. Heat Transfer*, **113**, 657–662.

Grigoropoulos, C. P., Emery, A. F., and Wipf, E., 1990, "Heat Transfer in Thin Silicon Films by Laser Line Sources," *Int. J. Heat Mass Transfer*, **33**, 797–803.

Grigoropoulos, C. P., Rostami, A. A., Xu, X., Taylor, S. L., and Park, H. K., 1993, "Localized Transient Surface Reflectivity Measurements and Comparison to Heat Transfer Modeling in Thin Film Laser Annealing," *Int. J. Heat Mass Transfer*, **36**, 1219–1229.

Hatano, M., Moon, S., Lee, M., and Grigoropoulos, C. P., 2000, "Excimer Laser-Induced Temperature Field in Melting and Resolidification of Silicon Thin Films," *J. Appl. Phys.*, **87**, 36–43.

Hawkins, W. G., and Biegelsen, D. K., 1983, "Origin of Lamellae in Radiatively Melted Silicon Films," *Appl. Phys. Lett.*, **42**, 358–360.

Herlach, D. M., 1994, "Non-equilibrium Solidification of Undercooled Metallic Melts," *Mater. Sci. Eng.*, **R12**, 177–272.

Im, J. S., Crowder, M. A., Sposili, R. S. *et al.*, 1999, "Controlled Super-Lateral Growth of Si Films for Microstructural Manipulation and Optimization," *Phys. Status Solidi (a)*, **166**, 603–617.

Im, J. S., Gupta, V. V., and Crowder, M. A., 1998, "On Determining the Relevance of Athermal Nucleation in Rapidly Quenched Liquids," *Appl. Phys. Lett.*, **72**, 662–664.

Im, J. S., Kim, H. J., and Thompson, M. O., 1993, "Phase Transformation Mechanisms Involved in Excimer Laser Crystallization of Amorphous Silicon Films," *Appl. Phys. Lett.*, **63**, 1969–1971.

Im, J. S., and Sposili, R. S., 1996, "Crystalline Si Films for Integrated Active-Matrix Liquid-Crystal Displays," *Mater. Res. Bull.*, **21**, 39–48.

Im, J. S., Sposili, R. S., and Crowder, M. A., 1997, "Single-Crystal Si Films for Thin-Film Transistor Devices," *Appl. Phys. Lett.*, **70**, 3434–3436.

Ishihara, R., and Matsumura, M., 1997, "Excimer-Laser-Produced Single-Crystal Silicon Thin-Film Transistors," *Jap. J. Appl. Phys.*, **36**, 6167–6170.

Ishikawa, K., Ozawa, M., Ho, C.-H., and Matsumura, M., 1998, "Excimer-Laser-Induced Lateral Growth of Silicon Thin-Films," *Jap. J. Appl. Phys.*, **37**, 731–736.

Jackson, K. A., and Kurtze, D. A., 1985, "Instability in Radiatively Melted Silicon Films," *J. Cryst. Growt.*, **71**, 385–390.

Jellison, G. E. Jr., Lowndes, D. H., Mashburn, D. N., and Wood, R. F., 1986, Time-Resolved Reflectivity Measurements of Silicon and Germanium Using a Pulsed Excimer KrF Laser Heating Beam," *Phys. Rev. B*, **34**, 2407–2417.

Jellison, G. E. Jr., Lowndes, D. H., and Wood, R. F., 1983, "A Detailed Examination of Time-Resolved Pulsed Raman Temperature Measurements of Laser Annealed Silicon," *Proc. Mater. Res. Soc.*, **13**, 35–40.

Kant, R., and Deckert, K. L., 1991, "Laser Induced Heating of a Multilayered Medium Resting on a Half-Space. Part II. – Moving Source," *J. Heat Transfer*, **113**, 12–20.

Kawamura, S., Sakurai, J., Nakano, M., and Takagi, M., 1982, "Recrystallization of Si on Amorphous Substrates by Doughnut-Shaped CW Ar Laser Beam," *Appl. Phys. Lett.*, **40**, 394–397.

Kim, H. J., and Im, J. S., 1996a, "Optimization and Transformation Analysis of Grain-Boundary-Location-Controlled Si Films," *Proc. Mater. Res. Soc.*, **697**, 401–406.

1996b, "New Excimer-Laser-Crystallization Method for Producing Large-Grained and Grain Boundary-Location-Controlled Si Films for Thin Film Transistors," *Appl. Phys. Lett.*, **68**, 1513–1517.

Kluge, M. D., and Ray, J. R., 1989, "Velocity versus Temperature Relation for Solidification and Melting of Silicon: A Molecular-Dynamics Study," *Phys. Rev. B*, **39**, 1738–1746.

Knapp, J. A., and Picraux, S. T., 1983, "Growth of Si on Insulator Using Electron Beams," *J. Cryst. Growth*, **63**, 445–452.

Kodas, T. T., Baum, T. H., and Comita, P. B., 1987, "Surface Temperature Rise in Multilayered Solids Induced by a Focused Laser Beam," *J. Appl. Phys.*, **61**, 2749–2753.

Kodera, H., 1963, "Diffusion Coefficients of Impurities in Silicon Melt," *Jap. J. Appl. Phys.*, **2**, 212–219.

Kraus, H. G., 1987, "Optical Spectral Radiometric/Laser Reflectance Method for Noninvasive Measurement of Weld Pool Surface Temperatures," *Opt. Eng.*, **26**, 1183–1190.

Lee, M., Moon, S., and Grigoropoulos, C. P., 2001a, "In-situ Visualization of Interface Dynamics during the Double Laser Recrystallization of Amorphous Silicon Thin Films," *J. Cryst. Growth*, **226**, 8–12.

2001b, "Ultra-Large Lateral Grain Growth by Double Laser Recrystallization of a-Si Films," *Appl. Phys. A*, **73**, 317–322.

Lee, M., Moon, S., Hatano, M., and Grigoropoulos, C. P., 2000, "Relationship between Fluence Gradient and Lateral Grain Growth in Spatially Controlled Excimer Laser Crystallization of Amorphous Silicon Films," *J. Appl. Phys.*, **88**, 4994–4999.

Lemons, R. A., and Bosch, M. A., 1982, "Microscopy of Si Films During Laser Melting," *Appl. Phys. Lett.*, **40**, 703–707.

Leonard, J. P., and Im, J. S., 2001, "Stochastic Modeling of Solid Nucleation in Supercooled Liquids," *Appl. Phys. Lett.*, **78**, 3454–3456.

Liu, T. M., and Oldham, W. G., 1983, "Channeling Effect of Low Energy Boron Implant in (100) Silicon," *IEEE Electron. Device Lett.*, **4**, 59–62.

Lompré, L. A., Liu, J. M., Kurz, H., and Bloembergen, N., 1983, "Time-Resolved Temperature Measurement of Picosecond Laser Irradiated Silicon," *Appl. Phys. Lett.*, **43**, 168–170.

Mansuripur, M., Connell, G. A. N., and Goodman, J. W., 1982, "Laser-Induced Local Heating of Multilayers," *Appl. Opt.*, **21**, 1106–1117.

Miaoulis, I. N., and Mikic, B. B., 1986, "Heat Source Power Requirements for High-Quality Recrystallization of Thin Silicon Films for Electronic Devices," *J. Appl. Phys.*, **59**, 1658–1666.

Moon, S.-J., Lee, M., and Grigoropoulos, C. P., 2002, "Heat Transfer and Phase Transformations in Laser Annealing of Thin Si Films," *J. Heat Transfer*, **124**, 253–264.

Moon, S., Lee, M., Hatano, M., and Grigoropoulos, C. P., 2000, "Interpretation of Optical Diagnostics for the Analysis of Laser Crystallization of Amorphous Silicon Films," *Micro. Thermoph. Eng.*, **4**, 25–38.

Mullins, W. W., and Sekerka, R. F., 1964, "Stability of a Planar Interface During Solidification of a Dilute Binary Alloy," *J. Appl. Phys.*, **35**, 444–451.

Murakami, K., Eryu, O., Takita, K., and Masuda, K., 1987, "Explosive Crystallization Starting from an Amorphous-Silicon Surface Region during Long-Pulse Laser Irradiation," *Phys. Rev. Lett.*, **59**, 2203–2206.

Nemanich, R. J., Biegelsen, D. K., and Hawkins, W. G., 1983, "Aligned, Coexisting Liquid and Solid Regions in Pulsed and CW Laser-Annealed Si," *Phys. Rev. B*, **27**, 7817–7819.

Ng, K. K., and Lynch, W. T., 1987, "The Impact of Intrinsic Series Resistance on MOSFET Scaling," *IEEE Trans. Electron Devices*, **34**, 503–511.

Nissim, Y. I., Lietoila, A., Gold, R. B., and Gibbons, J. F., 1980, "Temperature Distributions Produced in Semiconductors by a Scanning Elliptical or Circular CW Laser Beam," *J. Appl. Phys.*, **51**, 274–279.

Oh, C.-H., Ozawa, M., and Matsumura, M., 1998, "A Novel Phase-Modulated Excimer-Laser Crystallization Method of Silicon Thin Films," *Jap. J. Appl. Phys.*, **37-Part 2**, L492–L497.

Park, H. K., Xu, X., Grigoropoulos, C. P. *et al.*, 1992, "Temporal Profile of Optical Transmission Probe for Pulsed Laser Irradiation of Amorphous Silicon Films, *Appl. Phys. Lett.*, **61**, 749–751.

Preston, J. S., Sipe, J. E., and Van Driel, H. M., 1986a, "Phase Diagram of Laser Induced Melt Morphologies on Silicon," *Proc. Mater. Res. Soc.*, **51**, 137–142.

1986b, "Laser-Induced Morphological Phase-Transitions on Silicon Surfaces," *J. Opt. Soc. Am. B*, **3**, 156–157.

Preston, J. S., Sipe, J. E., Van Driel, H. M., and Luscombe, J., 1989, "Optical Absorption in Metallic–Dielectric Microstructures, *Phys. Rev. B*, **40**, 3931–3941.

Preston, J. S., Van Driel, H. M., and Sipe, J. E., 1987, "Order–Disorder Transitions in the Melt Morphology of Laser-Irradiated Silicon," *Phys. Rev. Lett.*, **58**, 69–72.

1989b, "Pattern Formation during Laser Melting of Silicon," *Phys. Rev. B*, **40**, 3942–3953.

Sands, T., Washburn, J., Gronsky, R. *et al.*, 1984, "Near-Surface Defects Formed during Rapid Thermal Annealing of Preamorphized and BF_2^+-Implanted Silicon," *Appl. Phys. Lett.*, **45**, 982–987.

Sedwick, T. O., 1981, "A Simple Optical Pyrometer for In-situ Temperature Measurement during CW Laser Annealing", *Proc. Mater. Res. Soc.*, **1**, 147–153.

Siegel, R., and Howell, J. R., 1992, *Thermal Radiation Heat Transfer*, 3rd edn, Washington D.C., Taylor and Francis.

Sipe, J. E., Young, J. F., Preston, J. S., and Van Driel, H. M., 1983, "Laser-Induced Periodic Surface Structure. I. Theory," *Phys. Rev. B*, **27**, 1141–1157.

Slack, G., 1964, "Thermal Conductivity of Pure and Impure Silicon, Silicon Carbide and Diamond," *J. Appl. Phys.*, **35**, 3460–3466.

Slaoui, A., Foulon, F., and Siffert, P., 1990, "Excimer Laser Induced Doping of Phosphorus into Silicon," *J. Appl. Phys.*, **67**, 6197–6201.

Sposili, R. S., and Im, J. S., 1996, "Sequential Lateral Solidification of Thin Silicon Films on SiO₂," *Appl. Phys. Lett.*, **69**, 2864–2866.

Stich, I., Car, R., and Parrinnello, M., 1989, "Bonding and Disorder in Liquid Silicon," *Phys. Rev. Lett.*, **63**, 2240–2243.

Stiffler, S. R., and Thompson, M. O., 1988, "Supercooling and Nucleation of Silicon after Laser Melting," *Phys. Rev. Lett.*, **60**, 2519–2522.

Stolk, P. A., Polman, A., and Sinke, W. C., 1993, "Experimental Test of Kinetic Theories for Heterogeneous Freezing in Silicon," *Phys. Rev. B*, **47**, 5–13.

Stultz, T. J., and Gibbons, J. F., 1981, "The Use of Beam Shaping to Achieve Large-Grain CW Laser-Recrystallized Polysilicon on Amorphous Substrates," *Appl. Phys. Lett.*, **39**, 498–500.

Tai, Y. C., Mastrangelo, C. H. and Muller, R. S., 1988, "Thermal Conductivity of Heavily Doped Low-Pressure Chemical Vapor Deposited Polycrystalline Silicon Films," *J. Appl. Phys.*, **63**, 1442–1447.

Thompson, M. O., Galvin, G. J., Mayer, J. W. *et al.*, 1984, "Melting Temperature and Explosive Crystallization of Amorphous Silicon during Pulsed Laser Irradiation," *Phys. Rev. Lett.*, **52**, 2360–2363.

Usami, A., Ando, M., Tsunekane, M. and Wada, T., 1992, "Shallow-Junction Formation on Silicon by Rapid Thermal Diffusion of Impurities from a Spin-on-Source," *IEEE Trans. Electron Devices*, **39**, 105–110.

Weiner, K. H., Carey, P. G., McCarthy, A. M., and Sigmon, T. W., 1993, "An Excimer-Laser-Based Nanosecond Thermal Diffusion Technique for Ultra-Shallow pn Junction Fabrication," *Microelectron Eng.*, **20**, 107–130.

Wolf, S. and Tauber, R. N., 1986, *Silicon Processing for the VLSI Era*, Vol. 1, New York, Lattice Press, p. 251.

Wood, R. F., and Geist, G. A., 1986a, "Theoretical Analysis of Explosively Propagating Molten Layers in Pulsed-Laser-Irradiated a-Si," *Phys. Rev. Lett.*, **57**, 873–876.

1986b, "Modeling of Nonequilibrium Melting and Solidification in Laser-Irradiated Materials," *Phys. Rev. B*, **34**, 2606–2620.

Wood, R. F., Geist, G. A., and Liu, C. L., 1996, "Two-Dimensional Modeling of Pulsed Laser-Irradiated a-Si and Other Materials," *Phys. Rev. B*, **53**, 15863–15870.

Xu, X., Taylor, S. L., Park, H. K., and Grigoropoulos, C. P., 1993, "Transient Heating and Melting Transformations in Argon-Ion Laser Irradiation of Polysilicon Films," *J. Appl. Phys.*, **73**, 8088–8096.

Xu, L., Grigoropoulos, C. P., and King, T.-J., 2006, "High Performance Thin Silicon Film Transistors Fabricated by Double Laser Crystallization," *J. Appl. Phys.*, **99**, 034508.

Young, J. F., Preston, J. S., Van Driel, H. M., and Sipe, J. E., 1983, "Laser-Induced Periodic Surface Structure. II. Experiments on Ge, Si, Al, and Brass," *Phys. Rev. B*, **27**, 1155–1172.

Zhang, X., Ho, J.-R., and Grigoropoulos, C. P., 1996, "Ultrashallow p⁺-Junction Formation in Silicon by Excimer Laser Doping – A Heat and Mass Transfer Perspective," *Int. J. Heat Mass Transfer*, **39**, 3835–3844.

8   Laser-induced surface modification

8.1   Hydrodynamic stability of transient melts

Laser melting of metals has revealed interesting topographical-development phenomena, even though laser-beam uniformity was maintained by applying a large-area excimer-laser pulse of nominally uniform intensity distribution (Kelly and Rothenberg, 1985). Van de Riet (1993) studied the roughening of metal surfaces after ablation by excimer-laser pulses ($\lambda = 308$ nm, $t_{\mathrm{pulse}} = 28$ ns) under 10^{-7} Pa pressure. When the thin surface layer of a metal is melted by a nanosecond laser pulse, surface tension tends to reduce regions of high curvature (roughness) and viscosity acts to dampen the resulting oscillations. Melting of gold by excimer-laser pulses ($\lambda_{\mathrm{exc}} = 248$ nm) at low energy densities (Bennett *et al.*, 1995) led to development of surface topography characterized by droplet and ridge formations as shown in Figure 8.1.

The formation of patterns on laser-melted and resolidified metals exhibits a characteristic two-dimensional periodicity (Figure 8.2) that lends itself to a stability analysis. The topography was attributed to a hydrodynamic response to phase change occurring at the surface of the target. The momentum equation balances the melt acceleration with stresses and body forces:

$$\rho \frac{\mathrm{D}\vec{u}}{\mathrm{D}t} = \mathrm{div}(\Sigma) + \vec{F}, \tag{8.1}$$

where Σ is the stress tensor and $\vec{F}$ the body-force vector.

Gravity is negligible as a body force acting on the molten layer in nanosecond pulsed-laser melting, even though it may be relevant in melts generated by continuous-wave (CW) or long-pulse lasers. For nanosecond laser pulses, an important body force may arise from the density change upon melting, in addition to the inertial force caused by the sudden expansion of the target. Besides the vapor recoil pressure which is exerted by the departing particles, gas and plasma pressure impart normal forces on the melt surface. However, normal forces cannot produce motion of the melt, unless there is a variation in the lateral distribution of the normal force. Capillarity contributes to surface-normal forces as seen in the Young–Laplace pressure balance:

$$P_1 - P_v = \sigma_{\mathrm{ST}}\left(\frac{1}{R_{\mathrm{c},1}} + \frac{1}{R_{\mathrm{c},2}}\right) = \frac{2\sigma_{\mathrm{ST}}}{R_{\mathrm{c,inv}}}, \tag{8.2}$$

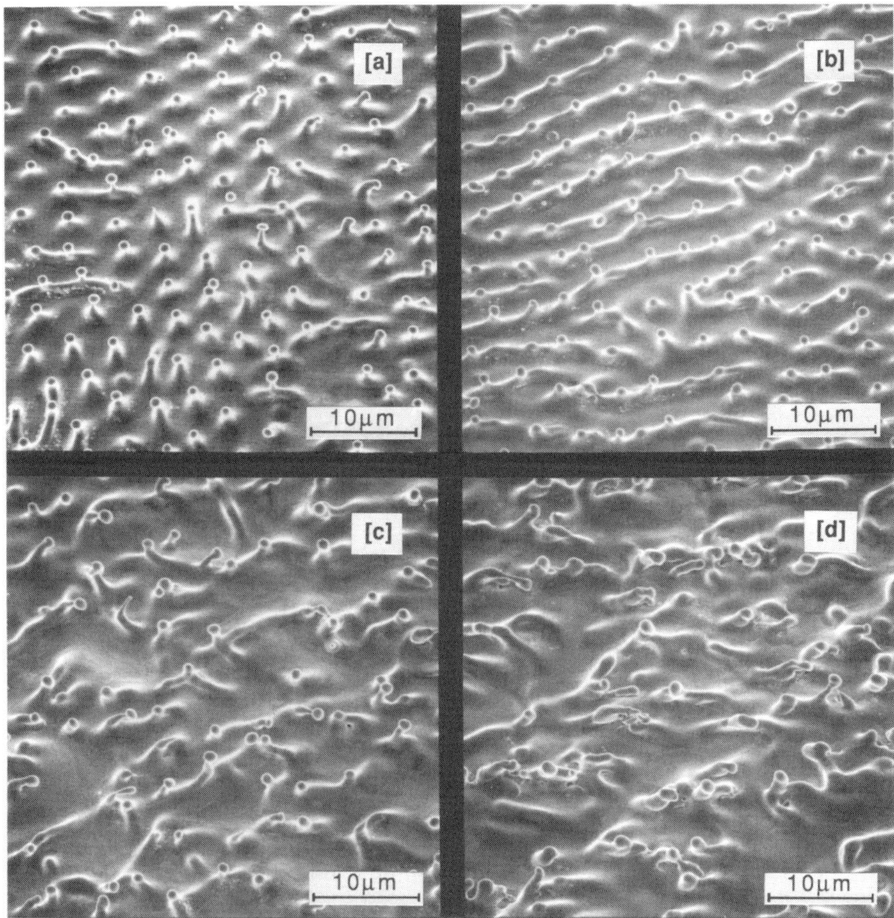

Figure 8.1. Scanning electron micrographs showing the influence of the laser-beam angle of incidence on the steady-state surface topography at a constant excimer-laser fluence, $F = 1.0$ J/cm^2: (a) $\theta_i = 5°$ produces an abundance of droplets with reduced ridge formations; (b) $\theta_i = 15°$ produces highly columnar ridge formations; (c) for $\theta_i = 25°$ a reduction in the number of droplets and loss of ridge columns is observed; and (d) $\theta_i = 45°$ produces higher surface agitation with reduced number of droplets. From Bennett *et al.* (1995), reproduced with permission from the American Institute of Physics.

where $R_{c,inv}$ is the invariant curvature of the surface. If $z = z_{surf}(x, y)$ is the equation of the surface and z_{surf} is assumed small everywhere,

$$\frac{2}{R_{c,inv}} = -\left(\frac{\partial^2 z_{surf}}{\partial x^2} + \frac{\partial^2 z_{surf}}{\partial y^2}\right). \tag{8.3}$$

Perturbation theory is applied to determine the dispersion relation for two-dimensional waves applied on a laser-induced melt pool that is assumed subject to constant acceleration, $-a_1\vec{s}_z$. In a coordinate system anchored to the surface, the pressure P is related to

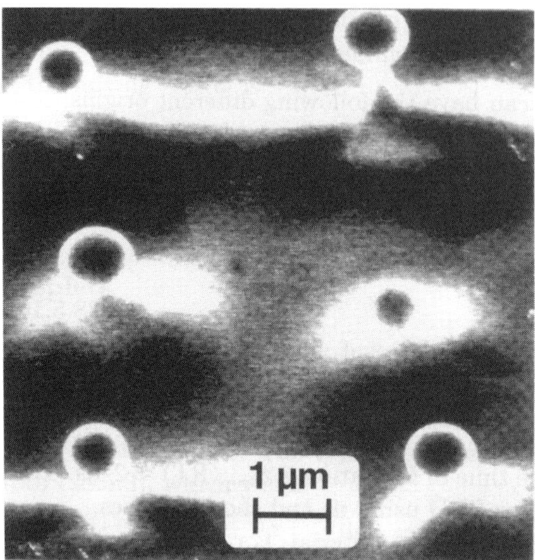

Figure 8.2. A high-magnification scanning-electron-micrograph image of surface droplets shown in Figure 8.1 ($F = 1.0$ J/cm^2, $\theta_i = 5°$), demonstrating characteristic length scales for droplet growth. From Bennett *et al.* (1995), reproduced with permission from the American Institute of Physics.

the true pressure P^* with respect to a fixed reference frame:

$$P = P^* + \rho \vec{z} \cdot \vec{s}_z. \tag{8.3}$$

The conservation equations for incompressible, inviscid flow are

$$\text{Continuity:} \quad \nabla \cdot \vec{u} = 0, \tag{8.4a}$$

$$\text{Momentum:} \quad \frac{\partial(\vec{u})}{\partial t} + \vec{u} \cdot \nabla \vec{u} = -\frac{\nabla P}{\rho}. \tag{8.4b}$$

The velocity and pressure fields are decomposed into the mean-value component plus a perturbation term:

$$\vec{u} = \bar{\vec{u}} + \vec{u}', \qquad P = \bar{P} + P'. \tag{8.5}$$

Since the conservation equations (8.4) have to be satisfied for the mean fields, the following equations hold for the perturbation fields:

$$\text{Continuity:} \quad \nabla \cdot \vec{u}' = 0, \tag{8.6a}$$

$$\text{Momentum:} \quad \frac{\partial(\vec{u}')}{\partial t} = -\frac{\nabla P'}{\rho}. \tag{8.6b}$$

Products between perturbation velocities constitute higher-order terms and have been neglected in (8.6b). An arbitrary disturbance in the fluid may be represented by a composition of normal modes, described by

$$\begin{Bmatrix} z_{\text{surf}} \\ u' \\ v' \\ P' \end{Bmatrix} = \begin{Bmatrix} A \\ \hat{u}(z) \\ \hat{v}(z) \\ \hat{P}(z) \end{Bmatrix} \exp[\mathrm{i}(\kappa_w y - \omega t)], \tag{8.7}$$

where the perturbation-velocity vector is decomposed into the (x, y) components: $\vec{u}' = (u', v')$. The continuity and momentum equations for the perturbation fields can be written as

$$\text{Continuity:} \quad \frac{\mathrm{d}\hat{u}}{\mathrm{d}z} + \mathrm{i}\kappa_w \hat{v} = 0, \tag{8.8a}$$

$$\text{Momentum:} \quad \begin{Bmatrix} -\mathrm{i}\omega\hat{u} \\ -\mathrm{i}\omega\hat{v} \end{Bmatrix} = \begin{Bmatrix} (\mathrm{d}\hat{P}/\mathrm{d}z)/\rho \\ \mathrm{i}\kappa_w \hat{P}/\rho \end{Bmatrix}. \tag{8.8b}$$

From the above equations,

$$\frac{\mathrm{d}^2 \hat{P}}{\mathrm{d}z^2} = -\mathrm{i}\rho\omega \frac{\mathrm{d}\hat{u}}{\mathrm{d}z} = -\rho\kappa_w \omega\hat{v} = \kappa_w^2 \hat{P}. \tag{8.9}$$

Consequently,

$$\hat{P}(z) = B' \cosh(-\kappa_w z) + C' \sinh(\kappa_w z). \tag{8.10}$$

Utilizing the no-slip boundary condition $\hat{u}(z = -z_{\text{int}}) = 0$, it is deduced that $B'/C' = \coth(\kappa_w z_{\text{int}})$. The constant C' is determined using the boundary condition $\partial z_{\text{int}}/\partial t = u'|_{z\to 0}$, from which it is found that $C' = A\omega^2 \rho/\kappa_w$. Equation (8.10) then gives

$$\hat{P}(z) = (A\omega^2 \rho/\kappa_w)[\coth(\kappa_w d_1) \cosh(-\kappa_w z) + \sinh(\kappa_w z)]. \tag{8.11}$$

Across the liquid–vapor interface, the true pressure change is

$$\left.(P_1^* - P_v^*)\right|_{z=z_{\text{int}}} \approx -\frac{\partial^2 z_{\text{int}}}{\partial y^2}. \tag{8.12}$$

The true pressure on the liquid is

$$P_1^* = \left.(\bar{P} + P' - \rho\vec{z}\cdot\vec{a}_1)\right|_{z=z_{\text{int}}}. \tag{8.13}$$

Recognizing that the mean pressure is $\bar{P} = P_v^*$, the following equation is derived for the pressure disturbance:

$$P'(z_{\text{int}}) + \rho z_{\text{int}} a_1 = -\sigma_{\text{ST}}(\partial^2 z_{\text{int}}/\partial y^2). \tag{8.14}$$

Utilizing (8.7), the above equation yields

$$\coth(\kappa_w d_1)\omega^2 = \left(\sigma_{\text{ST}}\kappa_w^3 - \rho a_1\kappa_w\right)/\rho. \tag{8.15}$$

If a_1 is sufficiently large, ω becomes an imaginary number and according to Equation (8.7) the disturbance will grow in time, giving rise to instability. If viscosity is taken into account, the momentum equation for the velocity disturbance becomes

$$\frac{\partial(\vec{u}')}{\partial t} = -\frac{\nabla P'}{\rho} + \nu_{\text{kin}} \nabla^2 \vec{u}'. \tag{8.16}$$

According to Pendleton *et al.* (1993), the dispersion relation for a layer of semi-infinite thickness becomes

$$\omega^2 + (\omega_{\text{im}})^2 = \frac{\sigma_{\text{ST}} \kappa_w^3}{2\rho}, \tag{8.17}$$

with

$$\omega_{\text{im}} = \frac{1}{\tau} = \frac{\mu_{\text{dyn}} \kappa_w^2}{2\rho}, \tag{8.18}$$

where τ is the damping time for the wavenumber κ_w. For waves of wavelength $\lambda = 1\ \mu\text{m}$ on a gold melt, $\tau \sim 0.2$ ns. Continuity dictates that the solid–liquid-interface velocity, u_{int}, is related to the material velocity of the melt, u_1, by the following expression:

$$u_1 = u_{\text{int}} \left(1 - \frac{\rho_s}{\rho_1} \right)_{\text{int}}. \tag{8.19}$$

Correspondingly, the acceleration of the melt is

$$a_1 = a_{\text{int}} \left(1 - \frac{\rho_s}{\rho_1} \right)_{\text{int}}. \tag{8.20}$$

Considering that for most metals $\rho_s > \rho_1$, it is inferred that the molten layer experiences an acceleration toward the solid bulk that contributes to a destabilizing body force together with the thermal expansion due to the rapid energy deposition. Figure 8.3 displays calculations of the thermal conditions and the interface velocity in the gold target obtained via the enthalpy method. The inertia force resulting from the volumetric expansion/contraction upon melting/freezing is approximately of the same order as the force due to volumetric expansion during the heating cycle. It is also noted that the phase-change-driven acceleration of the melt acts throughout the entire melting and solidification sequence, while the cooling process is relatively slower and consequently produces a comparatively weaker force. By taking the extrema of the dispersion relation, the most dangerous disturbance wavelength λ_D can be found by solving the following algebraic equation:

$$\left(3\Gamma_m x_D^2 - 1\right)\sinh(2x_D) + 2\left(\Gamma_m x_D^2 - 1\right) = 0. \tag{8.21}$$

In the above, $\Gamma_m = \sigma_{\text{ST}}/(\rho_1 a_1 d_1^2)$, $x_D = \kappa_{w,D} d_1$, and $\lambda_D = 2\pi/\kappa_{w,D}$. This Rayleigh–Taylor type of stability analysis predicts the observable $\sim$4-μm periodicity in topography development.

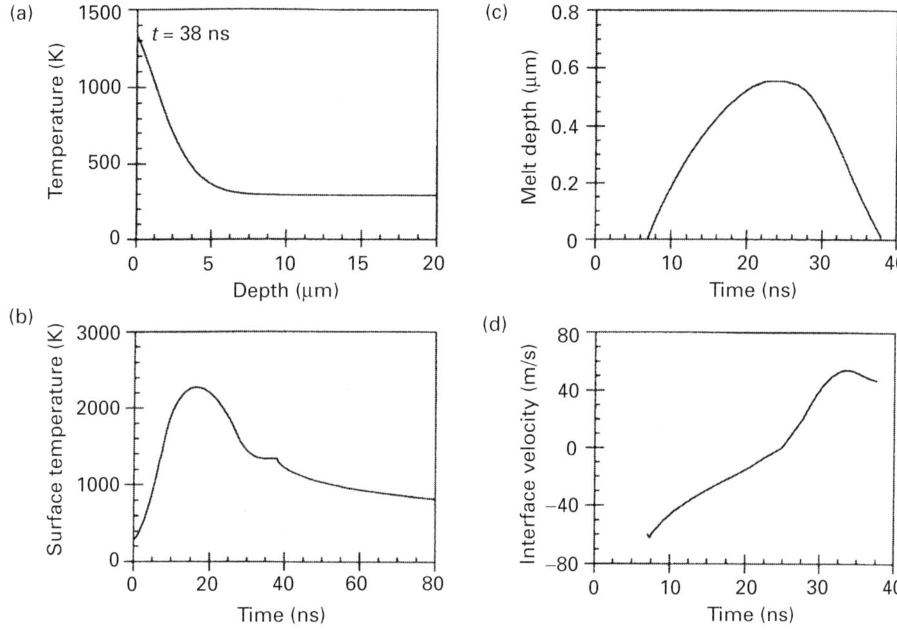

Figure 8.3. Numerically calculated thermal conditions of the gold target during sputtering at $F = 1.0$ J/cm^2: (a) the temperature distribution in the target upon completion of melt resolidification; (b) the temporal behavior of the surface temperature; (c) the temporal behavior of the melt depth; and (d) the temporal behavior of the liquid–solid interface velocity. From Bennett *et al.* (1995), reproduced with permission from the American Institute of Applied Physics.

8.2 Capillary-driven flow

8.2.1 Introduction

Whereas inertia forces due to the increase in density upon melting are destabilizing as in the experiment described above, a number of other factors may be contributing to the generation of surface topography. Periodic wave structures may be induced on the surface of semiconductors, metals, polymers, dielectric materials, and liquids. This phenomenon was observed for a variety of laser-beam parameters for CW beams or Q-switched pulses at intensities even below the damage threshold. A theory explaining the scaling of periodic patterns with the incident laser-light wavelength based on the concept of radiation "remnants" scattered from irregular surface structures was proposed by Sipe *et al.* (1983). Other authors attributed the formation of surface structures of periods longer than the incident laser-light wavelength to melt-flow instabilities. For example, Tokarev and Konov (1994) presented a theoretical study of thermocapillary waves induced by variations in surface shear stress during laser melting of metals and semiconductors. The spatial variation of the laser-beam intensity profile gives rise to surface-temperature and mass-concentration gradients leading to surface deformation. This is the physical principle in an application developed and implemented in the computer-hard-disk industry whereby a pulsed laser is used to create a landing zone on the hard-disk surface (Ranjan *et al.*, 1991). The laser-generated features are typically of diameter a few tens of micrometers and of crater shape or exhibit a central protrusion and a peripheral rim of height only a few tens of nanometers. It has been shown that the laser texturing drastically improves the tribological behavior and performance of the hard disks.

It is apparent that, in the case of a laser beam of Gaussian spatial intensity profile, the energy absorption and the resulting temperature field are not uniform across the irradiated spot. The shape of the surface of the molten pool is determined by the thermal expansion of the material, the density change upon melting and the variation of the surface tension on the surface of the liquid. When the temperature coefficient of surface tension $d\sigma_{ST}/dT < 0$, the surface is depressed. For laser intensities producing significant evaporation, the recoil of the particles leaving the surface may produce similar results. The situation may be even more complex. For example, in surface alloying, the dependence of the surface tension on the species concentration may yield

$$\frac{d\sigma_{ST}}{dT} = \frac{\partial\sigma}{\partial T} + \frac{\partial\sigma}{\partial C}\frac{\partial C}{\partial T} > 0.$$

In this case, the surface will extend outward at the center. Balandin *et al.* (1995) studied the flow of oxidized thin iron films that were melted by frequency-doubled ($\lambda = 532$ nm) Nd : YAG pulses of duration 20 ns, inducing transient surface temperatures and radial gradients up to 4000 K and 5×10^8 K/m, respectively. It was shown that the flow of the liquid was mainly driven by a rapidly changing gradient of the surface tension caused by a time-varying distribution of temperature and the surface concentration of dissolved active oxygen atoms. The influence of surface chemistry on the topography produced was examined by Schwarz-Selinger *et al.* (2001) in their study of the processing of crystalline Si by Q-switched, frequency-doubled ($\lambda = 532$ nm) Nd : YAG laser pulses of FWHM 1 ns. For low fluences near the melt threshold, surfaces carrying native oxides tend to produce positive bumps, in contrast with the crater shapes that are generated on surfaces treated by hydrofluoric acid prior to processing (Figure 8.4).

8.2.2 An approximate analytical model

Consider that the laser-pulse length is much larger than the time scale for the propagation of fluid momentum through the thickness of the melt layer, i.e. $t_{\text{pulse}} \gg d_l^2/\nu_{\text{kin}}$, where d_l and ν_{kin} are the thickness and kinematic viscosity of the melt. For an incompressible, steady-state flow the continuity and momentum equations are reduced to

$$\nabla \cdot \vec{u} = 0, \qquad (8.22a)$$

$$\nabla^2\vec{u} = 0. \qquad (8.22b)$$

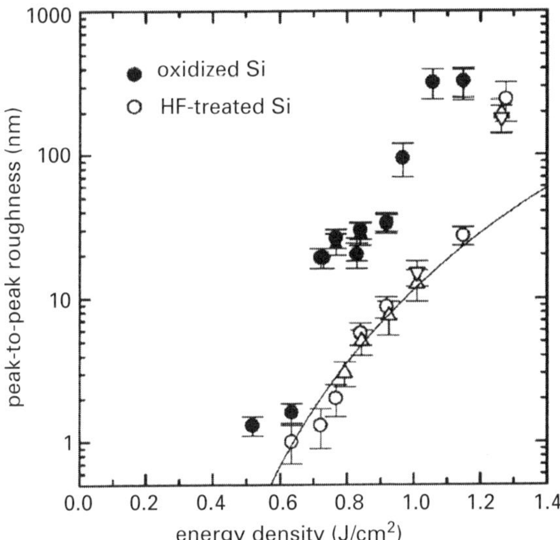

Figure 8.4. The peak-to-peak roughness of the dimples measured with AFM as a function of surface oxidation. Filled symbols: surface covered by a typical native oxide of Si; open symbols: surface treated with dilute hydrofluoric acid (HF) prior to laser texturing. In both cases, $2w \sim 3\ \mu$m. The solid line is a fit to the data for HF-treated Si using the functional form $\Delta z = C(F - F_{\text{th}})^4$, where F_{th} is the threshold for melting and the constant C is adjusted to fit the data, $C = 30$ nm cm^8 J^{-4}. From Schwarz-Selinger *et al.* (2001), reproduced with permission from the American Physical Society.

The boundary conditions on $\vec{u}$ are that the fluid flow vanishes at the solid–liquid interface, $\vec{u}|_{z=-d_1} = 0$, and that at the surface, $z = 0$,

$$\mu_{\text{dyn}}\left(\frac{\partial u_r}{\partial z} + \frac{\partial u_z}{\partial r}\right)\bigg|_{z=0} = \frac{\partial \sigma_{\text{ST}}}{\partial r}. \tag{8.23}$$

It is further assumed that derivatives in the radial direction are much smaller than derivatives in the z-direction. The momentum equation is then simplified to

$$\frac{\partial^2 u_r}{\partial z^2} = 0, \qquad \frac{\partial^2 u_z}{\partial z^2} = 0. \tag{8.24}$$

Integrating the first of these equations twice gives the radial component of the velocity as a function of depth:

$$u_r = \frac{1}{\mu_{\text{dyn}}} \frac{\partial \sigma_{\text{ST}}}{\partial r}(d_1 + z). \tag{8.25}$$

Under the assumption of constant viscosity, (8.22a) and (8.25) give

$$u_z = -\int_{-d_1}^{0} \frac{1}{r}\frac{\partial(ru_r)}{\partial r}\,\mathrm{d}z = \frac{1}{2\mu_{\text{dyn}}r}\frac{\partial}{\partial r}\left(rd_1^2\frac{\partial\sigma_{\text{ST}}}{\partial r}\right). \tag{8.26}$$

Since the change in the fluid morphology is the integral of u_z over time, the average surface temperature excursion $\langle T_{su} - T_m \rangle$ and melt depth $\langle d_l^2 \rangle$ are defined as follows:

$$\langle T_{su} - T_m \rangle = \int_{T_{su} > T_m} (T_{su} - T_m)\mathrm{d}t, \qquad (8.27a)$$

$$\langle d_l^2 \rangle = \int_{T_{su} > T_m} d_l^2 (T_{su} - T_m)\mathrm{d}t / \langle T_{su} - T_m \rangle. \qquad (8.27b)$$

The equation describing the change in the morphology produced by the laser pulse by thermocapillary-driven flow is

$$\Delta z = -\frac{1}{2\mu_{dyn}r} \frac{\partial \sigma_{ST}}{\partial T} \frac{\partial}{\partial r} \left(r\langle d_l^2 \rangle \frac{\partial \langle T_{su} - T_m \rangle}{\partial r} \right). \qquad (8.28)$$

In the limit of a short laser pulse, such that $t_{pulse} \ll d_l^2 / v_{kin}$, it is assumed that the momentum decays exponentially in time:

$$u_r = u_o \exp\left(-\frac{2v_{kin}}{d_l^2} t \right), \qquad (8.29)$$

where u_o satisfies the integrated momentum conservation:

$$\rho \int u_o \, \mathrm{d}z = \int \frac{\partial \sigma_{ST}}{\partial r} \mathrm{d}t. \qquad (8.30)$$

As in the previous derivation, the change in morphology is

$$\Delta z = \int \int \frac{1}{r} \frac{\partial}{\partial r} \left[r u_o \exp\left(-\frac{2v_{kin}t}{d_l^2} \right) \right] \mathrm{d}z \, \mathrm{d}t. \qquad (8.31)$$

On combining this expression with (8.27) and (8.30), it can be seen that Equation (8.28) is still valid, albeit with a different definition for $\langle d_l^2 \rangle$:

$$\langle d_l^2 \rangle = 2v_{kin} \int_0^{\infty} \exp\left(-\frac{2v_{kin}t}{d_l^2} \right) \mathrm{d}t. \qquad (8.32)$$

The melt depth scales as $d_l \sim (1 - R)(F - F_{th})/L_{sl}$, where F_{th} is the threshold fluence for melting and L_{sl} is the latent heat of melting, while the temperature excursion is given by

$$\langle T_{su} - T_m \rangle \sim (1 - R)(F - F_{th})d_l / K,$$

where K is the thermal conductivity. Replacing differentials of r by the radius of the laser spot w gives

$$\Delta z \sim (1 - R)^4 \frac{\partial \sigma_{ST}}{\partial T} \frac{(F - F_{th})^4}{\mu_{dyn} L_{sl}^3 K w^2}. \qquad (8.33)$$

The functional form $\Delta z \sim C(F - F_{th})^4$ fits the crater-depth data well. On taking $d_l = 100\,\mathrm{nm}$ and $v_{kin} = 2.4 \times 10^{-7}\,\mathrm{m^2/s}$, the predicted $d_l^2/(2v_{kin}) \approx 20$ ns, implying that the short-pulse limit is appropriate for the pulse length 1 ns. Assuming that lateral heat flow is negligible, the radial dependence of the melt depth can be calculated via a one-dimensional melting model and used as input to (8.27) and (8.32). The comparison of the measured topography with the numerical solution is displayed in Figure 8.5.

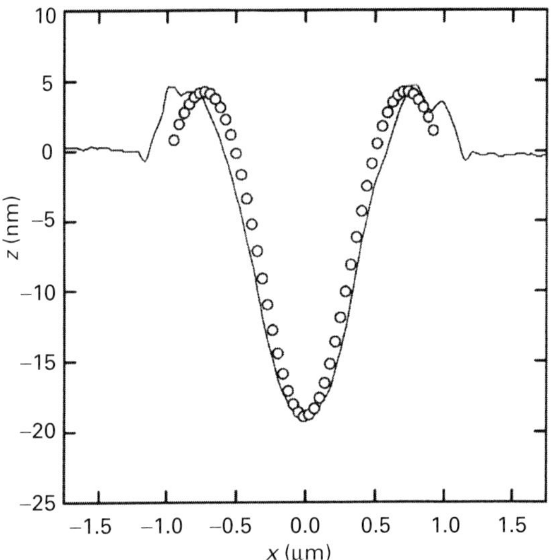

Figure 8.5. A comparison of the morphology of a laser dimple ($F_0 = 1.1$ J/cm^2, $2w = 2.6$ μm) created on HF-treated Si as measured by AFM (solid line) and as predicted (open circles). From Schwarz-Selinger *et al.* (2001), reproduced with permission from the American Physical Society.

8.2.3　Computational modeling of capillary-driven surface modification

The equations governing heat, mass, and momentum transfer, under the assumption of incompressible flow with the appropriate boundary and initial conditions, are mass conservation

$$\nabla \cdot \vec{u} = 0, \tag{8.34}$$

momentum conservation

$$\rho \frac{\mathrm{D}\vec{u}}{\mathrm{D}t} = -\nabla P + \mu_{\mathrm{dyn}} \nabla^2 \vec{u}, \tag{8.35}$$

initial condition

$$\vec{u}(t = t_0) = 0, \tag{8.36a}$$

and boundary conditions

$$\vec{u}(\vec{r} = \vec{r}_{\mathrm{int}}) = 0, \tag{8.36b}$$

i.e. a no-slip condition at the solid–liquid interface, and

$$\vec{t}_{\mathrm{surf}} = \nabla \sigma_{\mathrm{ST}} - (P_\sigma + P_v)\vec{n} \tag{8.36c}$$

at the free surface. The shear force at the melt surface is specified by the Marangoni effect. Normal-shear forces arise from the Young–Laplace pressure P_σ and the recoil pressure P_v.

The heat-conduction equation models energy conservation:

$$\frac{\mathrm{D}T}{\mathrm{D}t} = \alpha \, \nabla^2 T. \tag{8.37}$$

The initial condition is

$$T(t = 0) = T_\infty. \tag{8.38a}$$

The boundary conditions are

$$T(\vec{r} \to \infty) = T_\infty, \tag{8.38b}$$

$$-K \left. \frac{\partial T}{\partial n} \right|_{\mathrm{sur}} = I_{\mathrm{las}}(r, t), \tag{8.38c}$$

where $I_{\mathrm{las}}(r, t)$ is the transient laser intensity impinging on the surface.

The diffusion in a constant-density fluid with constant thermal properties is governed by the equation

$$\frac{\mathrm{D}C}{\mathrm{D}t} = D \, \nabla^2 C. \tag{8.39}$$

The initial condition is

$$C(t = 0) = C_0. \tag{8.40a}$$

The boundary conditions are

$$C(\vec{r} \to \infty) = C_0, \tag{8.40b}$$

$$-\rho D \left. \frac{\partial C}{\partial n} \right|_{\mathrm{sur}} = j_{\mathrm{ev}}, \tag{8.40c}$$

where the solute mass vapor flux, j_{ev}, is given by the Hertz–Knudsen–Langmuir formula. Surface tension is assumed to be a function of both temperature, T, and mass surface concentration. For pure liquids, the surface tension is known to depend linearly on the temperature. On the basis of thermodynamic arguments, it can be shown that the dependence of surface tension on concentration is logarithmic. The system of equations, subject to the specified initial and boundary conditions, is solved numerically for an axisymmetric domain using the finite-element method. The predicted surface-deformation profiles (Bennett *et al.*, 1997) shown in Figure 8.6 are consistent with the measured data (Tam *et al.*, 1996).

8.2.4   Experimental probing

Photothermal deflection

Photothermal displacement (PTD) techniques have been developed to measure slight displacements due to laser heating by using a probing beam (Mandelis, 1992). However, conventional PTD schemes require high-frequency modulation of the heating beam in order to detect a minute deflection signal due to a small temperature change or small deformation on the sample surface in the thermoelastic regime (Olmstead *et al.*, 1983). Moreover, conventional PTD setups are not technically practical for detecting the bump

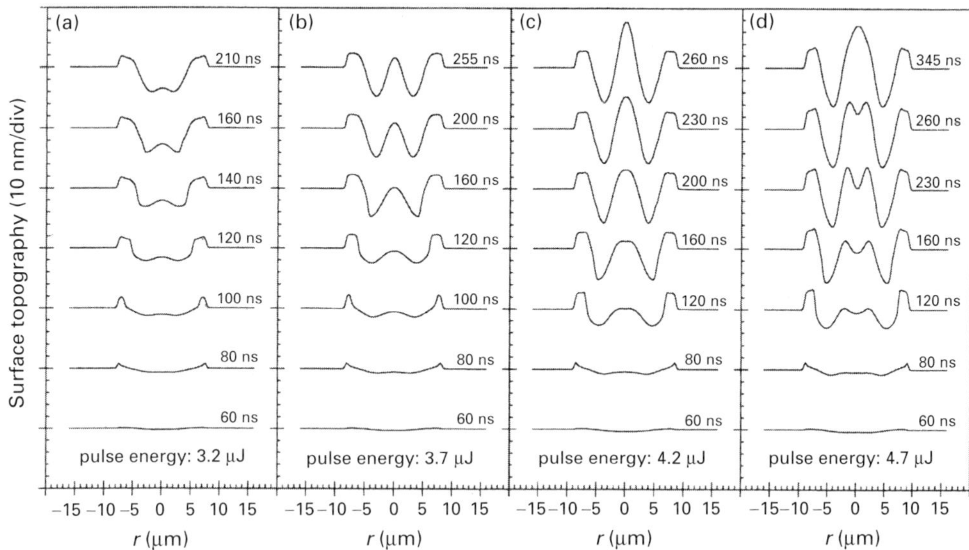

Figure 8.6. A summary of transient surface topography for various pulse energies. From Bennett *et al.* (1997), reproduced with permission from the American Society of Mechanical Engineers.

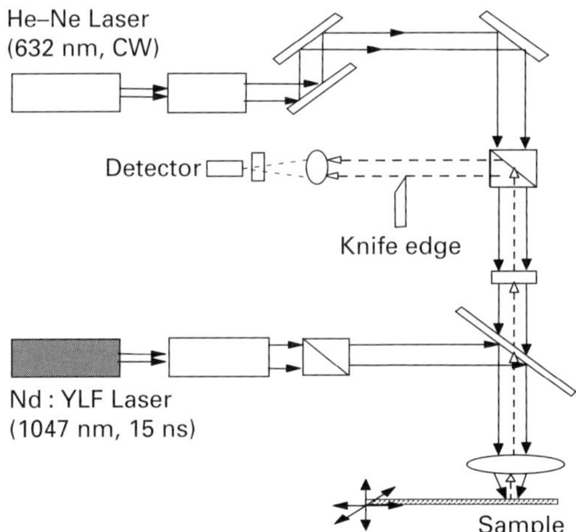

Figure 8.7. The experimental setup of the photothermal-displacement method. From Chen *et al.* (1998), reproduced with permission from the American Institute of Physics.

growth dynamics with the required stability and alignment reproducibility. Chen *et al.* (1998) developed a PTD setup to study the transient melting and surface deformation of materials upon single-pulsed-laser heating. This setup provides a technique that is robust and of high resolution both in time and in space for measuring transient deformation on the material surface (Figure 8.7). When a single pulse is fired onto the target, the surface experiences thermoelastic expansion, melting, and deformation, leading to changes in the

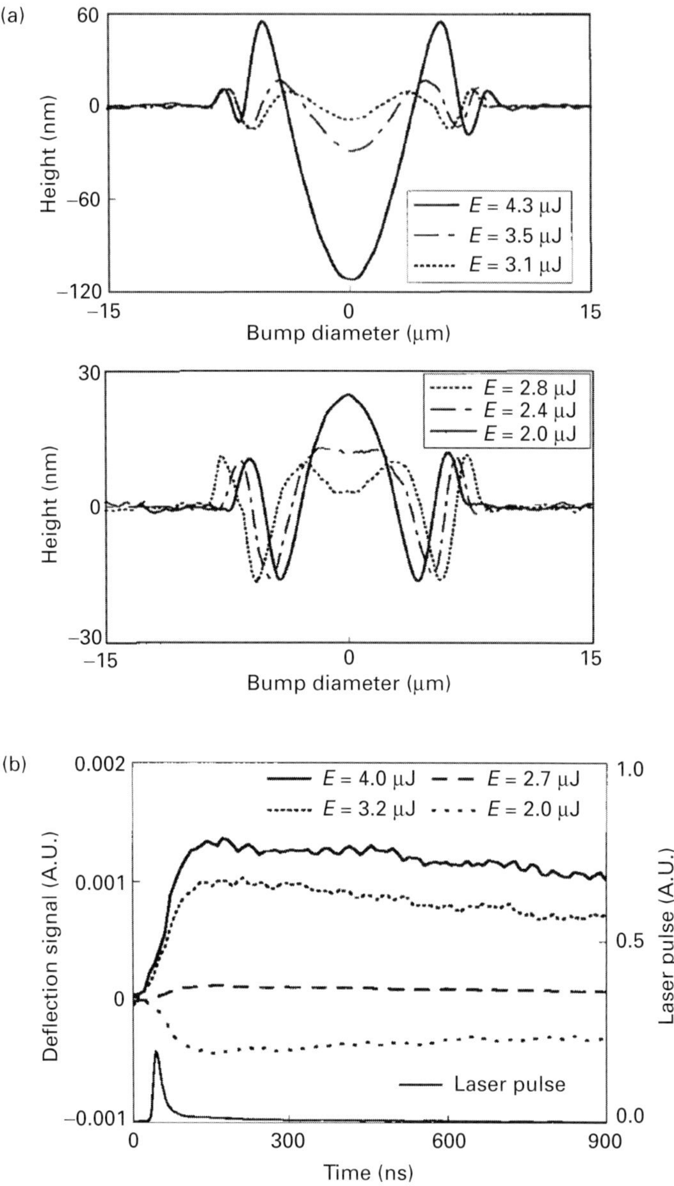

Figure 8.8. Transient deflection signals at various laser-heating energies. The deflection signal is enhanced due to the crater formation, and weakened for the sombrero case. From Chen *et al.* (1998), reproduced with permission from the American Institute of Physics.

reflected He–Ne-laser beam. The detected signal comprises contributions stemming from changes in optical properties upon melting as well as from the evolving topography. By detecting the deflection signal of the probe He–Ne-laser beam, the transient deformation of the surface can be tracked with nanosecond time resolution (Figure 8.8). A variant of this method was applied by Shiu *et al.* (1999b) for *in situ* probing of the formation of miniature glass-surface features that will be discussed in Section 8.3.

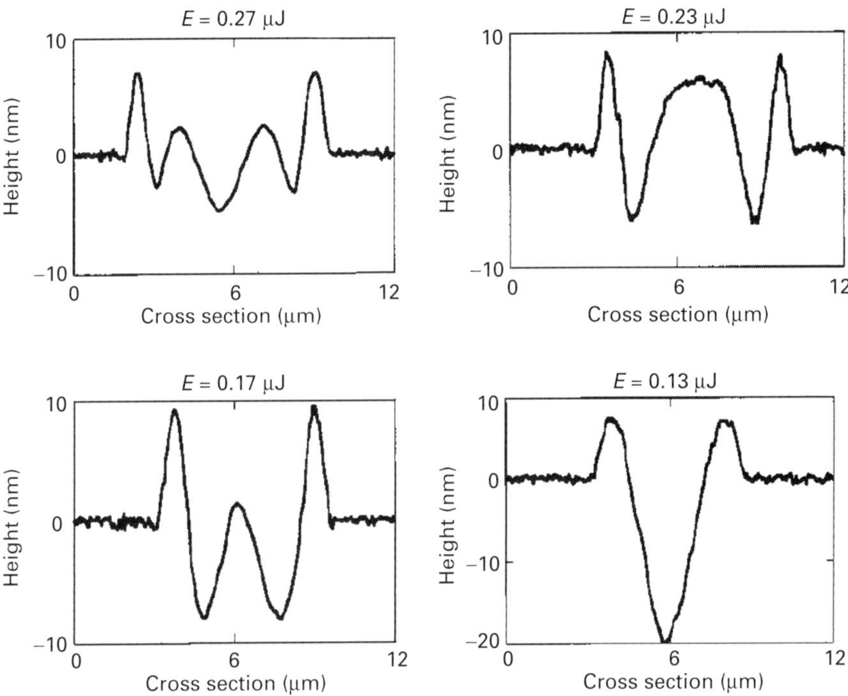

Figure 8.9. The laser-energy dependence of the bump shape (cross section, measured by AFM) in green-laser texturing. From Chen *et al.* (2000), reproduced with permission from the American Society of Mechanical Engineers.

Optical visualization

A laser flash-photograpy (LFP) system was utilized to visualize the dynamic process of surface deformation (Chen *et al.*, 2000). The pulse energy is adjusted to create smooth bumps. The bumps have diameters of about 5 μm and rim heights varying from 5 to 20 nm (Figure 8.9). Visualization of the entirety of bump growth by the laser flash deflection microscope has been conducted for various pulse energies as shown in Figure 8.10.

Photoelectron imaging

Figure 8.11 shows a schematic diagram of the high-speed double-frame time-resolving photoelectron microscope (Bostanjoglo *et al.*, 2000; Nink *et al.*, 1999) for imaging pulsed-laser surface-modification processes. A transmission electron microscope (TEM) was modified by replacing the standard electron gun by a UV-laser pulse-driven photoelectron-emission gun, consisting of a zirconium-covered hairpin-shaped rhenium wire as a photo-emitter, a focusing electrode, and a grounded anode with an integrated aluminum mirror to direct the laser pulses onto the emitter tip. The cathode-driving laser was a two-step Q-switched and frequency-quadrupled Nd : YAG laser, producing two 8-ns pulses with a selectable spacing of 20 ns to 2 ms. The double pulse was focused onto the electron-emitter tip and produced two photoelectron pulses that were

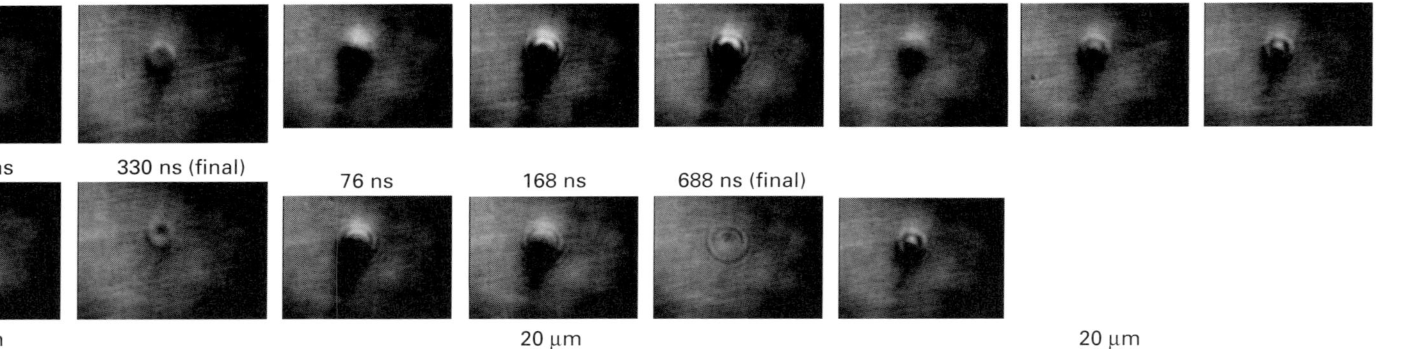

Figure 8.10. A sequence of images of bumps with green-laser pulse energies of (a) 0.13 μJ (V type), (b) 0.23 μJ (sombrero type), and (c) 0.27 μJ (double-rim type) produced by a laser-flash-deflection microscope. From Chen *et al.* (2000), reproduced with permission from the American Society of Mechanical Engineers.

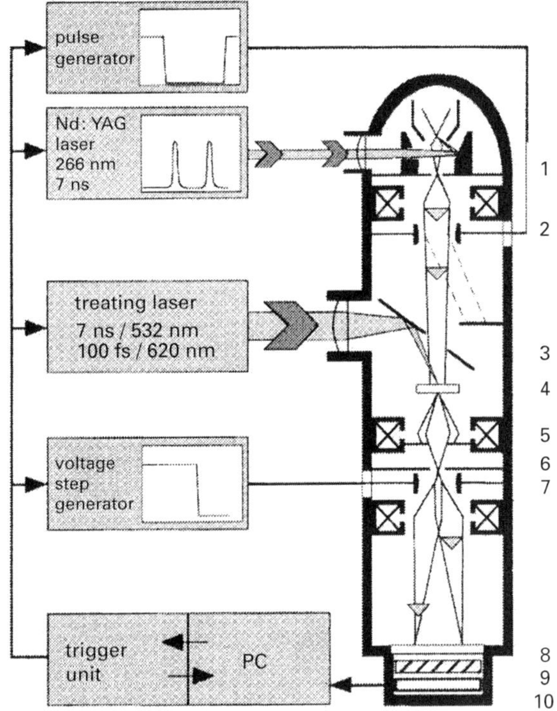

Figure 8.11. High-speed transmission electron microscopy with a pulsed laser for specimen treating: (1) laser-driven photoelectron gun, (2) beam blanker, (3) lasers for treating the specimen, (4) specimen, (5) objective lens, (6) field aperture, (7) frame shifter, (8) fiber plate transmission screen, (9) MCP electron image intensifier, and (10) CCD sensor. From Bostanjoglo *et al.* (2000), reproduced with permission from Elsevier.

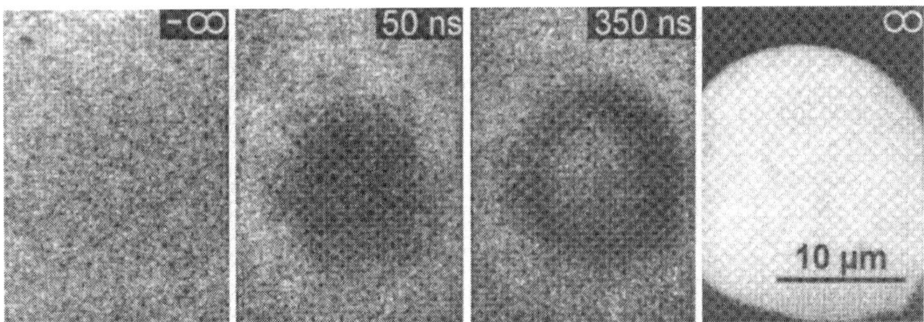

Figure 8.12. Short exposure images of the flow in a NiP melt produced by a 7-ns pulse. The moment of exposure is counted from the maximum of the laser pulse ("−∞" before, "∞" 10 s after the laser pulse.) From Bostanjoglo *et al.* (2000), reproduced with permission from Elsevier.

accelerated and illuminated the specimen. A second frequency-doubled Nd : YAG laser operating at $\lambda = 532$ nm was used for the *in situ* treatment of the specimen. The electron images produced by the double electron exposure were recorded by a multi-channel plate (MCP)-intensified CCD camera. According to Bostanjoglo and Weingärtner (1997), the rise in sample temperature due to the photoelectron pulses was of the order of 10 K while the imaging spatial resolution for the 4-ns exposure was about 1.3 μm. Figure 8.12 shows typical short-exposure TEM bright-field images of a melt produced by laser pulses that depict the previously mentioned interplay between the inward flow produced at 50 ns and the outward flow at 350 ns that finally opens a hole. It is expected that at higher energies local chemical reactions and phase transitions producing steep gradients of the concentration of the surface active atoms or movement of broken oxides on the melt surface might occur.

8.3   Glass-surface modification

Glass surfaces can be modified permanently by CO_2-laser irradiation that couples efficiently with the Si—O absorption band. An application of this process has been the texturing of alternative hard-disk substrates (e.g. Kuo *et al.*, 1997; Tam *et al.*, 1996). A net increase in volume is observed in the glass substrate if the induced temperature remains much less than that required to vaporize the material, implying that thermal expansion/contraction of the material during and after heating is an important factor.

8.3.1   The glass-transition temperature

During the glass-forming (cooling) process, the liquid begins to deviate from its equilibrium state when the available thermal energy is insufficient to provide fast enough atomic movements for the system to stay in equilibrium. The deviation from the equilibrium state can be characterized by the glass-transition temperature (Scherer, 1986, Varshneya, 1994; Zallen, 1983). As the cooling proceeds, the atomic mobilities become lower and lower and eventually no perceptible rearrangement of the structure occurs over practical amounts of time as the material reaches a glassy (solid) state. How early the departure from the equilibrium liquid line begins depends on the cooling rate. During fast cooling, the structure has less time to rearrange itself so as to stay on the equilibrium line, and, therefore, deviates from equilibrium at a higher temperature. The phenomenon of laser heating and cooling of glass is described using the volume–temperature diagram (Figure 8.13). Glass is typically manufactured in a furnace with a slow cooling rate (of the order of 0.1 °C/s), and can be represented with a transition temperature T_g^*. Initially, the glass is considered to be in a state represented by point 1. Laser heating with microsecond pulse duration brings the glass to the liquid state. The cooling rate of 10^6 °C/s forces volumetric contraction past a transition temperature T_g until eventually the "permanent" state 2 is reached.

8.3.2   An analytical model

In a first approximation, the heat-affected zone (Figure 8.14) is defined as the region in the heated glass where the peak temperature exceeds the glass-transition temperature

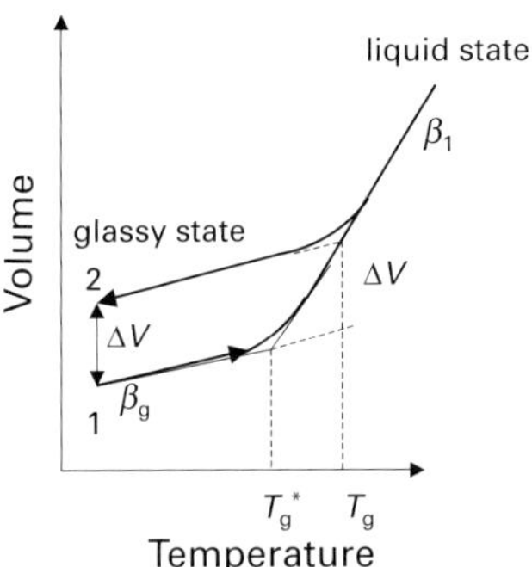

Figure 8.13. The volume–temperature diagram of a glass-forming liquid. The glass is heated by a CO_2 laser from point 1 to full liquid and retracts to point 2 after faster cooling.

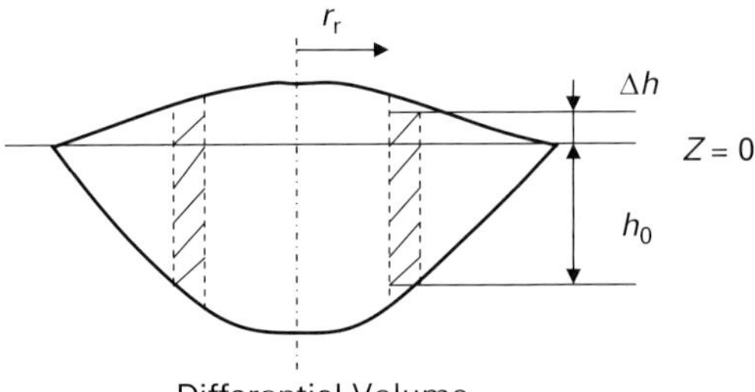

Figure 8.14. A net volume increase ΔV is induced by the permanent structural change resulting from the faster cooling rate, and the affected zone V_0 is the locus region in the heated glass where the peak temperature exceeds the new glass-transition temperature T_g. In this case $\Delta h_b / h_0$ is equal to $\Delta V / V_0$ because radial symmetry is assumed.

T_g. A uniform cooling rate is assumed over the heated area, with a single T_g. Furthermore, the affected zone is surrounded by material that expands and contracts freely at temperatures below T_g. Owing to the low thermal conductivity of glass, heat loss and thermal diffusion during the laser pulse are neglected. The net increase in volume can be written as

$$\Delta V = (\beta_l - \beta_g)V_{\mathrm{HAZ}}(T_g - T_g^*), \tag{8.41}$$

where β_l and β_g are the volume thermal-expansion coefficients in the liquid and glassy states, respectively, and V_{HAZ} is the volume of the heat-affected zone. Assuming axial symmetry, the above relation reduces to

$$\Delta h(r) = (\beta_l - \beta_g)h_b(r)(T_g - T_g^*), \tag{8.42}$$

where $\Delta h(r)$ and $h_b(r)$ represent the bump height and depth in the radial direction, where the maximum temperature exceeds the new glass-transition temperature T_g. For a Gaussian laser beam, the energy absorbed per unit time and unit volume is of the form

$$Q_{ab}(r, z) = \gamma(1 - R)\frac{2P_1}{\pi w^2}\exp\left(-\frac{2r^2}{w^2}\right)\exp(-\gamma z). \tag{8.43}$$

The peak temperatures occur at the surface right upon completion of the laser pulse, at $t = t_{pulse}$:

$$T_{pk}(r) = \frac{Q_{ab}t_{pulse}}{\rho C_p} = \left(\frac{\gamma(1 - R)}{\rho C_p}\right)F\exp\left(-\frac{2r^2}{w^2}\right), \tag{8.44}$$

where $F = 2P_1 t_{pulse}/(\pi w^2)$. Since the temperature decays exponentially in the z-direction during heating, the bump height is

$$h_b(r) = \frac{1}{\gamma}\ln\left(\frac{T_{pk}(r)}{T_g}\right). \tag{8.45}$$

By combining Equation (8.42) with Equation (8.43) we obtain

$$\Delta h_b(r) = \frac{\beta_1 - \beta_g}{\gamma}(T_g - T_g^*)\ln\left(\frac{T_{pk}}{T_g}\right). \tag{8.46}$$

Substituting from Equation (8.44) for T_{pk} gives

$$\Delta h_b(r) = C_g\ln\left[\frac{F}{F_{th}}\exp\left(-\frac{2r^2}{w^2}\right)\right] = h_{b,max} - C_g\left(\frac{2r^2}{w^2}\right), \tag{8.47}$$

where $h_{b,max}$, the maximum bump height at $r = 0$, is

$$h_{b,max} = C_g\ln\left(\frac{F}{F_{th}}\right), \tag{8.48}$$

and

$$C_g = \frac{\beta_1 - \beta_g}{\gamma}(T_g - T_g^*), \tag{8.49}$$

$$\frac{1}{F_{th}} = \frac{\gamma(1 - R)}{\rho C_p}\frac{1}{T_g}. \tag{8.50}$$

The diameter of the bump can also be predicted utilizing Equation (8.44):

$$d_b = \sqrt{2}w\sqrt{\ln\left(F\frac{\gamma(1 - R)}{\rho C_p}\frac{1}{T_g}\right)} = \sqrt{2}w\sqrt{\ln\left(\frac{F}{F_{th}}\right)}. \tag{8.51}$$

Figure 8.15 displays AFM images and cross sections of bumps generated in soda-lime glass subjected to CO_2-laser pulses of wavelength $\lambda = 10.6\,\mu\text{m}$ with a $1/e^2$ beam

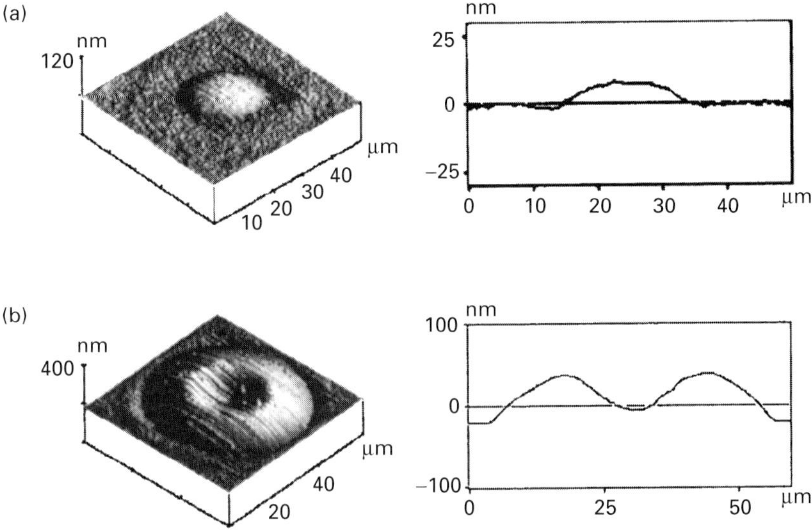

Figure 8.15. Atomic-force-microscopy images and cross sections for soda-lime glass: (a) $E = 18\,\mu\text{J}$ results in a dome-shaped bump and (b) $E = 56\,\mu\text{J}$ initiates a central dimple on the surface. From Shiu *et al.* (1999a), reproduced with permission from the American Institute of Physics.

radius at the sample of $31\,\mu\text{m}$. The constants C_{g} and F_{th} were obtained by fitting experimental data and then used for the prediction of the surface topography as shown in Figure 8.16.

8.3.3 Detailed numerical modeling

A detailed model for the prediction of the laser-induced surface topography in glass materials has been developed by Bennett and Li (2001). Above the glass-transition temperature glass undergoes a structural change whose rate depends on the thermodynamic and fictive temperature in the glass and is assumed to be of the form

$$\frac{\mathrm{d}T_{\text{g}}}{\mathrm{d}t} = -\frac{T_{\text{g}} - T}{\tau(T_{\text{g}}, T)}, \tag{8.52}$$

where $\tau(T_{\text{g}}, T)$ is the relaxation time constant. If the relaxation time constant is known, Equation (8.52) can be integrated forward in time and the fictive temperature throughout the heat-affected zone is found as part of the transient solution. Assuming that the relaxation time can be obtained from bulk glass viscosity data, the Vogel–Fulcher–Tammann equation for viscosity is considered:

$$\log_{10}(\mu_{\text{dyn}}) = A_{\mu_{\text{dyn}}} + \frac{B_{\mu_{\text{dyn}}}}{T - C_{\mu_{\text{dyn}}}}. \tag{8.53}$$

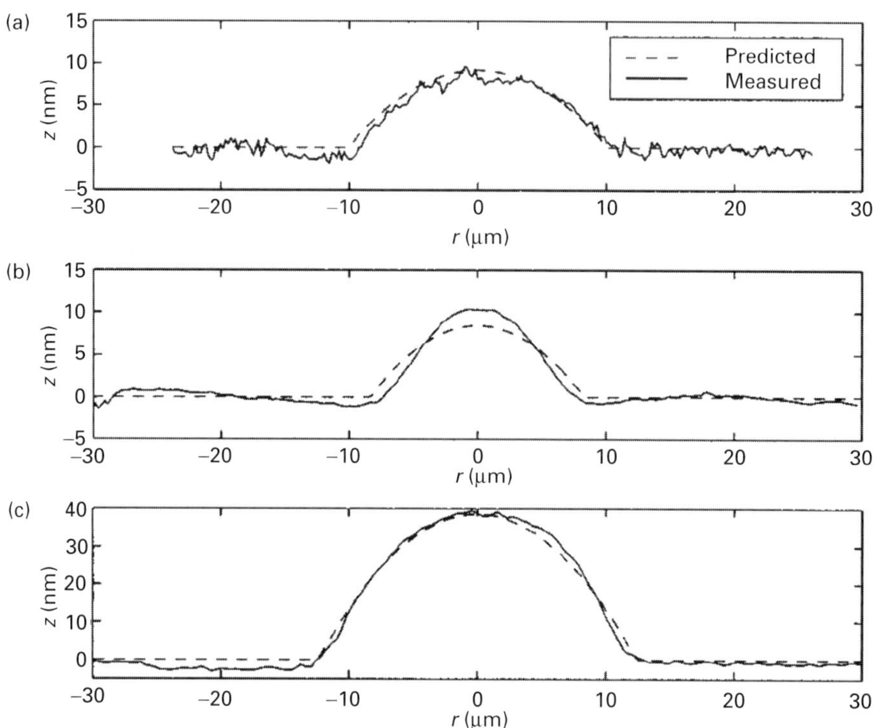

Figure 8.16. A comparison between predicted (dashed line, using C and F_{th} obtained from Figure 8.15) and measured (solid line) bump profiles: (a) soda-lime, $E = 18\,\mu J$; (b) Corning 7059, $E = 20\,mJ$; and (c) MemCor, $E = 18\,\mu J$. From Shiu *et al.* (1999a), reproduced with permission from the American Institute of Physics.

The relaxation time constant, τ is then evaluated:

$$\log_{10}(100s/\tau) = \frac{B_{\mu_{dyn}}(T - T_g^*)}{(T_g^* - C_{\mu_{dyn}})(T - C_{\mu_{dyn}})}. \tag{8.54}$$

In a cylindrical system of coordinates, the strain field $\varepsilon_{str}^{T} = \{\varepsilon_{str,r}, \varepsilon_{str,z}, \gamma_{str,rz}, \varepsilon_{str,\theta}\}$ is computed by considering the contributions of elastic strain, plastic strain, and temperature strain: $\varepsilon_{str} = \varepsilon_{str}^{e} + \varepsilon_{str}^{p} + \varepsilon_{str}^{o}$. A linear relation is applied to evaluate the dependence of the temperature strain on the thermodynamic and dynamic glass-transition temperature:

$$\varepsilon_r^{o} = \varepsilon_z^{o} = \varepsilon_\theta^{o} = \beta_T(T - T_0) + \beta_{T_g}(T_g - T_g^*), \tag{8.55}$$

where β_T and β_{T_g} are the coefficient of thermal expansion and an additional component resulting from the departure of the dynamic from the normal glass-transition temperature. Figure 8.17 gives a quantitative description of the relationship between thermodynamic temperature and the dynamic glass-transition temperature along the centerline of the laser-heated zone. At $1\,\mu s$, and while the thermodynamic temperature has exceeded 1000 K, the dynamic glass-transition temperature is still at the bulk value

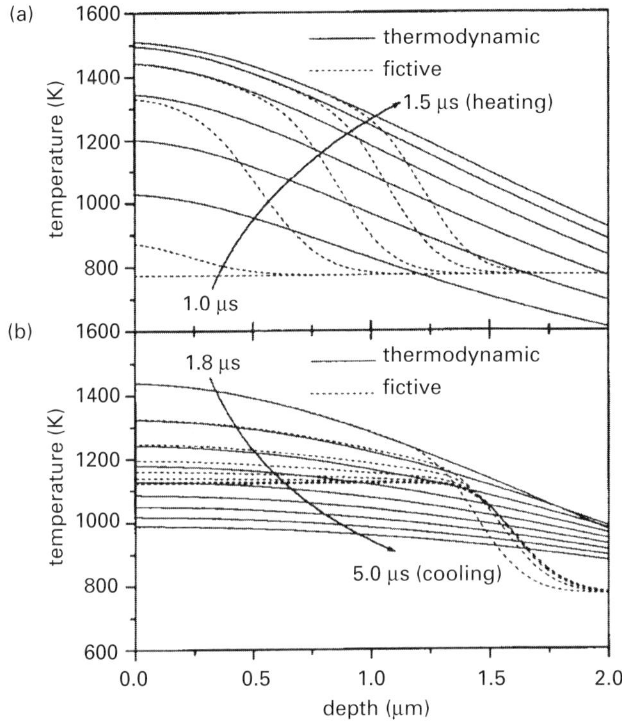

Figure 8.17. The relationship between thermodynamic temperature (solid lines) and fictive temperature (dashed lines) along the centerline of the laser-heated zone in an aluminosilicate glass. A CO_2 laser pulse of wavelength $\lambda = 10.6$ µm, $1/e^2$ radius 13.5 µm, and pulse energy 3.0 µJ, with a Gaussian temporal profile with FWHM pulse width $t_{pulse} = 0.7$ µs and peak intensity at $t_{max} = 1$ µs is utilized. (a) For heating, depth profiles are given at 100-ns intervals between 1.0 and 1.5 µs. (b) For cooling, depth profiles are given at 400-ns intervals between 1.8 and 5.0 µs. From Bennett and Li (2001), reproduced with permission from the American Institute of Physics.

of 773 K. The transition from a glassy to a liquid state occurs over a range of temperature exceeding 1100 K and, for temperatures above 1300 K, the glass-transition temperature is essentially identical to the thermodynamic temperature. The transient evolution of the surface is due to several types of subsurface changes. The first is the reversible thermoelasticity that causes the surface to swell during heating and contract during cooling. The second effect is nonreversible plastic motion of the near-surface volume that has been heated to above the glass-transition temperature. Volumetric plastic changes can occur due to structural changes arising from the departure from the glass-transition temperature. The transient surface topography is depicted in Figure 8.18. As the surface approaches the final shape, the thermoelastic contribution is removed and the topography is entirely due to plastic strain over the region that has been heated to above the glass-transition temperature. Figure 8.19 shows a comparison of the predicted and measured final bump height obtained by fitting $\beta_{T_g} = 2.5 \times 10^{-5}$ to the experimental data.

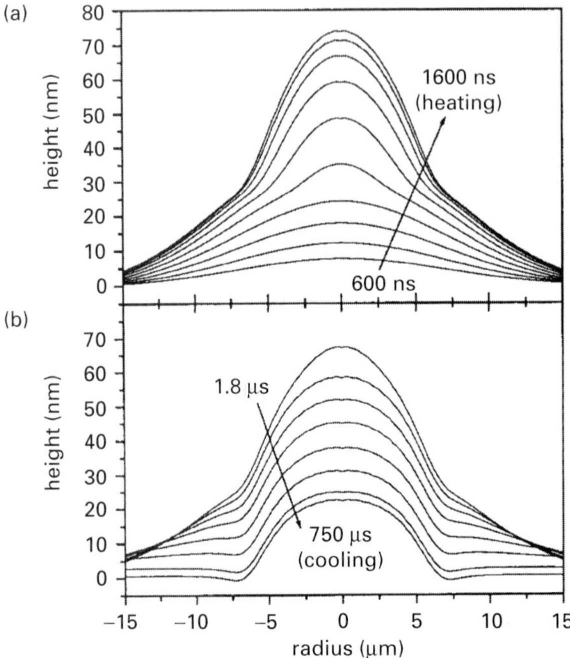

Figure 8.18. Profiles of glass-surface topography during the thermal cycle for the conditions of Figure 8.17. (a) During heating, topography is shown at 100-ns intervals between 600 and 1600 ns. (b) During cooling, topography is shown for 1.8 μs, 3.2 μs, 8.4 μs, 12.8 μs, 25.0 μs, 50.0 μs, 100 μs, and 750 μs. From Bennett and Li (2001), reproduced with permission from the American Institute of Physics.

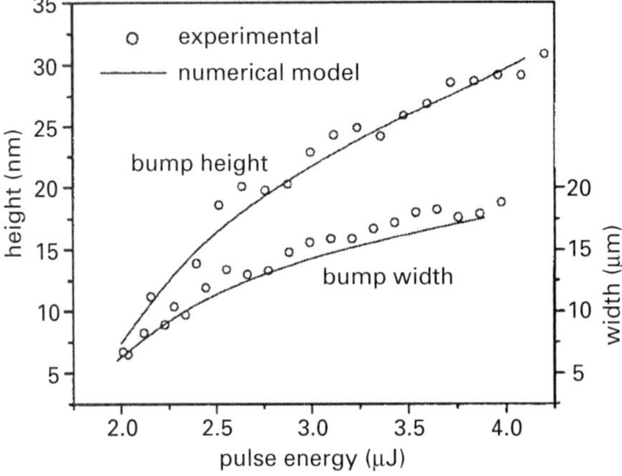

Figure 8.19. A summary comparison between model results and experimental measurements of laser-induced texture bump dimensions as a function of pulse energy. From Bennett and Li (2001), reproduced with permission from the American Institute of Physics.

References

Balandin, V. Yu., Otte, D., and Bostanjoglo, O., 1995, "Thermocapillary Flow Excited by Focused Nanosecond Laser Pulses in Contaminated Thin Liquid Iron Films," *J. Appl. Phys.*, **78**, 2037–2048.

Bennett, T. D., Grigoropoulos, C. P., and Krajnovich, D. J., 1995, "Near-Threshold Laser Sputtering of Gold," *J. Appl. Phys.*, **77**, 849–868.

Bennett, T. D., Krajnovich, D. J., Grigoropoulos, C. P., Baumgart, P., and Tam, A. C., 1997, "Marangoni Mechanism in Pulsed Laser Texturing of Magnetic Hard Disks," *J. Heat Transfer*, **119**, 589–598.

Bennett, T. D., and Li, L., 2001, "Modeling Laser Texturing of Silicate Glass," *J. Appl. Phys.*, **89**, 942–950.

Bostanjoglo, O., and Weingärtner, M., 1997, "Pulsed Photoelectron Microscope for Imaging Laser-Induced Nanosecond Processes," *Rev. Sci. Instrum.*, **68**, 2456–2460.

Bostanjoglo, O., Elschner, R., Mao, Z., Nink, T., and Weingärtner, M., 2000, "Nanosecond Electron Microscopes," *Ultramicroscopy*, **81**, 141–147.

Chen, S. C., Cahill, D. G., and Grigoropoulos, C. P., 2000, "Melting and Surface Deformation in Pulsed Laser Surface Micro-modification of NiP Disks," *J. Heat Transfer*, **122**, 107–112.

Chen, S. C., Grigoropoulos, C. P., Park, H. K., Kerstens, P., and Tam, A. C., 1998, "Photothermal Displacement Measurement of Transient Melting and Surface Deformation during Pulsed Laser Heating," *Appl. Phys. Lett.*, **73**, 2093–2095.

Kelly, R., and Rothenberg, J. E., 1985, "Laser Sputtering 3. The Mechanism of the Sputtering of Metals at Low Energy Densities," *Nucl. Instrum. Meth. Phys. Res. B*, **7–8**, 755–763.

Kuo, D., Vierk, S. D., Rauch, O., and Polensky, D., 1997, "Laser Texturing on Glass and Glass–Ceramic Substrates," *IEEE Trans. Magnetics*, **33**, 94–99.

Mandelis, A., ed., 1992, *Progress in Photothermal & Photoacoustic Science & Technology*, New York, Elsevier.

Nink, T., Galbert, F., Mao, Z. L., and Bostanjoglo, O., 1999, "Dynamics of Laser Pulse-Induced Melts in Ni–P Visualized by High-Speed Transmission Electron Microscopy," *Appl. Surf. Sci.*, **138–139**, 439–443.

Nink, T., Mao, Z. L., and Bostanjoglo, O., 2000, "Melt Instability and Crystallization in Thin Amorphous Ni–P Films," *Appl. Surf. Sci.*, **154–155**, 140–145.

Olmstead, M. A., Amer, N. M., Kohn, S., Fournier, D., and Boccara, A. C., 1983, "Photothermal Displacement Spectroscopy: An Optical Probe for Solids and Surfaces," *Appl. Phys. A*, **32**, 141–158.

Ranjan, R., Lambeth, D. N., Tromel, M., Goglia, P., Li, Y., 1991, "Laser Texturing for Low-Flying-Height Media," *J. Appl. Phys.*, **69**, 5745–5747.

Scherer, G. W., 1986, *Relaxation in Glass and Composites*, San Diego, CA, Academic Press.

Schwarz-Selinger, Th., Cahill, D. G., Chen, S. C., Moon, S.-J., and Grigoropoulos, C. P., 2001, "Micron-Scale Modifications of Si Surface Morphology by Pulsed-Laser Texturing," *Phys. Rev. B*, **64**, 155 323–155 329.

Shiu, T.-R., Grigoropoulos, C. P., Cahill, D. G., and Greif, R., 1999a, "Mechanism of Bump Formation on Glass Substrates during Laser Texturing," *J. Appl. Phys.*, **86**, 1311–1318.

Shiu, T.-R., Grigoropoulos, C. P., and Greif, R., 1999b, "Measurement of Transient Glass Surface Deformation during Laser Heating," *J. Heat Transfer*, **121**, 1042–1048.

Sipe, J. E., Young, J. F., Preston, J. S., and van Driel, H. M., 1983, "Laser-Induced Periodic Surface Structure," *Phys. Rev. B*, **27**, 1141–1158.

Tam, A. C., Pour, I. K., Nguyen, T. *et al.*, 1996, "Experimental and Theoretical Studies of Bump Formation During Laser Texturing of Ni–P Substrates," *IEEE Trans. Magnetics*, **32**, 3771–3773.

Tokarev, V. N., and Konov, V. I., 1994, "Suppression of Thermocapillary Waves in Laser Melting of Metals and Semiconductors," *J. Appl. Phys.*, **76**, 800–805.

van de Riet, E., Nillesen, C. J. C. M., and Dieleman, J., 1993, "Reduction of Droplet Emission and Target Roughening in Laser Ablation and Deposition of Metals," *J. Appl. Phys.*, **74**, 2008–2012.

Varshneya, A. K., 1994, *Fundamentals of Inorganic Glasses*, San Diego, CA, Academic Press.

Zallen, R., 1983, *The Physics of Amorphous Solids*, New York, Wiley.

9   Laser processing of organic materials

S. Mao and C. P. Grigoropoulos

9.1   Introduction

Laser ablation of polymers and other molecular materials constitutes the basis for a range of well-established applications, such as matrix-assisted laser desorption–ionization (MALDI) (Hillenkamp and Karas, 2000), laser surgery (Niemz, 2003) including the widely used laser-assisted *in situ* keratomileusis (LASIK) technique, surface microfabrication and lithography (Lankard and Wolbold, 1992), and pulsed laser deposition (PLD) of organic coatings (Bäuerle, 2000; Chrisey and Hubler, 1994). Interaction of UV laser pulses with an organic substance typically results in photothermal and/or photochemical processes in the irradiated material. Generally, photothermal processes, which produce heat in the sample, dominate when the laser photon energy is small, whereas photochemical processes occur when the laser photon energy is larger than the chemical-bond energies of the molecules.

For lasers operating at near-IR wavelengths, photothermal processes usually play a major role. With deep-UV (wavelength shorter than $\sim$200 nm) laser irradiation, in which the photon energy is larger than the typical energy of the chemical bonds of molecules, photochemical processes are usually responsible for the onset of ablation. For laser ablation of organic materials with wavelengths between the near IR and deep UV, photothermal and photochemical processes are often interrelated (Ichimura *et al.*, 1994). On the other hand, the characteristics of material ejection depend on the nature of the ablation process (Srinivasan, 1986; Georgiou *et al.*, 1998). For instance, ablation of organic materials results in little lateral damage in the sample when the photochemical processes are dominant. In contrast, debris due to lateral damage of the laser-ablated sample is generally observed when photothermal processes prevail (Linsker *et al.*, 1984; Garrison and Srinivasan, 1985). Figure 9.1 shows a comparison between clean laser ablation and ablation with significant lateral damage. Evidently, the physical processes involved in laser ablation of organic materials are complex, and detailed understanding of the underlying mechanisms remains a challenge.

9.2   Fundamental processes

The complexity of the interplay between photothermal and photochemical processes involved in laser ablation of polymers and other molecular materials presents a

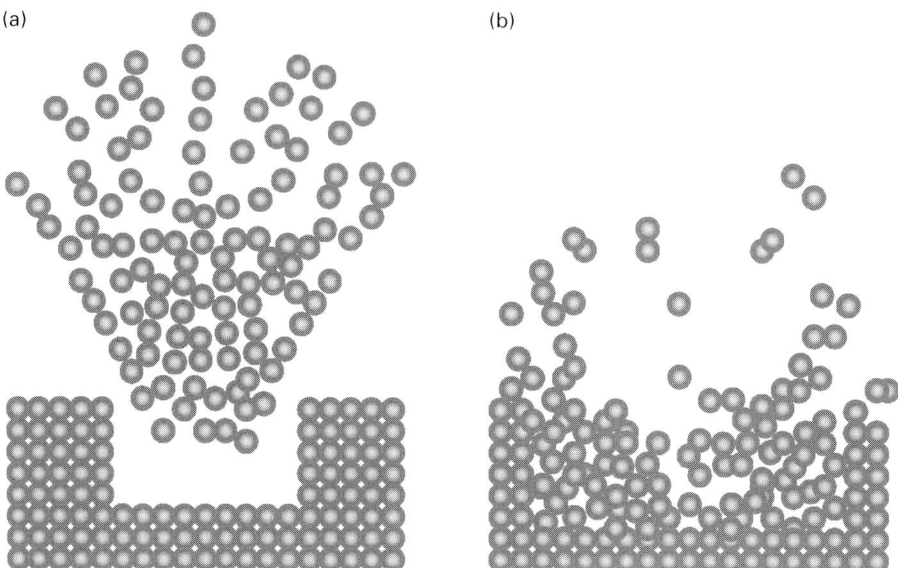

Figure 9.1. Clean ablation (shown on the left) and ablation with damage (shown on the right), adapted from Garrison and Srinivasan (1985) with permission from the Institute of Applied Physics.

challenge for both theoretical and modeling studies of the phenomena. The general processes typically start with excitation of optically active states in the organic sample, followed by thermalization of deposited laser energy, bond-breaking chemical reactions, fragmentation of the excited molecules, ejection of a volume of the molecular material, and propagation of a pressure wave into the material. There have been significant experimental studies aimed at elucidating the fundamental processes of laser ablation of polymers and organic molecules. For example, the influences of laser pulse duration (Dreisewerd *et al.*, 1996; Cramer *et al.*, 1997), laser fluence and wavelength (Dreisewerd *et al.*, 1995; Berkenkamp *et al.*, 2002), laser-beam incidence angle (Westman *et al.*, 1994), the number of successive laser pulses (Koubenakis *et al.*, 2001), and molecular volatility (Yingling *et al.*, 2001) have been investigated. In addition to molecular yields commonly detected using time-of-flight mass spectrometry, ablation characteristics such as cluster ejection (Handschuh *et al.*, 1999; Heitz and Dickinson, 1999) and time-resolved plasma evolution have also been studied.

The photochemical mechanism assumes that the absorption of photons by electrons results immediately in the breaking of a chemical bond, leading to molecular dissociation. In contrast, the photothermal model assumes that the absorption of laser photon energy by electrons is rapidly equilibrated into the phonon system, resulting in rapid local heating of the interacting volume. Subsequently molecular dissociation occurs through thermally activated decomposition process at elevated temperature. The critical difference between the two mechanisms is whether incident photons are initially converted to heat prior to dissociation, or directly induce the dissociation while bypassing the formation of a thermal reservoir. Although the difference appears conceptually

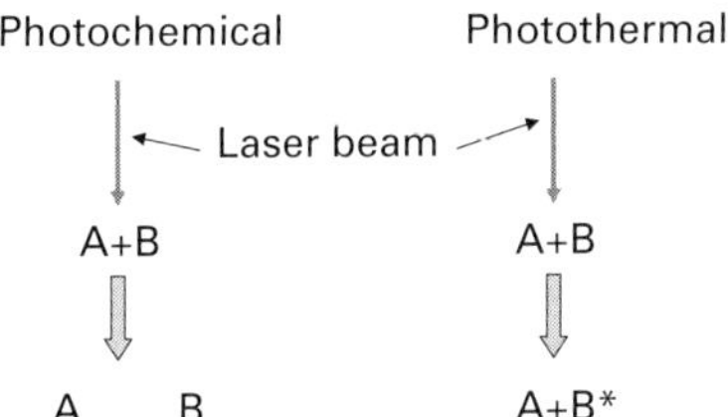

Figure 9.2. Main differences between photochemical and photothermal effects.

simple, it has proven to be exceedingly difficult to resolve experimentally. Figure 9.2 shows schematically the primary difference between the photochemical and photothermal processes. Assuming an AB molecule, the photochemical process would break the bonding between A and B, thus directly dissociating the molecule into two parts. In contrast, the photothermal process would initially heat the molecule to an excited state without bond-breaking.

9.2.1   Photothermal processes

From the early observations of laser ablation of polymers (Kawamura *et al.*, 1982), it was found that UV laser light can initiate a series of decomposition steps via deposition of energy into chromophores as well as electronic excitation, which is rapidly converted into vibrational heating of the solid. The experimentally measured single-pulse etch depth is presented in Figure 9.3.

Experiments have indicated that there is no significant difference in thermal behavior among different wavelengths for laser irradiation of polymers (Yeh, 1986). For example, for solid poly(ethylene terephthalate), polymethyl methacrylate (PMMA), polycarbonate, and polyimide, the absorbed energy density at the ablation threshold varies only slightly with the wavelength. It is understood that photons of different wavelengths are absorbed by different electronic states of the molecules, and electronic-to-vibrational energy transfer leads to thermal decomposition and ablation. When the laser pulse fluence is below a threshold value, almost all the laser energy deposited into the target is converted into thermal energy; when the threshold laser fluence is exceeded, most of the excess energy would be carried away by ablation products.

According to a model proposed by Cain *et al.* (1992), the ablation is assumed to take place via a pseudo-zeroth-order thermal degradation. The absorbed laser energy is converted through nonradiative decay mechanisms to an equivalent temperature. The corresponding target-surface-recession process can be described by the following Arrhenius equation:

$$R \sim \frac{dX_{ab}}{dt} \sim k_0 \exp\left(-\frac{E_a}{k_B T_s}\right), \tag{9.1}$$

where X_{ab} is the position of the receding surface, R is the recession speed (or etch rate), E_a is the activation energy for decomposition, k_0 is an Arrhenius pre-exponential constant, T_s is the surface temperature, which is generally a function of time and position, and

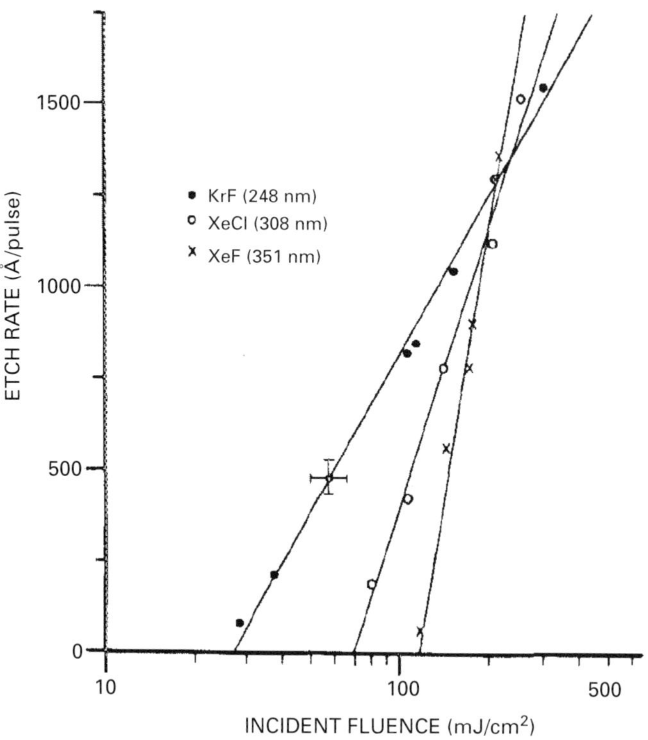

Figure 9.3. The fluence dependence of the measured single-pulse etch depth, adapted from Brannon *et al.* (1985) with permission from the Institute of Applied Physics.

k_B is the Boltzmann constant. The total ablation depth is obtained by integrating the instantaneous etch rate over time:

$$X_{ab} \sim \int k_0 \exp\left(-\frac{E_a}{k_B T_s}\right) dt. \tag{9.2}$$

In the above, the transient surface temperature $T_s = T_s(t)$ is considered proportional to the laser fluence, $T_s \sim F$, yielding the following expressions for the ablation depth and etch rate:

$$X_{ab} \sim k_0 \exp\left(-\frac{B}{F}\right) \Delta t, \tag{9.3}$$

$$R \sim k_0 \exp\left(-\frac{B}{F}\right). \tag{9.4}$$

Experiments also suggested that the etch rate is environment-independent (Brannon *et al.*, 1985). However, when multiple laser pulses are directed onto organic materials regarded as weak absorbers (such as PMMA), the ablation or etch rate should be reconsidered because of the incubation effect. This is because the first pulse of laser light alters the polymer chemically so that its response to the second pulse would be different. Weakly absorbing polymers can undergo substantial non-ablative degradation and become more absorptive. The incubation effect is particularly pronounced if no polymer is removed by the first pulse.

For laser-produced ablation products, there is also a threshold behavior, although the threshold is not as well defined as that for the ablation or etch rate. It is generally observed that more small fragments, including diatomic molecules, are produced at high fluences at the expense of large fragments. For instance, solid products of laser-ablated polyimide were analyzed by using IR spectrometry on samples collected on KBr substrates in vacuum (Yeh, 1986). At low fluences the solid products consist of fragments of a wide range of sizes, with some of them still exhibiting the parent polyimide structure. Above a certain threshold laser fluence, the absorption signature of the parent polyimide disappears.

Efforts at estimating the surface temperature rise have been made by balancing the laser input energy and the internal energy of the polymer (Dyer and Sidhu, 1985; Gorodetsky *et al.*, 1985), on the assumption that most of the energy absorbed is confined within the irradiated area because the thermal diffusion length of polymers during laser interaction is less than the photon-absorption depth. A more accurate estimate of the temperature rise requires knowledge of the laser pulse energy absorbed, mass density and specific heat as a function of temperature. A rough estimate of the temperature rise for most polymers is of the order of 1000 K at the threshold fluence for laser ablation. At such a high temperature, thermal decomposition, rapid thermal expansion, and phase transformation can all induce bond-breaking that leads to ablation.

9.2.2 Photochemical processes

Owing to the complexity of the processes involved in photochemical reactions, understanding the photochemical processes during laser ablation remains a challenge. In a typical photochemical process, the photon-absorption event starts with the excitation of molecules followed by breaking of the molecule's chemical bonds, forming molecules typically with a smaller number of atoms. The resulting photoproducts occupy a larger volume and create a local pressure excess inside the irradiated volume. Because the process of energy transfer (corresponding to immediate molecular dissociation) in this case is considerably faster than the vibrational relaxation process (typically on a time scale of picoseconds) (Garrison and Srinivasan, 1985; Dlott, 1990), the electronic-to-vibrational energy transfer (thermal effect) can be neglected before material ejection.

Detailed discussion of the photochemical models in the context of molecular relaxation pathways is given in Bäuerle (2000). Conventionally, there is a threshold energy density of the laser pulse above which ablation occurs. For a total number density of chromophores N, the number density of broken bonds is

$$N_{\mathrm{B}}(z) \approx N \frac{\sigma I(x)}{h\nu}, \tag{9.5}$$

where $I(x)$ is the intensity at a depth x and σ is the cross section for bond-breaking in response to the photon energy $h\nu$. According to the Beer–Lambert law, the attenuated laser intensity and fluence at depth x into the target can be described by

$$I(x) = I_0 e^{-\gamma x} \quad \text{and} \quad F(x) = F_0 e^{-\gamma x}, \tag{9.6}$$

where γ is the absorption coefficient. At the ablation depth X_{ab}, the threshold number density of broken bonds is

$$N_{B,th} \approx N \frac{\sigma I_0 e^{-\gamma X_{ab}}}{h\nu}, \tag{9.7}$$

corresponding to a threshold local intensity at that location of

$$N_{B,th} \approx N \frac{\sigma I_{0,th}}{h\nu}. \tag{9.8}$$

Assuming that a certain threshold fluence must be exceeded in order for ablation to occur, I should then be at least equal to the threshold value, and I_0 must be increased over $I_{0,th}$ by a factor of $\exp(\gamma X_{ab})$ so that single-pulse ablation occurs at the desired depth X_{ab}. This exponential factor compensates for the exponential decrease of the fluence as light penetrates into the target. A simple relationship between light absorption and the single-pulse ablation depth can therefore be established:

$$X_{ab} = \frac{1}{\gamma} \ln\left(\frac{I_0}{I_{0,th}}\right) = \frac{1}{\gamma} \ln\left(\frac{F}{F_{th}}\right). \tag{9.9}$$

However, a threshold behavior is not exclusive to photochemical processes. Rapidly increasing ablation rates are observed, for example, in thermal processes through the so-called explosive-vaporization process, invoking a threshold-type behavior. Consequently, the validity of the above relation should not preclude thermal processes. Comprehensive studies of the effect of the photochemical processes of laser ablation of polymers and organic molecular materials have been conducted using molecular-dynamics simulations. A mesoscopic coarse-grained breathing-sphere model was applied to the modeling of photochemical processes during laser ablation of organic films. The key point of the model is to treat each molecule (or group of atoms) in the system as a single particle that has the true translational degrees of freedom but an approximate internal degree of freedom (Zhigilei *et al.*, 1997, 1998). The internal degree of freedom allows spheres to change their sizes and controls the rate of energy transfer from the excited molecules to their surroundings. The main advantage of the model is its ability to study the dynamics of the system at the mesoscopic length scale, a regime that is not accessible with either atomistic or continuum computational methods.

As an example, molecular-dynamics simulations were used to investigate the effect of photochemical processes on molecular ejection mechanisms during 248-nm laser irradiation of organic solids. Photochemical reactions were found to release additional energy into the irradiated sample and decrease the average cohesive energy, therefore decreasing the value of the ablation threshold. The yield of emitted fragments becomes significant only above the ablation threshold. Below the threshold, only the most volatile photoproduct is ejected and only in a very small amount, whereas the remainder of the photoproducts are trapped inside the sample. The occurrence of photochemical decomposition processes and subsequent chemical reactions changes the temporal and

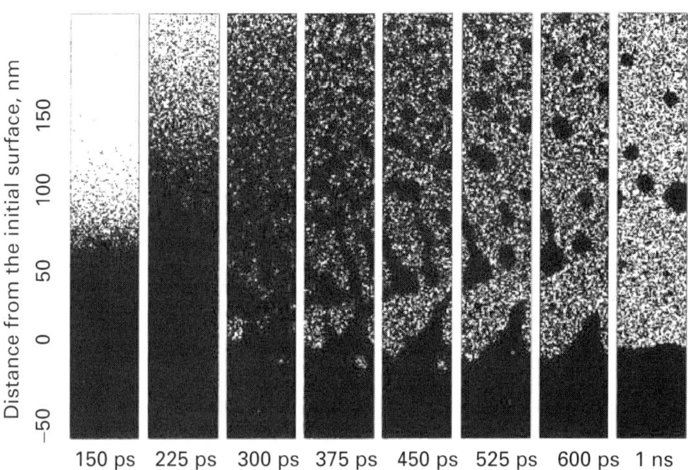

Figure 9.4. Snapshots from the molecular-dynamics simulation of laser irradiation of a molecular solid with pulse duration of 150 ps and fluence of 61 J/m^2, adapted from Zhigilei and Garrison (2000) with permission from the Institute of Applied Physics.

spatial energy-deposition profile, which is different from the case of pure photothermal ablation. The chemical reactions create an additional local pressure buildup and, as a result, generate a strong and broad inward-propagating acoustic pressure wave. The strong pressure wave in conjunction with the temperature increase in the absorbing region causes the ejection of hot massive molecular clusters. These massive clusters later disintegrate in the plume into smaller clusters and monomers due to ongoing chemical reactions. The ejection and disintegration of large clusters result in higher material-removal rates as well as a higher plume density. This photochemical process is shown schematically in Figure 9.4, where snapshots were obtained from molecular-dynamics simulations.

The results from molecular-dynamics simulations are in good agreement with experiments that provide microscopic perspectives of the photochemical processes. The simulation also showed that the ablation threshold is decreased and the material-removal rate is increased when the photochemical process is taken into account (Yingling and Garrison, 2007). This conclusion agrees well with the observation that the threshold value increases with the wavelength in experiments on laser ablation of polymers (Brannon *et al.*, 1985). As is expected, in general, the longer the wavelength, the weaker the photochemical mechanism.

9.2.3   Interplay between photochemical and photothermal processes

There are two main mechanisms playing roles in laser ablation of polymers and other organic molecular materials. Typical characteristics of photochemical processes would be (i) the lack of damage to the remaining material (i.e. there is no melting), (ii) the nonequilibrium characteristics regarding the translational, rotational, and vibrational energies of ablation products, and (iii) the observed difference in terms of species obtained in ablation at UV versus IR wavelengths. Contradictory evidence could be offered by citing (i) the dependence of the ablation rate on pulse length and intensity, and (ii) the Arrhenius behavior of the ablation rate near the "threshold" for observable material removal. It is likely that the nature of the process is system (i.e. laser and sample)-dependent with no clear distinction (Dyer, 2003). The photochemical mechanism usually dominates when the wavelength is short (e.g. <200 nm) and the photothermal mechanism is more important for longer wavelengths. For the case of wavelengths in between, the process is often a combination of photochemical and photothermal processes, and the interplay between them is responsible for the onset of ablation as well as subsequent plume dynamics. Although the quantum yield for photo-induced bond-breaking is low, photons can efficiently excite chain-scission sites. A reasonable hypothesis on the coupled ablation process could therefore involve photo-initiated scission followed by thermal unzipping, although this is hard to elucidate experimentally. A combined photothermomechanical process could be presented in the form of the following equation:

$$\frac{\partial N_B(x,t)}{\partial t} = \eta \frac{\sigma I(x,t)}{h\nu} N + k_0 N \exp\left(-\frac{\Delta E_a - \zeta S_{zz}(x,t)}{k_B T(x,t)}\right), \qquad (9.10)$$

where η is the quantum yield, S_{zz} the dilatational stress component, and ζ a factor quantifying the strength of the mechanical coupling. The first term in the above indicates the direct defect formation, whereas the second models the thermal generation of defects that is mediated by the thermoelastic stress field that is induced by the laser excitation.

It is well established that photochemical reactions can play a major role in deep-UV ablation of polymers and other molecular materials such as biological tissues. Measurements of the *single-shot* ablation rate of polyimide by Küper *et al.* (1993) using a quartz microbalance system indicated a thermal-like behavior for ablation at 248, 308, and 351 nm, manifested by an S-type dependence of the ablation rate with respect to the fluence. In contrast, the ablation behavior at 193 nm exhibited a sharp threshold reminiscent of photochemical processes. Consistent with this work were the *in situ* experimental measurements by Brunco *et al.* (1992). These investigators utilized thin-film Ni–Si thermistors and applied conductive-heat-transfer analysis to infer that the peak temperature in thin PI films irradiated by 248-nm excimer-laser light at the threshold fluence of 36 mJ/cm^2 was 1660 ± 100 K.

In general, the role of the photochemical reactions in cases when photochemical processes are closely coupled with thermal ones, as for irradiation at 248 nm, has not been understood fully. To distinguish and pinpoint experimentally the photochemical and photothermal processes, each stage of the excitations and relaxations prior to ablation needs to be checked. Although it is difficult to isolate one mechanism from another experimentally, it is easier to consider numerically only one mechanism or the combination of the two. The molecular-dynamics simulation based on a mesoscopic coarse-grained breathing-sphere model as discussed in the previous section was able to distinguish some of the features involved in the interplay of the two mechanisms. In model calculations, the photothermal event is realized by depositing the photon energy into the kinetic energy of internal vibrations of excited molecules, and the photochemical event is realized by allowing a sphere to break into several smaller spheres representing the

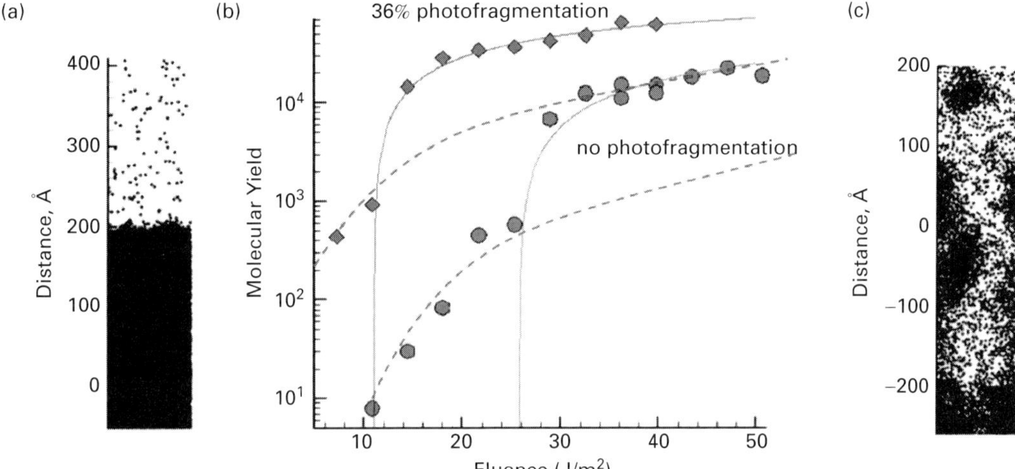

Figure 9.5. A representation of molecular yields versus laser fluence for chlorobenzene solids resulting from photothermal processes (i.e. no photofragmentation) and from 36% photochemical and 64% photothermal processes. The snapshots of the plume in the near-surface region representing (a) desorption and (b) ablation regimes were taken from simulations of a cholorobenzene sample at 2.53 and 4.7 mJ/cm^2. The laser beam impinges on the sample from the top and zero distance denotes the original surface. Adapted from Yingling and Garrison (2007) with permission from Elsevier.

fragments of the excited molecule. By means of this model, the effect of the combination of photothermal and photochemical mechanisms underlying laser ablation of PMMA was numerically investigated. Figure 9.5 shows a general dependence of molecular yield on fluence for both photothermal and photochemical processes. At low laser fluences, mostly monomers are ejected from the surface according to a thermal-desorption model (Figure 9.5(a)). As the laser fluence is increased, the surface region starts to melt, and evaporation from the liquid surface increases. The dramatic increase in the total amount of the ejected material at a certain fluence corresponds to the ablation threshold. The ablation regime is described by an explosive disintegration of material and the homogeneous decomposition of the plume into a mixture of molecular clusters of various sizes and individual molecules (see Figure 9.5(c)). The occurrence of photochemistry influences the yield and lowers the ablation threshold, indicating the change in molecular ejection mechanism (see Figure 9.5(b)).

9.3 Applications

Polymers and other organic materials have many important applications. In general, the applications of laser processing of polymers can be grouped into two main categories (Phipps, 2007). In the first one, the structures produced in the material are of interest, and in the second, the application is based on laser-ablation products. For instance, as a mature technology, structuring of polymers has been used in industry for the production of nozzles for inkjet printers and to prepare via holes in multi-chip modules through polyimide.

As an example of emerging applications, biodegradable polymeric materials have had a significant impact on medical technology, greatly enhancing the efficacy of many existing drugs and enabling the construction of entirely new therapeutic modalities (La Van *et al.*, 2003). These degradable materials have been employed in many areas of therapeutic medicine, from commonly used resorbable surgical sutures (Behravesh *et al.*, 1999) to controlled-release drug-delivery vehicles (Allen and Cullis, 2004) and implantable cell-support scaffolds (Middleton and Tipton, 2000). The development of next-generation drug-delivery and tissue-engineering devices based on biodegradable polymers is contingent on the fashioning of features on a scale analogous to the size of cells and organelles. Then UV lasers can modify the surface of a material through direct chemical bond-breaking and minimize the heat effects on the surrounding material. The use of solid-state femtosecond lasers allows micromachining of polymers with precise resolution since the influence of heat conduction around the machined area during the process can be ignored. This can be accomplished primarily because of the short time scale of laser–material interaction in which the excitation energy is highly confined. Therefore, femtosecond UV-laser ablation is very appealing for micro-patterning biodegradable polymers, which have been utilized to generate high-resolution micro-patterns on biodegradable polymers (Aguilar *et al.*, 2005).

9.3.1   Laser fabrication of polymers

Excimer-laser ablation is a useful fabrication tool suitable for surface structuring of polymers and other molecular solids due to their high UV absorption and possible non-thermal ablation behavior. Polymers typically display high absorption of UV irradiation. The photothermal and photochemical processes lead to a dissociation of the polymer and subsequently local ejection of material. The occurrence of photochemical reactions is responsible for a clean, well-defined geometry of the ablated zone. The lateral dimensions of the removed polymer are determined by the laser-beam size, and the depth of the hole depends on the laser intensity. Typically the lateral resolution is limited to a few micrometers while the ablation rate (depth per pulse) can be a few tens of nanometers. These properties make polymers a very suitable material for excimer-laser ablation.

On the other hand, thin films of polymers can be deposited by pulsed-laser ablation, given the demonstrated ability of laser pulses to etch away the near-surface layer. A significant amount of work on the formation of polymer films by pulsed-laser evaporation was conducted by Hansen and Robitaille (1988a, 1988b). In their work, laser wavelengths ranging from 193 to 1064 nm and pulse energy densities of 0.01–2 J/cm^2 were utilized to investigate various polymer systems, such as polyethylene, polycarbonate, polyimide, and PMMA. A strong correlation between the film quality and the optical density at the excitation wavelength was observed. Films produced by UV-laser irradiation have been found to be of considerably higher quality. More interestingly, it is possible to synthesize polymer–metal composite thin films by pulsed-excimer-laser co-ablation.

Figure 9.6. An image of 40–50-μm features through 50-μm-thick polyimide. Adapted from www.oxfordlasers.com.

Since any conventional attempt to introduce metals into an already-formed polymer (e.g. by use of ion implantation) cracks the polymer, it is essential that metals be included in the formation stage of the polymer films. The method of laser co-ablation might not be applicable to the situation wherein the vaporization threshold for the metal is significantly higher than that for the polymer. In such a situation, it may be necessary to split the laser beam and adjust separately the energy density on the metal and polymer surfaces. Figure 9.6 shows a microstructure obtained from UV-laser drilling into a polyimide sample.

9.3.2   MALDI

MALDI is a laser-assisted soft ionization process that produces molecular ions from large nonvolatile molecules, such as biopolymers and synthetic polymers, with minimum fragmentation. This technique has become a major and powerful tool for analysis of nonvolatile organic compounds (Karas *et al.*, 1987). The growth of this technique has been further accelerated by innovative mass-spectrometry technologies.

The basic MALDI process involves multiple intermediate steps. Initially, analytes (e.g. polymers or peptides) are dissolved in a solvent such as water, followed by admixture of a compound (called a matrix) such as *trans*-cinnamic acid that has a good

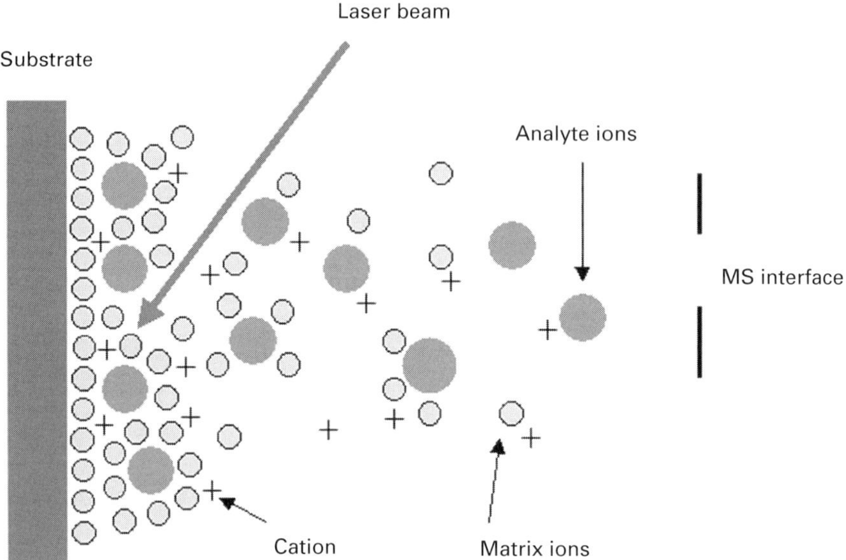

Figure 9.7. A schematic diagram of the MALDI technique.

UV-absorption property. Usually about 10^4 times more UV absorber than polymer is used to form the matrix. In the next step, the solvent is evaporated and a layer of UV-absorbing compound is formed in an airtight sample chamber. A laser beam is then directed onto the compound sample, and the laser energy is absorbed by the matrix, with a portion of the energy being passed to the polymer molecules. The matrix material also reacts with the polymers in such a way that the polymer becomes charged ions. This process is depicted schematically in Figure 9.7. The ions formed (including analyte ions and matrix ions) are subsequently analyzed by mass spectroscopy.

Despite its complex characteristics, simulations have been able to offer some insight into the fundamental aspects of MALDI. Molecular-dynamics simulation studies (Zhigilei and Garrison, 1999) predicted the existence of two distinct regimes of molecular ejection – surface desorption at low laser fluences and volume expulsion (phase explosion) at higher fluences. This prediction is supported by a number of recent experimental observations. In MALDI experiments performed by Dreisewerd *et al.* (1995), the detection threshold for the ejection of a large analyte molecule, bovine insulin, was found to be significantly higher than that for neutral matrix molecules (sinapic acid). A reasonable assumption is that large and heavy analyte molecules can be ejected only in the ablation regime, through their entrainment into the expanding plume of matrix molecules.

9.3.3   Laser–tissue interactions

Laser–tissue interactions constitute a unique application in the broader field of laser ablation of organic materials. Clinical applications of lasers require an understanding of the fundamental mechanisms that govern laser–tissue interactions. Several books have

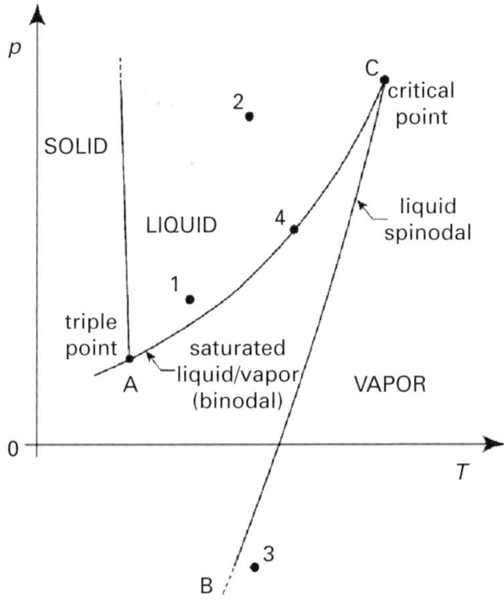

Figure 9.8. The path taken through the p versus T projection of the thermodynamic phase diagram for confined boiling ($1 \to 2 \to 6$) and for tissue ablation involving a phase explosion ($1 \to 2 \to 3 \to 4 \to 5 \to 6$). The actual path followed depends on the rate of energy deposition, number density of heterogeneous nuclei, and mechanical strength of the tissue matrix relative to the saturation vapor pressure corresponding to the ambient spinodal temperature. Adapted from Vogel and Venugopalan (2003) with permission from the American Chemical Society.

served as valuable resources in the field of laser–tissue interactions (Welch and van Gemert, 1995; Niemz, 2003; Waynant, 2002). There are also a few review articles on the general science and applications (Hillenkamp *et al.*, 1980; Oraevsky *et al.*, 1991; van Leeuwen *et al.*, 1995; Walsh, 1995; Vogel and Venugopalan 2003), as well as on particular aspects of laser–tissue interactions. For instance, UV-laser ablation of polymers and biological tissues was discussed in Srinivasan (1986) and Srinivasan and Braren (1989), the photophysics of laser–tissue interactions was examined in Boulnois (1986), and the thermal process during laser–tissue interactions was the topic of McKenzie (1990).

Important to the understanding of laser–tissue interaction are the properties of biological tissues, such as their composition and morphology, mechanical strength, and thermal denaturation. Soft biological tissues can be considered as a type of material consisting of cells that reside in and attach to an extracellular matrix (ECM). By mass most soft tissues are composed of water (55%–99%) and collagen (0–35%). The ratio of ECM components to the total tissue mass varies significantly among tissue types. The interaction of collagen and the other ECM elements with water is important when considering energy-transport processes within the tissue under laser irradiation. Secondly, the mechanical properties of biological tissues influence laser ablation significantly, insofar as the elasticity and strength of the tissue would modulate the kinetics and dynamics of the ablation process. In addition, thermal denaturation of ECM proteins resulting from pulsed-laser irradiation is of great importance because it affects the dynamics of the ablation process and governs the extent of thermal damage produced in the remaining tissue.

On the other hand, optical-absorption properties of tissues are governed by the electronic, vibrational, and rotational structures of the constituent biomolecules. The dominant tissue chromophores in the UV are proteins, DNA, and melanin, since the absorption of UV-laser irradiation by water at room temperature is negligible. Moreover, the thermal and mechanical transients generated by pulsed-laser ablation are substantial and may result in significant alteration of the tissue's optical properties. Therefore, dynamic change of optical properties in tissues needs to be taken into account in most cases.

There are three main processes (surface vaporization, normal boiling, and phase explosion) responsible for material removal during laser ablation of tissues. Initially, surface vaporization (along the binodal line) takes place (as shown in Figure 9.8), and, in the course of the formation and growth of vapor bubbles, normal boiling occurs. However, when the rate of deposition of energy into the tissue is high enough to superheat a subsurface layer to the spinodal decomposition temperature, phase explosion dominates the ablation process and a significant volume of the tissue is abruptly ejected from the sample; the superheated layer is transformed into a multiphase mixture.

References

Aguilar, C. A., Lu, Y., Mao, S. S., and Chen, S., 2005, "Direct Micro-patterning of Biodegradable Polymers using Ultraviolet and Femtosecond Lasers," *Biomaterials*, **26**, 7642–7649.

Allen, T. M., and Cullis, P. R., 2004, "Drug Delivery Systems: Entering the Mainstream," *Science*, **303**, 1818–1822.

Bäuerle, D., 2000, *Laser Processing and Chemistry*, Berlin, Springer-Verlag.

Behravesh, E., Yasko, A. W., Engle, P. S., and Mikos, A. G., 1999, "Synthetic Biodegradable Polymers for Orthopaedic Applications," *Clin. Orthopaedics*, **367(S)**, S118–S129.

Berkenkamp, S., Menzel, C., Hillenkamp, F., and Dreisewerd, K., 2002, "Measurements of Mean Initial Velocities of Analyte and Matrix Ions in Infrared Matrix-Assisted Laser Desorption Ionization Mass Spectrometry," *J. Am. Soc. Mass Spectrometry*, **13**, 209–220.

Boulnois, J. L., 1986, "Time–Irradiance Reciprocity Relationship as a Model for Laser–Tissue Interactions," *Lasers Med. Sci.*, **6**, 168.

Brannon, J. H., Lankard, J. R., Baise, A. I., Burns, F., and Kaufman, J., 1985, "Excimer Laser Etching of Polyimide," *J. Appl. Phys.*, **58**, 2036–2043.

Brunco, D. P., Thompson, M. O., Otis, C. E., and Goodwin, P. M., 1992, "Temperature Measurements of Polyimide during KrF Excimer Laser Ablation," *J. Appl. Phys.*, **72**, 4344–4350.

Cain, S. R., Burns, F. C., Otis, C. E., and Braren, B., 1992, "Photothermal Description of Polymer Ablation: Absorption Behavior and Degradation Time Scales," *J. Appl. Phys.*, **72**, 5172–5178.

Chrisey, D. B., and Hubler, G. K., eds., 1994, *Pulsed Laser Deposition of Thin Films*, New York, Wiley-Interscience.

Cramer, R., Haglund, R. F. Jr., and Hillenkamp, F., 1997, "Matrix-Assisted Laser Desorption and Ionization in the O—H and C=O Absorption Bands of Aliphatic and Aromatic Matrices: Dependence on Laser Wavelength and Temporal Beam Profile," *Int. J. Mass Spectrometry*, **169**, 51–67.

Dlott, D. D., 1990, "Ultrafast Vibrational Energy Transfer in the Real World: Laser Ablation, Energetic Solids, and Hemeproteins," *J. Opt. Soc. Am. B*, **7**, 1638–1652.

Dreisewerd, K., Schürenberg, M., Karas, M., and Hillenkamp, F., 1995, "Influence of the Laser Intensity and Spot Size on the Desorption of Molecules and Ions in Matrix-Assisted Laser-Desorption Ionization with a Uniform Beam Profile," *Int. J. Mass Spectrometry*, **141**, 127–148.

1996, "Matrix-Assisted Laser Desorption/Ionization with Nitrogen Lasers of Different Pulse Widths," *Int. J. Mass Spectrometry*, **154**, 171–178.

Dyer, P. E., 2003, "Excimer Laser Polymer Ablation: Twenty Years on," *Appl. Phys. A*, **77**, 167–163.

Dyer, P. E., and Sidhu, J., 1985, "Excimer Laser Ablation and Thermal Coupling Efficiency to Polymer-Films," *J. Appl. Phys.*, **57**, 1420–1422.

Garrison, B. J., and Srinivasan, R., 1985, "Laser Ablation of Organic Polymers – Microscopic Models for Photochemical and Thermal Processes," *J. Appl. Phys.*, **57**, 2909–2914.

Georgiou, S., Koubenakis, A., Lassithiotaki, M., and Labrakis, J., 1998, "Formation and Desorption Dynamics of Photoproducts in the Ablation of van der Waals films of Chlorobenzene at 248 nm," *J. Chem. Phys.*, **109**, 8591–8600.

Gorodetsky, G., Kazyaka, T. G., Melcher, R. J., and Srinivasan, R., 1985, "Calorimetric and Acoustic Study of Ultraviolet-Laser Ablation of Polymers," *Appl. Phys. Lett.*, **46**, 828–830.

Handschuh, M., Nettesheim, S., and Zenobi, R., 1999, "Laser-Induced Molecular Desorption and Particle Ejection from Organic Films," *Appl. Surf. Sci.*, **137**, 125–135.

Hansen, S. G., and Robitaille, T. E., 1988a, "Formation of Polymer Films by Pulsed Laser Evaporation," *Appl. Phys. Lett.*, **52**, 81–83.

1988b, "Arrival Time Measurements of Films Formed by Pulsed Laser Evaporation of Polycarbonate and Selenium," *J. Appl. Phys.*, **84**, 2122–2129.

Heitz, J., and Dickinson, J. T., 1999, "Characterization of Particulates Accompanying Laser Ablation of Pressed Polytetrafluorethylene (PTFE) Targets," *Appl. Phys. A*, **68**, 515–523.

Hillenkamp, F., and Karas, M., 2000, "Matrix-Assisted Laser Desorption/Ionization, an Experience," *Int. J. Mass Spectrometry*, **200**, 71–77.

Hillenkamp, F., Pratesi, R., and Sacchi, C. A., eds., 1980, *Lasers in Biology and Medicine*, New York, Plenum Press, p. 37.

Himmelbauer, M., Arenholz, E., and Baüerle, D., 1996, "Single-Shot UV-Laser Ablation of Polyimide with Variable Pulse Lengths," *Appl. Phys. A*, **63**, 87–90.

Ichimura, T., Mori, Y., Shinohara, H., and Nishi, N., 1994, "Photofragmentation of Chlorobenzene – Translational Energy Distribution of the Recoiling Cl Fragment," *Chem. Phys.*, **189**, 117–125.

Karas, M., Bachmann, D., Bahr, U., and Hillenkamp, F., 1987, "Matrix-Assisted Ultraviolet-Laser Desorption of Nonvolatile Compounds," *Int. J. Mass Spectrom. Ion Processes*, **78**, 53–68.

Kawamura, Y., Toyoda, K., and Namba, S., 1982, "Effective Deep Ultraviolet Photoetching of Poly(methyl methacrylate) by an Excimer Laser," *Appl. Phys. Lett.*, **40**, 374–375.

Koubenakis, A., Labrakis, J., and Georgiou, S., 2001, "Pulse Dependence of Ejection Efficiencies in the UV Ablation of Bi-component van der Waals Solids" *Chem. Phys. Lett.*, **346**, 54–60.

Küper, S., Brannon, J., and Brannon, K., 1993, "Threshold Behavior in Polyimide Photoablation: Single-Shot Rate Measurements and Surface-Temperature Modeling," *Appl. Phys. A*, **56**, 43–50.

Lankard, J. R. Sr., and Wolbold, G., 1992, "Excimer Laser Ablation of Polyimide in a Manufacturing Facility," *Appl. Phys. A*, **54**, 355–359.

La Van, D. A., McGuire, T., and Langer, R., 2003, "Small-Scale Systems for *in vivo* Drug Delivery," *Nature Biotechnol.*, **21**, 1184–1191.

Linsker, R., Srinivasan, R., Wynne, J. J., and Alonso, D. R., 1984, "Far-Ultraviolet Laser Ablation of Atherosclerotic Lesions," *Lasers Surg. Medicine*, **4**, 201–206.

McKenzie, A. L., 1990, "Physics of Thermal Processes in Laser Tissue Interaction," *Phys. Med. Biol.*, **35**, 1175–1209.

Middleton, J. C., and Tipton, A. J., 2000, "Synthetic Biodegradable Polymers as Orthopedic Devices," *Biomaterials*, **21**, 2335–2346.

Niemz, M. H., 2003, *Laser–Tissue Interactions: Fundamentals and Applications*, 3rd edn, Berlin, Springer-Verlag.

Novis, Y., Pireaux, J. J., Brezini, A. *et al.*, 1988, "Structural Origin of Surface Morphological Modifications Developed on Poly(ethylene terephthalate) by Excimer Laser Photoablation," *J. Appl. Phys.*, **64**, 365–370.

Oraevsky, A. A., Esenaliev, R. O., and Letokhov, V. S., 1991, "Pulsed Laser Ablation of Biological Tissue: Review of the Mechanisms," in *Laser Ablation, Mechanisms and Applications*, edited by J. C. Miller and R. F. Haglund, New York, Springer, pp. 112–122.

Phipps, C., 2007, *Laser Ablation and its Application*, New York, Springer.

Raimondi, F., Abolhassani, S., Brütsch, R. *et al.*, 2000, "Quantification of Polyimide Carbonization after Laser Ablation," *J. Appl. Phys.*, **88**, 3659–3666.

Srinivasan, R., 1986, "Ablation of Polymers and Biological Tissue by Ultraviolet Lasers," *Science*, **234**, 559–565.

1992, "Etching Polymer Films with Ultraviolet Laser Pulses of Long (10–400 μs) Duration," *J. Appl. Phys.*, **72**, 1651–1653.

Srinivasan, R., and Braren, B. 1989, "Ultraviolet Laser Ablation of Organic Polymers," *Chem. Rev.*, **89**, 1303–1316.

Srinivasan, R., Braren, B., and Casey, K. G., 1990, "Nature of Incubation Pulses in the Ultraviolet Laser Ablation of Polymethyl Methacrylate," *J. Appl. Phys.*, **68**, 1842–1847.

Srinivasan, R., Hall, R. R., Loehle, W. D., Wilson, W. D., and Albee, D. C., 1995, "Chemical Transformations of the Polyimide Kapton Brought about by Ultraviolet Laser Radiation," *J. Appl. Phys.*, **78**, 4881–4887.

van Leeuwen, T., Jansen, E. D., Motamedi, M., Borst, C., and Welch, A. J., 1995, "Pulsed Laser Ablation of Soft Tissue," in *Optical–Thermal Response of Laser-Irradiated Tissue*, edited by A. J. Welch and M. J. C. van Gemert, New York, Plenum Press, pp. 709–763.

Vogel, A., and Venugopalan, V., 2003, "Mechanisms of Pulsed Laser Ablation of Biological Tissues," *Chem. Rev.*, **103**, 577–644.

Walsh, J. T., 1995, "Pulsed Laser Angioplasty: A Paradigm for Tissue Ablation, in *Optical–Thermal Response of Laser-Irradiated Tissue*; Welch, A. J., and van Gemert, M. J. C., Eds., New York, Plenum Press, p. 865.

Waynant, R. W., ed., 2002, *Lasers in Medicine*, Boca Raton, FL, CRC Press.

Welch, A. J., and van Gemert, M. J. C., eds., 1995, *Optical–Thermal Response of Laser-Irradiated Tissue*, New York, Plenum Press.

Westman, A., Huth-Fehre, T., Demirev, P. *et al.*, 1994, "Matrix-Assisted Laser Desorption Ionization Dependence of the Ion Yield on the Laser Beam Incidence Angle," *Rapid Commun. Mass Spectrometry*, **8**, 388–393.

Yeh, J. T. C., 1986, "Laser Ablation of Polymers," *J. Vac. Sci. Technol. A*, **4**, 653–658.

Yingling, Y. G., and Garrison, B. J., 2007, "Incorporation of Chemical Reactions into UV Photochemical Ablation of Coarse-Grained Material," *Appl. Surf. Sci.*, **253**, 6377–6381.

Yingling, Y. G., Zhigilei, L. V., Garrison, B. J. *et al.*, 2001, "Laser Ablation of Bicomponent Systems: A Probe of Molecular Ejection Mechanisms," *Appl. Phys. Lett.*, **78**, 631–633.

Zhigilei, L. V., and Garrison, B. J., 1999, "Molecular Dynamics Simulation Study of the Fluence Dependence of Particle Yield and Plume Composition in Laser Desorption and Ablation of Organic Solids," *Appl. Phys. Lett.*, **74**, 1341–1343.

2000, "Microscopic Mechanisms of Laser Ablation of Organic Solids in the Thermal and Stress Confinement Irradiation Regimes," *J. Appl. Phys.*, **88**, 1281–1298.

Zhigilei, L. V., Kodali, P. B. S., and Garrison, B. J., 1997, "Molecular Dynamics Model for Laser Ablation and Desorption of Organic Solids," *J. Phys. Chem. B*, **101**, 2028–2037.

1998, "A Microscopic View of Laser Ablation," *J. Phys. Chem. B*, **102**, 2845–2853.

10 Pulsed-laser interaction with liquids

10.1 Rapid vaporization of liquids on a pulsed-laser-heated surface

10.1.1 Background

The laser-beam interaction with materials in liquid environments exhibits unique characteristics in a variety of technical applications. The explosive vaporization of liquids induced by short-pulsed laser irradiation is utilized in laser cleaning to remove microcontaminants (Park *et al.*, 1994) and in medical laser surgery. Physical understanding of superheated liquids and liquid-to-vapor phase transitions has been sought in order to achieve better control of such applications. The transient development of the bubble-nucleation process and the onset of phase change were monitored by simultaneous application of optical-reflectance and -scattering probes (Yavas *et al.*, 1993). The numerical heat-conduction calculation also shows that the solid surface achieves temperatures of tens of degrees of superheat (Yavas *et al.*, 1994). Real-time measurement of the surface temperature transient in the course of the laser-induced vaporization process is needed, since the surface temperature is an important parameter in heterogeneous nucleation. The kinetics of heterogeneous bubble nucleation and the growth dynamics have long been a subject of intense research interest (Skripov, 1974; Stralen and Cole, 1979; Carey, 1992).

Enhanced pressure is produced in the interaction of short-pulsed laser light with liquids (Sigrist and Kneubühl, 1978). The efficient coupling of laser light into pressure is of interest in many technical and medical areas, such as laser cleaning to remove surface contaminants, laser tissue ablation, corneal sculpturing (Vogel *et al.*, 1990), and gall-stone fragmentation (Teng *et al.*, 1987). It is also widely used for surface treatment of metals and nondestructive testing of materials. The enhancement of pressure in a liquid or at its interface with another medium has been attributed to plasma formation, bubble cavitation, and rapid evaporation. Plasma formation occurs only for high-intensity laser irradiation involving ablation or optical breakdown. Collapsing bubbles generate a strong impulse of pressure (Rayleigh, 1917). High-speed photography has shown that the cavitation bubbles near a solid boundary introduce liquid-jet formation (Vogel *et al.*, 1989) that is responsible for cavitation damage (Tomita *et al.*, 1984). Many technical applications are based on the generation of pressure by implosive bubble collapse, such as ultrasonic cleaning. In the ablation of a liquid film by a short-pulsed laser, the pressure production is ascribed to the explosive growth of bubbles brought about by instantaneous heating (Do *et al.*, 1993). Vapor explosion and emission of a pressure wave by rapid evaporation of liquids is known to be destructive in many industrial accidents, such as nuclear-reactor failure.

Quantification of pressure in liquid pools subjected to intense short-pulsed-laser irradiation has been achieved by various methods. These include the use of hydrophone or piezoelectric transducers (Leung *et al.*, 1992; Schoeffmann *et al.*, 1988), fast photography (Vogel *et al.*, 1989), interferometry (Ward and Emmony, 1991a, 1991b), and optical probing (Doukas *et al.*, 1991). Several optical methods rely on measurements of the shock-wave velocity and the shock-wave velocity–pressure relations for the determination of pressure (Doukas *et al.*, 1991; Harith *et al.*, 1989). In a lower-energy regime wherein the velocity is not supersonic, such a calibration is not applicable. Since high pressure is needed to create shock waves, the detection limit of this kind of technique is typically in the kilobar range. In the case of water, a Mach number of 1.35 corresponds to a pressure of 5 kbar according to the shock-wave and thermodynamic relation (Rice and Walsh, 1957). Hydrophone or piezoelectric transducer probes are of limited use due to ringing and lack of bandwidth, leading to probable underestimation of pressure. In the short-pulsed-laser generation of pressure when the frequency is well above the megahertz range, piezoelectric transducers need to be placed close to the sample in order to avoid severe attenuation of high-frequency waves in the liquid. This causes undesirable disturbance in the pressure field and a false signal due to direct absorption of laser light by the sensor element. The photoacoustic sensing technique, which is based on the probe-beam-deflection method, has shown its ability to overcome these difficulties.

Pressure amplitude ranging from a few atmospheres up to hundreds of kilobars is generated upon laser interaction with absorbing liquids or transparent liquids near an absorbing solid boundary. The reported pressure values depend on the laser intensity, optical properties of liquids and/or absorbing solid, characteristics of the pressure wave, and detection geometry. Hence a more universal parameter, namely, the conversion efficiency from the optical energy delivered by a laser pulse into acoustic energy has been utilized. The parameter ranges from 10^{-4} in the case of thermoelastic expansion (Lyamshev and Naugol'nykh, 1981; Lyamshev and Sedov, 1981) to 0.3 in the case of optical breakdown (Teslenko, 1977). It has been shown that the optical breakdown or surface ablation and subsequent plasma formation are the most efficient mechanisms for pressure generation (Sigrist, 1986). However, these are not desirable in many technical applications due to their destructive nature. The explosive vaporization of liquids, on the other hand, can generate pressure efficiently without inflicting permanent damage on the sample. It had been shown earlier that the additional evaporation of liquid causes the peak pressure to be up to an order of magnitude higher than that due to the thermal expansion alone (Sigrist, 1978). Theoretical description is difficult due to the formation of a superheated liquid, development of nucleation centers, and thermal instability of the evaporation front (or bubble).

The bubble size and growth rate are important physical quantities for understanding the rapid phase-change process. When the laser pulse irradiates the solid surface, tiny bubbles nucleate on the surface and then grow. The bubbles may eventually coalesce, in which process both the overall number density and the shape of the bubble clusters or aggregates are expected to change. Therefore, the presence of multiple bubbles complicates theoretical calculation of the growth rate. Optical techniques based on Mie scattering were applied to measure the size of spherical bubbles (Hansen, 1985; Holt and Crum, 1990). Barber and Putterman (1992) also utilized Mie scattering to measure the radius of a single, isolated micrometer-size sonoluminescing bubble. Direct visualization of a cavitation bubble and the emanating pressure field has been conducted using a Mach–Zehnder interferometric technique (Ward and Emmony, 1991a, 1991b; Alloncle *et al.*, 1994). A high-sensitivity differential interferometer was introduced by Rovati *et al.* (1993) to measure the pulsation of millimeter-size gas bubbles. Nevertheless, the small (sub-micrometer) length and short time scales render direct *in situ* measurement of the size of nanosecond-pulsed-laser-induced bubbles difficult.

10.1.2   Temperature measurement

Accurate measurement of the temperature of the short-pulsed-laser-irradiated target is essential for understanding the laser-beam interaction with materials and assessing the optimal operating parameters in many technical laser applications such as pulsed-laser micromachining. Efforts to quantify the temperature development during laser-irradiation processes include temperature estimation by using thermocouples (Dyer and Sidhu, 1985), use of pyroelectric crystals (Gorodetsky *et al.*, 1985), probing the absorptance change of dye (Lee *et al.*, 1992), and using Ni–Si thermistors (Brunco *et al.*, 1992).

The solid sample utilized for carrying out the temperature measurement (Park *et al.*, 1996a, 1996b) is shown in Figure 10.1. The optical-temperature-sensor layer is a thin polycrystalline silicon (p-Si) film whose optical properties vary with temperature. By utilizing the formalism of the characteristic transmission matrix, the lumped-structure reflectivity and transmissivity can be obtained. The calculated temperature profile is utilized to compute the theoretical reflectivity response. Figure 10.2 shows the results from measurements of the transient temperature and bubble growth at atmospheric pressure for the water–chromium interface for two excimer-laser fluences. As bubbles form and grow on the surface, the frontside specular reflectance exhibits distinct transient behavior, as can be seen in the top panels. The increase in reflectance following the excimer-laser pulse is caused by the formation of a thin layer of small embryonic bubbles (Yavas *et al.*, 1993), while the trailing decrease is due to scattering losses caused by enlarged bubbles. Finally, the slow recovery is due to bubble collapse. Nucleation manifests itself as an increase in the front-side reflectance. The degree of superheating required for the nucleation of embryos is significantly lower than that for the formation of bubbles. The dynamics of bubble growth can be understood as follows. Microscopic embryos are nucleated as soon as the surface reaches the boiling condition (the "nucleation threshold"). Bubbles can grow in size if sufficient heat is supplied from the surface. When their radii exceed the limit $R_b \gg \lambda/(2\pi n)$, i.e. about 0.06 μm ($\lambda = 488$ nm and $n = 1.34$), scattering losses become much more appreciable and the specular reflectance decreases. From the temperature measurements, bubble growth takes place only if the surface temperature is about 100 K above the boiling temperature for water and methanol at atmospheric pressure. The bubble-growth speed at the liquid-inertia-controlled limit

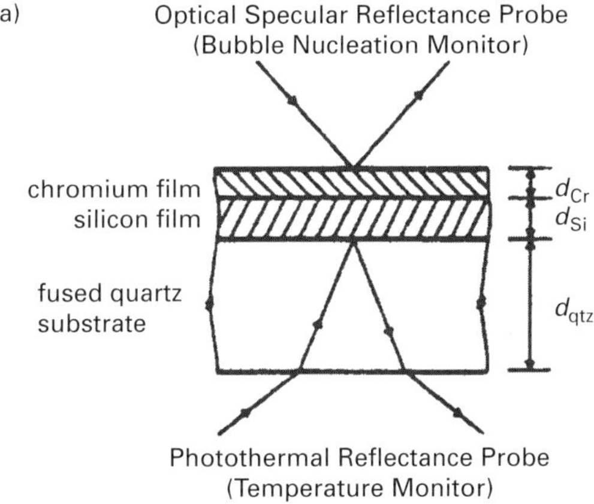

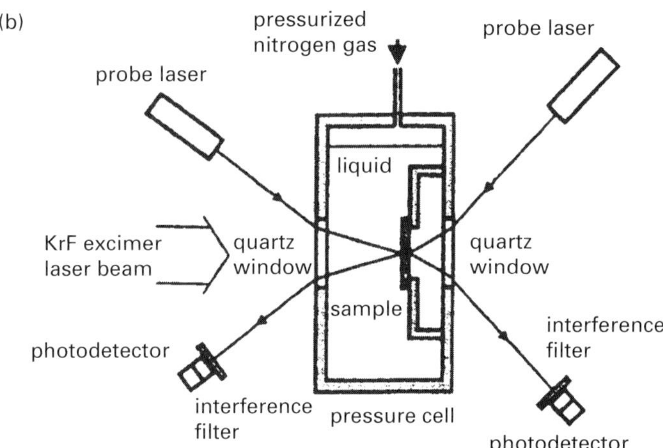

Figure 10.1. A schematic diagram of the experimental setup: (a) the optical detection scheme and (b) an overall view. The sample is composed of a 0.15-μm-thick Cr film and 0.35-μm-thick p-Si to maximize the optical response due to interference. From Park *et al.* (1996a), reproduced with permission from the American Institute of Physics.

can be estimated by equating the kinetic energy of the flow to the work performed by pressure forces, $dR_b/dt = (2/3)(P_v - P_\infty/\rho)^{1/2}$ (Carey, 1992). This estimate provides an upper bound for the growth velocity (Prosperetti and Plesset, 1978). The peak surface temperature for water at the bubble-growth threshold for each ambient pressure is measured and plotted in Figure 10.3. The solid line indicates the saturation curve and the dashed line the limit of superheating (Avedisian, 1985). The area between the saturation curve and the limit of the superheating curve is the region of metastable states. It is observed that the threshold temperature increases with pressure until it merges with the saturation curve at 2.2 MPa. The degree of superheating required for the bubble growth on a nanosecond time scale decreases with pressure and becomes zero near 2.2 MPa.

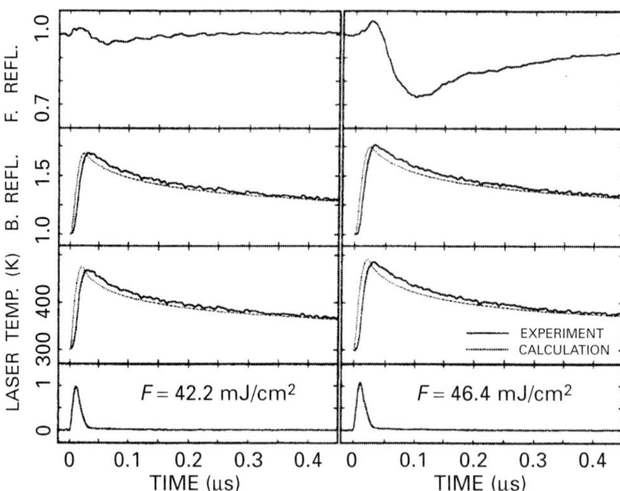

Figure 10.2. The experimental reflectance curves (solid lines) for water are shown in both top panels for the front-side reflectance and in both second panels for the back-side photothermal reflectance. The dotted lines in the second and third panels are the calculated transient reflectivity response and surface temperature, respectively. Shown as solid lines in the third panels are resultant surface-temperature traces from the measured reflectances. The bottom panels show the excimer-laser pulses, $F = 42.2$ mJ/cm^2 (left) and 46.4 mJ/cm^2 (right). From Park *et al.* (1996a), reproduced with permission from the American Institute of Physics.

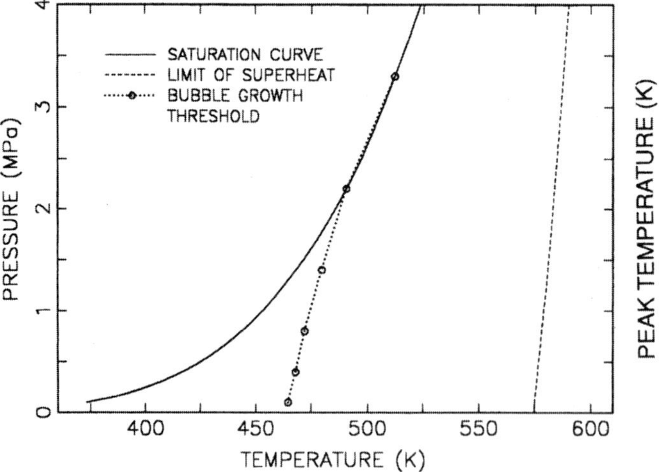

Figure 10.3. The bubble-growth-threshold temperatures and liquid pressures are identified as symbols o and connecting dotted lines in the *P–T* diagram. From Park *et al.* (1996b), reproduced with permission from the American Society of Mechanical Engineers.

10.1.3 Pressure production

The transient pressure generated by the vaporization of water on a solid surface is observed experimentally using a commercial piezoelectric transducer of bandwidth 100 MHz and the photoacoustic (PA) probe-beam-deflection method (Park *et al.*, 1996c). The experimental setup is shown in Figure 10.4(a). A PA probe-beam-deflection probe is also shown. The technique is based on detecting the refractive-index gradient in water induced by the change of density. The change of density can be created by temperature or pressure gradients. While crossing a probe beam, a pressure wave introduces a refractive-index gradient across the waist of the probe beam. Therefore, the probe beam deflects toward the optically denser side, producing the so-called "mirage effect."

The experimental setup for the optical-reflectance probe together with the PA probe-beam-deflection method is shown in Figure 10.4(b). The PA probe-beam-deflection technique provides a fast and contactless way of detecting acoustic-wave propagation. When the probe diameter is much smaller than the wavelength of the PA pulse, the deflection angle φ_{def} is (Tam, 1986)

$$\varphi_{\mathrm{def}} \cong \frac{l}{n_0} \frac{\partial n}{\partial r}, \tag{10.1}$$

where n is the refractive index of the medium, with an equilibrium value of n_0, l the interaction length of the photoacoustic pulse with the probe beam, and r the radial coordinate (Figure 10.5). The refractive-index gradient is proportional to the time derivative of the acoustic wave, so the probe-beam-deflection signal is a measure of the time derivative of the pressure pulse at the probe-beam position, i.e.

$$\varphi_{\mathrm{def}}(t) \propto \frac{\partial P}{\partial t}. \tag{10.2}$$

In the detection scheme with a knife edge blocking the lower half of the probe beam (Figure 10.5), the signal received by a photodetector, I_{d}, using Equation (10.1), is written as

$$I_{\mathrm{d}} = \mathrm{Gain} \times \left[1 + \mathrm{erf}\left(\sqrt{2} \frac{\pi w \varphi_{\mathrm{def}}}{\lambda} \right) \right], \tag{10.3}$$

where Gain is the gain constant which can be determined by measuring the initial signal when the deflection is zero, $I_{\mathrm{d}} = \mathrm{Gain}$, and L is the distance from the focal plane to the knife edge. Therefore, the deflection angle is

$$\varphi_{\mathrm{def}} = \frac{\lambda}{\sqrt{2\pi} w} \, \mathrm{erf}^{-1}\left(1 - \frac{I_{\mathrm{d}}}{\mathrm{Gain}} \right), \tag{10.4}$$

where erf^{-1} is the inverse error function and the deflection angle toward the sample is given a positive sign. Figure 10.6(a) shows a typical deflection signal detected by the photodetector. The trace shown in Figure 10.6(b) is the deflection angle φ_{def} computed using Equation (10.4). The deflection angle is easily converted to $\partial P/\partial t$ using Equation (10.1), the $\partial n/\partial P$ value (0.000 15 MPa^{-1} for water at $\lambda = 632.8$ nm; Schiebener *et al.*, 1990), and the speed of sound. Finally, the pressure $P(t)$ can be obtained by integrating with respect to time (Figure 10.6(c)).

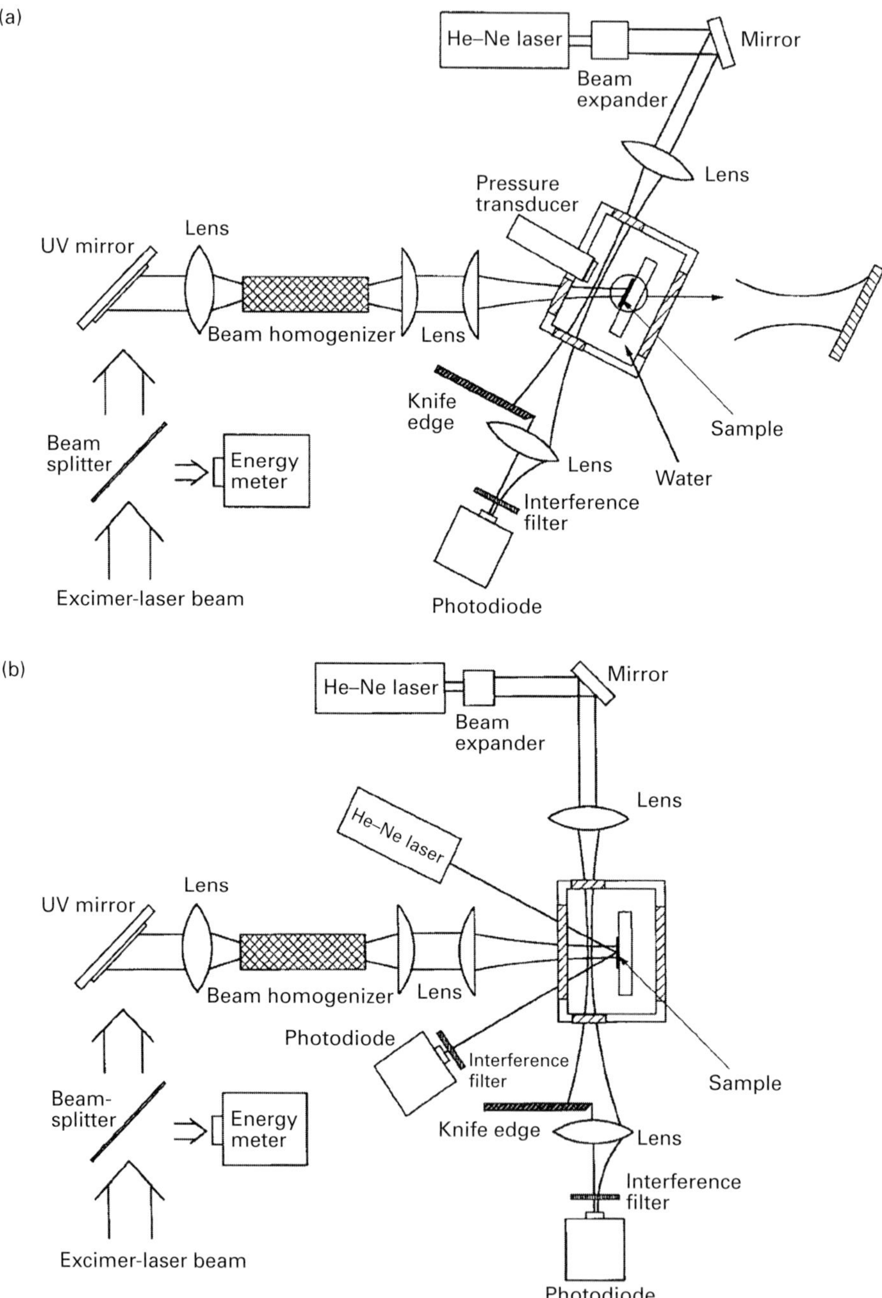

Figure 10.4. Schematic diagrams of the experimental setup with (a) the piezoelectric transducer and the photoacoustic-deflection probe, and (b) the optical-reflectance probe and the photoacoustic-deflection probe. The diameter of the deflection probe beam is exaggerated. From Park *et al.* (1996c), reproduced with permission from the American Institute of Physics.

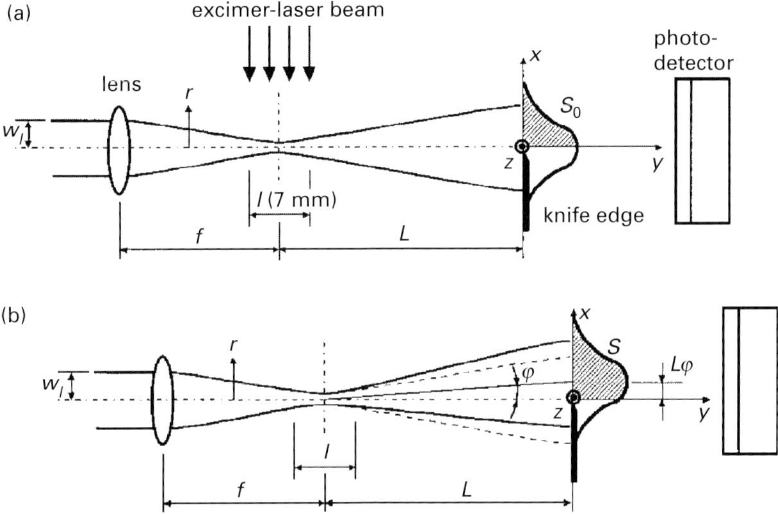

Figure 10.5. Geometries of (a) the undeflected and (b) the deflected probe beam with a Gaussian intensity profile. From Park *et al.* (1996c), reproduced with permission from the American Institute of Physics.

The possible pressure-production mechanisms during laser-induced vaporization of liquids on solid surfaces are thermal expansion, rapid bubble growth and collapse, and plasma formation. For the laser energy levels considered in this work (<0.1 J/cm^2), the last mechanism is not likely to contribute. When laser-induced heating occurs, both liquid and solid experience thermoelastic expansion and thermal stress. The stress wave produced by thermoelasticity can be calculated from the equations of motion, the stress–strain relations, and the imposed boundary conditions (Bushnell and McCloskey, 1968).

It has been shown that a vibrating body transmits energy in the form of acoustic radiation to the surrounding medium (Stokes, 1868). Bubbles have since been recognized as a source of acoustic emission by Strasberg (1956). He found that the volume pulsations of bubbles can generate high sound pressures. When a bubble entrained in a liquid undergoes expansion or contraction, acoustic pressure is radiated from the bubble surface. Pressure can be generated also during the adiabatic collapse of bubbles (Rayleigh, 1917; Strasberg, 1956). A factor to be accounted for is related to the effects due to multiple bubbles. It has been shown experimentally that the pressure induced by multiple-bubble collapse is weaker than that caused by collapse of a single bubble (Crum, 1994). Tomita *et al.* (1984) discovered that the maximum impulsive pressure quickly decreases with reduction in spacing between bubbles due to significant interactions and subsequent loss of spherical symmetry.

When the laser fluence is low, i.e. below the vaporization threshold of about 50 mJ/cm^2, the pressure is generated thermoelastically. Upon exceeding this threshold, another contribution of pressure is attributed to bubble growth, as shown in Figure 10.7(a). The observed signal is a superposition of the thermal-expansion-induced pressure and the bubble-growth-induced pressure. A similar trend has been observed (Zweig *et al.*, 1993).

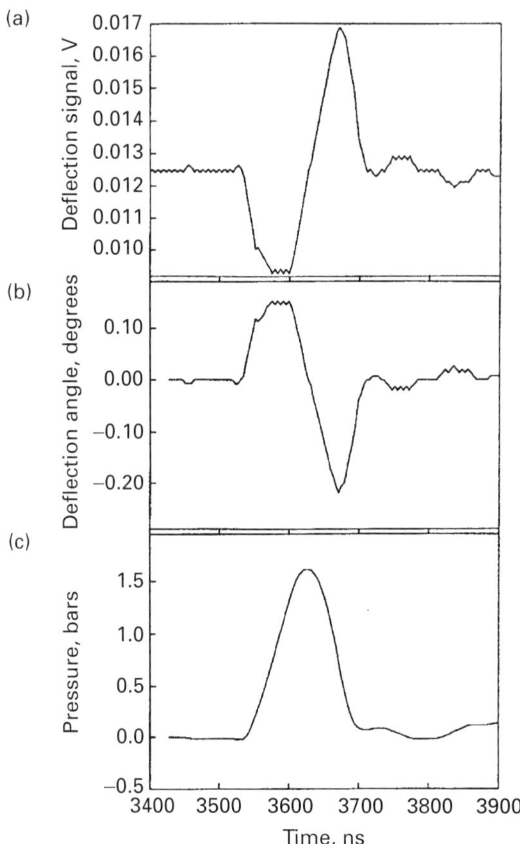

Figure 10.6. (a) The deflection signal received by the photodetector for water irradiated with the excimer-laser fluence of 46.1 mJ/cm^2. Also shown are (b) the converted deflection angle, φ_{def}, and (c) the pressure. From Park *et al.* (1996c), reproduced with permission from the American Institute of Physics.

Thermoelastic pressure is generated by the temperature rise and has narrow temporal width because its duration scales with the time required for the temperature rise, i.e. the excimer-laser pulse width. On the other hand, pressure generated by bubble growth has a longer temporal duration, scaling with the bubble-growth time (Figure 10.7(b)).

10.1.4 Kinetics of vapor phase change

Michelson interferometry is the best-known technique to be based on the interference effect of two mutually coherent beams. As mentioned above, the nanosecond-laser-induced bubble-growth process can be approximated by invoking the picture of a growing thin vapor film. If the surface roughness of the film (or size of the individual bubbles in the early stage) is much smaller than the wavelength of the probe laser beam, the interference technique can be used to detect the thickness of the vapor film (or effective thickness of the bubble layer). The difference between the refractive indices of water vapor and water causes a net change in the optical path length of an interfering beam as

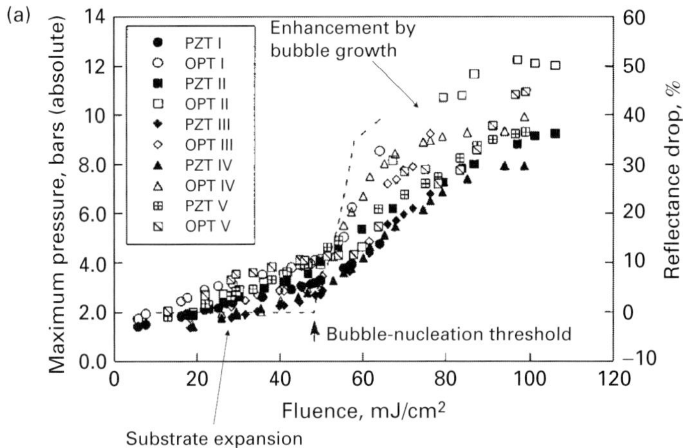

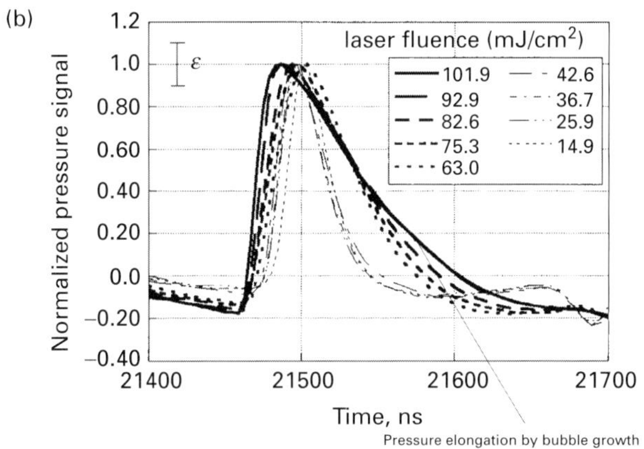

Figure 10.7. The pressure-pulse amplitudes are plotted in the left-hand figure as a function of the excimer-laser fluence. The data produced by the transducer are represented by the symbols labeled "PZT" and the data obtained by the deflection probe are represented by the symbols labeled "OPT." The experiments were done five times, and corresponding data points are labeled as indicated by the Roman numerals. The amplitude of the drop in optical specular reflectance is also plotted with the dashed line. The bubble-nucleation threshold is marked by the arrow. The right-hand figure gives the pressure-pulse traces that are normalized by the peak pressue. The dashed curves indicate the pressure pulses in the thermoelastic regime and the solid traces indicate the pressure pulses above the bubble-nucleation threshold. It can be seen that the bubble generation and vapor-phase formation extend the duration of the released pressure. From Park *et al.* (1996c), reproduced with permission from the American Institute of Physics.

the vapor film grows. A schematic diagram of the experimental setup is shown in Figure 10.8. The measured "effective film thickness" of the bubble layer and the corresponding reflectance signal are displayed in Figures 10.9(a) and (b). The early stage of bubble growth is largely dominated by an inertia-controlled process. The growth rate of the effective vapor film in the initial stage ranges from 0.5 to 1 m/s. These values are considered as lower limits of the bubble-growth rate since the effective layer thickness does not directly represent bubble size.

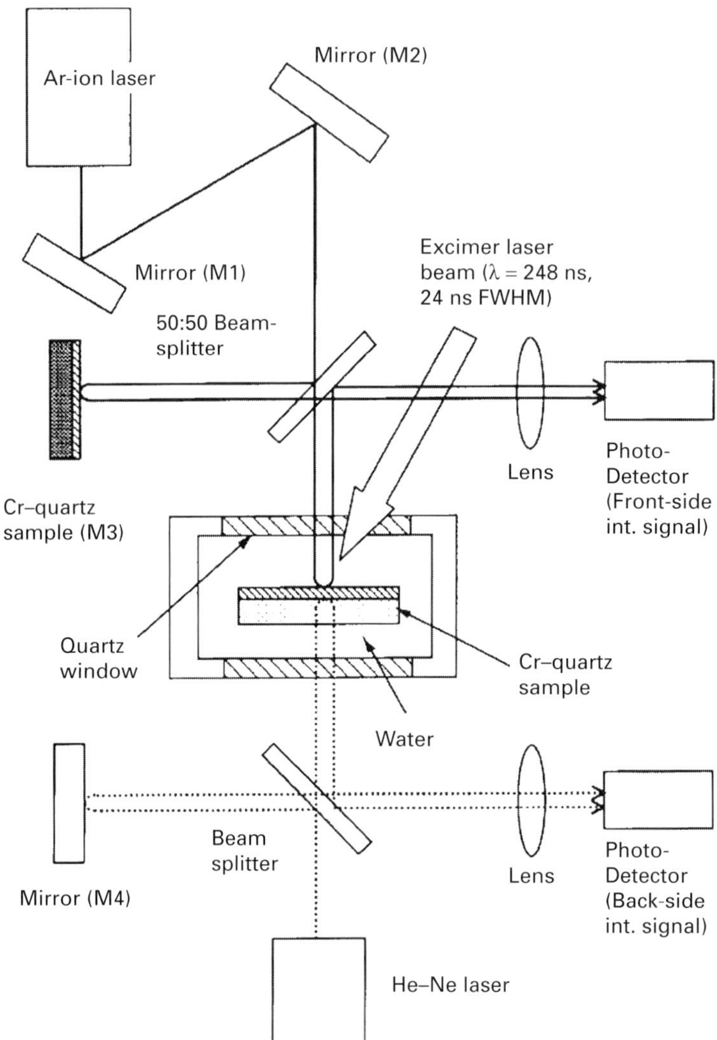

Figure 10.8. A schematic diagram of the experimental setup using two Michelson interferometers (front-surface probe and back-surface probe) for measuring the change in optical path length due to bubble formation on a Cr surface and for monitoring the vibration of the solid sample, respectively. From Kim *et al.* (2001), reproduced with permission from Elsevier.

10.2   Pulsed-laser interaction with absorbing liquids

10.2.1   Backgound

Owing to the short time scale and noncontact nature of the high-density energy transfer, the interaction of a short laser pulse with an absorbing-liquid medium has been one of the important issues in the study of rapid phase change and bubble dynamics. An acoustic pulse can be generated either by a purely thermoelastic mechanism or by the combined

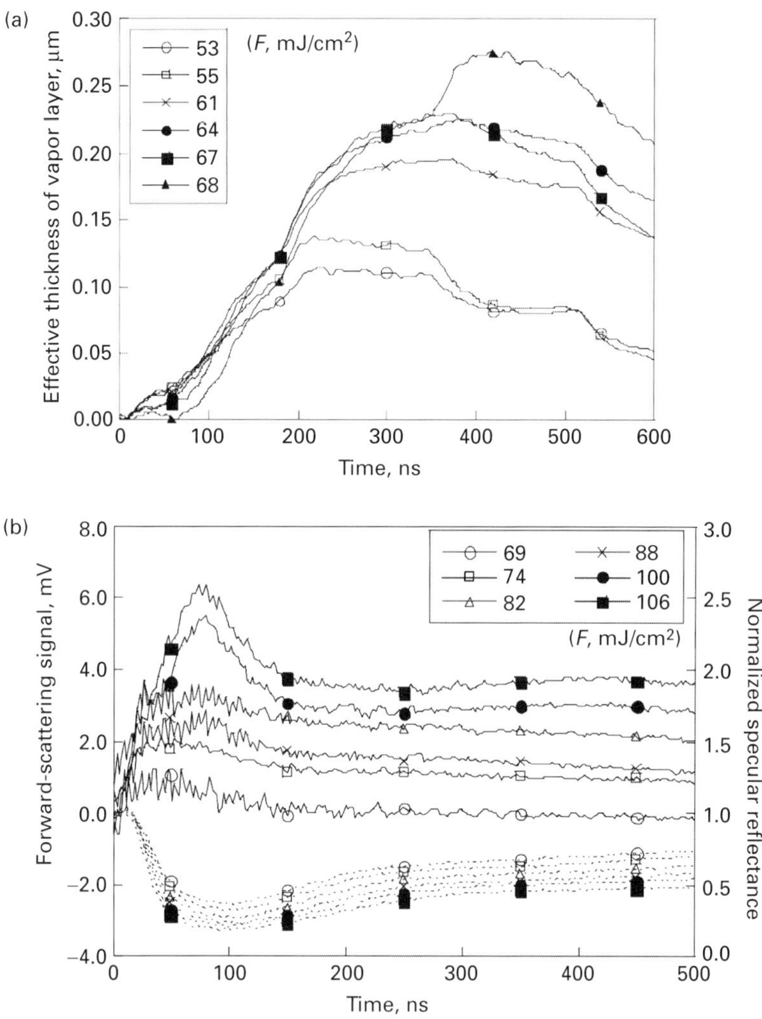

Figure 10.9. (a) The effective thickness of the bubble layer measured by recording the change in optical path length at various laser fluences and (b) the corresponding normal-incidence reflectance signal on the vapor-formed surface. From Kim *et al.* (2001), reproduced with permission from Elsevier, and Kim *et al.* (2001), reproduced with permission from Elsevier.

effect of that together with other nonlinear processes such as liquid evaporation, dielectric breakdown, and electrostriction (Bunkin and Komissarov, 1973; Lyamshev and Naugol'nykh, 1981; Lyamshev and Sedov, 1981). This photoacoustic energy-transfer mechanism can be employed in a number of applications, including a variety of biomedicalones (Teng *et al.*, 1987; Cross *et al.*, 1988; Vogel *et al.*, 1990; Asshauer *et al.*, 1994; Oraevsky *et al.*, 1995; Paltauf *et al.*, 1996). Results of studies on laser-induced biological-tissue removal suggest that the low threshold for tissue ablation can be ascribed to the mechanical effects (spallation) caused by short-pulsed heating under stress-confinement conditions (Oraevsky *et al.*, 1995; Paltauf and Schmidt-Kloiber, 1996). It has also been

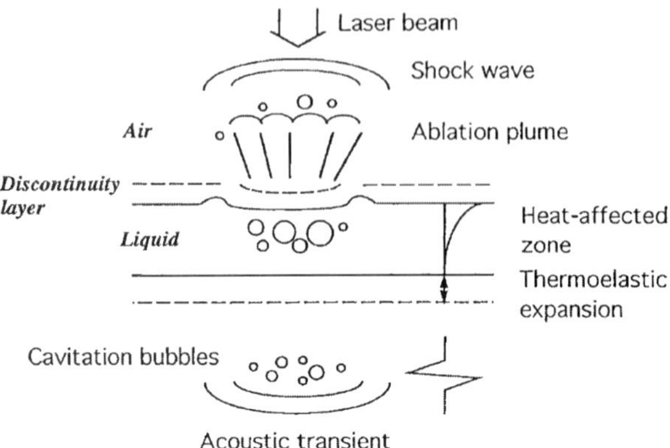

Figure 10.10. A diagram indicating various phenomena involved in the pulsed-laser-induced ablation of absorbing liquids.

argued that the physical mechanism underlying the biological-tissue ablation bears a strong resemblance to that of the absorbing-liquid ablation. Another potential application is in the pulsed-laser removal of surface contaminants via the ablation of a deliberately deposited liquid film, such as in the CO_2-laser-based cleaning tool (Héroux *et al.*, 1996). In this case, however, the ablation mechanism for a thin liquid film formed on a solid surface has to be considered since the mechanism might be significantly different from that of the ablation of bulk liquids.

Irradiation of absorbing-liquid surfaces by a laser pulse can drive a number of inter-related phenomena, as depicted in Figure 10.10. The energy absorption increases the liquid temperature and the subsequent thermal expansion of the heated volume produces thermoelastic-stress waves. The vapor phase can be formed by further temperature increase or by a local pressure drop, while acoustic excitation can be induced by the dynamics of the vapor plume and cavities in the liquid. Thus, the pulsed-laser ablation of absorbing liquids can take place through several different physical processes. If the heating rate is gradual so that the laser-pulse duration is long compared with the acoustic relaxation time (the time taken for an acoustic wave to cross the heated zone), thermal evaporation due to the temperature rise under constant ambient pressure is usually observed, i.e. typical "slow" boiling phenomena. However, in the case of "rapid" heating by a short-pulsed laser, the effect of thermoelastic excitation of the acoustic field becomes a dominating factor in the ablation process at low laser fluences: ablation is driven by the tensile component of the thermoelastic stress without a substantial temperature increase ("cold" ablation). This mechanism is explained by the metastability of liquids subjected to negative stress of considerable magnitude (Fisher, 1948; Batchelor, 1967). The significance of the thermoelastic-stress generation stems from the fact that low hydrodynamic pressure below the equilibrium vapor pressure can create cavities filled with either vapor or noncondensable gases from pre-existing gas pockets. The dynamics of the cavity growth and collapse can play a major role in the ablation process and the acoustic-pulse generation. It has been demonstrated that cavitation bubbles near a free surface can produce a high-speed jet in the liquid (Blake *et al.*, 1987). Oraevsky *et al.* (1995) also observed that substantial liquid-stream ejection is followed by the collapse of large cavitation bubbles in the ablation of an aqueous solution of K_2CrO_4 by a Q-switched Nd : YAG laser. Accordingly, the acoustic-pulse generation and material removal by the pulsed-laser irradiation of a liquid surface do not just depend on the thermal-energy balance in a simple way as in the case of "slow" heating of the liquid. More importantly, the phenomena are affected by the optically excited stress field, the dynamics of the vapor cavity formed, and the evolution of the ablation plume ejected into the ambient air.

When a large amount of energy is deposited during a short interval at high laser fluences, substantial superheating of the liquid above the equilibrium vaporization temperature is achieved. If the temperature of the superheated liquid becomes comparable to the thermodynamic critical temperature, "explosive vaporization" takes place due to the instantaneous increase of the homogeneous bubble-nucleation rate in the superheated liquid. The metastability behavior of the liquid due to large superheating and the bubble-nucleation kinetics are critical in this case. A strong shock wave can also be formed in the ambient air by the expansion of a high-temperature and high-pressure ablation cloud over the laser spot.

10.2.2 Optical generation of acoustic transients

The transient heating by a laser pulse induces thermal expansion of the absorbing-liquid volume, which initially pushes the surrounding liquid medium. This momentum supply into the surrounding medium and the formation of a rarefaction zone inside the heated volume are responsible for the compressive and tensile stresses in the medium, respectively. The acoustic pulse at a certain location in the medium can be either compressive or tensile, depending on the arrival time of the stress wave generated in the absorption zone. While the thermal expansion directed into the medium (observer) generates positive stress (a compression wave), the outward expansion directed toward the free surface generates negative stress (a rarefaction wave). The early analytical works in one-dimensional planar and spherical geometries showed that the acoustic wave comprises a leading positive component that is followed by a negative component originated by the outward expansion (Gournay, 1966; Hu, 1969). Three-dimensional analyses, however, demonstrated that the acoustic pulse may have a more complicated shape than a simple compression–rarefaction pulse (Kasoev and Lyamshev, 1977; Karabutov *et al.*, 1979).

Simple estimates of the magnitude of the pressure pulse generated by the thermoelastic mechanism are possible for two limiting cases. In the short-pulse limit (acoustic-penetration depth d_{aco} $\ll$ optical-penetration depth d_{abs}, i.e. $c_s t_{pulse} \ll 1/\gamma$), the volume that absorbs laser energy cannot expand during the laser-pulse heating and the entire zone rapidly enters the compression stage. Let the beam-spot area be A and the characteristic length of the energy-absorption zone be $1/\gamma$ so that the total laser-energy-absorbing volume is $V = A/\gamma$. In this limiting case, the thermal expansion of the volume is

$$\Delta V = \frac{E_{abs}\beta}{\rho C_p},$$
(10.5)

where E_{abs} represents the absorbed laser energy, β the volumetric thermal-expansion coefficient, ρ the density, and C_p the specific heat at constant pressure.

By virtue of the definition of the speed of sound in a liquid,

$$c_s^2 = \left(\frac{\partial P}{\partial \rho}\right)_s = -\frac{1}{\rho^2}\left(\frac{\partial P}{\partial v}\right)_s, \tag{10.6}$$

the pressure increase due to the volume expansion is given by

$$\Delta P \approx \rho c_s^2 \frac{\Delta V}{V}. \tag{10.7}$$

Substitution of Equation (10.5) into the above equation yields

$$\Delta P \approx \frac{c_s^2 \beta}{C_p} \frac{E_{abs}}{V} = \Gamma_G \gamma F, \tag{10.8}$$

where the constant $\Gamma_G = c_s^2 \beta / C_p$ is the so-called Grüneisen parameter. This parameter quantifies the ability of the medium to convert internal energy into pressure production. The long-pulse limit corresponds to the case when the laser-pulse duration is much longer than the acoustic-pulse travel time ($c_s t_{\text{pulse}} \gg 1/\gamma$). In this limit, the acoustic pulse travels in the absorption zone with finite speed during the laser-pulse heating. From the momentum balance, the pressure rise is estimated:

$$\Delta P \approx \rho c_s U_{\text{exp}} = \rho c_s \frac{\Delta V}{A t_{\text{pulse}}} = \frac{c_s \beta}{C_p} \frac{E_{abs}}{A t_{\text{pulse}}} = \frac{c_s \beta}{C_p} \frac{F_{abs}}{t_{\text{pulse}}}, \tag{10.9}$$

where U_{exp} denotes the characteristic velocity of volumetric expansion. For a fixed-laser fluence and absorption coefficient, short laser pulse heating converts the electromagnetic energy into acoustic energy in a more efficient way. The limiting value of the maximum pressure approaches $\Gamma_G \gamma F_{abs}$ as the pulse duration is reduced. Equations (10.8) and (10.9) also show that the magnitude of the thermoelastic pressure for a fixed laser-pulse duration and laser fluence is initially linearly proportional to the absorption coefficient, but eventually approaches a constant value as the absorption coefficient becomes large.

The temperature field induced by pulsed-laser heating can be modeled via conductive heat transfer, assuming that the laser energy absorption in the liquid is quantified by volumetric heat generation that decays exponentially from the liquid surface. Since the laser-beam-spot dimensions are much larger than the optical and thermal penetration depths, the temperature field $T(x, t)$ can be obtained using the one-dimensional heat-conduction equation. On the time scale of interest (up to a few microseconds), the temperature field calculated by neglecting heat diffusion is a good approximation for typical liquids of low thermal conductivity. According to this simplification, the temperature rise in the irradiated volume is directly proportional to the energy absorbed during the laser-pulse duration. After the laser pulse, the temperature field is considered constant. Hence, the following expression is obtained for the temperature field by neglecting thermal diffusion:

$$\theta(x, t) = T - T_\infty = \frac{1 - R}{\rho C_p} \gamma \exp(-\gamma x) \int_0^t I(t) dt. \tag{10.10}$$

The stress field induced by surface heating of a semi-infinite liquid can be calculated by using a theoretical model based on the thermoelastic expansion of the liquid subjected to transient heating (the so-called "thermal shock"). The governing equation for modeling the one-dimensional thermoelastic material displacement u_d is (Boley and Weiner, 1967)

$$\frac{1}{c_s^2}\frac{\partial^2 u_d}{\partial t^2} = \frac{\partial^2 u_d}{\partial x^2} - \beta\frac{\partial\theta}{\partial x}. \tag{10.11}$$

The initial condition is

$$\left.\frac{\partial u_d}{\partial t}\right|_{x,t=0} = u_d(x,0) = 0, \tag{10.12}$$

and the boundary conditions for a free surface (no normal stress) and a rigid boundary are

$$\left.\frac{\partial u_d}{\partial x}\right|_{x=0,t} = \beta\theta(0,t) \tag{10.13}$$

and

$$u_d(0,t) = 0, \tag{10.14}$$

respectively. The general solution of the above equation is obtained by applying the Laplace-transform method:

$$u_d = \int_0^t f(\xi)g(t-\xi)d\xi. \tag{10.15}$$

In the above equation, the function $f(t)$ is determined from the temporal shape of the laser pulse as

$$f(t) \equiv \frac{1-R}{\rho C_p}\gamma\int_0^t I(t)dt, \tag{10.16}$$

so that

$$\theta(x,t) = f(t)\exp(-\gamma x). \tag{10.17}$$

In the study by Kim $et\ al.$ (1998), the temporal shape of the excimer-laser pulse was approximated to be triangular ($t_{\max} = 17$ ns, $t_{\text{pulse}} = 48$ ns). Hence, the temperature field in the liquid is expressed as

$$\theta(x,t) = \frac{1-R}{\rho C_p}\gamma F \exp(-\gamma x)$$

$$\times\begin{cases}\dfrac{t^2}{t_{\text{pulse}}t_{\max}} & (0 < t < t_{\text{peak}}),\\[2ex]\dfrac{t_{\max}}{t_{\text{pulse}}} + \dfrac{(t-t_{\text{pulse}})(2t_{\text{pulse}} - t_{\max} - t)}{t_{\text{pulse}}(t_{\text{pulse}} - t_{\max})} & (t_{\text{peak}} < t < t_{\text{pulse}}),\\[2ex]1 & (t_{\text{pulse}} < t).\end{cases} \tag{10.18}$$

The function $g(t)$ is obtained for the free- and rigid-boundary cases as

$$g(t) = \begin{cases} \dfrac{\beta c_s}{2}\{\exp[\gamma(c_s t - x)] - \exp[-\gamma(c_s t + x)]\} & (t < x/c_s), \\ -\dfrac{\beta c_s}{2}\{\exp[-\gamma(c_s t - x)] + \exp[-\gamma(c_s t + x)]\} & (t > x/c_s), \end{cases} \tag{10.19}$$

and

$$g(t) = \begin{cases} \dfrac{\beta c_s}{2}\{\exp[\gamma(c_s t - x)] - \exp[-\gamma(c_s t + x)]\} & (t < x/c_s), \\ \dfrac{\beta c_s}{2}\{\exp[-\gamma(c_s t - x)] - \exp[-\gamma(c_s t + x)]\} & (t > x/c_s), \end{cases} \tag{10.20}$$

respectively. Once the displacement u_d is given, the normal stress σ_{xx} is given by the equation

$$\frac{\sigma_{xx}}{\rho c_s^2} = \frac{\partial u_d}{\partial x} - \beta\theta. \tag{10.21}$$

The one-dimensional assumption in the above formulation is valid in the region close to the absorbing surface if the cross section of the laser beam is larger than the optical penetration depth and the characteristic wavelength of the excited sound wave ($d_{abs} \ll x \ll d_{aco}$). If the wave propagation exhibits three-dimensional effects, the above model is not appropriate.

If the liquid surface is covered by a quartz plate with acoustic impedance much greater than that of the liquid, the boundary condition (10.14) applies. In this case, the thermal expansion is confined only to the downward direction into the liquid and the acoustic wave is purely compressive. In contrast, the wave launched from the free surface has both compressive and tensile phases. The overall pulse width approximately matches the laser-pulse duration since the temperature change that causes compression–rarefaction occurs during this interval. As shown in Equation (10.18), the maximum (surface) temperature is determined by γF for a fixed laser-pulse duration. The effect of varying laser fluence and absorption coefficient for a constant value of γF (i.e. constant temperature) on the thermoelastic-stress generation is represented in Figure 10.11 (in this case, the maximum temperature rise is 12 K). Evidently, the amplitude and width of the acoustic pulse increase as larger laser fluence (weaker absorption) is used.

The acoustic-transient generation accompanying the change in the thermodynamic state of the liquid medium can involve various physical mechanisms. First, rapid expansion of the vapor cavity by thermal evaporation of bubble nuclei can produce acoustic pulses in the liquid. Second, surface evaporation with violent vapor-plume ejection imparts substantial recoil pressure on the liquid. Third, the collapse of an empty or vapor cavity generated by a tensile-stress field can induce a pressure wave of large amplitude. The relative importance of the vapor-cavity expansion in acoustic-pulse generation depends on the absorbed energy density. If the heating is strong, i.e. if forced vaporization occurs before the thermoelastic compression pulse is formed, the action of vapor-cavity expansion should be taken into consideration. However, if the heating is moderate, the effect of vaporization is secondary and the thermoelastic mechanism is dominant.

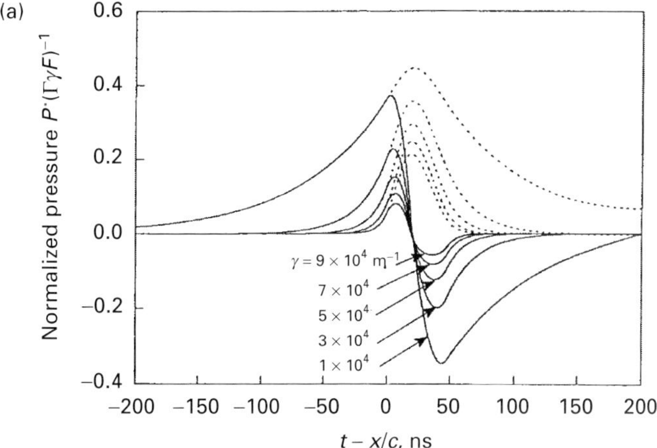

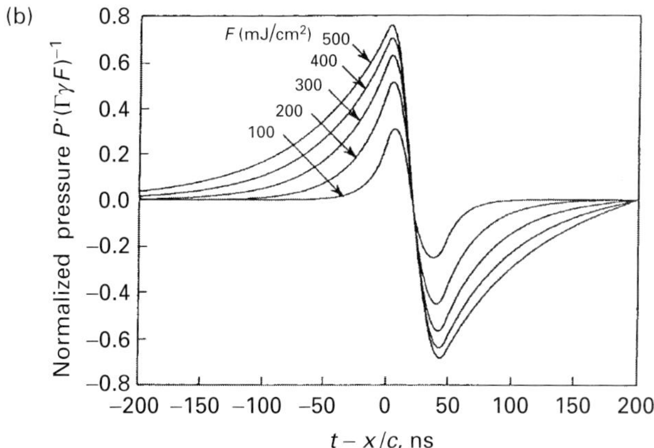

Figure 10.11. Temporal profiles of normalized thermoelastic stress induced by an excimer-laser irradiation of K_2CrO_4 solution (a) for various absorption coefficients and (b) for various concentrations and a fixed $\gamma F = 5.0 \times 10^7$ J/m^3. From Kim *et al.* (1998), reproduced with permission from Springer-Verlag. In (a) the solid lines represent the pressure pulse under free-boundary conditions and the dotted lines represent that under rigid-boundary conditions.

The liquid ablation induced by laser irradiation of a free surface involves surface vaporization as well as volumetric evaporation due to the bubble growth in the liquid. If the absorbed energy density is large enough, the vapor plume attains a large amount of kinetic energy and exerts recoil momentum upon leaving the surface. Because the surface-ablation mechanism and the consequent acoustic-pulse formation are very complicated, a thorough understanding of the process has not yet been achieved (especially for when the surface temperature is close to the critical temperature). If the temperature rise is moderate (below the critical temperature), the kinetic theory of vaporization may be applied to determine the vapor flux from the surface.

If a liquid medium is subjected to tensile stress, cavitation bubbles can be formed in the liquid. When these cavities collapse, liquid rushes toward the center of the cavity,

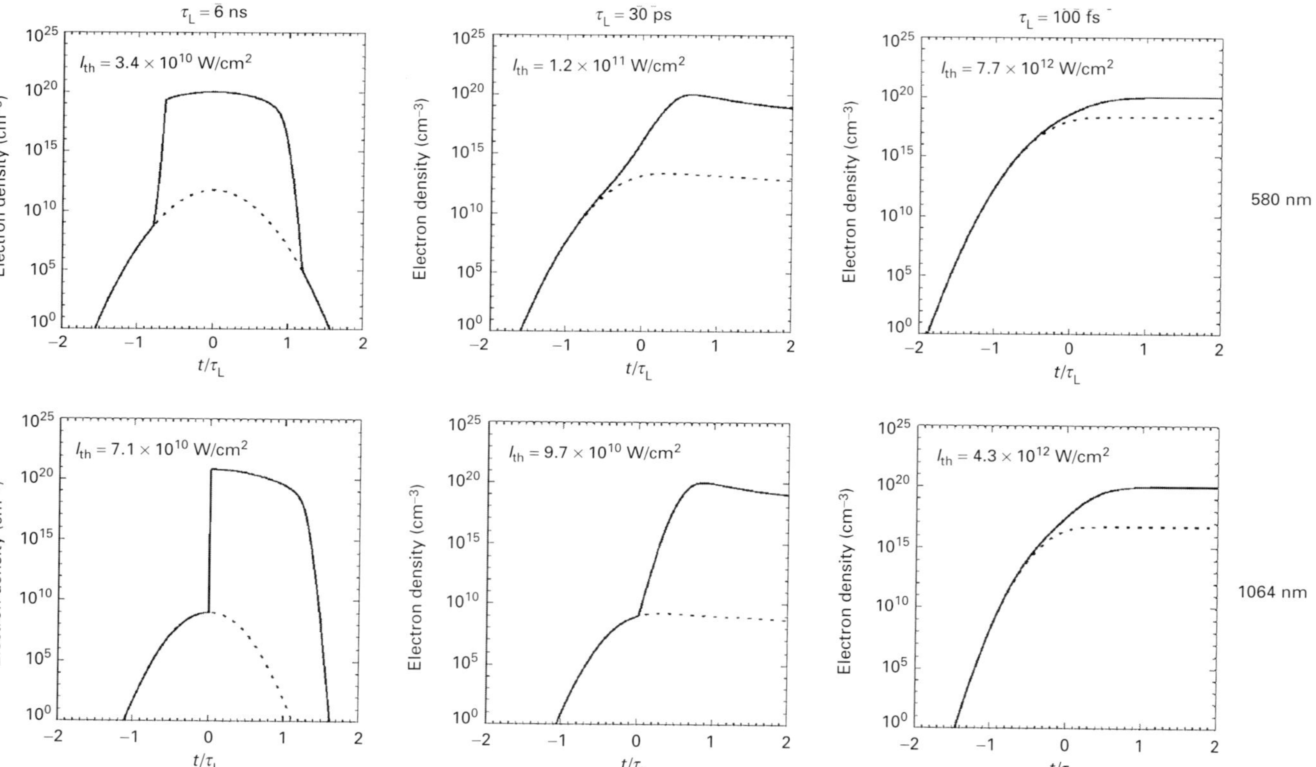

Figure 10.15. The evolution of the free-electron density of the breakdown threshold for various pulse durations (from left to right) and wavelengths (from top to bottom). The calculations were performed for breakdown in pure water with $N_{cr} = 10^{20}$ cm^{-3} and a spot size of 5 μm. For each pulse duration, the threshold irradiance I_{th} at which the curves were calculated is indicated in the plot. Besides the total free-electron concentration (solid curve), the concentration due to multiphoton absorption alone (dotted curve) is plotted as a function of time. The time axis has been normalized with respect to the laser pulse duration τ_L. From Noack and Vogel (1999), reproduced with permission from the IEEE.

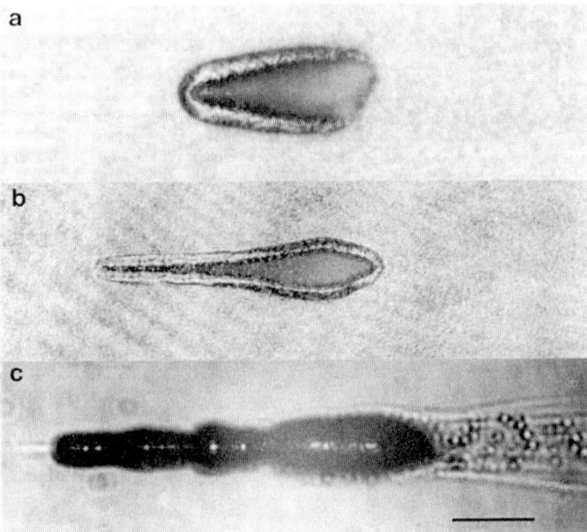

Figure 10.16. Optical breakdown at superthreshold irradiance: (a) pulse duration $\tau_L = 6$ ns, wavelength $\lambda = 1064$ nm, focusing angle $\theta = 22°$, $E = 8.2$ mJ, and $E/E_{th} = 60$, picture taken $\Delta t = 10$ ns after the start of the laser pulse; (b) $\tau_L = 30$ ps, $\lambda = 1064$ nm, $\theta = 14°$, $E = 740\,\mu$J, $E/E_{th} = 150$, $\Delta t = 8$ ns; and (c) $\tau_L = 100$ fs, $\lambda = 580$ nm, $\theta = 16°$, $E = 35\,\mu$J, $E/E_{th} = 200$, and $\Delta t = 2\,\mu$s. The laser light producing breakdown was incident from the right. The bar represents a length of 100 mm. After the nanosecond and picosecond pulses, (a) and (b), the breakdown region was delineated. One can see the luminescent plasma as well as the cavitation bubble and the shock wave produced by plasma expansion, but no other changes are observed in the surrounding liquid. In contrast, 2 μs after the femtosecond-laser pulse, in (c) refractive-index changes are visible in the laser-beam path upstream, the cavitation bubble indicating that the liquid has been heated by the laser pulse. The refractive-index changes were made visible by slightly defocusing the image. A contribution of acoustic transients to the observed refractive-index changes was excluded by taking the photographs after the transients had propagated out of the irradiated region. From Noack and Vogel (1999), reproduced with permission from the IEEE.

the focal volume is

$$I(r, z, t) = \frac{P_0}{\pi w^2(z)} \exp\left[\left(\frac{r}{w_0}\right)^2\right] \exp\left[-4\ln 2\left(\frac{t - zn/c}{t_{\mathrm{FWHM}}}\right)^2\right]. \qquad (10.26)$$

This would make the electron density a function of r, z and t, namely $N_e = N_e(r, z, t)$, but the two-dimensionality, assuming radial symmetry, would not be captured by the rate equation. To incorporate spatial effects into the rate equation, the diffusion term of Equation (10.22) is replaced by a constant diffusivity coefficient, D_e, and cylindrical diffusion terms. The electron-density rate equation then becomes

$$\frac{\partial N_e(r, z, t)}{\partial t} = \left(\frac{\partial N_e}{\partial t}\right)_{\mathrm{mp}} + \eta_{\mathrm{casc}} N_e + D_e\left[\frac{1}{r}\frac{\partial}{\partial r}\left(r\frac{\partial N_e}{\partial r}\right) + \frac{\partial^2 N_e}{\partial z^2}\right] - \eta_{\mathrm{rec}} N_e^2.$$

$$(10.27)$$

An upper limit on N_e is physically imposed by the coupling efficiency of the laser energy into the plasma. When the plasma frequency, ω_p, equals the frequency of the

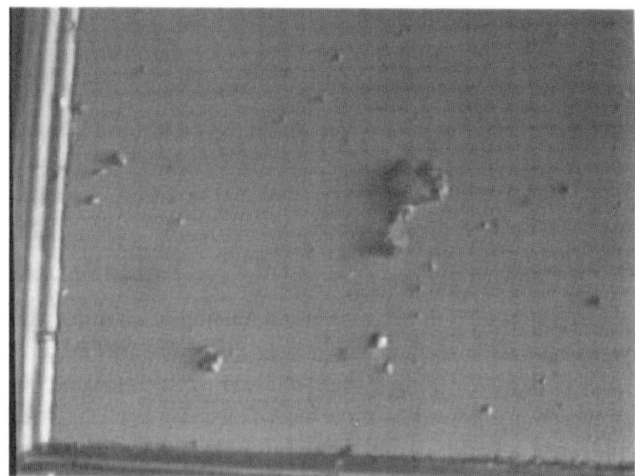

86 mJ/cm² (before)

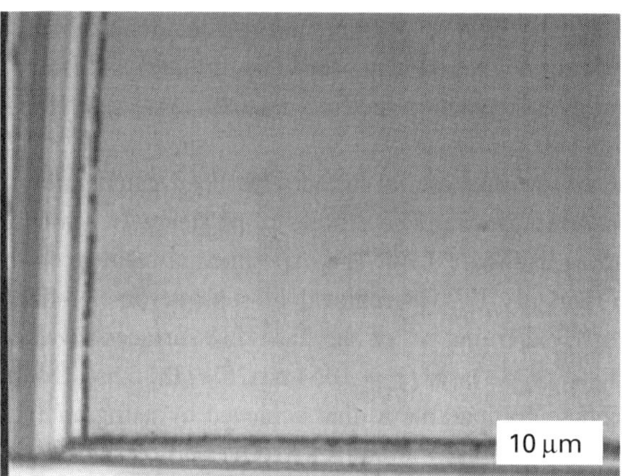

86 mJ/cm² (after)

Figure 11.5. Optical-microscope photographs of a NiP hard-disk surface contaminated with 0.3-µm-sized (average) alumina (Al_2O_3) particles and the same spots after ten steam Nd : YAG laser pulses ($\lambda = 1064$ nm) at $F = 86$ mJ/ cm² ($\theta_i = 40°$). From She *et al.* (1999), reproduced with pemission by the American Institute of Physics.

Nd : YAG laser ($\lambda = 1064$ nm) beams carry photon energies of 5.01 and 1.16 eV, respectively. The magnitudes of many chemical (covalent, ionic, hydrogen, etc.) bond energies are smaller than the photon energy of excimer-laser light. Typical ablation thresholds of organic materials subjected to UV radiation are much lower than those of inorganic substrates (Si, metals, etc.). Related issues have been examined in Chapter 10. The following discussion pertains to the removal of particles from solid surfaces.

11.4.1   Dry laser cleaning

In the LC process, the laser pulse energy is absorbed by a solid surface during a very short time. Since the temperature increase and the subsequent thermal expansion of the solid material have finite magnitudes at the same time, an inertia force of large magnitude is applied to contaminants sitting on the surface. The order of magnitude of this force can be estimated by assuming that the heat-affected zone (thermal penetration depth $d_{\text{th}} \sim \sqrt{\alpha t_{\text{pulse}}}$) expands by an amount $\beta d_{\text{th}} \Delta T$, where ΔT is the temperature excursion. The order of magnitude of the inertia force then becomes

$$f_{\text{i}} \sim \frac{4}{3} \pi R_{\text{pa}}^3 \beta d_{\text{th}} \, \Delta T / t_{\text{pulse}}^2. \tag{11.12}$$

When a particle on a surface is subjected to laser radiation, both the particle and the substrate undergo transient expansions, $z_{\text{pa}}(t)$ and $z_{\text{su}}(t)$, respectively. The dynamic deformation is given by

$$\delta_{\text{sep}}(t) = z_{\text{su}}(t) + z_{\text{pa}}(t) - z_{\text{f}}(t) + \delta_0. \tag{11.13}$$

In the above, $z_{\text{f}}(t)$ is the particle displacement and δ_0 the initial deformation parameter, given by

$$\delta_0 = \frac{1}{8} \left(\frac{2 R_{\text{pa}} h_{\text{LvW}}^2}{\varepsilon_0^4 E_{\text{Y}}^{\text{eff2}}} \right)^{\frac{1}{3}}, \tag{11.14}$$

where ε_0 is the least distance between the particle surface and the flat substrate, which is of the order of 3–4 Å (Derjaguin *et al.*, 1975). Neglecting energy loss due to plastic deformation and sound generation, the acceleration due to the elastic force is, according to Lu *et al.* (2000a, 2000b),

$$\frac{4}{3} \pi R_{\text{pa}}^3 \rho_{\text{pa}} \frac{\text{d}^2 z_{\text{f}}}{\text{d}t^2} = \frac{4}{3} \sqrt{R_{\text{pa}}} E_{\text{Y}}^{\text{eff}} \left[\delta_{\text{sep}}(t)^{3/2} - \delta_0^{3/2} \right], \tag{11.15}$$

where ρ_{pa} is the density of the particle.

The temperature field in the substrate under and around the particle is in general three-dimensional, can be complex, and may deviate from the direct scaling of the incident laser intensity via Beer's law due to near-field focusing effects. Luk'yanchuk (2003) presented a theoretical model for the evaluation of these effects as a function of the particle size, the complex refractive indices of both the particle and the substrate, and the incident laser wavelength. Theoretical calculations of the near-field intensity distribution indicate that there is a significant enhancement near the contacting point. As noted before, the optical properties of the substrate play a significant role in determining the intensity distribution. Accordingly, the incident intensity at the contacting point underneath the particle is higher in the case of the higher-reflectivity Si than for a transparent glass substrate. Work by Mosbacher *et al.* (2001), Fourrier *et al.* (2001), and Münzer *et al.* (2001) has also established the role of near-field optical enhancement effects. Correspondingly, the induced temperature right underneath the particle is expected to be higher than that predicted without considering the modulation of the intensity due to the presence of the particle and the ensuing near-field effects. In fact, the departure is sufficiently high to induce damage and thus serve as a near-field scheme (e.g. Huang *et al.*, 2002) for deliberate nanoprocessing. This of course constitutes a practical limitation to the utilization of LC. Laser-cleaning thresholds based on the local ablation of substrate

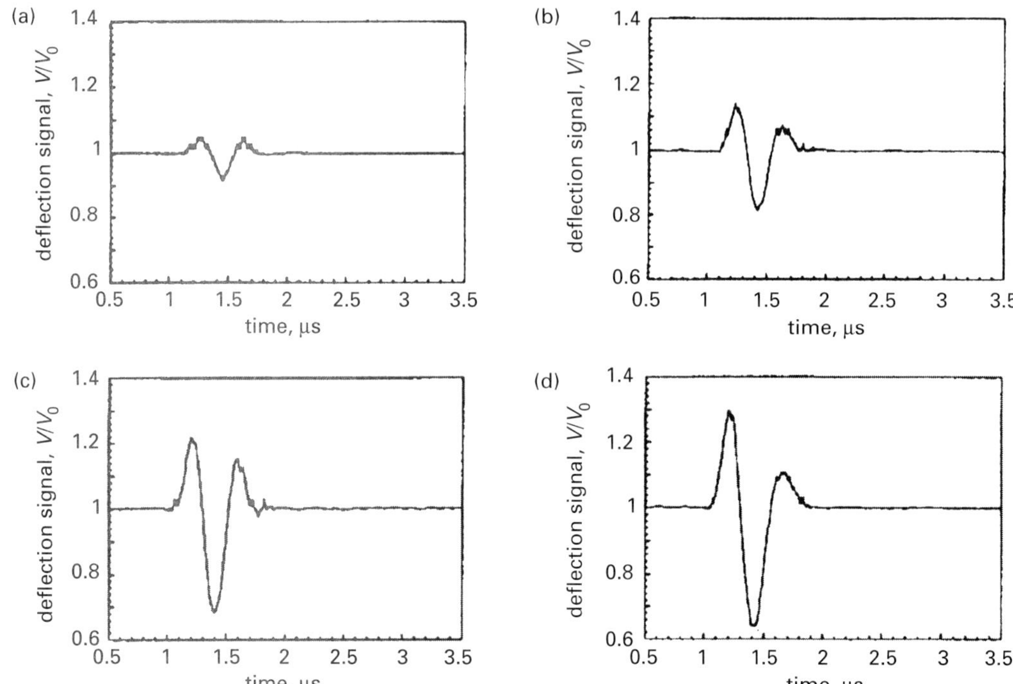

Figure 11.8. The temporal deflection signal profile measured with the photoacoustic-deflection method at several Nd : YAG laser fluences: (a) 20, (b) 27, (c) 30, and (d) 52 mJ/cm^2. The laser wavelength is $\lambda = 355$ nm. From She *et al.* (1999), reproduced with pemission by the American Institute of Physics.

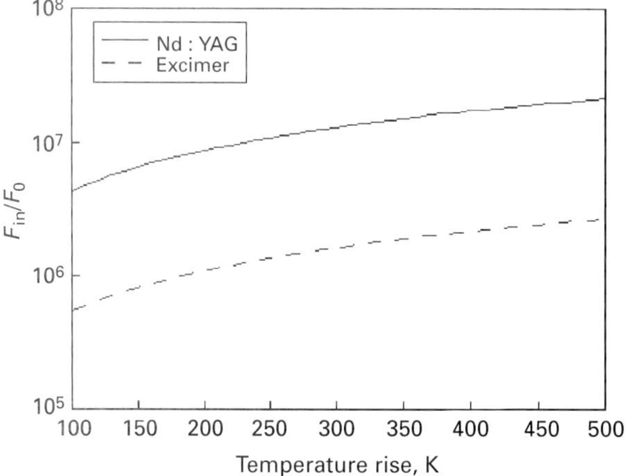

Figure 11.9. The normalized inertia force acting on a 1-μm-sized alumina particle, calculated by use of Equation (11.12).

acting on a sub-micrometer contaminant might not be large enough to overcome the adhesion force. The superior efficiency of steam LC shown in the previous sections indicates that explosive vaporization at the liquid–solid interface plays a dominant role in the LC process. The physical process of contaminant (sub-micrometer particles) removal is explained by invoking strong superheating of liquid and subsequent bubble nucleation and growth. When bubble nuclei are formed and further grow either at the solid–liquid interface or in the superheated-liquid layer, the pressure inside a bubble becomes very high in order to overcome the surface-tension force. Steam LC is then achieved by this locally concentrated high pressure around the particle–substrate-contact region. The temporal variation of the surface temperature has been calculated considering conductive heat transfer, neglecting the phase-change effect in a first-order approximation of the actual interface temperature between a solid surface and a superheated liquid. The calculated transient temperature shows that the liquid is already superheated above the nominal boiling temperature at the laser fluence of 26 mJ/cm^2 for the wavelength $\lambda = 355$ nm. At 60 mJ/cm^2, superheating to close to the critical temperature is predicted, validating the expectation of explosive vaporization. Another possible effect of introducing a thin liquid film is the reduction of the total adhesion force between contaminants and surfaces brought about by relaxation of capillary adhesion. In the explosive-vaporization process, a high degree of superheating is obtained by nanosecond laser heating of a solid surface immersed in a bulk liquid (see Chapter 10). Even though different dynamics of liquid vaporization and ablation is expected in the case of a thin liquid film, a thermodynamically similar nonequilibrium rapid phase transition takes place during the LC process.

Explosive vaporization accompanies acoustic excitation resulting from the bubble dynamics. The acoustic transient in the rapid vaporization process (in bulk liquid) has been measured quantitatively in the far field by a piezoelectric pressure transducer and by a photoacoustic beam-deflection probe as described in the Section 10.1. Results shown in Figure 10.7 indicate that the pressure launched into the liquid exceeds the linear pressure caused by thermoelastic expansion of the target for laser fluences exceeding the so-called "bubble-nucleation threshold." This bubble-nucleation threshold is certainly defined by the observation method. Experimental work by Yavas *et al.* (1997) utilizing a sensitive plasmon probe indicated that nucleation of bubbles as small as 20 nm in diameter occurs on the surface at lower fluences than those detected by optical reflectance. The required degree of superheating for nucleation of such nano-bubbles would be only 10 K above the saturation temperature that corresponds to ambient atmospheric pressure. In accordance with the above-mentioned optical deflection measurement, the peak pressure amplitude was found to lie in the range $\sim$1–5 MPa. An interferometric probe detected the dynamics of an "effective" bubble layer whose growth velocity is estimated to be 0.5–1 m/s, as derived from Figure 10.9(a). This work also revealed that the growth velocity is much higher than the collapse velocity. Accordingly, it is inferred that the emission of cavitation pressure at the end of the bubble-collapse stage is not so significant as that due to the initial bubble expansion, which is consistent with the results of the acoustic-transient measurement. On the basis of the experimental results, the following conclusions can be drawn concerning the acoustic enhancement due to liquid vaporization.

- Rapid vaporization of liquid yields an enhanced acoustic pulse that is of larger magnitude than the thermoelastic stress.
- The acoustic augmentation comes from rapid vapor-phase expansion during the initial stage of bubble growth, unlike in the ultrasonic cleaning technique, in which the cleaning mechanism relies on acoustic bubble cavitation. No significant cavitation pressure develops during the LC process.

References

Afif, M., Girardeau-Montaut, J. P., Tomas, C., 1996, "*In situ* Surface Cleaning of Pure and Implanted Tungsten Photocathodes by Pulsed Laser Irradiation," *Appl. Surf. Sci.*, **96–98**, 469–473.

Arnold, N., 2003, "Theoretical Description of Dry Laser Cleaning," *Appl. Surf. Sci.*, **208–209**, 15–22.

Arnold, N., Schrems, G., and Baüerle, D., 2004, "Ablative Thresholds in Laser Cleaning of Substrates from Particulates," *Appl. Phys. A*, **79**, 729–734.

Belov, N. N., 1989, "Laser Cleaning of an Optical Surface," *Opt.-Mekh. Promyshlennost*, **56**, 35–38.

Bowling, R. A., 1988, "A Theoretical Review of Particle Adhesion," in *Particles on Surfaces 1: Detection, Adhesion, and Removal*, K. L. Mittal, ed., New York, Plenum Press, pp. 129–142.

Derjaguin, B. V., Muller, V. M., and Toporov, Y. P., 1975, "Effect of Contact Deformations on the Adhesion of Particles," *J. Colloid Interface Sci.*, **53**, 314–326.

Engelsberg, A. C., 1991, "Removal of Surface Contaminants by Irradiation from a High-energy Source," US Patent 5 024 968.

1995, "Laser-Assisted Cleaning Proves Promising," *Precision Cleaning*, May, 35–42.

Fourrier, T., Schrems, G., Mühlberger, T., 2001, "Laser Cleaning of Polymer Surfaces," *Appl. Phys. A*, **72**, 1–6.

Halfpenny, D. R., and Kane, D. M., 1999, "A Quantitative Analysis of Single Pulse Ultraviolet Dry Laser Cleaning," *J. Appl. Phys.*, **86**, 5541–6646.

Hamaker, H. C., 1937, "The London–Van der Waals Attraction between Spherical Particles," *Physica*, **IV**, 5194–5197.

Héroux, J. B., Boughaba, S., Ressejac, I., Sacher, E., and Meunier, M., 1996, "CO_2 Laser-Assisted Removal of Submicron Particles from Solid Surfaces," *J. Appl. Phys.*, **79**, 2857–2862.

Huang, S. M., Hong, M. H., Luk'yanchuk, B. S. *et al.*, 2002, "Pulsed Laser-Assisted Surface Structuring with Optical Near-field Enhanced Effects," *J. Appl. Phys.*, **92**, 2495–500.

Imen, K., Lee, S. J., and Allen, S. D., 1991, "Laser-Assisted Micron Scale Particle Removal," *Appl. Phys. Lett.*, **58**, 203–205.

Kelley, J. D., and Hovis, F. E., 1993, "A Thermal Detachment Mechanism for Particle Removal from Surfaces by Pulsed Laser Irradiation," *Microelectron. Eng.*, **20**, 159–170.

Krupp, H., 1967, "Particle Adhesion Theory and Experiment," *Adv. Colloid Interf. Sci.*, **1**, 111–239.

Lee, L.-H., 1991, *Fundamentals of Adhesion*, New York, Plenum Press.

Lee, S. J., Imen, K., and Allen, S. D., 1992, "CO_2 Laser Assisted Particle Removal Threshold Measurements," *Appl. Phys. Lett.*, **61**, 2314–2316.

Leenaars, A. F. M., 1988, "A New Approach to the Removal of Sub-Micron Particles from Solid (Silicon) Substrates," in *Particles on Surfaces 1: Detection, Adhesion, and Removal*, K. L. Mittal, ed., New York, Plenum Press, pp. 361–372.

Lifshitz, E. M., 1956, "Theory of Molecular Attractive Forces between Solids," *Sov. Phys. JETP*, **2**, 73–83.

Lu, Y. F., Song, W. D., Ye, K. D. *et al.*, 1997, "Removal of Submicron Particles from Nickel–Phosphorous Surface by Pulsed Laser Irradiation," *Appl. Surf. Sci.*, **120**, 317–322.

Lu, Y. F., Takai, M., Komuro, S., Shiokawa, T., and Aoyagi, Y., 1994, "Surface Cleaning of Metals by Pulsed Laser Irradiation in Air," *Appl. Phys. A*, **59**, 281–288.

Lu, Y. F., Zheng, Y. W., and Song, W. D., 2000a, "Laser Induced Removal of Spherical Particles from Silicon Wafers," *J. Appl. Phys.*, **87**, 1534–1538.

2000b, "Characterization of Ejected Particles during Laser Cleaning," *J. Appl. Phys.*, **87**, 549–552.

Luk'yanchuk, B. S., 2003, *Laser Cleaning*, Singapore, World Scientific.

Magee, T. J., and Leung, C. S., 1991, "Scanning UV Removal of Contaminants from Semiconductor and Optical Surfaces," in *Particles on Surfaces, 3: Detection, Adhesion and Removal*, K. L. Mittal, ed., New York, Plenum Press.

Mann, K., Wolff-Rottke, B., and Müller, F., 1996, "Cleaning of Optical Surfaces by Excimer Laser Radiation," *Appl. Surf. Sci.*, **96–98**, 463–468.

Mittal, K. L., ed., 1988, *Particles on Surfaces 1: Detection, Adhesion, and Removal*, New York, Plenum Press.

Mosbacher, M., Dobler, V., Boneberg, J., and Leiderer, P., 2000, "Universal Threshold for the Steam Laser Cleaning of Submicron Spherical Particles," *Appl. Phys. A*, **70**, 669–672.

Mosbacher, M., Münzer, H.-J., Zimmermann, J. *et al.*, 2001, "Optical Field Enhancement Effects in Laser-Assisted Particle Removal," *Appl. Phys. A*, **72**, 41–44.

Münzer, H.-J., Mosbacher M., Bertsch, M. *et al.*, 2001, "Local Field Enhancement Effects for Nanostructuring of Surfaces," *J. Microscopy*, **202**, 129–35.

Park, H. K., Grigoropoulos, C. P., Leung, W. P., and Tam, A. C., 1994, "A Practical Excimer Laser-based Cleaning Tool for Removal of Surface Contaminants," *IEEE Trans. Comp. Pack. Manuf. Technol. A*, **17**, 631–643.

She, M., Kim, D., and Grigoropoulos, C. P., 1999; "Liquid-Assisted Pulsed Laser Cleaning Using Near-infrared and Ultraviolet Radiation," *J. Appl. Phys.*, **86**, 6519–6524.

Tabor, D., 1977, "Surface Forces and Surface Interactions," *J. Colloid Interf. Sci.*, **58**, 2–13.

Tam, A. C., Leung, W. P., Zapka, W., and Ziemlich, W., 1992, "Laser-Cleaning Techniques for Removal of Surface Particulates," *J. Appl. Phys.*, **71**, 3515–3523.

Van den Tempel, M., 1972, "Interaction Forces between Condensed Bodies in Contact," *Adv. Colloid Interf. Sci.*, **3**, 137–159.

Visser, J., 1975, "Adhesion of Colloidal Particles," *Surf. Colloid Sci.*, **8**, 3–84.

Yavas, O., Schilling, A., Bischof, J., Boneberg, J., and Leiderer, P., 1997, "Bubble Nucleation and Pressure Generation during Laser Cleaning of Surfaces," *Appl. Phys. A*, **64**, 331–339.

Wesner, D. A., Mertin, M., Lupp. F., and Kreutz, E. W., 1996, "Cleaning of Copper Traces on Circuit Boards with Excimer Laser Radiation," *Appl. Surf. Sci.*, **96–98**, 479–483.

Zapka, W., Asch, K., Keyser, J., and Meissner, K., 1989, European Patent EP 0-297-506-A2.

Zapka, W., Ziemlich, W., Leung, W. P., and Tam, A. C., 1993, "Laser Cleaning Removes Particles from Surfaces," *Microelectron Eng.*, **20**, 171–183.

12 Laser interactions with nanoparticles

12.1 Size effects on optical properties

12.1.1 The classical Mie solution for a spherical particle

Consider a plane x-polarized wave of amplitude E_0 and vacuum wavelength λ_{vac} incident upon a homogeneous, isotropic sphere of radius R_{pa}. Let the complex refractive index of the particle be $n_1^c = n_1 - ik_1$ and let n be the refractive index of the matrix medium, which is assumed to be transparent to the incident radiation. The size parameter is defined as

$$\chi_{\text{M}} = \frac{2\pi R_{\text{pa}}}{\lambda}, \tag{12.1}$$

where λ is the wavelength in the medium, i.e.

$$\lambda = \frac{\lambda_{\text{vac}}}{n}. \tag{12.2}$$

The scattering and extinction efficiencies, defined as the ratios of the scattered energy flow and the energy taken away from the beam versus the incident flux on the particle are

$$Q_{\text{M;sca}} = \frac{2}{\chi_{\text{M}}^2} \sum_{l=1}^{\infty} (2l + 1)(|a_{l\text{M}}|^2 + |b_{l\text{M}}|^2), \tag{12.3a}$$

$$Q_{\text{M;ext}} = \frac{2}{\chi_{\text{M}}^2} \sum_{l=1}^{\infty} (2l + 1)\text{Re}(a_{l\text{M}} + b_{l\text{M}}). \tag{12.3b}$$

The absorption efficiency, i.e. the ratio of the absorbed power versus the energy incident on the particle is simply found by taking $Q_{\text{M;abs}} = Q_{\text{M;ext}} - Q_{\text{M;sca}}$. Assuming equal magnetic permeabilities for the particle and the matrix medium, the complex coefficients $a_{l\text{M}}$ and $b_{l\text{M}}$, commonly referred to as the Mie coefficients, are given by

$$a_{l\text{M}} = \frac{m^c \psi_l(m^c \chi_{\text{M}})\psi_l'(\chi_{\text{M}}) - \psi_l(\chi_{\text{M}})\psi_l'(m^c \chi_{\text{M}})}{m^c \psi_l(m^c \chi_{\text{M}})\xi_l'(\chi_{\text{M}}) - \xi_l(\chi_{\text{M}})\psi_l'(m^c \chi_{\text{M}})}, \tag{12.4a}$$

$$b_{l\text{M}} = \frac{\psi_l(m^c \chi_{\text{M}})\psi_l'(\chi_{\text{M}}) - m^c \psi_l(\chi_{\text{M}})\psi_l'(m^c \chi_{\text{M}})}{\psi_l(m^c \chi_{\text{M}})\xi_l'(\chi_{\text{M}}) - m^c \xi_l(\chi_{\text{M}})\psi_l'(m^c \chi_{\text{M}})}, \tag{12.4b}$$

where the complex coefficient $m^c = n_1^c/n$ and ψ_l and ξ_l are the lth-order Ricatti Bessel functions. Note that the scattering efficiency may exceed unity, since the scattering includes diffraction effects that are modified by interference.

12.1.2 Rayleigh scattering: sphere small compared with wavelength

Utilizing power-series expansion of spherical Bessel functions on the general Mie solution, the following expressions are correct to $O(\chi_M^4)$:

$$Q_{\text{M;ext}} = 4\chi_M \, \text{Im}\left\{ \frac{(m^c)^2 - 1}{(m^c)^2 + 2}\left[1 + \frac{\chi_M^2}{15}\left(\frac{(m^c)^2 - 1}{(m^c)^2 + 2} \right) \frac{(m^c)^4 + 27(m^c)^2 + 38}{2(m^c)^2 + 3} \right] \right\}$$
$$+ \frac{8}{3}\chi_M^4 \, \text{Re}\left\{ \left[\frac{(m^c)^2 - 1}{(m^c)^2 + 2} \right]^2 \right\}, \tag{12.5a}$$

$$Q_{\text{M;sca}} = \frac{8}{3}\chi_M^4 \left| \frac{(m^c)^2 - 1}{(m^c)^2 + 2} \right|^2. \tag{12.5b}$$

If $|m^c|\chi_M \ll 1$ and assuming that

$$\frac{4\chi_M^3}{3} \, \text{Im}\left[\frac{(m^c)^2 - 1}{(m^c)^2 + 2} \right] \ll 1,$$

$$Q_{\text{M;abs}} = 4\chi_M \, \text{Im}\left[\left| \frac{(m^c)^2 - 1}{(m^c)^2 + 2} \right| \right]. \tag{12.6}$$

Consequently, the absorption cross section, $C_{\text{M;abs}} = \pi R_{\text{pa}}^2 Q_{\text{M;abs}}$, is proportional to the volume of the particle. On the other hand, the scattering cross section is proportional to the square of the volume of the particle. In the Rayleigh regime, the scattering cross section is much smaller than the absorption cross section. Equation (12.6) is rewritten in terms of the dielectric constant, ε^c, of the particle material and the permittivity of the matrix medium, ε_{mat}:

$$Q_{\text{M;abs}} = 4\chi_M \, \text{Im}\left[\frac{\varepsilon^c - \varepsilon_{\text{mat}}}{\varepsilon^c + 2\varepsilon_{\text{mat}}} \right]. \tag{12.7}$$

Resonance behavior is to be expected should $\varepsilon^c = \varepsilon_R - i\varepsilon_{\text{Im}} = -2\varepsilon_{\text{mat}}$, which necessarily implies that $\varepsilon_R = -2\varepsilon_{\text{mat}}$, since the matrix medium is considered non-absorbing. The appearance of the so-called Fröhlich peak is apparent in the spectral behavior of the absorption efficiency of gold nanoparticles with $R_{\text{pa}} = 5$ nm shown in Figure 12.1.

12.1.3 Effective-medium theory

Assuming a dilute dispersion of spherical nanoparticles of complex dielectric constant ε^c in a medium of permittivity ε_{mat}, the Maxwell-Garnet (1904) prediction of the average

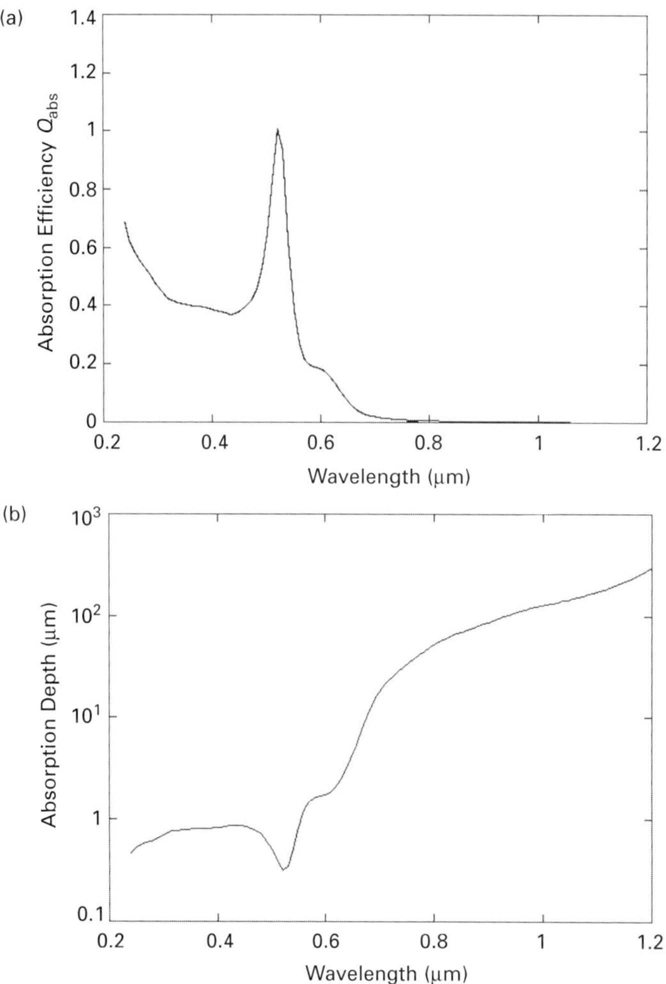

Figure 12.1. (a) The wavelength dependence of the absorption efficiency for a single gold particle of radius $R_{\mathrm{pa}} = 5$ nm. (b) The dependence of the optical penetration depth in a toluene ($n = 1.5$) suspension of 5-nm gold particles.

dielectric constant of the composite medium is

$$\varepsilon_{\mathrm{av}}^{c} = \varepsilon_{\mathrm{m}}\left[1 + \frac{3F_{\mathrm{v}}\left(\dfrac{\varepsilon^{c} - \varepsilon_{\mathrm{mat}}}{\varepsilon^{c} + \varepsilon_{\mathrm{mat}}}\right)}{1 - F_{\mathrm{v}}\left(\dfrac{\varepsilon^{c} - \varepsilon_{\mathrm{mat}}}{\varepsilon^{c} + \varepsilon_{\mathrm{mat}}}\right)}\right], \tag{12.8}$$

where F_{v} is the nanoparticle volumetric function. On the other hand, Bruggeman's expression is

$$F_{\mathrm{v}}\frac{\varepsilon^{c} - \varepsilon_{\mathrm{av}}^{c}}{\varepsilon^{c} + \varepsilon_{\mathrm{av}}^{c}} + (1 - F_{\mathrm{v}})\frac{\varepsilon_{\mathrm{m}} - \varepsilon_{\mathrm{av}}^{c}}{\varepsilon_{\mathrm{m}} + \varepsilon_{\mathrm{av}}^{c}} = 0. \tag{12.9}$$

The prediction of the effective-medium theory for a suspension of gold nanoparticles in toluene shown in Figure 12.1(b) agrees very well with the Rayleigh behavior.

12.1.4 Size-correction effects

Clearly, the refractive-index concept has to be modified for ultra-small particles wherein the mean free path can be dominated by collisions with the free boundary. Assuming that electrons are diffusely reflected at the boundary, the Drude damping term can be written as

$$\zeta = \zeta_{\text{bulk}} + \frac{v_{\text{F}}}{L}, \tag{12.10}$$

where ζ_{bulk} is the damping constant for the bulk-material counterpart and v_{F} is the electron velocity at the Fermi surface. The characteristic length may be taken as $L = 4R_{\text{pa}}/3$. In metals, near the plasma frequency $\omega^2 \gg \zeta^2$. Consequently, the imaginary part of the Drude dielectric function (Equation (1.115b)) is well approximated by

$$\varepsilon_{\text{Im}}(\omega, R) = \frac{\omega_{\text{p}}^2}{\omega^3}\zeta = \varepsilon_{\text{Im;bulk}} + \frac{3}{4}\frac{\omega_{\text{p}}^2}{\omega^3}\frac{v_{\text{F}}}{R_{\text{pa}}}. \tag{12.11}$$

Once the Mie solution has been obtained, the temperature distribution inside the particle can be evaluated by calculating the internal volumetric heat-generation term. In a polar system of coordinates, assuming azimuthal symmetry,

$$Q_{\text{M;abs}}(r, \theta) = \frac{4\pi n k_1 I_0}{\lambda_{\text{vac}}}|\vec{E}(r, \theta)/\vec{E}_0|^2,$$

where I_0 is the laser-beam intensity. The factor $|\vec{E}/\vec{E}_0|^2$ is the spatial source distribution, i.e. the squared magnitude ratio of the electric field $\vec{E}(r, \theta)$ with respect to the incident electric field $\vec{E}_0$. Longtin *et al.* (1995) showed that the distortion of the laser-beam profile with respect to time depends on the diffusional dimensionless parameter $t_{\text{pulse}}\alpha_{\text{pa}}/R_{\text{pa}}^2$, where α_{pa} is the particle thermal diffusivity. For high values of this parameter (i.e. long laser pulses), the temperature profile relaxes and becomes uniform across the particle. However, as the pulse duration becomes shorter, the effects of the nonuniform laser energy deposition are distinct in terms of their yielding spatially varying temperature profiles.

12.2 **Melting of nanoparticles**

12.2.1 Size-dependent depression of the melting point

It has been shown that nanoparticles possess properties different from those of their bulk counterparts. An explanation for this difference in thermophysical behavior is based on the ratio between surface atoms and inner-phase atoms for nanoparticle systems. In a gold particle of diameter 4 nm, 40% of the atoms are in fact surface atoms. The ratio of atoms at the surface, N_{su}, and the total number of atoms, N, of a particle is inversely proportional to the radius of the particle, R_{pa}: $N_{\text{su}}/N \propto 1/R_{\text{pa}}$. The energy of atoms at the surface differs from the energy of inner atoms. The small radius of curvature and finite size of the particle emphasize the influence of the surface atoms. Thermodynamic properties such as the melting and boiling points drastically differ from the bulk properties. The

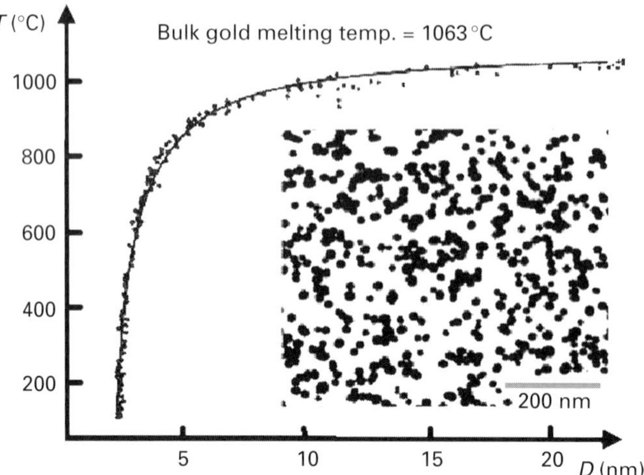

Figure 12.2. Depression of the melting point of gold nanoparticles as a function of the particle diameter. From Buffat and Borel (1976), reproduced with permission from the American Physical Society.

reduction of the melting temperature with decreasing particle size is of interest. This *thermodynamic size effect* was reported and experimentally investigated by Gladkich *et al.* (1966), Buffat and Borel (1976), and Peppiatt and Sambles (1975). According to the phenomenological model and the experimental observations presented by Buffat and Borel (1976), the melting temperature of gold particles significantly decreases when the diameter is smaller than 5–7 nm. Figure 12.2 depicts the depression of the melting point of gold nanoparticles that was deduced via diffraction of high-energy electrons.

The model by Kofman *et al.* (1994) attempts to explain the effect of the curvature on the melting-point reduction. The free energy of a solid flat surface of area A and containing N atoms is

$$g_{G_{s,f}} = N\mu_{che;s} + A\sigma_{ST;sv}. \tag{12.12}$$

Now, assuming that N_l atoms are in the liquid state,

$$g_{G_{sl,f}} = (N - N_l)\mu_{che,s} + N_l\mu_{che,l} + A\tilde{\Gamma}. \tag{12.13}$$

The parameter $\tilde{\Gamma}$ depends on the thickness of the liquid layer, d_l, and a length scale that is characteristic of the short-range interactions:

$$\tilde{\Gamma} = \sigma_{ST;sl} + \sigma_{ST;lv} + \tilde{\Sigma}e^{-(d_l/\xi_{char})}, \tag{12.14}$$

where ξ_{char} is of the order of a few ångström units and

$$\tilde{\Sigma} = \sigma_{ST;sv} - (\sigma_{ST;sl} + \sigma_{ST;lv}).$$

Consider next a spherical particle of radius R_{pa} containing N atoms. The free energy of a solid particle is

$$g_{G_s} = N\mu_{\text{che,s}} + 4\pi R_{\text{pa}}^2 \sigma_{\text{ST;sv}}. \tag{12.15}$$

In the case of surface melting with a thin melt-layer thickness, d_1,

$$g_{G_{sl}} = (N - N_1)\mu_{\text{che,s}} + N_1\mu_{\text{che,l}}$$
$$+ 4\pi R_{\text{pa}}^2 \left[\sigma_{\text{ST;sl}}\left(\frac{R_{\text{pa}} - d_1}{R_{\text{pa}}}\right)^2 + \sigma_{\text{ST;lv}} + \tilde{\Sigma}_{\text{su}}e^{-(d_1/\xi_{\text{char}})} \right], \tag{12.16}$$

where

$$\tilde{\Sigma}_{\text{su}} = \sigma_{\text{ST;sv}} - \left[\sigma_{\text{ST;lv}} + \sigma_{\text{ST;sl}}\left(\frac{R_{\text{pa}} - d_1}{R_{\text{pa}}}\right)^2 \right].$$

The departure of the melting temperature from the bulk melting temperature is found by minimizing $\Delta g_G = g_{G;sl} - g_{G;s}$, and utilizing the relation

$$N_1(\mu_{\text{che,l}} - \mu_{\text{che,s}}) = V_1\rho L_{\text{sl}}\frac{\Delta T_{\text{m}}}{T_{\text{m}}}, \tag{12.17}$$

where V_1 is the volume of the liquid layer and L_{sl} is the latent heat of melting. The computed temperature excursion is

$$\frac{\Delta T_{\text{m}}}{T_{\text{m}}} = \frac{2\sigma_{\text{ST;sl}}}{\rho L_{\text{sl}}(R_{\text{pa}} - d_1)}(1 - e^{-(d_1/\xi_{\text{char}})}) + \frac{\tilde{\Sigma}_{\text{su}}R_{\text{pa}}^2}{\rho L_{\text{sl}}\xi_{\text{char}}(R_{\text{pa}} - d_1)^2}e^{-(d_1/\xi_{\text{char}})}. \tag{12.18}$$

In the case of a sharp interface ($\xi_{\text{char}} \to 0$), the temperature depression is identical to what results on considering a transition from entirely solid to entirely liquid:

$$\frac{\Delta T_{\text{m}}}{T_{\text{m}}} = \frac{2}{\rho L_{\text{sl}}R_{\text{pa}}}(\sigma_{\text{ST;sv}} - \sigma_{\text{ST;lv}}), \tag{12.19}$$

that is the so-called Pawlow relation that implies equality of the densities of the solid and liquid phases. The depression of the melting point as a function of particle size has several interesting applications in the fabrication of electronic circuits on sensitive substrates that will be outlined in Chapter 13.

12.2.2 Ultrafast-laser interactions with nanoparticles

Consider a dilute monodispersion of metal nanoparticles embedded in a lossless (transparent) matrix medium (Del Fatti *et al.*, 1999). Let the complex dielectric constant of the nanoparticles be $\varepsilon^c(\omega) = \varepsilon_R(\omega) - i\varepsilon_{\text{Im}}(\omega)$ and the real dielectric index of the matrix be $\varepsilon_{\text{mat}}(\omega)$. According to the effective-medium theory, the absorption coefficient of the dispersion is

$$\gamma(\omega) = 9F_v\varepsilon_{\text{mat}}^{3/2}\frac{\omega}{c}\frac{\varepsilon_{\text{Im}}}{(\varepsilon_R + 2\varepsilon_{\text{mat}})^2 + (\varepsilon_{\text{Im}})^2}. \tag{12.20}$$

Because of the confinement in a dielectric material, the absorption coefficient is enhanced at a certain frequency ω_0, at which $\varepsilon_R(\omega_0) + 2\varepsilon_{mat} = 0$, which is the condition for a surface plasmon resonance, i.e. a peak of collective electron oscillation within the nanoparticle.

The dielectric constant of the metal nanoparticles is written as $\varepsilon^c = \varepsilon^c_{bulk} + \varepsilon^c_{free}$, where ε^c_{bulk} and ε^c_{free} represent the contributions of the interband transitions and the intraband term that is due to conduction-band electrons. A modified Drude expression can be utilized to represent the latter term:

$$\varepsilon^c_{free}(\omega) = -\frac{\omega_p^2}{\omega(\omega - is_{sca,free}(\omega))}, \tag{12.21}$$

where ω_p is the bulk plasma frequency and $s_{sca,free}$ the electron-scattering rate in the nanoparticle that is given below in a modification of Equation (12.10):

$$s_{sca,free}(\omega) \approx \frac{1}{\tau(\omega)} + \tilde{s}(\omega)\frac{v_F}{R_{pa}}. \tag{12.22}$$

In the above, $\tau(\omega)$ is the relaxation time due to intrinsic electron–phonon and electron–electron scattering, and v_F is the electron Fermi velocity. The proportionality factor is that derived by Hache *et al.* (1986),

$$\tilde{s}(\omega) = \frac{1}{\hbar\omega}\int_0^\infty \left(\frac{E}{E_F}\right)^{\frac{3}{2}} \left(\frac{E+\hbar\omega}{E_F}\right)^{\frac{1}{2}} f_e(E)[1 - f_e(E+\hbar\omega)]dE, \tag{12.23}$$

where f_e is the electron-distribution function. According to the above relations, as the temperature increases, so do $\tilde{s}(\omega)$ and $s_{sca,free}(\omega)$, thereby increasing the contribution to the nonlinear response with respect to ultrafast-laser excitation. By utilizing Equations (12.21)–(12.23), the absorption coefficient of the dispersion in the neighborhood of the plasmon resonance is found to be

$$\gamma(\omega) = \frac{A\omega^2\omega_0^4 s_{sca}}{\left(\omega^2 - \omega_0^2\right)^2 + \left(\omega_0^2 s_{sca}/\omega\right)^2}, \tag{12.24}$$

with the resonance frequency given by

$$\omega_0 = \frac{\omega_p}{\sqrt{\varepsilon_{bulk;R}(\omega_0) + 2\varepsilon_{mat}}}. \tag{12.25}$$

The overall electron-scattering rate is

$$s_{sca} = s_{sca,free} + \frac{\omega_0^3}{\omega_p^2}\varepsilon_{bulk;Im}(\omega_0). \tag{12.26}$$

Time-resolved pump-and-probe experiments were conducted by Del Fatti *et al.* (1999) to investigate this nonlinear behavior. Silver nanoparticles of $R_{pa} \sim 6.4$ nm, embedded in a 1 : 1 P_2O_5–BaO matrix with $F_v \sim 10^{-4}$ were subjected to Ti : Al_2O_3 femtosecond-laser pulses. Figure 12.3(a) shows the differential transmission $\Delta T/T$ as a function of

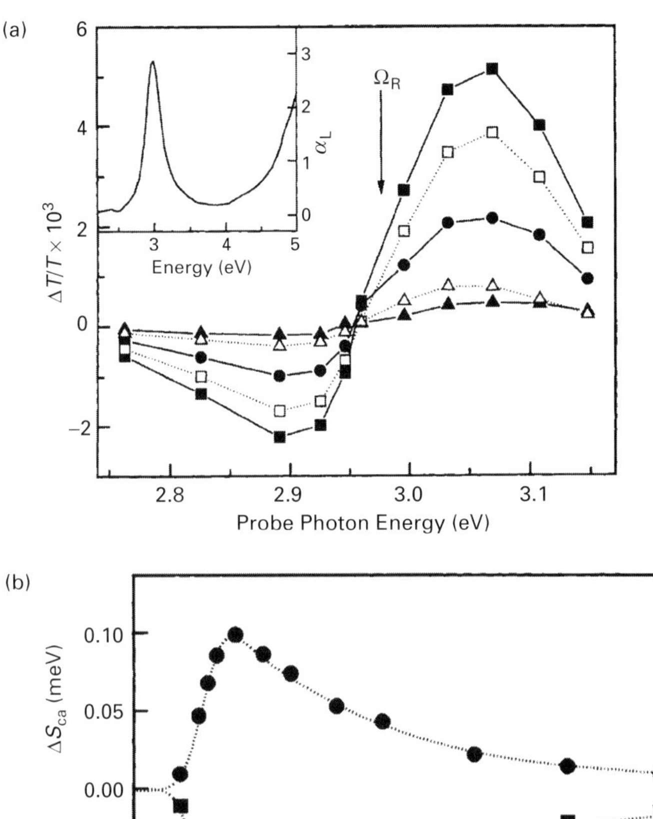

Figure 12.3. (a) The transmission change $\Delta T/T$ for probe time delays of $t_{\mathrm{D}} = -0.15$ ps (triangles), 0 ps (circles), 0.15 ps (squares), 0.45 ps (open squares), and 2 ps (open triangles) around the surface plasmon resonance (SPR) in Ag nanoparticles with $R_{\mathrm{pa}} \sim 6.4$ nm. The inset shows the absorption spectrum of the sample around the surface plasmon resonance of 2.98 eV. (b) The measured time dependence of the SPR frequency shift $\Delta\omega_0$ and broadening $\Delta\gamma$ in Ag nanoparticles with $R_{\mathrm{pa}} \sim 4.9$ nm for a pump fluence of 180 mJ/cm^2. The dashed lines correspond to convolutions of the pump–probe correlation function with an exponential response function with $\tau_{\mathrm{e-ph}} = 800$ fs. (c) The size dependence of the effective electron–phonon coupling time $\tau_{\mathrm{e-ph}}$ measured in Ag nanoparticle samples. The dotted line is a guide to the eye. From Del Fatti *et al.* (1999), reproduced with permission from Springer-Verlag.

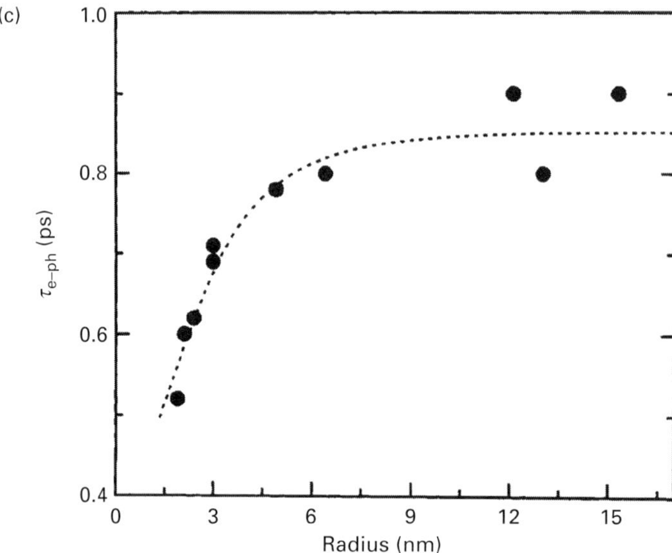

Figure 12.3. (*cont.*)

the probe-photon energy. The asymmetrical shape of the $\Delta T/T$ curves that experience a change of sign at a red-shifted ω_0 indicates a broadening of the surface plasmon resonance. Figure 12.3(b) depicts the resonance shift and broadening for a particle of $R_{\mathrm{pa}} = 4.9$ nm. Figure 12.3(c) indicates that the effective electron–phonon relaxation time $\tau_{\mathrm{e-ph}}$ deduced from these measurements is a function of the nanoparticle size and moves toward bulk values as R_{pa} increases. The relaxation times of electron–phonon coupling and phonon–phonon coupling depend upon the nature of the matrix material as shown by Mohamed *et al.* (2001). As the laser power increases, the metal nanoparticles may permanently deform, thus modifying the aggregate optical properties of the medium (Kaempfe *et al.*, 1999).

The melting behavior of nanoparticles upon femtosecond-laser excitation has been investigated by Link *et al.* (2000). It was found that ultrafast-laser-irradiation of defect-free gold nanorods induces internal defects that are dominated by twins and stacking faults as shown in Figure 12.4 that displays the nanoparticle structural transformation at laser energy below that required for complete melting. These defects act as precursors to the conversion of {110} facets into the more stable {100} and {111} facets. The defect formation and local internal melting are followed by surface reconstruction minimizing the surface free energy and diffusion that tends to produce symmetrical nanodots. This sequence is in contrast to the thermal process, in which the melting is initiated at the nanoparticle surface. Complete melting of the nanorods in a colloidal solution and their transformation into spherical nanoparticles was observed under multiple femtosecond-laser pulses. For laser pulses of lower than threshold energy, the number of pulses required to achieve this increased significantly as the laser energy decreased. On the other hand, for laser energies above the threshold for complete melting, the required number of pulses remained constant.

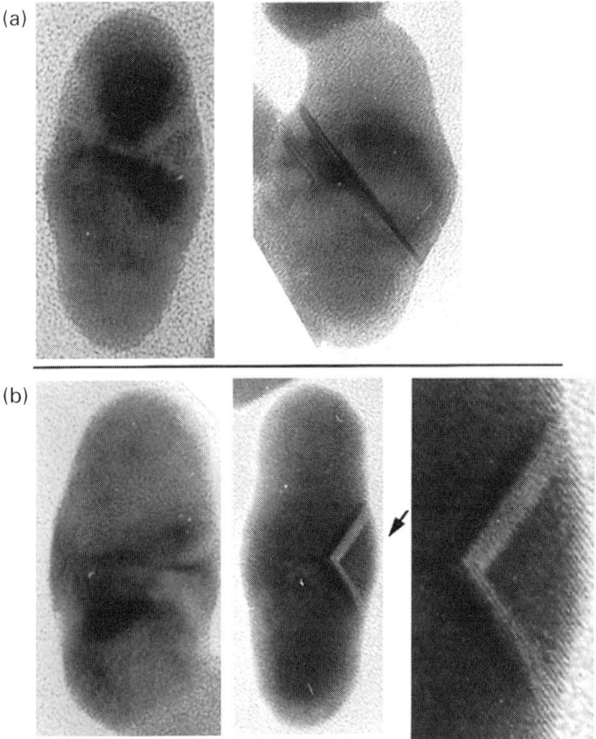

Figure 12.4. A high-resolution TEM image of gold nanorods after exposure to nanosecond-laser pulses with a fluence of 250 mJ/cm^2 (20 μJ per pulse): (a) nanorods with stacking faults and (b) twinned particles. Similar defect structures are found in spherical gold nanoparticles after complete shape transformation from nanorods, whereas the as-prepared nanorods are found to be defect-free. From Link *et al.* (2000), reproduced with permission from the American Chemical Society.

12.3   Laser-induced production of nanoparticles

Unique and tailored properties of nanoparticles are observed below 10 nm, where quantum effects become dominant. To take advantage of these properties, it is important to fabricate particles of given size and size distribution. In this regard, laser-ablation techniques offer some distinct flexibility.

12.3.1   Ablation of solid targets

As the laser-ablation plume is ejected, condensation of the vapor is a mechanism for producing nanoparticles. The ambient pressure and gas composition are critical factors in the condensation process. In experimental results it has typically been found that the size of the nanoparticles produced ranges from several to tens of nanometers. Via collisions, larger particles can be formed, but the probability of this is low. As noted previously, mechanical exfoliation, phase explosion, and hydrodynamic instabilities

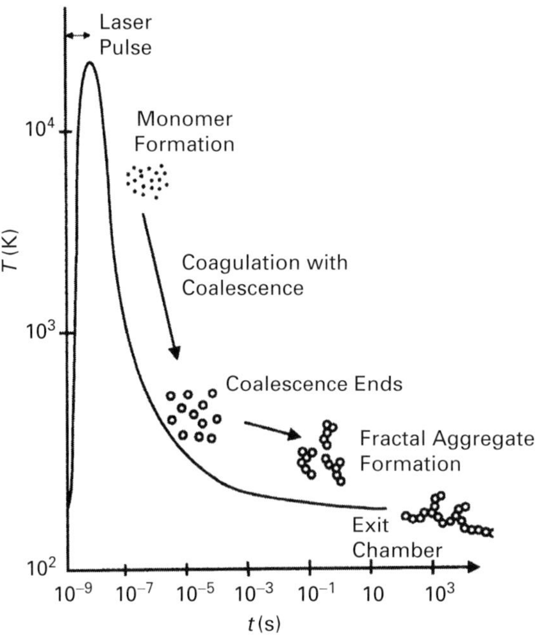

Figure 12.5. A schematic diagram of the formation of oxide nanoparticles. Monomer formation occurs when the gas has cooled to the point at which oxide dissociation ends. Coalescence continues during a period of rapid quenching and ends at the same time for various materials. Hence particles of similar size are produced; these form aggregates with a fractal dimension of 1.7–1.9 by cluster–cluster collision. From Ullmann *et al.* (2002), reproduced with permission from Springer-Verlag.

lead to ablation of larger particles. The size of the nanoparticles can be explained qualitatively via collision–coalescence arguments (Ullmann *et al.*, 2002) as shown in Figure 12.5. During the first stage of the ablation, vaporized material forms monomers, which possibly react with the background gas. When the temperature drops, nucleation occurs and nanoparticles begin to form. The rapid expansion of the high-temperature vapor produces much higher formation rates than those encountered in flame reactors. Once the temperature falls below the point at which coalescence ceases, the size of primary nanoparticles is set. Further collisions over much longer time scales lead to the formation of nanoparticle chain aggregates.

The nanoparticles are typically collected on neighboring or remote substrates and analyzed by transmission electron microscopy (TEM), determination of specific surface area by the method of Brunauer, Emett and Teller, X-ray diffraction, Raman spectroscopy, SEM, STM, etc. A typical setup for producing ultrafine oxide nanoparticles at high rates (1.5 g/min per kW of CO_2 laser power) is shown in Figure 12.6(a) (Gaertner and Lydtin, 1994). As can be seen in Figures 12.6(b) and (c), the compression of the particle distribution is effected by regulating the flowing gas velocity and the chamber pressure.

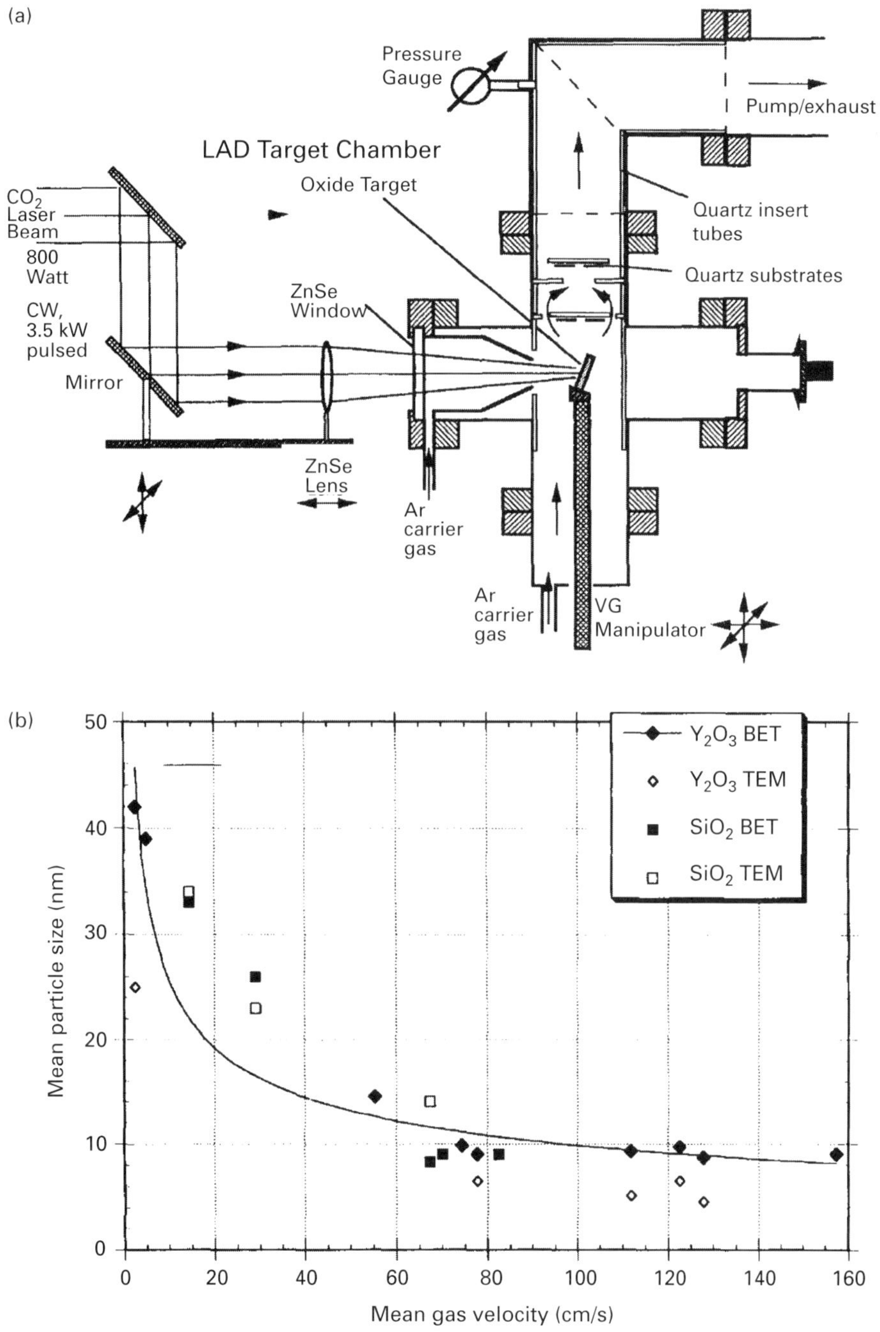

Figure 12.6. (a) The experimental setup for deposition of ultrafine particles by CO_2 laser ablation; (b) the relationship between mean particle size and mean gas velocity; and (c) the mean particle size as a function of chamber pressure. From Gaertner and Lydtin (1994), reproduced with permission from Pergamon.

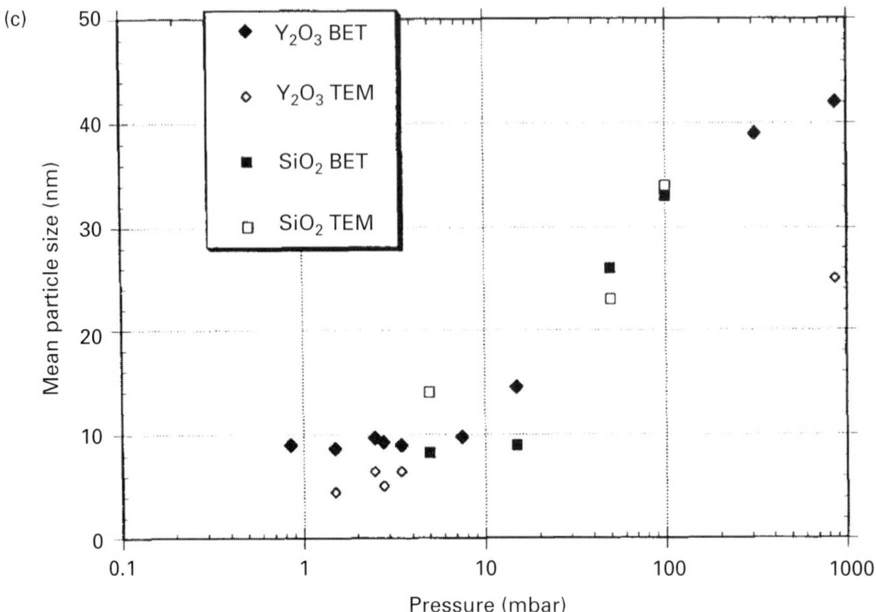

Figure 12.6. (*cont.*)

High-purity Si nanodots can be produced by ablation in a low-pressure inert gas. Figure 12.7(a) shows the DMA-based setup employed by Seto *et al.* (2001). First, Si is vaporized by focused laser-beam irradiation. The laser plasma is formed by excitation of Si and He. Polydisperse Si nanoparticles are generated by sudden cooling. Some of the particles become charged during multiphoton excitation processes in the laser plasma. After the classification by DMA though the application of a transverse electric field, the size of the nanoparticles is controlled. At this point, most of the particles bear a single charge, but some particles have double charges (with the same mobility). Finally, isolated nanodots are formed by supersonic deposition of the nanoparticle beam onto the substrate. By measuring the current from the nanoparticle beam, the density on the surface could be controlled. Results are depicted in Figures 12.7(b)–(d).

12.3.2   Ablation in aqueous solutions

Colloidal nanoparticle suspensions can be synthesized via ablation of a solid target immersed in a liquid. Mafuné *et al.* (2001) demonstrated gold nanoparticle formation via ablation of a solid target in an aqueous solution of sodium dedecyl sulfate (SDS) with a Nd : YAG laser beam ($\lambda = 1.064$ nm). The initial rapid embryonic nucleation is followed by slow growth due to the tendency to terminate the growth due to the formation of SDS coating on the particle. Correspondingly, the size of the nanoparticles was found to decrease with increasing surfactant concentration and decreasing laser power. Figure 12.8(a) shows the formation of nanoparticles of diameter 1–15 nm. Subsequent irradiation by the frequency-doubled harmonic at $\lambda = 532$ nm could dissociate the

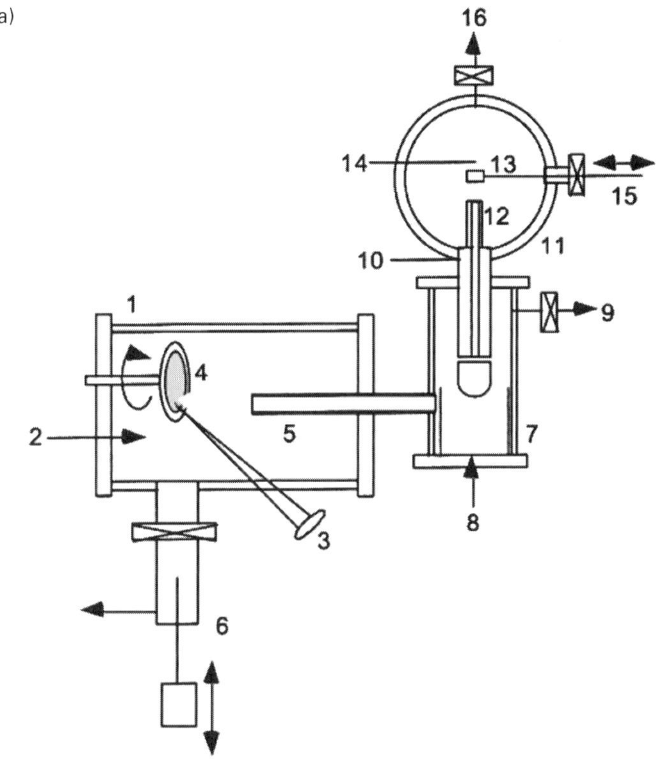

Figure 12.7. (a) A schematic diagram of the apparatus for synthesis of silicon single dots. In the laser-ablation chamber (1), the laser beam is focused (3) onto the silicon target (4) in the helium background gas (2). The target is rotated at 8 rpm. Generated nanoparticles are sampled (5) by the gas stream and introduced into the DMA (7). The size of the nanoparticles is controlled by the balance of sheath-gas flow (8) and applied voltage (10). The size-selected nanoparticle beam is focused (12) onto the substrate (TEM grid (13)). Current from the nanoparticles is measured by an electrometer (14). The whole system is pumped by a turbomolecular pump, a mechanical booster pump, and a rotary pump. Si targets (4) and substrates (13) are exchanged through load–lock systems (6, 15). (b) A dark-field TEM image of the nanodots classified at 7 nm by the DMA. The scale bar is 20 nm. In total 35 dots are shown in the area of 550 nm × 400 nm. (c) The distribution of the surface-area-equivalent diameter of nanodots measured by TEM. The first peak, a, is close to the diameter of the setting value of the DMA (7 nm). The second peak, b, at about 10 nm is considered to be that of doubly charged particles, which have the same electrical mobility as that of 7-nm particles. (d) A high-resolution TEM image of single nanodots of diameter A, 10 nm, B, 7 nm, and C, 5 nm. The lattice parameters of these dots were about 0.3 nm and close to that of the [111] face of silicon nanowire. From Seto *et al.* (2001), reproduced with permission from the American Chemical Society.

nanoparticles and shrink both the mean size and the spread (Figure 12.8(b)). Note that the $\lambda = 1.064$-nm beam is not absorbed efficiently in the nanoparticles (Figure 12.1). In a continuation study Mafuné *et al.* (2003) demonstrated successful formation of stable, surfactant-free platinum nanoparticles of median diameter 7.4 nm in pure water.

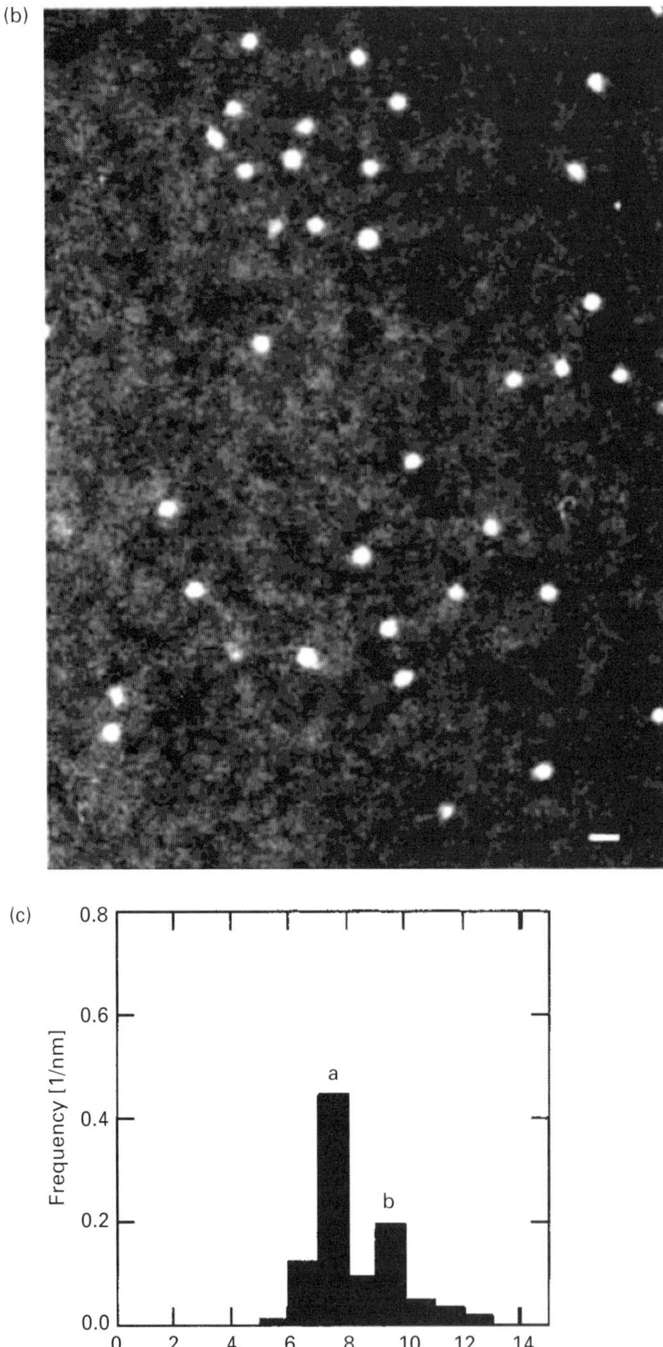

Figure 12.7. (*cont.*)

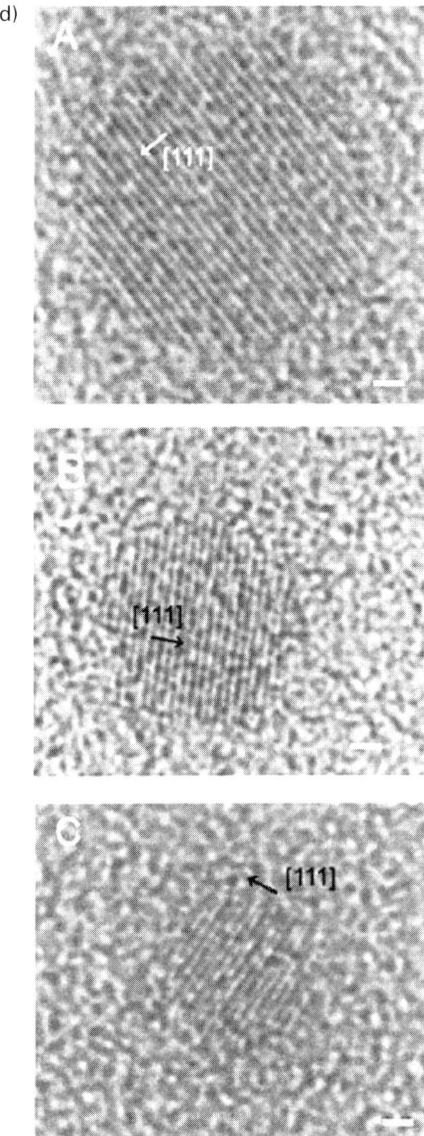

Figure 12.7. (*cont.*)

12.3.3　Ablation of consolidated particles

A new process of nanoparticle synthesis, called laser ablation of consolidated microparticles (LACM), has been developed using pulsed-laser ablation of consolidated microparticles. Microparticles of metals, including Cu, Al, and Ag, are consolidated by a cold isobaric press with pressures up to a few hundred MPa before laser irradiation.

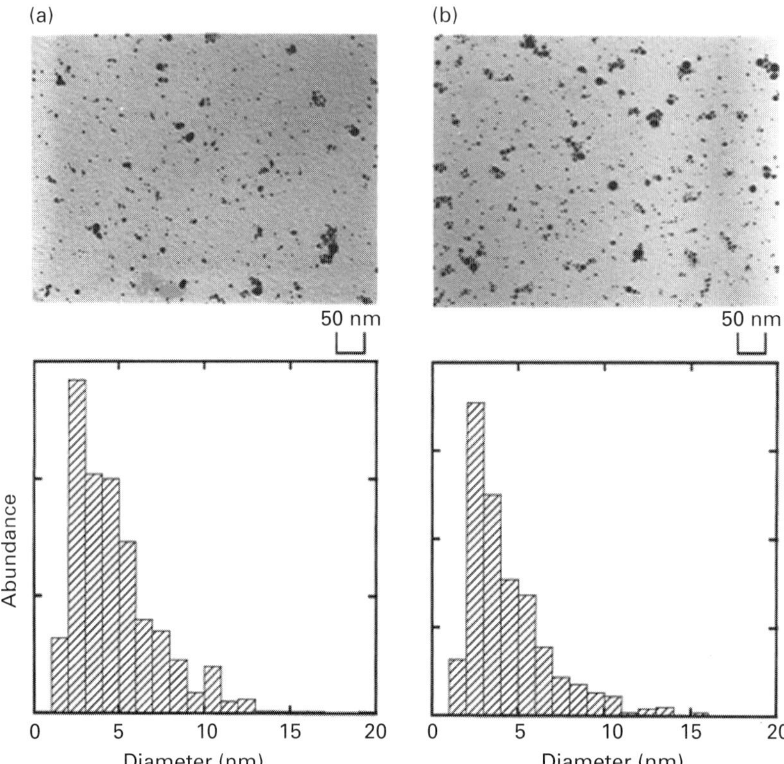

Figure 12.8. Electron micrographs and size distributions of the gold nanoparticles produced by 1064-nm-laser ablation at 80 mJ/pulse in a 0.01 M aqueous solution of sodium dodecylsulfate (SDS). Panels (a) and (b) show the electron micrographs of the gold nanoparticles remaining in the top layer of the solution before and after centrifugation, respectively. Panel (c) shows an electron micrograph and the size distribution of the gold nanoparticles produced by 1064-nm-laser ablation at 80 mJ/pulse in a 0.01 M aqueous solution of SDS (the same condition as in (a)) and subsequent laser irradiation at 532 nm (50 mJ/pulse for 60 min). From Mafuné *et al.* (2001), reproduced with permission from the American Chemical Society.

Nanoparticles are then synthesized in air or inert gases by high-power pulsed-laser ablation of the microparticles, e.g. using a Q-switched Nd : YAG laser (Figure 12.9). It has been shown that the degree of compaction plays a significant role in determining the size of the nanoparticles produced, as do the laser fluence and collector position. Results from photoacoustic-deflection probing and nanosecond time-resolved visualization indicate that the novel process attains an increased efficiency of coupling of laser energy with the target. *In situ* visualization of the Cu-microparticle explosion and nanoparticle synthesis (Figure 12.10) indicates that the threshold laser fluence for particle ablation is substantially lower than that required for bulk-metal ablation. Consequently, the results confirm that a photomechanical mechanism

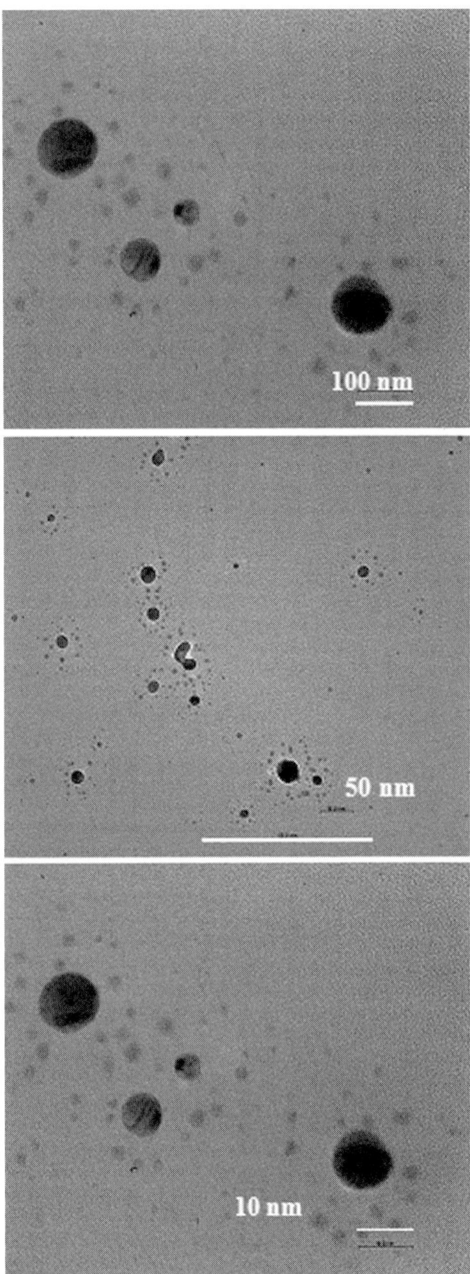

Figure 12.9. Transmission-electron-microscopy images of silver nanoparticles generated from 8-μm-sized microparticles (laser fluence 2 J/cm^2, laser wavelength 355 nm, consolidation pressure 40 MPa; the ambient gas is air). Courtesy of Dongsik Kim of POSTECH University, Korea.

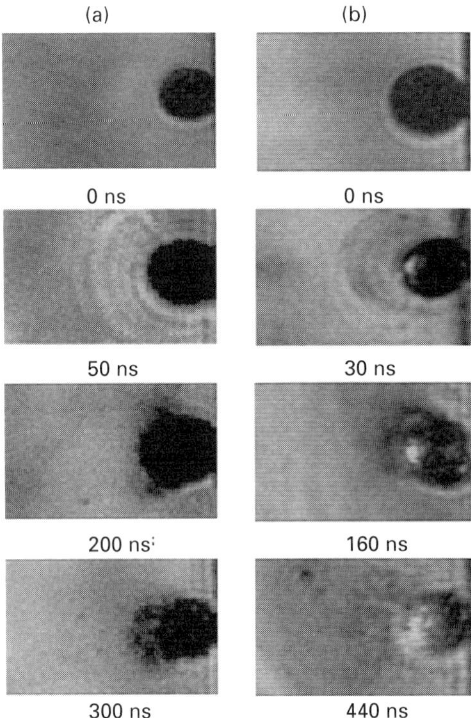

Figure 12.10. Time-resolved shadowgraph images displaying ablation of a Cu microparticle (diameter 48 μm) during the synthesis of CuO nanoparticles; (a) wavelength 355 nm, fluence 1 J/cm^2; and (b) wavelength 532 nm, fluence 1.8 J/cm^2. Courtesy of Dongsik Kim of POSTECH University, Korea.

associated with the shock-wave generation plays a significant role in the ablation process.

References

Arcidiacono, S., Bieri, N. R., Poulikakos D., and Grigoropoulos, C. P, 2004, "On the Coalescence of Gold Nanoparticles," *Int. J. Multiphase Flow*, **30**, 979–994.

Bohren, C. F., and Huffman, D. R., 1983, *Absorption and Scattering of Light by Small Particles*, New York, John Wiley.

Buffat, P., and Borel, J. P., 1976, "Size Effect on the Melting Temperature of Gold Particles," *Phys. Rev. A*, **13**, 2287–2298.

Del Fatti, N., Flytzanis, C., and Vallee, F., 1999, "Ultrafast Induced Electron-Surface Scattering in a Confined Metallic System," *Appl. Phys. B*, **68**, 433–437.

Flüeli, M., Buffat, P. A., and Borel, J.-P., 1988, "Real Time Observation by High Resolution Electron Microscopy (HREM) of the Coalescence of Small Gold Particles in the Electron Beam," *Surf. Sci.*, **202**, 343–353.

Gaertner, G. F., and Lydtin, H., 1994, "Review of Ultrafine Particle Generation by Laser Ablation from Solid Targets in Gas Flows," *Nanostr. Mater.*, **4**, 559–568.

Gladkich, N. T., Niedermayer, R., and Spiegel, K., 1966, "Nachweis großer Schmelzpunkt-serniedrigungen bei dünnen Metallschichten," *Phys. Status Solidi.* (b), **15**, 181–192.

Hache, F., Ricard, D., and Flytzanis, C., 1986, "Optical Nonlinearities of Small Metal Particles – Surface Mediated Resonance and Quantum Size Effects," *J. Opt. Soc. Am. B*, **3**, 1647–1655.

Kaempfe, M., Rainer, T., Berg, K. J., Seifert, G., and Graener, H., 1999, "Ultrashort Laser Pulse Induced Deformation of Silver Nanoparticles in Glass," *Appl. Phys. Lett.*, **74**, 1200–1202.

Kofman, R., Cheyssac, P., Aouaj, A. *et al.*, 1994, "Surface Melting Enhanced by Curvature Effects," *Surf. Sci.*, **303**, 231–246.

Link, S., Wang, Z. L., and El-Sayed, M. A., 2000, "How Does a Gold Nanorod Melt?," *J. Phys. Chem. B*, **104**, 7867–7870.

Longtin, J. P., Qiu, T. Q., and Tien, C.-L., 1995, "Pulsed Laser Heating of Highly Absorbing Particles," *J. Heat Transfer*, **117**, 785–788.

Mafuné, F., Kohno, J., Takeda, Y., and Kondow, T., 2001, "Formation of Gold Nanoparticles in Aqueous Solution of Surfactant," *J. Phys. Chem. B*, **105**, 5114–5120.

2003, "Formation of Stable Platinum Nanoparticles by Laser Ablation in Water," *J. Phys. Chem. B*, **107**, 4218–4223.

Maxwell-Garnet, J. C., 1904, "Colours in Metal Glasses and Metallic Films," *Phil. Trans. Royal Soc. A*, **203**, 385–420.

Mohamed, M. B., Ahmadi, T. S., Link, S. *et al.*, 2001, "Hot Electron and Phonon Dynamics of Gold Nanoparticles Embedded in a Gel Matrix," *Chem. Phys. Lett.*, **343**, 55–63.

Peppiatt, S. J., and Sambles, J. R., 1975, "Melting of Small Particles. 1. Lead," *Proc. Roy. Soc. London A – Math. Phys. Eng. Sci.*, **345**, 387–390.

Seto, T., Kawakami, Y., Suzuki, N., Hirasawa, M., and Aya, N., 2001, "Laser Synthesis of Uniform Silicon Single Nanodots" *Nano Lett.*, **1**, 315–318.

Ullmann, M., Friedlander, S. K., and Schmidt-Ott, A., 2002, "Nanoparticle Formation by Laser Ablation," *J. Nanoparticle Res.*, **4**, 499–509.

Zhu, H., and Averback, R. S., 1996, "Sintering Process of Two Nanoparticles: A Study by Molecular Dynamics Simulations," *Phil. Mag. Lett.*, **73**, 27–33.

13 Laser-assisted microprocessing

13.1 Laser chemical vapor deposition

Consider a target material immersed in a reactive ambient medium. An incident laser beam may excite and dissociate the reactant molecules. Consequently, excited molecules or radicals diffuse to the solid surface and may interact with the target surface, resulting in etching or deposition. These processes are thoroughly discussed in Bäuerle (2000). Figure 13.1 gives a schematic illustration of the laser-induced chemical-processing systems utilizing either direct beam incidence onto the substrate or processing via a beam propagating in a direction parallel to the substrate. *Thermal* or *pyrolytic* chemical laser processing is characterized by the rapid dissipation of the excitation energy into heat. In this case, the particular excitation mechanisms are not significant and the processing rate is chiefly determined by the induced temperature distribution. However, the physicochemical processes involved may be drastically different from the conventional "thermal-processing" treatment. This is highlighted by the extremely confined laser-beam radiant energy and hence the temperature-distribution localization to high peak temperatures and steep temperature gradients. Furthermore, pulsed-laser-induced heating rates can be very fast, reaching 10^{12} K/s, even 10^{15} K/s, leading to a regime where the chemical reaction deviates greatly from equilibrium. Consequently, one may expect the formation of new phases, microstructures, and morphologies through novel chemical-reaction pathways. On the other hand, *photochemical* or *photolytic* laser chemical-processing conditions apply when the thermalization of the excitation energy is slow. As noted in Bäuerle (1998), this condition frequently applies for chemical reactions of excited molecules among themselves or with the substrate surface, photoelectron transfer and chemisorption of species on solid surfaces, and photochemical species desorption. Pure photothermal and photochemical conditions may be regarded as limiting cases of laser-induced chemical processing. Often both processes contribute to the overall reaction rate, implying that one is dealing with a so-called *photophysical* process.

Laser chemical vapor deposition (LCVD) is an extremely versatile materials-synthesis technique that enables the formation of technologically attractive microstructures of well-defined dimensions in a single-step maskless process (Bäuerle, 2000; Ibbs and Osgood, 1989). Several applications have been pursued, including the fabrication of contacts, circuit lines, and interconnects, and the repair of lithographic masks. It also

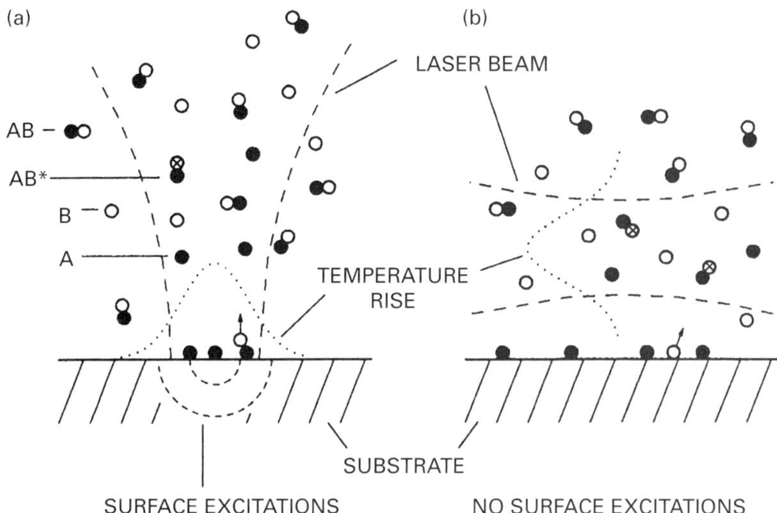

Figure 13.1. Illustrations of laser-induced chemical processing (LCP) at perpendicular (case a) and parallel laser-beam incidence (case b). Thermal (photothermal, pyrolytic) LCP at perpendicular incidence is in general performed with precursors AB that do not absorb the laser radiation. Nonthermal (photochemical, photolytic) LCP is based on selective excitation/ dissociation of precursor molecules and/or the substrate. In a purely photochemical process, the laser-induced temperature rise can be ignored. From Bäuerle (1998), reproduced with permission from Wiley-Interscience.

offers unique capabilities for the formation, coating, and patterning of nonplanar, three-dimensional objects.

The decomposition of precursor molecules in LCVD can be activated either pyrolytically or nonthermally (photolytic LCVD) or by a combination thereof (i.e. photophysical LCVD). In pyrolytic deposition, laser-induced heating of the substrate is employed to decompose the gases above it. In photolytic processes, direct dissociation of the parent gas is accomplished by the optical excitation of bound–free transitions. The type of process activation determines the morphology of the deposit while the deposition rate is a function of the laser power, intensity, pulse duration, wavelength, and substrate material. Major advantages of pyrolytic LCVD are the synthesis of high-purity materials and fast deposition rates, but an obvious drawback is that it cannot be used if the substrate softens or melts prior to gas decomposition upon exposure to laser radiation. In contrast, photolytic LCVD can be used for low-temperature deposition onto fragile substrates, including organic materials or complex III–V compounds. Furthermore, photolytic deposition is less sensitive to the substrate-surface conditions and composition than is pyrolytic deposition. However, deposition rates in photolytic processes are 2–4 orders of magnitude lower than those associated with pyrolytic deposition. Furthermore, typical laser scanning speeds are in the range 0.001–0.1 μm/s for photolytic deposition, compared with $O(10\,\mu\text{m/s})$ speeds for pyrolytic deposition. The lateral feature resolution could be 100 μm for photolytic direct writing, but the deposition might not be well localized because the intensity threshold is very low, allowing spurious effects.

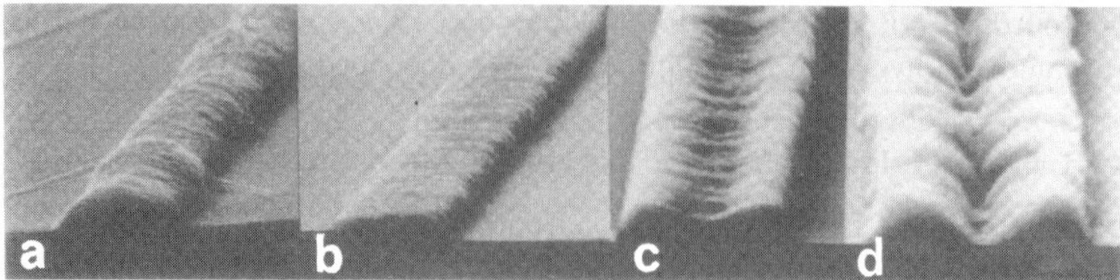

Figure 13.2. Pyrolytic microdeposition of Si lines from SiH_4 gas upon argon-ion-laser irradiation at $\lambda = 488$ nm. The substrate is silicon. From Bäuerle (2000), reproduced with permission from Springer-Verlag.

Photophysical LCVD achieved by a twin- or single-laser-beam configuration seeks to combine the advantages of both deposition schemes by initiating nucleation photochemically and progressing thereafter according to a conventional thermal script. Precursor molecules are usually either (a) halogen compounds, hydrocarbons, and silanes or (b) alkyls, carbonyls, and organometallic complexes. The first class of molecules possesses electronic transitions in the ultraviolet (UV) or deep-UV range and is normally utilized in pyrolytic LCVD, even though the temperatures required are relatively high. Conversely, the second category possesses electronic transitions at visible or UV wavelengths at which laser sources can be matched to effect photophysical LCVD. However, high-intensity UV laser radiation can initiate thermal chemistry; or alternatively, if the pyrolytic laser source is operating in the visible or near-UV range, photodissocation of weakly bound complexes may occur. Selection of the precursor material, or parent gas, is instrumental in developing a viable nanodeposition process.

Upon flowing a parent-gas (e.g. silane for deposition of Si)–carrier-gas mixture into the process chamber, gas molecules impinging on the substrate – some of which are physisorbed – are exposed to focused laser radiation. For substrates that absorb the incident laser light, the induced temperature distribution will cause thermal dissociation of gas molecules at or near the surface, which can then form stable nuclei. In the case of transparent substrates, nucleation can be initiated by the atomic products of adsorbed molecules dissociated by the laser-beam radiation. A fraction of the photo-fragments created by gas-phase dissociation within the focal zone condenses preferentially on the nuclei generated within the irradiated area.

Parent-gas molecules not directly exposed to the laser radiation or to high temperatures as a result of laser–substrate interactions will not crack and therefore not be deposited. However, the depositing atoms (dissociation products of the parent gas) that are physisorbed onto the substrate outside the area occupied by the nuclei will undergo surface diffusion and ultimately will be captured by existing nuclei, form new nuclei or re-evaporate. When the diffusion length of a physisorbed atom (adatom) is determined by the length scale of the laser-irradiated zone, atom capture by the intentionally formed nuclei should be most probable and selective growth will occur. Achieving favorable growth conditions involves controlling the size of the laser-affected zone, the density of parent-gas molecules, and the kinetic energy of the depositing species.

An example of pyrolytic micropatterning is given in Figure 13.2 that depicts Si lines deposited from SiH_4 onto Si wafers (Bäuerle, 2000). As the laser power increases, the

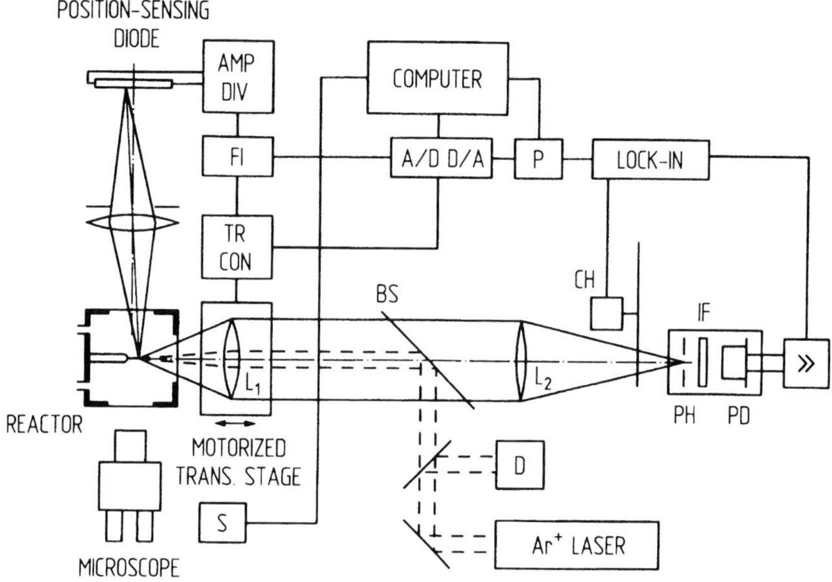

Figure 13.3 The experimental setup employed for *in situ* temperature measurements during LCVD of fibers: AMP DIV, amplifier and analog divider; BS, beamsplitter; CH, light chopper; D, power meter; FI, electronic filter; P, peak detector; PD, Si photodiode; PH, pinhole; S, switch; and TR CON, translation control. From Doppelbauer and Bäuerle (1986).

morphology of the lines changes to concave. These changes can be attributed to the laser-induced temperature distribution, the transport of reactants and products in the gas phase, and the surface diffusion of species. The LCVD technique can be implemented for the fabrication of fibers, crystal rods of virtually any length, and even three-dimensional microstructures. Figure 13.3 shows a schematic diagram of the setup for *in situ* temperature measurements during the steady growth of fibers (Doppelbauer and Bäuerle, 1986). Growth of ultrahigh-quality single-crystalline Si whiskers is depicted in Figure 13.4. It is remarkable to observe the high deposition rates achieved by LCVD shown in Figure 13.5. Owing to the fact that the heat-affected zone is confined to the proximity of the wafer, it is possible to apply higher parent-gas partial pressures than typically used in traditional chemical vapor deposition (CVD).

Various LCVD-based approaches can be employed for the rapid prototyping of three-dimensional microstructures of complex geometries. When only one beam is used, as is typically the case, LCVD generates rods parallel or nearly parallel to the incident beam. In the case of growth of transparent alumina, the rod absorbs light relatively weakly in a volumetric fashion rather than in a shallow surface layer. Figure 13.6(a) shows the dual, overlapping-beam principle (Lehmann and Stuke, 1995). No growth can occur for either beam alone. However, a sufficiently high and uniform temperature field is formed in the overlapping focal region that is well defined. By moving the sample in any direction at a speed comparable to the growth rate ($\sim$10 μm/s), arbitrary, free-standing structures can be produced. Figures 13.6(b) and (c) depict an impressive example of the application of LCVD to fabrication of photonic band-gap (PGB) structures (Wanke *et al.*, 1997).

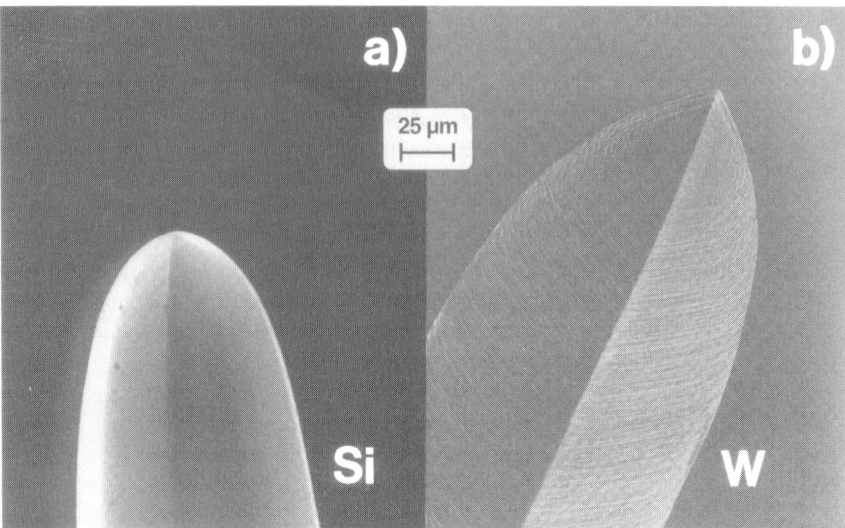

Figure 13.4. Scanning-electron-microscopy pictures of the tips of laser-grown single-crystalline fibers: (a) Si grown from SiH_4 and (b) W grown from WF_6 plus H_2. From Bäuerle *et al.* (1983), reproduced with permission from Springer-Verlag.

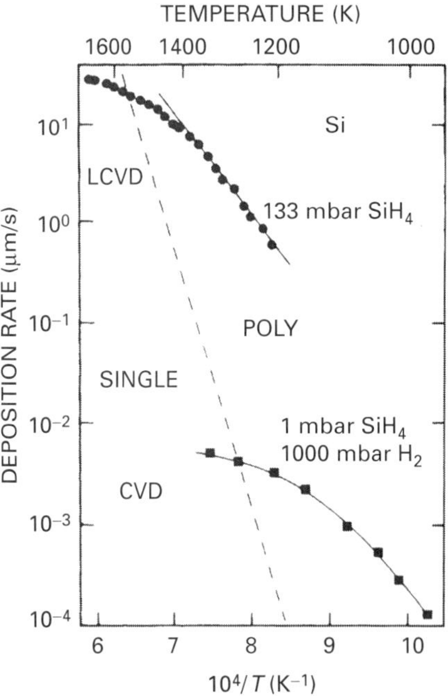

Figure 13.5. An Arrhenius plot for the growth of Si from SiH_4 by LCVD and standard CVD. Regions of single-crystalline and polycrystalline growth are shown (separated by a dashed line); the intersection points of the LCVD and CVD curves are at 1555 and 1262 K, respectively. From Bäuerle (2000), reproduced with permission from Springer-Verlag.

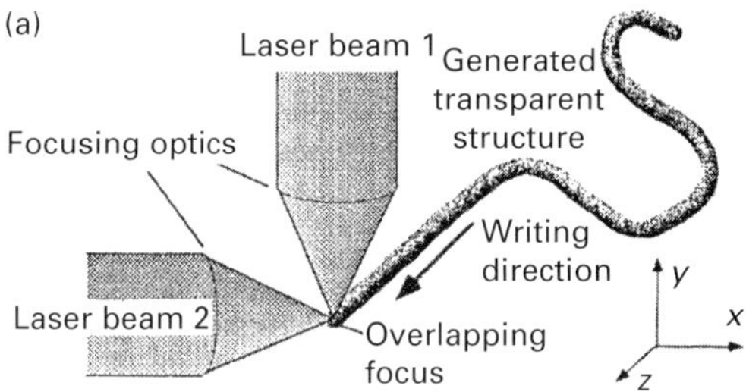

Figure 13.6. (a) The three-dimensional direct-writing principle. (b) The PGB structure. The periodic structure consists of 15 rows of perpendicularly arranged aluminum oxide rods 40 mm in diameter, with a periodicity of 133 μm and rod lengths of 3000 μm. (c) A partly disassembled aluminum oxide rod structure showing the aluminum oxide rods (diameter 40 μm) and their organization (periodicity 133 μm). From Lehmann and Stuke (1995) and Wanke *et al.* (1997), reproduced with permission from the AAAS.

13.2 Laser direct writing

13.2.1 Laser-induced forward transfer

The ability to deposit patterns, spots, and lines with sub-millimeter resolution may have applications in microelectronics as well as in the opto-electronics fabrication industries.

Figure 13.6. (*cont.*)

Conventional methods of surface patterning are CVD, plasma CVD, and sputtering, which by their nature have no spatial selectivity, requiring subsequent etching steps to define micropatterns. In this regard, the laser-induced forward-transfer (LIFT) technique utilizes pulsed lasers to remove thin-film material from a transparent support and deposit the ejected fragment onto a suitable substrate. The thin film that is deposited onto a quartz plate is transferred by using a single laser pulse onto the receiving substrate that is usually placed parallel to the source thin film. The LIFT process was first shown by Bohandy *et al.* (1986, 1988) to produce 50-μm-wide Cu lines by using single nanosecond excimer-laser pulses (193 nm) under high vacuum (10^{-6} mbar). Among several studies, Fogarassy *et al.* (1989a, 1989b) reported the microdeposition of 100-μm-wide patterns of superconducting thin films using nanosecond ArF and Nd : YAG lasers. The dynamics of the laser-ablation transfer of thin coatings effected by near-IR ($\lambda = 1064$ nm) 23-ps laser pulses was studied by Lee *et al.* (1992) via an optical microscope with picosecond temporal resolution. This investigation showed that the velocity of the ejected material corresponded to a Mach number of 0.75 and that the use of picosecond-laser pulses resulted in a reduction of the laser-fluence threshold by one order of magnitude, compared with the use of 100-ns-laser pulses.

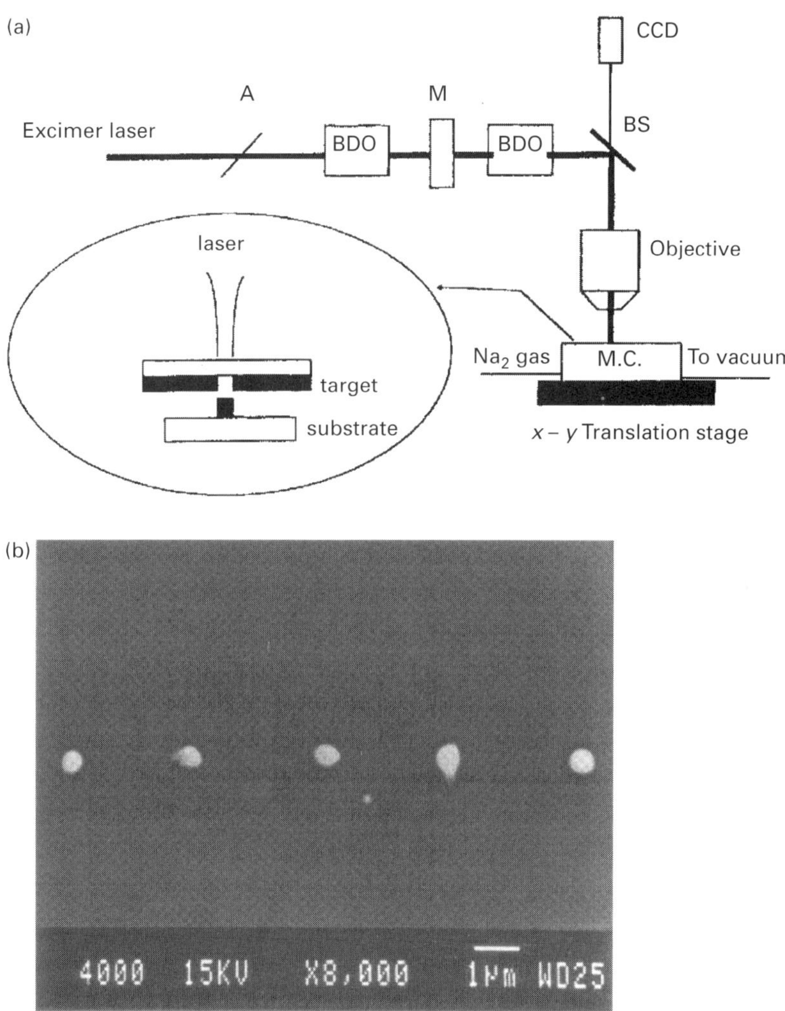

Figure 13.7. (a) A schematic diagram of the optical setup for excimer-laser microdeposition: A, attenuator; BBO, beam-delivery optics; M, mask; BS, image/beamsplitter; CCD, camera; and MC, microdeposition cell. (b) A scanning electron micrograph of isolated Cr dots deposited onto glass by femtosecond-laser microdeposition. The target source was 400-Å-thick Cr. The UV-illuminated area was 4 μm × 4 μm and the energy density was 100 mJ/cm². From Zergioti *et al.* (1998a), reproduced with permission from Elsevier.

The work of Zergioti *et al.* (1998a, 1998b) demonstrated direct deposition of Cr and In_2O_3 microstructures via sub-picosecond KrF excimer-laser radiation in the LIFT mode (Figure 13.7(a)). The short pulse length and consequently limited thermal diffusion lowered the material-removal threshold compared with that for nanosecond pulses, thereby enabling deposition of high-resolution features (Figure 13.7(b)). By controlling the ambient pressure at 0.1 Torr, complex holographic patterns could be fabricated

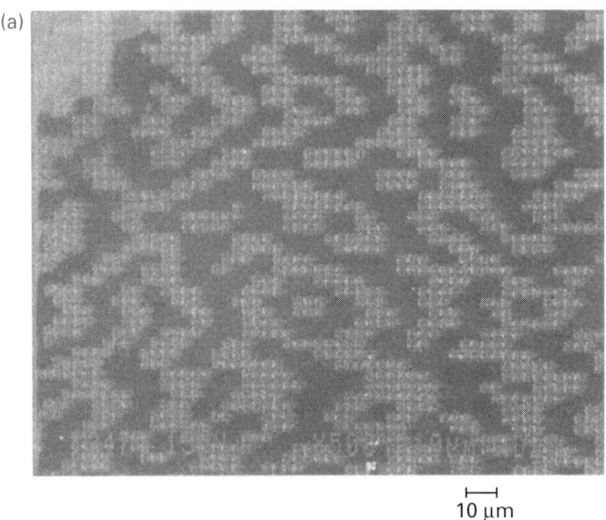

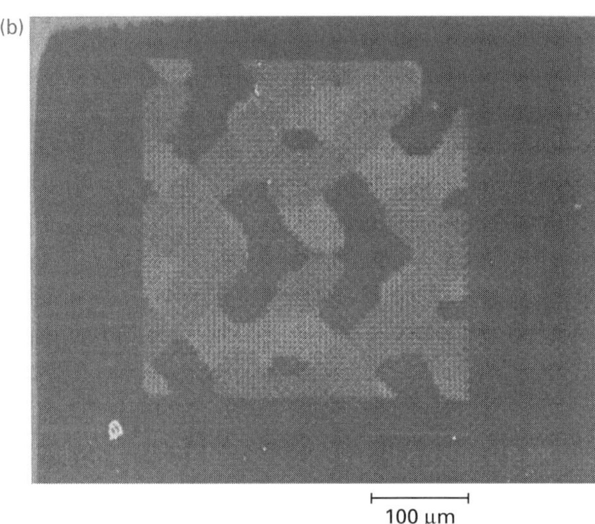

Figure 13.8. (a) A scanning electron micrograph of a computer-generated holographic pattern produced by microdeposition of CR onto silicon (100). The target source was 400-Å-thick Cr. The pattern consists of 64 × 64 pixels with pixel size 3 μm × 3 μm. (b) A scanning electron micrograph of a computer-generated multilevel structure of Cr on glass. The pixel size of the pattern was 5 μm × 5 μm. From Zergioti *et al.* (1998a), reproduced with permission from Elsevier.

(Figures 13.8(a) and (b)). Time-resolved Schlieren imaging (Koundourakis *et al.*, 2001; Papazoglou *et al.*, 2002) revealed that the sub-picosecond, LIFT-mode film ejection occurs in a forward manner with small lateral angular spread (Figure 13.9(a)). In contrast, direct front-surface ablation (Figure 13.9(b)) exhibits higher divergence. On the other hand, nanosecond lasers tend to produce a dispersed particle cloud (Figure 13.9(c)) that is deposited onto the receiving substrate in irregular aggregates.

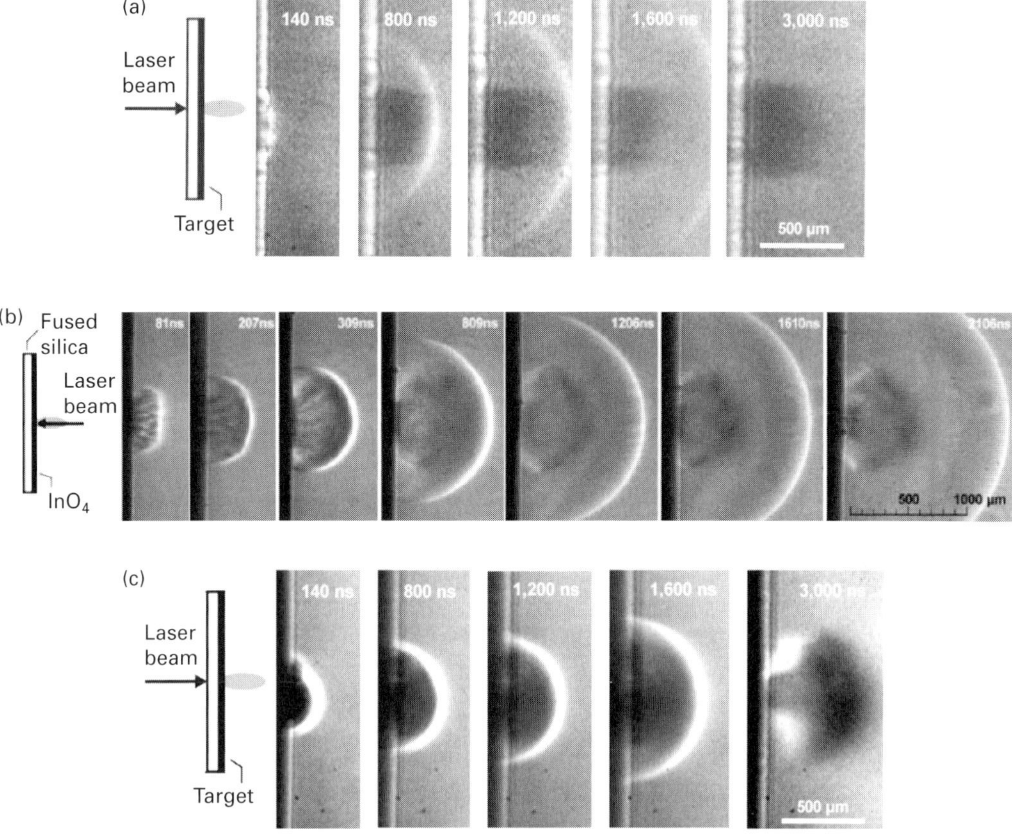

Figure 13.9. (a) Shadowgraph images of InO_x film under the LIFT process at various delay times. The laser fluence was 410 mJ/cm². The detected material ejection is highly directional with an angular divergence of 3°. (b) Shadowgraph images of the InO_x film under backward laser irradiation at various delay times. The laser fluence was 430 mJ/cm². The velocity of the ejected-material front is ∼430 ± 19 m/s and the blast-wave velocity is ∼500 ± 25 m/s. The detected material ejection is quite divergent, with an angular divergence of ∼30°. (c) The nanosecond LIFT of Cr film is shown to highlight its difference from picosecond-laser ablation. From Papazoglou *et al.* (2002), reproduced with permission from the American Institute of Physics.

13.2.2 MAPLE-assisted direct writing

In matrix-assisted pulsed-laser evaporation (MAPLE) (Chrisey *et al.*, 2003), an organic compound is dissolved in a matrix material that is generally a volatile solvent such as alcohol to form a solution. The solution is frozen at −100 °C to produce a solid target. Upon pulsed-laser irradiation, solvent molecules vaporize and thermal energy is transferred to the organic molecules that are desorbed from the target surface without significant decomposition to form a film on a receiving substrate while the solvent is pumped out. A key characteristic of this gentle deposition process is that the organic molecule remains intact. The MAPLE direct-write (DW) technique further incorporates the LIFT material-removal process (Piqué *et al.*, 1999). As shown in Figure 13.10(a),

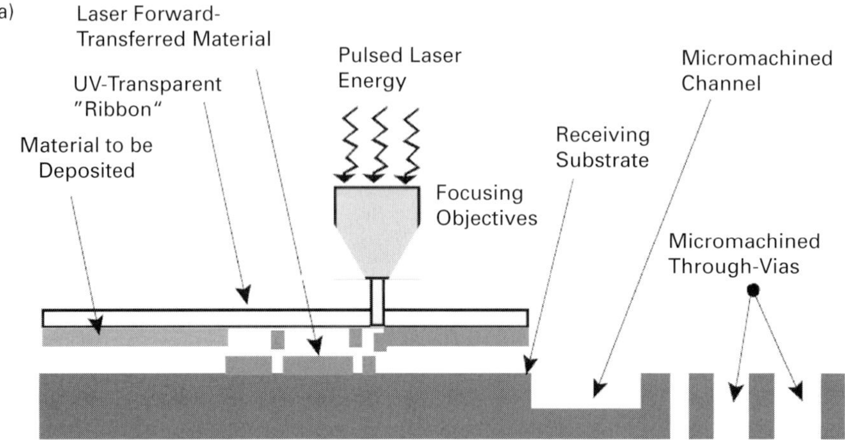

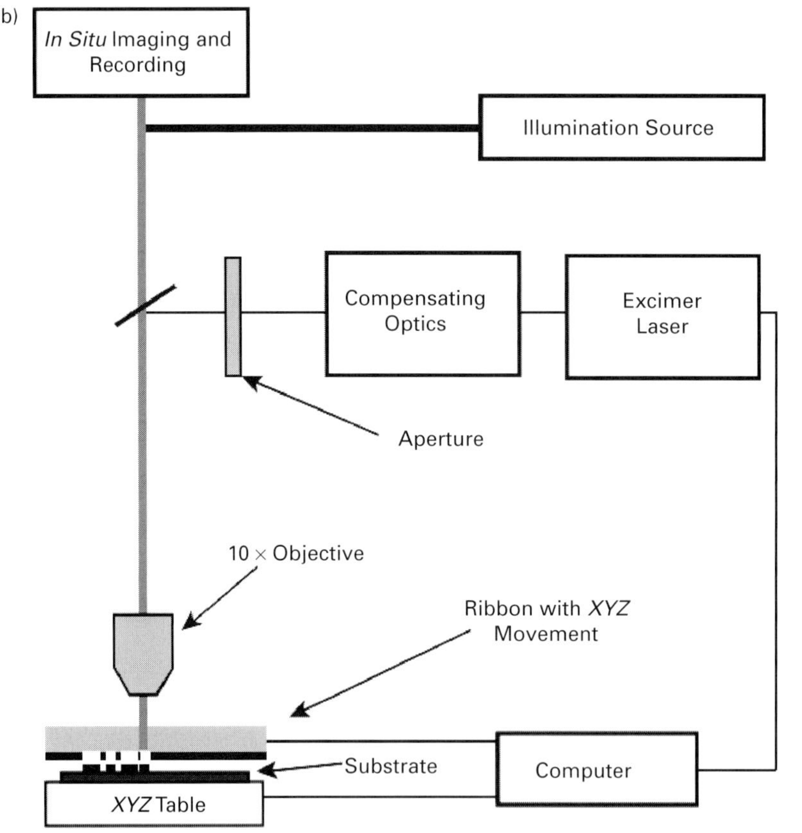

Figure 13.10. (a) A schematic representation of the MAPLE direct-writing (DW) process. (b) A schematic diagram of the MAPLE DW apparatus. From Piqué *et al.* (1999), reproduced with permission from Springer-Verlag.

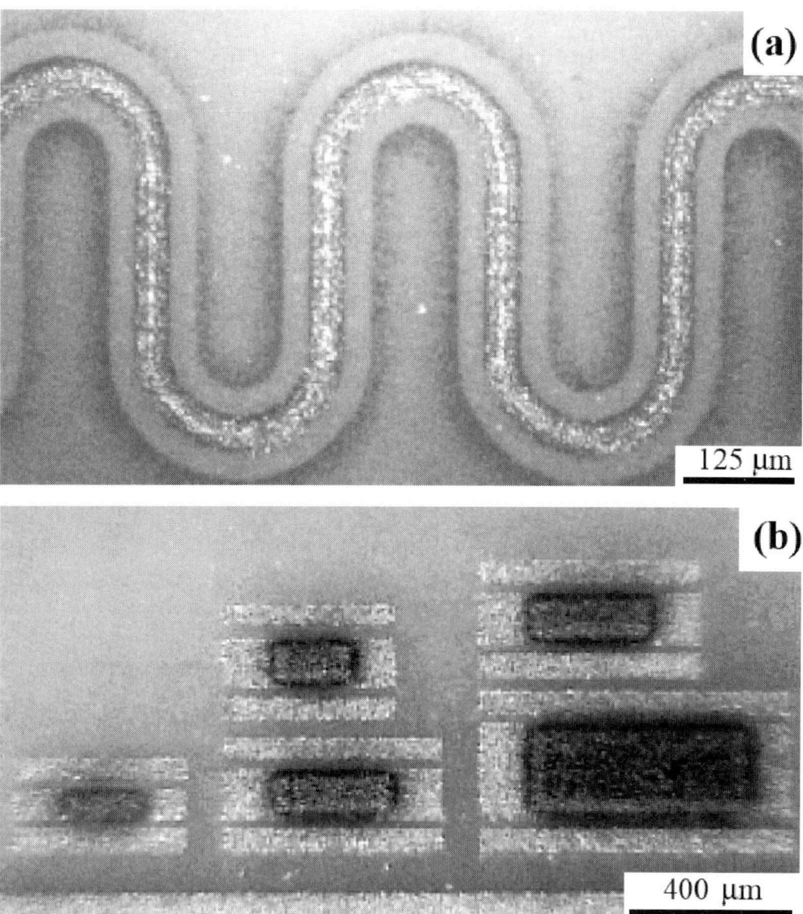

Figure 13.11. (a) Lines deposited by LIFT on an RO4003 circuit board. The Au line is approximately 30 mm wide after a final laser-trimming step performed along both sides of the line. (b) An optical micrograph of nichrome coplanar resistors made by LIFT on RO4003. From Piqué *et al.* (1999), reproduced with permission from Springer-Verlag.

the donor transparent substrate (the "ribbon") is coated with a powder of the material to be deposited in a photosensitive polymer or organic binder. Figure 13.10(b) shows a schematic diagram of the computer-driven MAPLE DW apparatus. Examples of deposited conductor lines are shown in Figure 13.11. By combining microdeposition and micromachining, Piqué *et al.* (2004) achieved fabrication of mesoscale elctromechanical sources, such as primary Zn–Ag$_2$O and secondary Li-ion microbatteries (see Figure 13.12).

13.3.2 Hybrid microprinting and laser sintering of nanoparticle suspensions

The emergence of consistent methods for manufacturing ultrafine particles has come about with the aim of utilizing their remarkable thermophysical properties that may be

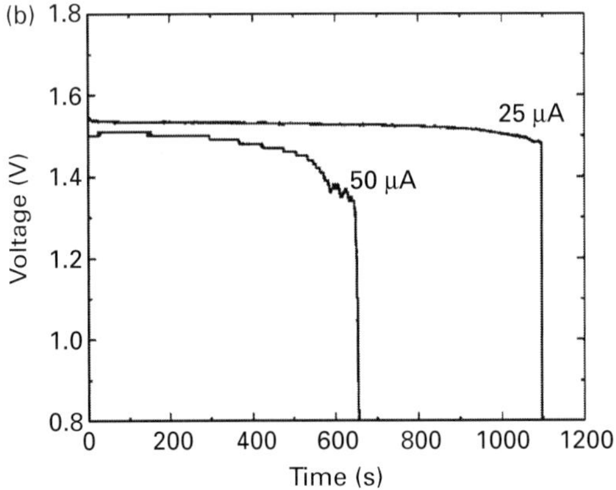

Figure 13.12. (a) An optical micrograph of a planar Ag_2O/Zn alkaline microbattery made by laser direct writing. (b) Discharge behaviors from two planar alkaline microbatteries, operating at 25 and 50 mA, as a function of time. The total electrode mass is $\sim$250 mg in both cases. From Piqué *et al.* (2004), reproduced with permission from Springer-Verlag.

significantly different from those of their bulk counterparts. In an order-of-magnitude sense, particles of diameters well into the sub-micrometer region and on the nanometer scale possess significant individual characteristics that differentiate them from bulk materials. As shown in Chapter 12, gold nanosized particles possess a drastically lower melting temperature (approximately 400 °C) than that of bulk gold (approximetaly 1300 °C).

Gold nanoparticles are prepared by a two-phase reduction method. The average nanocrystal size is 1–3 nm (Figure 13.13) and the size is coarsely tunable by adjusting the ratio of capping groups to metal salt, whereas size-selective precipitation is employed to narrow the initial size distribution. Monolayer-protected gold nanoparticles are suspended in alpha-terpineol at a proportion of 10% by weight. The reflectivity of printed nanoparticle film starts to increase and the electrical resistivity starts to decrease at a curing temperature of 130–140 °C. According to observations, the surface monolayer

(a)

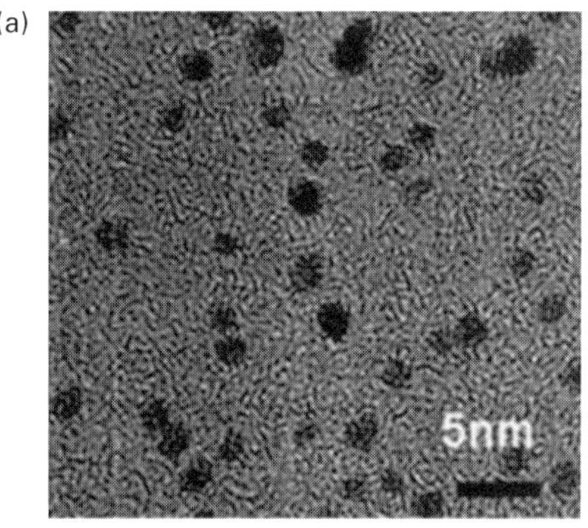

(b)

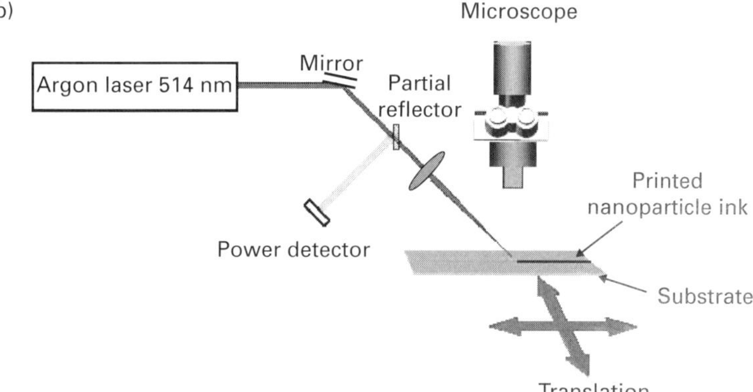

Figure 13.13. (a) A TEM image of the synthesized nanoparticles and (b) a schematic diagram of the laser-sintering setup. From Ko *et al.* (2007a), reproduced with permission from the American Institute of Physics.

(hexanethiol) is detached from the nanoparticle at 130–140 °C. Micro-lines of nano-ink were deposited onto a polyimide film by the generation of micro-droplets using the piezoelectric drop-on-demand (DOD) printing system. The DOD jetting system includes a back-pressure controller and a purging system. A detailed description of the experimental system is given in Chung *et al.* (2004, 2005) and Bieri *et al.* (2003). After inkjet printing of nanoparticle ink on polyimide film, a temporally modulated continuous Ar$^+$ laser ($\lambda = 514$ nm) was applied to evaporate residual solvent and sinter the nanocrystals to form low-resistivity conducting microstructures.

Several laser parameters (wavelength, pulse width, power density, beam-spot size etc.) can be tuned to obtain optimum processing. Highly effective energy coupling (absorption) can occur at the appropriate wavelength. On the other hand, the heat-affected zone and thermal damage to the substrate can be minimized by controlling the pulse width. In contrast, thermal damage extends over the entire sample for sintering on a hot plate. Local manipulation of the inkjet-printed features enables more versatile processing while the final feature obtained by sintering on a hot plate is mainly limited by

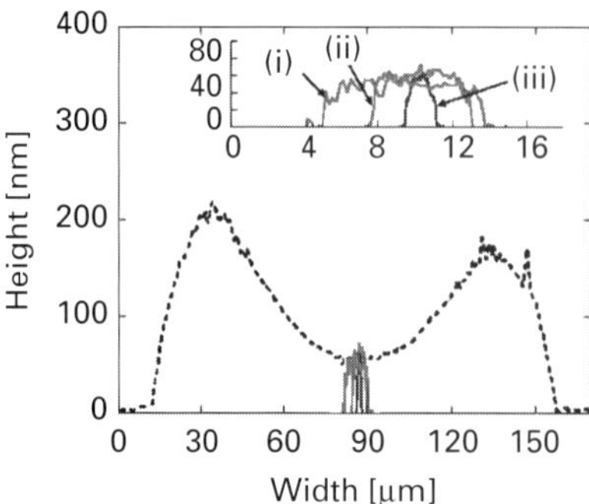

Figure 13.14. An AFM cross-sectional profiles of sintered metal nanoparticles produced by selective laser sintering (solid lines, (i) $\sim$10 μm, (ii) $\sim$3 μm, and (iii) $\sim$1 μm, shown in the inset and in the main plot) and on a hot plate as printed (dotted line, $\sim$100 μm). Note that the feature size can be reduced by two orders of magnitude. Also note the absence of the high rim structure caused by the ring-stain problem. From Ko (2006).

the initial inkjet printing. Selective laser sintering can enable enhancement of resolution and uniformity. In typical inkjet processing, most nanoparticles are deposited at the edge of the droplet, which is often referred to as the "ring-stain problem" (Deegan *et al.*, 1997). This nonuniformity can cause problems when another layer needs to be deposited on top of the film. However, since the central part is rather flat, by selective sintering and washing out leftover unsintered inkjet-printed nanoparticles, the resolution can be enhanced by one or two orders of magnitude since it is regulated by the size of the focused laser beam. Note the resolution enhancement of the original inkjet-printed line (the dotted line in Figure 13.14) after selective laser sintering (solid lines: 1–10 μm depending on the laser-beam size and laser power).

While selective laser sintering is used to make micro positive features with a low-power continuous Ar^+ laser, selective laser ablation is used to define micrometer to sub-micrometer negative features. Small negative features are critical in high-quality transistors because the switching speed and output current directly depend on the channel size. The core concept of selective laser ablation is based on the finding that the nanoparticle film can be ablated at much lower laser energy than bulk film (at least an order of magnitude lower (Ko *et al.*, 2006)). This low ablation threshold adds significant processing capability because ablation can be done at low laser energy, thereby minimizing thermal effects and enabling the use of sensitive substrates. Very clean ablation profiles are obtained, which is almost impossible without using ultrafast (picosecond or femtosecond lasers). Depending on the fluence, the vaporized plume or plasma exerts a recoil pressure, expelling outward or even splashing the melt pool, which tends to resolidify at the periphery of the crater (Figure 13.15(b)). This elevated topography may cause shorts in multilayer devices. Minute features can be fabricated with proper focusing optics (sub-micrometer features in Figure 13.15(g)). This could be explained by invoking the combined effects of melting-temperature depression, lower conductive heat-transfer loss, strong absorption of the incident laser beam, and relatively weak

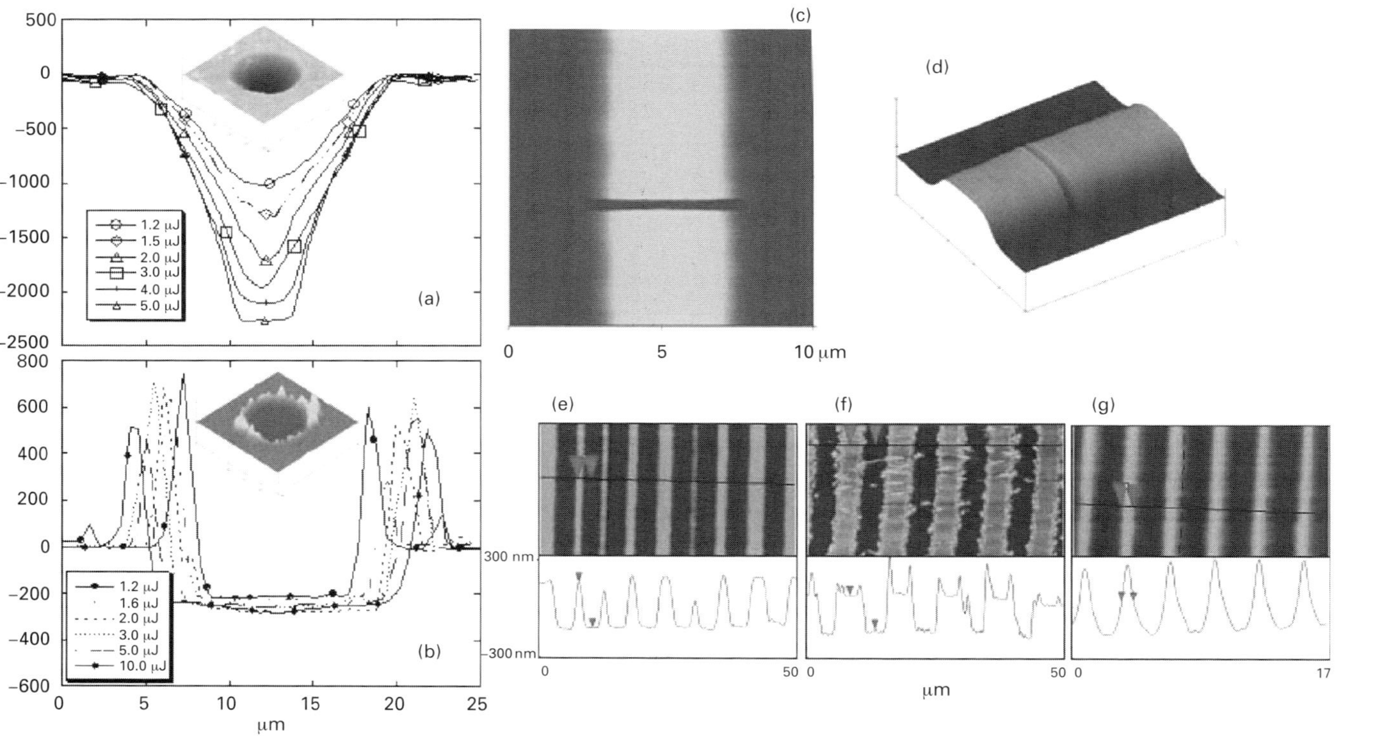

Figure 13.15. Atomic-force-microscopy (AFM) scanned images of ablation profiles produced by single-shot irradiation with a 20× objective lens at various energies of samples prepared at (a) 120°C (before sintering) and (b) 160°C (after sintering). Both inset figures are surface-morphology AFM images at incident laser energy 3 μJ. In (c) and (d) are shown AFM topology pictures of two electrodes separated by a sub-micrometer channel (∼800 nm) formed by laser ablation and sintering of metal nanoparticles. (e)–(g) Channel formation on the upper layer in multilayer structures: AFM depth and cross-sectional images of gold positive line features made from (e) an unsintered film of nanoparticles by laser ablation with a 20× infinity-corrected long working lens, (f) a sintered film of nanoparticles (bulk gold film) by laser ablation with a 20× lens, and (g) an unsintered film of nanoparticles by laser ablation with a 50× lens. Note that the positive features are always cleaner and well defined in laser ablation of unsintered nanoparticle film and at much lower ablation energy. From Ko *et al.* (2006), reproduced with permission from the American Institute of Physics.

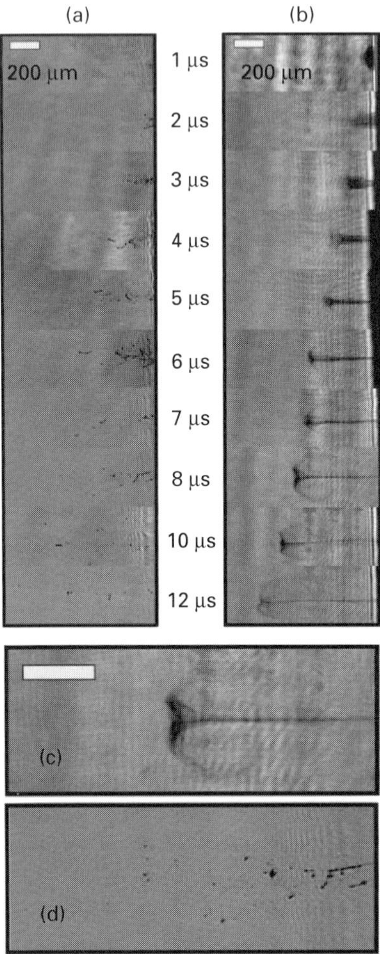

Figure 13.16. Time resolved shadowgraph images of the ablation-plume ejection from (a) sintered nanoparticle film and (b) unsintered nanoparticle film. The images were taken for 1–12 μs and processed by background subtraction for better contrast. Magnified views at 8 μs for (c) unsintered nanoparticle film and (d) sintered nanoparticle film are shown. Inset scale bars correspond to 200 μm. A laser energy of 8.5 μJ with beam radius 12.5 μm is used for ablation of sintered film, (a) and (d), and 4.2 μJ is used for ablation of unsintered film, (b) and (c). From Ko (2006).

bonding between nanoparticles during laser irradiation. Even smaller (sub-100-nm) features can be defined by coupling laser pulses to near-field scanning optical microscope (NSOM) instruments.

On the other hand, the ablation of the unsintered nanoparticle film produces a mushroom-shaped plume as shown in Figures 13.16(b) and (c). The ejecta are in the form of nanoparticles as shown in Figure 13.17(b). In contrast the ablation of sintered nanoparticle film is in the form of a streak of micrometer-sized droplets (Figures 13.16(a) and (d)) that is corroborated by the collected microparticles shown in Figure 13.17(a).

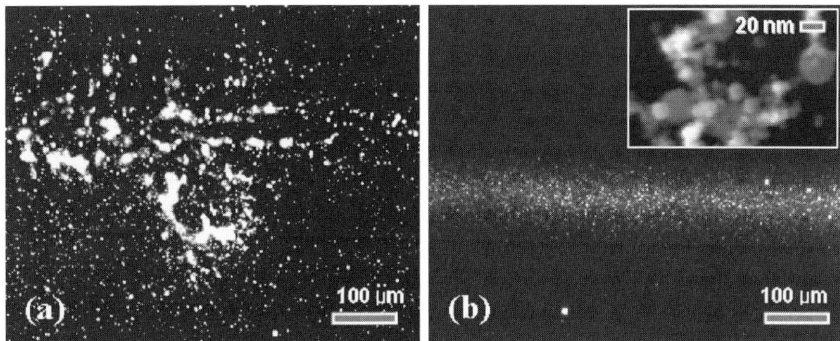

Figure 13.17. A dark-field microscope image of laser-ablation ejecta captured from (a) sintered gold thin film and (b) unsintered gold nanoparticle film. Ejecta from sintered gold film were of diameter several micrometers, whereas ejecta from unsintered nanoparticle film were aggregates of diameter 5–50 nm. The inset is a SEM image of nanoparticle aggregates. From Ko (2006).

The significant ablation-threshold difference between sintered and unsintered nanoparticle film can be used for selective-layer-ablation processing. In principle, selective ablation of a multilayer system can be done by placing the focal point of the laser exactly on the target-layer, expecting that the underlying layer would be outside the depth of focus. However, this approach is difficult since the target-layer thickness is of the order of several tens of nanometers and the thickness of the intermittent dielectric layer is also very small. Consequently, a very tight depth of focus would be needed in proportion to the multilayer-separation distance. In contrast, the differential ablation threshold between the laser-sintered and unsintered gold nano-ink enables effective and robust multilayer processing. The process entails the processing of structures in the "as-deposited" nanoparticle form, thereby avoiding damage to the underlying and sintered structures.

After inkjet printing of nanoparticle ink, continuous argon-ion laser ($\lambda = 514$ nm) irradiation was performed to sinter gold nano-ink, forming conducting lines. The sintering process depends on the intensity of the incident laser and the laser scanning speed. An electrical resistivity close to that of bulk gold is achieved (Figure 13.18(b)). The micro-conductor lines were used to fabricate crossover capacitors (Figure 13.18(c)). High-resolution, all-inkjet-printed field-effect transistors (FETs) could be fabricated benefiting from the recent development of air-stable carboxylate-functionalized polythiophene (Murphy *et al.*, 2005). All fabrication steps consist of inkjet printing of functional materials, and processing via laser irradiation. Furthermore, the entire cycle of fabrication and device characterization could be performed at room temperature, under ambient pressure, and in an air environment without resorting to expensive lithographic methods. The organic field-effect transistors (OFETs) fabricated had a typical bottom-gate/bottom-contact coplanar transistor configuration wherein the channel length was defined by selective laser sintering or ablation of the inkjet-printed metal nanoparticle ink. Figure 13.19 outlines the lithography-free sequence and displays images of the fabricated transistors (Ko, 2006) whose performance characteristics compare very well with those of devices fabricated using the same materials but lithographically.

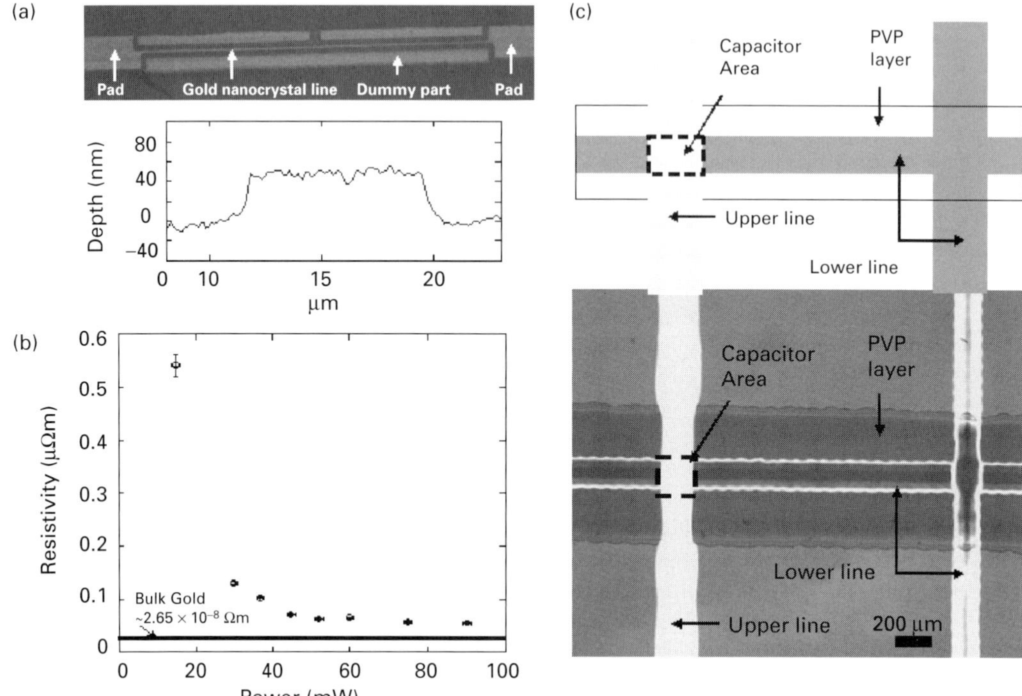

Figure 13.18. (a) A resistivity test structure inkjetted and processed by a Nd : YAG laser at 1.2 μJ with a 20× objective lens. After inkjetting and ablation, the sample was sintered by an Ar-ion laser at various powers to yield the resistivity data (b) shown in the graph, where the bottom solid line represents the bulk resistivity of gold. (c) Crossover capacitor schematics (top) and a test structure (bottom) on polyimide film. A PVP layer is sandwiched between the lower and upper lines. The lower line is under the PVP layer and the upper line is above the PVP layer. From Ko *et al.* (2007b), reproduced with permission from Elsevier.

13.3   Laser microstereolithography

Microstereolithography is a valuable tool, allowing relatively simple conversion of computer-aided-design models into truly three-dimensional microcomponents and devices possessing complex geometrical shapes. The basic scheme involves scanning a laser beam to solidify a photopolymer in a layer-by-layer manner. Arbitrary shapes can thus be produced by stacking successive layers. Continuous lasers have been utilized, demonstrating superb three-dimensional structure capabilities. On the other hand, femtosecond lasers are also implemented in microstereolithography, taking advantage of multiphoton absorption and limited diffusion to seek even finer feature resolution.

13.3.1   Photoinitiation and photopolymerization

As discussed in the review by Sun and Kawata (2004), polymerization is a chain reaction generating solid macromolecules from oligomers, i.e. unsaturated small

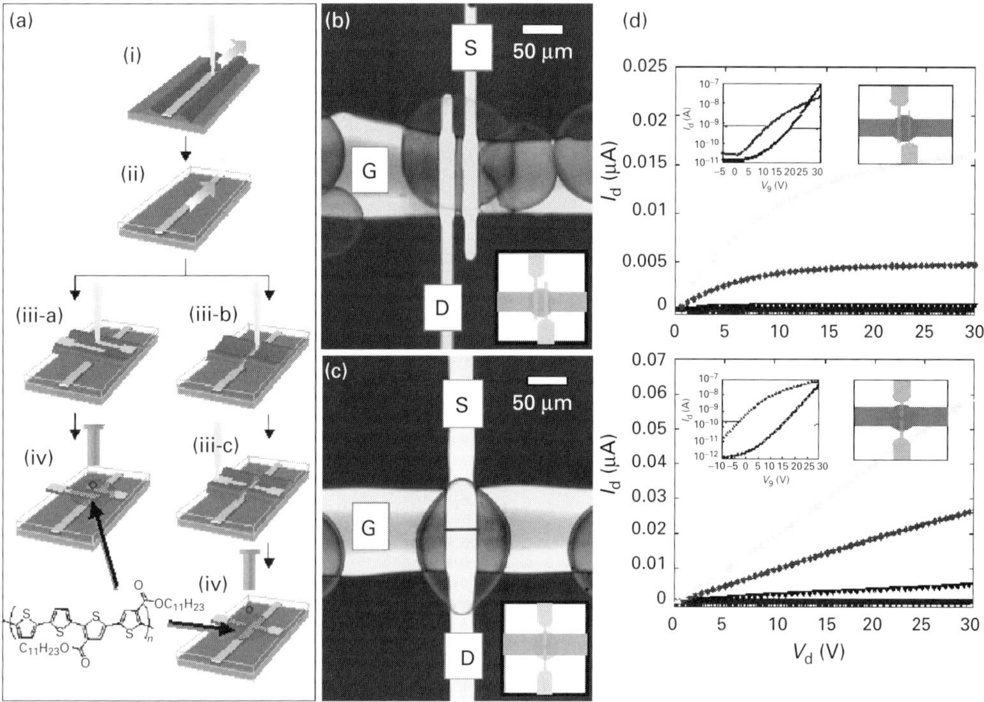

Figure 13.19. (a) Process steps for OFET fabrication by laser ablation and sintering of metal nanoparticle ink. (i) Nanoparticle ink is inkjet-printed onto either glass or plastic (exhibiting the ring-stain problem); a focused Ar-ion-laser beam scans the printed nanoparticle ink and selectively induces sintering. The dark line represents unsintered nanoparticle ink and the bright line represents sintered gold. (ii) Washout of unsintered nanoparticle ink and inkjet printing or spin-coating of a PVP dielectric layer. (iii) Inkjet printing of another structure on top of the PVP layer and either scanning with an Ar-ion-laser beam to induce selective sintering (iii-a) or use of a scan-pulsed Nd : YAG laser beam to induce selective ablation (iii-b) to define the channel. The Ar-ion-laser beam scans the ablated structure for selective sintering (iii-c). (vi) Washout of unsintered nanoparticle ink and inkjet printing of semiconducting polymer at the channel. The inset in the bottom of (a) shows the chemical structure of a polythiophene derivative containing electron-withdrawing substituents. (b) An OFET fabricated by nanoparticle selective laser sintering. (c) An OFET fabricated by a combination of nanoparticle laser ablation and sintering. The dielectric layer was spin-coated for (b) and (c). The two vertical lines are the source and drain electrodes and the underlying horizontal line is a gate line. Circles are inkjet-printed semiconducting polymer. Insets show schematics of the OFETs. Dotted lines are the edges of an original inkjet-printed line. Note how selective laser sintering can reduce the feature size and laser ablation can produce a very short channel. (d) Output and transfer characteristics of laser-processed OFETs. The top graphs correspond to the characteristics of a laser-sintered OFET ($L \sim 10$ μm, $W \sim 800$ μm) and the bottom graphs correspond to the characteristics of laser-ablated and -sintered OFETs ($L \sim 1.5$ μm, $W \sim 122$ μm). From Ko (2006).

molecules:

$$\text{Mon} \xrightarrow{\text{Mon}} \text{Mon}_2 \xrightarrow{\text{Mon}} \text{Mon}_3 \longrightarrow \cdots \longrightarrow \text{Mon}_{n-1} \xrightarrow{\text{Mon}} \text{Mon}_n. \tag{13.1}$$

The curing process can be divided into three steps: photo-initiation, propagation, and termination. Considering that the quantum yield of such a reaction is relatively low, it is necessary to introduce photo-initiators, i.e. low-mass molecules that are sensitive to light, giving radical species or cations:

$$\text{PI} \xrightarrow{h\nu} \text{PI}^{\text{exc}} \rightarrow \text{R}^{\bullet} \tag{13.2}$$

The above equation depicts the production of an active species, in this case a radical, upon the interaction of the photon with the photo-initiator and the production of an intermediate excited state. The radical then attacks the monomers and triggers the following chain reaction:

$$\text{R}^{\bullet} + \text{Mon} \rightarrow \text{RMon}^{\bullet} \xrightarrow{\text{Mon}} \text{RMonMon}^{\bullet} \cdots \rightarrow \text{RMon}_n^{\bullet}. \tag{13.3}$$

The termination of this sequence occurs through either of the following channels:

$$\begin{aligned} \text{RMon}_n^{\bullet} + \text{RMon}_m^{\bullet} &\rightarrow \text{RMon}_{m+n}\text{R}, \\ \text{RMon}_n^{\bullet} + \text{RMon}_m^{\bullet} &\rightarrow \text{RMon}_n + \text{RMon}_m. \end{aligned} \tag{13.4}$$

The role of photosensitizers is to absorb laser light and subsequently transfer energy to a photo-initiator:

$$\text{Sens} \xrightarrow{h\nu} \text{Sens}^{\text{exc}} \cdots \xrightarrow{\text{PI}} \text{PI}^{\text{exc}} \rightarrow \text{R}^{\bullet}. \tag{13.5}$$

The light absorption and photo-initiation processes are coupled by

$$\begin{aligned} \frac{\mathrm{d}C_{\text{PI}}}{\mathrm{d}t} &= -\varpi_{\text{qy;PI}}\gamma_{\text{PI}}^{\text{mol}}C_{\text{PI}}I, \\ \frac{\mathrm{d}I}{\mathrm{d}z} &= -\gamma_{\text{PI}}^{\text{mol}}C_{\text{PI}}I, \end{aligned} \tag{13.6}$$

where C_{PI}, $\varpi_{\text{qy;PI}}$, and $\gamma_{\text{PI}}^{\text{mol}}$ are the photo-initiator concentration, quantum yield, and molecular absorption coefficient, respectively. Furthermore, the mass transfer of both the radicals and the photo-initiators as well as the released heat transfer can be considered according to the model presented by Fang *et al.* (2004).

13.3.2   Projection microstereolithography

The typical setup involves a tightly focused laser spot scanning over the photo-curable resin surface, constructing solid microstructures via light-induced photopolymeriza-tion. Beyond polymeric microstructures, fabrication of ceramics and metals with complex geometrical shapes has been demonstrated. In this case, UV-curable resin is mixed with fine powders (Zhang *et al.*, 1999). The serial nature of this process renders this sort of processing slow, requiring hours to fabricate a microstructure. An alternative approach is to project a pattern onto the liquid surface. This parallel approach still requires multiple masks. A faster approach could be implemented by

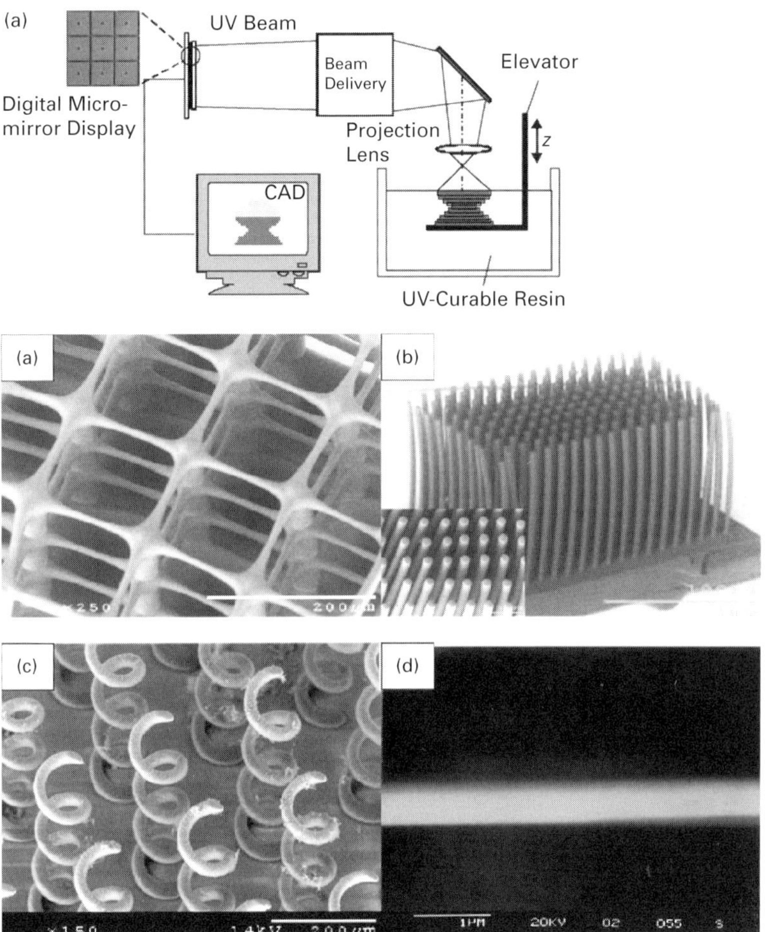

Figure 13.20. The figure on the left shows a schematic diagram of the projection-microstereolithography setup. The figure on the right depicts three-dimensional complex microstructures fabricated by the process: (a) a micro matrix with suspended beam diameter of 5 μm; (b) a high-aspect-ratio micro-rod array consisting of 21 × 11 rods of overall size 2 mm × 1 mm (the rod diameter and height are 30 μm and 1 mm, respectively); (c) a micro-coil array with coil diameter 100 μm and wire diameter 25 μm; and (d) a suspended ultrafine line of diameter 0.6 μm. From Sun *et al.* (2005), reproduced with permission from Elsevier.

using liquid-crystal-display (LCD) masks (Bertsch *et al.*, 1997). The LCD approach is limited by the switching speeds of tens of milliseconds, large pixel size, and rather low filling ratio. Furthermore, the low optical density of the refractive elements during the off mode limits the contrast of the transmitted pattern. Another issue is the high absorption of UV light during the on mode that limits the potential use of optimized UV-curable resins. A reflective type of display, the Digital Micromirror Device (DMD) was used by Sun *et al.* (2005) to produce three-dimensional microstructures with a minimum feature resolution of 0.6 μm via the introduction of UV dopants. Figure 13.20 gives the experimental concept and examples of the structures produced.

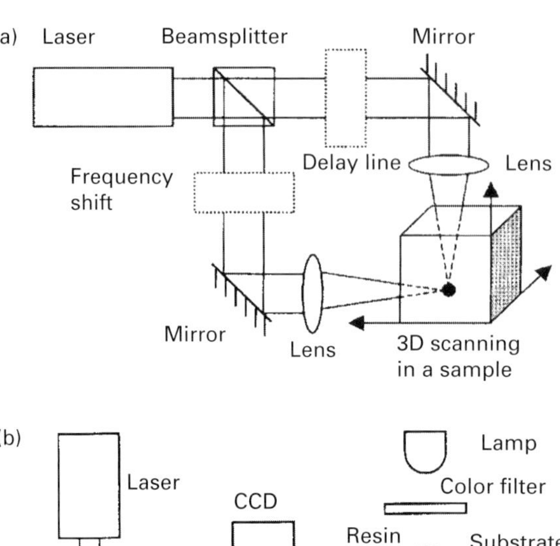

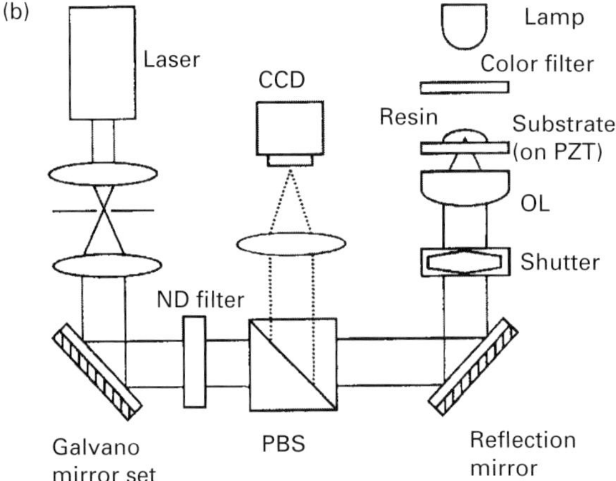

Figure 13.21. (a) A crossbeam two-photon two-color scanning laser microscopic system. Pulses from two beams split from an identical laser output should overlap in both time and temporal domains so that a two-photon-absorption process could be launched by simultaneously absorbing two photons. Removal of the frequency shifter gives rise to a degenerative two-photon fabrication system. (b) A two-photon direct-writing laser-microfabrication system. The Galvano mirror set is used for scanning the laser beam in the two horizontal dimensions, and along the longitudinal direction a PZT stage is used. The laser power was continuously adjusted by a neutral density (ND) filter. The polarization beamsplitter (PBS) lets the laser beam pass but reflects the illumination light to the CCD monitor for *in situ* monitoring of the fabrication process. OL, objective lens. From Sun and Kawata (2004), reproduced with permission from Springer-Verlag.

13.3.3 Two-photon photopolymerization and three-dimensional lithographic microfabrication

As noted in Chapter 6, two-photon (TP) excitation provides a means of activating chemical or physical processes with high spatial resolution in three dimensions, enabling the development of microfluidic and waveguide structuring of glass, three-dimensional fluorescence imaging, optical data storage, and lithographic microfabrication. Photopolymer systems exhibit low TP photosensitivity since the initiators have small TP absorption cross sections. Consequently, a practical limitation is that resins that normally polymerize at the wavelength λ polymerize at 2λ only at high intensities. Chemistry design is

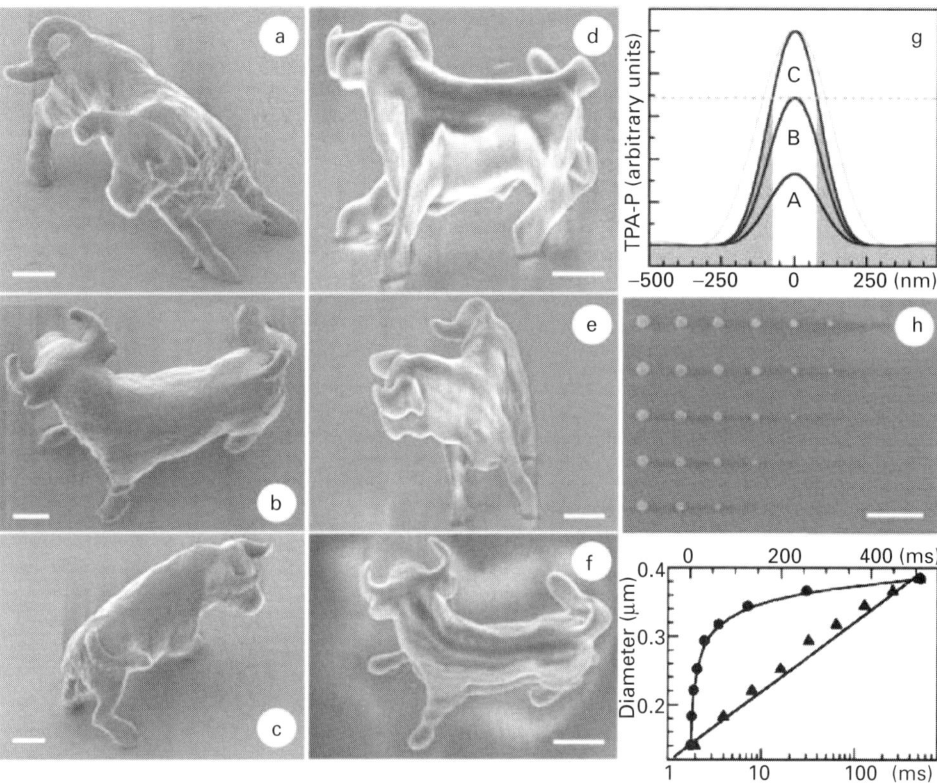

Figure 13.22. Microfabrication and nanofabrication at subdiffraction-limit resolution. A titanium : sapphire laser operating in mode-lock at 76 MHz and 780 nm with a pulse width of 150 fs was used as an exposure source. The laser was focused by an objective lens of high numerical aperture ($\sim$1.4). (a)–(c) A bull sculpture produced by raster scanning; the process took 180 min. (d)–(f) The surface of the bull was defined by two-photon absorption (TPA; that is, surface-profile scanning) and was then solidified internally by illumination under a mercury lamp, reducing the TPA scanning time to 13 min. (g) Achievement of subdiffraction-limit resolution, where A, B, and C, respectively, denote the laser-pulse energy below, at, and above the TPA polymerization threshold (dashed line). The line represents the range of single-photon absorption. TPA-P, TPA probability. (h) A scanning electron micrograph of voxels formed at various exposure times and laser-pulse energies. (i) The dependence of the lateral spatial resolution on the exposure time. The laser-pulse energy was 137 pJ. The same data are presented using both logarithmic (triangles; bottom axis) and linear (circles; top axis) coordinates, to show the logarithmic dependence and threshold behaviour of TPA photopolymerization. Scale bars, 2 μm. From Kawata *et al.* (2001), reproduced with permission from the Nature Publishing Group.

therefore needed in order to increase chromophore sensitivity and/or number density without aggregation. In this regard, Cumpston *et al.* (1999) reported TP absorption cross sections as high as $1.250 \times 10^{-50}\ \mathrm{cm^4}\,\mathrm{s}$ per photon and an enhanced TP sensitivity relative to that with UV initiators. Systems implementing TP processing via near-IR fentosecond-laser radiation are depicted in Figure 13.21. The flexibility and power of TP processes for three-dimensional lithographic structuring is highlighted in the widely referred-to example by Kawata *et al.* (2001) shown in Figure 13.22.

References

Bäuerle, D., 1998, "Laser-Induced Fabrication and Processing of Semiconductors: Recent Developments," *Phys. Status Solidi* (a), **166**, 543–554.

2000, *Laser Processing and Chemistry*, 3rd edn, Heidelberg, Springer-Verlag.

Bäuerle, D., Leyendecker, G., Wagner, D., Bauser, E., and Lu, Y. C., 1983, "Laser Grown Single-Crystals of Silicon," *Appl. Phys. A*, **30**, 147–149.

Bertsch, A., Yezequel, J. Y., and Andre, J. C., 1997, "Study of the Spatial Resolution of a New 3D Microfabrication Process: The Microstereophotolithography Using a Dynamic Mask-Generator Technique," *J. Photochem. Photobiol. A – Chem.*, **107**, 275–281.

Bieri, N. R., Chung, J., Haferl, S. E., Poulikakos, D., and Grigoropoulos, C. P., 2003, "Microstructuring by Printing and Laser Curing of Nanoparticle Solutions," *Appl. Phys. Lett.*, **82**, 3529–3531.

Bohandy, J., Kim, B. F., and Adrian, F. J., 1986, "Metal Deposition from a Supported Metal Film Using an Excimer Laser," *J. Appl. Phys.*, **60**, 1538–1539.

Bohandy, J., Kim, B. F., Adrian, F. J., and Jette, A. N., 1988, "Metal Deposition at 532 nm Using a Laser Transfer Technique," *J. Appl. Phys.*, **63**, 1158–1162.

Chrisey, D. B., Piqué, A., McGill, R. A. *et al.*, 2003, "Laser Deposition of Polymer and Biomaterial Films," *Chem. Rev.*, **103**, 553–576.

Choi, T.-Y., Poulikakos, D., and Grigoropoulos, C. P., 2004, "Fountain-Pen Based Laser Microstructuring with Gold Nanoparticle Inks," *Appl. Phys. Lett.*, **85**, 13–15.

Chung, J., Bieri, N. R., Ko, S., Grigoropoulos, C., and Poulikakos, D., 2004, "In-Tandem Deposition and Sintering of Printed Gold Nanoparticle Inks Induced by Continuous Gaussian Laser Irradiation," *Appl. Phys. A*, **79**, 1259–1261.

Chung, J., Ko, S., Grigoropoulos, C. P. *et al.*, 2005, "Damage-Free Low Temperature Pulsed Laser Printing of Gold Nanoninks on Polymers," *J. Heat Transfer*, **127**, 724–732.

Cumpston, B. H., Ananthavel, S. P., Barlow, S. *et al.*, 1999, "Two-Photon Polymerization Initiators for Three-Dimensional Optical Data Storage and Microfabrication," *Nature*, **398**, 51–54.

Deegan, R. D., Bakajin, O., Dupont, T. F. *et al.*, 1997, "Capillary Flow as the Cause of Ring Stains from Dried Liquid Drops," *Nature*, **389**, 827–829.

Doppelbauer, J., and Bäuerle, D., 1986, "Kinetic Studies of Pyrolytic Laser-Induced Chemical Processes," in *Laser Processing and Diagnostics II*, ed. D. Bäuerle, K. L. Kompa, and L. D. Laude, Les Ulis, Les Editions de Physique.

Fang, N., Sun, C., and Zhang, X., 2004, "Diffusion-Limited Photopolymerization in Scanning Micro-stereolithography," *Appl. Phys. A*, **79**, 1839.

Fogarassy, E., Fuchs, C., Kerherve, F., Hauchecorne, G., and Perriere, J., 1989a, "Laser-Induced Forward Transfer of High-T_c YBaCuO and BiSrCaCuO Superconducting Thin Films," *J. Appl. Phys.*, **66**, 457–459.

1989b, "Laser-Induced Forward Transfer: A New Approach for the Deposition of High T_c Superconducting Thin Films," *J. Mater. Res.*, **4**, 1082–1086.

Ibbs, K. G., and Osgood, R. M. (editors), 1989, *Laser Chemical Processing for Microelectronics*, Cambridge, Cambridge University Press.

Kawata, S., Sun, H.-B., Tanaka, T., and Takada, K., 2001, "Fine Features for Functional Microdevices," *Nature*, **412**, 697–698.

Ko, S. H., 2006, *Flexible Electronics Fabrication by Lithography-Free Low Temperature Metal Nanoparticle Laser Processing*, Ph.D. Dissertation, University of California, Berkeley.

Ko, S. H., Choi, Y., Hwang, D. *et al.*, 2006, "Nanosecond Laser Ablation of Gold Nanoparticle Films," *Appl. Phys. Lett.*, **89**, 141126.

Ko, S. H., Pan, H., Grigoropoulos, C. P. *et al.*, 2007a, "Air Stable High Resolution Organic Transistors by Selective Laser Sintering of Inkjet Printed Metal Nanoparticles," *Appl. Phys. Lett.*, **90**, 141103.

Ko, S. H., Chung, J., Pan, H., Grigoropoulos, C. P., Poulikakos, D., 2007b, "Fabrication of Multilayer Passive and Active Electric Components on Polymer Using Inkjet Printing and Low Temperature Laser Processing," *Sensors Actuators A*, **134**, 161–168.

Koundourakis, G., Rockstuhl, C., Papazoglou, D. *et al.*, 2001, "Laser Printing of Active Optical Microstructures," *Appl. Phys. Lett.*, **78**, 868–870.

Lee, I. Y. S., Tolbert, W. A., Dlott, D. D. *et al.*, 1992, "Dynamics of Laser Ablation Transfer Imaging Investigated by Ultrafast Microscopy," *J. Imag. Sci. Technol.*, **36**, 180–187.

Lehmann, O., and Stuke, M., 1995, "Laser-Driven Movement of Three-Dimensional Microstructures Generated by Laser Rapid Prototyping," *Science*, **270**, 1644–1646.

Murphy, A. R., Liu, J. S., Luscombe, C. *et al.*, 2005, "Synthesis, Characterization, and Field-Effect Transistor Performance of Carboxylate-Functionalized Polythiophenes with Increased Air Stability," *Chem. Mater.*, **17**, 4892–4899.

Papazoglou, D. G., Karaiskou, A., Zergioti, I., and Fotakis, C., 2002, "Shadowgraphic Imaging of the Sub-ps Laser-Induced Forward Transfer Process," *Appl. Phys. Lett.*, **81**, 1594–1596.

Piqué, A., Arnold, C. B., Kim, H., Ollinger, M., and Sutto, T. E., 2004, "Rapid Prototyping of Micropower Sources by Laser Direct-Write," *Appl. Phys. A*, **79**, 783–786.

Piqué, A., Chrisey, D. B., Auyeung, R. C. Y. *et al.*, 1999, "A Novel Laser Transfer Process for Direct Writing of Electronic and Sensor Materials," *Appl. Phys. A*, **69**, S279–S284.

Sun, H.-B., and Kawata, S., 2004, "Two-Photon Photopolymerization and 3D Lithographic Microfabrication," *Adv. Polym. Sci.*, **170**, 169–273.

Sun, C., Fang, N., Wu, D. M., and Zhang, X., 2005, "Projection Micro-stereolithography Using Digital Micro-mirror Dynamic Mask," *Sensors Actuators A*, **121**, 113–120.

Wanke, M. C., Lehmann, O., Müller, K., Wen, Q., and Stuke, M., 1997, "Laser Rapid Prototyping of Photonic Band-Gap Microstructures," *Science*, **275**, 1284–1286.

Zergioti, I., Mailis, S., Vainos, N. A. *et al.*, 1998a, "Microdeposition of Metals by Femtosecond Excimer Laser," *Appl. Surf. Sci.*, **127–129**, 601–605.

1998b, "Microdeposition of Metal and Oxide Structures Using Ultrashort Laser Pulses," *Appl. Phys. A*, **66**, 579–582.

Zergioti, I., Papazoglou, D. G., Karaiskou, A. *et al.*, 2003, "A Comparative Schlieren Imaging Study between ns and Sub-ps Laser Forward Transfer of Cr," *Appl. Surf. Sci.*, **208–209**, 177–180.

Zergioti, I., Karaiskou, A., Papazoglou, D. G., Kapsetaki, M., and Kafetzopoulos, D., 2005, "Femtosecond Laser Microprinting of Biomaterials," *Appl. Phys. Lett.*, **86**, 163902.

Zhang, X., Jiang, X. N., and Sun, C., 1999, "Micro-stereolithography of Polymeric and Ceramic Microstructures," *Sensors. Actuators A*, **77**, 149–156.

14 Nano-structuring using pulsed laser radiation

14.1 Introduction

Fundamental understanding of microscale phenomena has been greatly facilitated in recent years, largely due to the development of high-resolution mechanical, electrical, optical, and thermal sensors. Consequently, new directions have been created for the development of new materials that can be engineered to exhibit unusual properties at sub-micrometer scales. Surface engineering is a critical sub-field of nanotechnology because of the paramount importance of surface-interaction phenomena at the micro/nano-machine level. Nanofabrication of complex three-dimensional patterns cannot be accomplished with conventional thermo-chemo-mechanical processes. While laser-assisted processes have been effective in component microfabrication with basic dimensions in the few-micrometer range, there is an increasing need to advance the science and technology of laser processing to the nanoscale. Breakthroughs in various nanotechnologies, such as information storage, optoelectronics, and bio-fluidics, can be achieved only through basic research on nanoscale modification and characterization of surfaces designed to exhibit special topographical and compositional features at high densities.

Since their invention in the 1980s, scanning-microscopy techniques such as scanning tunneling microscopy (STM), atomic-force microscopy (AFM), scanning near-field optical microscopy, and further variants thereof, have not only become indispensable tools for ultrahigh-resolution imaging of surfaces and measurement of surface properties but also offered hitherto unexplored possibilities to locally modify materials at levels comparable to those of large atoms, single molecules, and atomic clusters. These nanometric investigation tools have been used extensively in numerous high-resolution nanostructuring studies, to manipulate single atoms, and also as effective all-inclusive nanofabrication tools. In view of rapid advances in nanotechnology, various applications and research fields are being envisioned, such as ultrahigh-density data storage, mask repair, Fresnel optics for X-rays, nanoelectronics, nanomechanics, and nanobiotechnology, all of which depend on the fabrication of features only 5–10 nm in lateral dimensions (Quate, 1997). These feature sizes are more than an order of magnitude less than present industrial ultraviolet lithography standards. Owing to the light diffraction that limits the size of the minimum resolvable feature to one-half of the wavelength of the radiation used, conventional optical lithography is not applicable at the nanoscale. Although other techniques, such as electron-beam, ion-beam, and X-ray lithography, provide means for high-resolution nanoprocessing, the excessive equipment cost, low processing speeds, and lack of materials suitable for fabricating various optical elements highlight the need for alternative and/or complementary process schemes.

Confinement of optical energy to small dimensions can be achieved by coupling laser radiation to near-field scanning optical microscopes (NSOMs). Historically, the concept of utilizing the optical near field was proposed by Synge in the 1920s (Synge, 1928) and experimentally examined first in the microwave range (Ash and Nicholls, 1972), but could not be verified initially in the visible wavelength range due to difficulty in fabricating the aperture and controlling the gap distance. Since the mid 1980s, aided by advances in scanning probe microscopy and fiber-pulling technology (Lewis *et al.*, 1984), scanning near-field optical microscopy was established as a powerful nanoscale characterization tool. Example NSOMs were demonstrated by Pohl *et al.* (1984), Betzig *et al.* (1991), and Betzig and Trautman (1992) as a means to achieve probing resolution far below the wavelength used, yet at optical frequencies. In addition to their use as nanoscale microscopy tools, NSOMs have been applied to various materials-processing tasks (Betzig *et al.*, 1991; Wegscheider *et al.*, 1995). Thin-metal-film ablation tests using optical near-field schemes have been carried out (Lieberman *et al.*, 1999; Nolte *et al.*, 1999). Since laser sources are available with various photon energies and pulse lengths, matching with the absorption properties of the substrate can be dictated in order to accomplish local chemical and structural modification with precision and accuracy, without detrimental thermo-mechanical side effects.

The main technical challenge stems from the difficulty in transmitting pulses of sufficient energy for nanostructuring through nanoscale openings. However, recent work has shown the potential of nanoscale laser energy deposition for precise surface modification. The surface topography, texture, crystalline structure, and chemical composition could be modified with lateral feature definition in the nanometer range by coupling nanosecond- and femtosecond-laser beams to NSOM probes in both apertured and apertureless configurations.

14.2 Apertureless NSOM nanomachining

Surface nanostructuring of various materials including gold, gold/palladium, poly(methyl methacrylate), and polycarbonate by illumination of a scanning probe microscope (SPM) tip with nanosecond-laser pulses has been demonstrated in the literature (Gorbunov and Pompe, 1994; Jersch and Dickmann, 1996; Huang *et al.*, 2002). Although laser-assisted STM-based nanoprocessing schemes have yielded resolutions of the order of 30 nm (Lu *et al.*, 1999; Boneberg *et al.*, 1998), restrictions on the electrical conductivity of both the material to be processed and the scanning probe tip as well as the ambient working conditions make these methods unsuitable for a wide range of electrically nonconductive and biological materials. Laser nanoprocessing schemes coupled with an AFM operated either in contact or in noncontact mode provide effective means for overcoming the aforementioned drawbacks and enable the development of nanoscopic features of lateral dimensions less than 20 nm (Jersch *et al.*, 1997, Lu *et al.*, 2001). In spite of the increasing interest in this nanostructuring scheme, the actual processing mechanism still remains unexplained. In most cases, AFM or STM configurations with the tip in actual contact or in very close proximity (e.g. 3–5 Å) with the sample surface were used, which complicated the identification of the basic modification mechanism. Despite the high enhancement factors proposed for the electric field intensity in the vicinity of the tip (Dickmann *et al.*, 1997), contact of the SPM tip with the sample surface due to thermal expansion (Huber *et al.*, 1998; Boneberg *et al.*, 1998) limits the reliability and resolution of these nanostructuring techniques.

14.2.1   The experimental setup

Figure 14.1 shows a schematic diagram of the experimental apparatus used by Chimmalgi *et al.* (2003, 2005a). Surface nanostructuring was accomplished with femtosecond pulses of FWHM duration 83 fs, wavelength 800 nm, and pulse energy in the range of a fraction of a microjoule to a few microjoules. In the nanostructuring experiments a constant AFM tip–sample separation distance of 3–5 nm was maintained with the aid of nanolithography software that disables tip-position feedback signals and enables tip movement along complex contours. Commercially available etched silicon and coated silicon nitride microprobes were used for both surface scanning and nanostructuring. Electric-field-intensity distributions were obtained with the finite-difference time-domain (FDTD) method using electromagnetic (EM) near-field simulation software (Tempest 6.0; Pistor, 2001). A significant enhancement of the field intensity is predicted. The electric field distribution at the tip exhibits the *lightning-rod effect*, which is more pronounced for longer wavelengths (Ohtsu, 1998). The lateral confinement of the field, with a FWHM of $\sim$10 nm, is believed to be the main factor for the high spatial resolution observed in the nanostructuring experiments. Simulations indicated practically zero enhancement of the electric field for S-polarization. The actual field-strength enhancement may differ from the theoretical predictions due to various influencing factors, such as approximations in the modeling of the actual tip shape, the increase in the effective tip–sample separation distance during processing due to the removal of material, surface roughness, the presence of native oxide films on the tip and sample surfaces, and *in situ* change of the apex radius of the scanning tip.

14.2.2   The temperature distribution

Heat-transfer analysis of ultrathin films subjected to ultra-short laser pulses presents significant challenges due to difficulties in modeling the dynamics of nonequilibrium heating and the lack of pertinent thermophysical data for ultrathin films (Smith *et al.*, 1999). The film thickness, deposition method, substrate material, and surface roughness are some of the factors directly affecting the thermophysical properties. Consequently, these properties for the thin films vary from their corresponding bulk values. Mechanisms such as surface and grain-boundary scattering tend to reduce the thermal conductivity

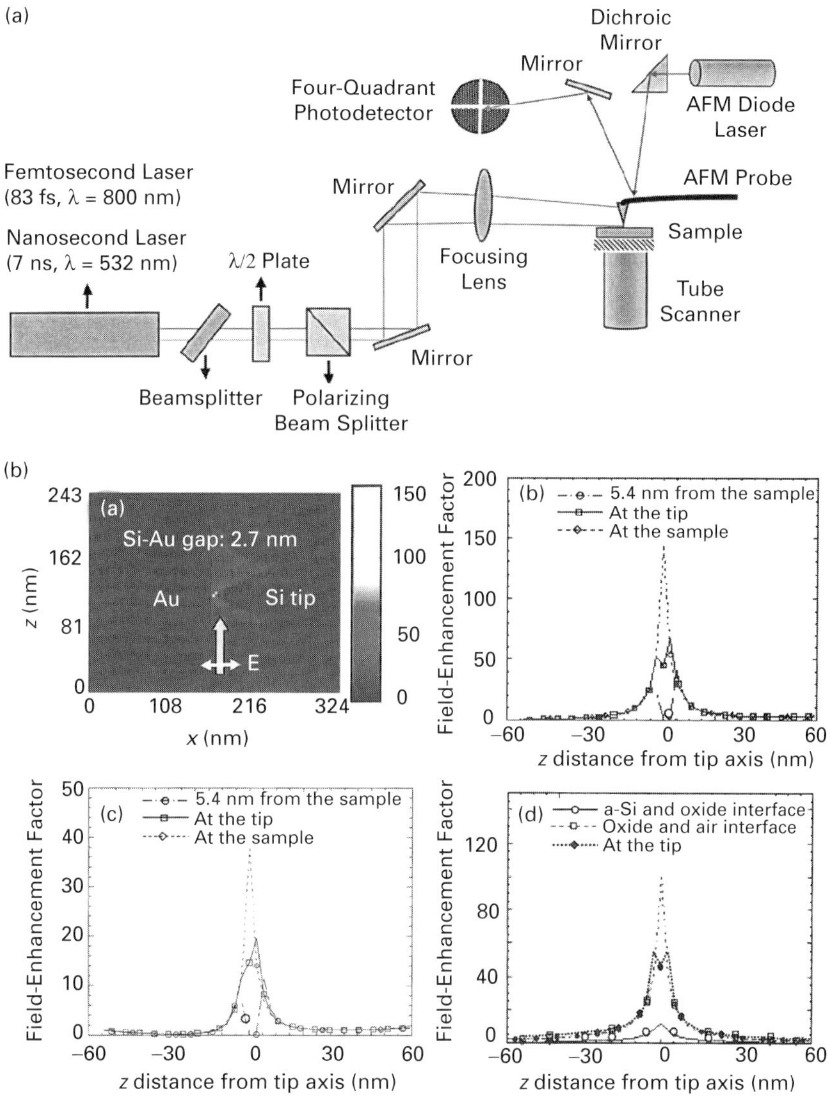

Figure 14.1. A schematic diagram of the apertureless near-field-scanning-optical-microscope experimental setup for surface nanostructuring. The normalized electric-field-intensity distributions shown are for the Si probe tip irradiated with a P-polarization femtosecond laser of wavelength $\lambda = 800$ nm. The tip-apex–sample-surface separation is 2.7 nm. From Chimmalgi *et al.* (2005a), reproduced with permission by the American Institute of Physics.

of ultrathin metallic films used in the nanostructuring experiments (Chen and Hui, 1999).

Temporal transient temperature distributions in a film subjected to ultra-short laser pulses can be obtained from an approximate three-dimensional analysis. The hyperbolic two-temperature model given below is simplified by neglecting lattice conduction and used to model the energy transfer and subsequent heating by an axisymmetric heat source with Gaussian temporal and spatial distributions:

$$C_{v,e}(T_e)\frac{\partial T_e}{\partial t} = -\left(\frac{q_{H,er}}{r} + \frac{\partial q_{H,er}}{\partial r} + \frac{\partial q_{H,ez}}{\partial z}\right)$$
$$- G(T_e - T_l) + Q_{ab}(r, z, t), \tag{14.1}$$

$$\tau_e\frac{\partial q_{H,er}}{\partial t} + q_{H,er} = -K_e\frac{\partial T_e}{\partial r}, \tag{14.2}$$

$$\tau_e\frac{\partial q_{H,ez}}{\partial t} + q_{H,er} = -K_e\frac{\partial T_e}{\partial z}, \tag{14.3}$$

$$C_{v,l}\frac{\partial T_l}{\partial t} = G_c(T_e - T_l). \tag{14.4}$$

In the above, the subscripts r and z indicate the radial and axial directions, e and l the electron and lattice systems. G is the coupling constant (e.g. (6.17a)), K_e the electron conductivity, and q the heat flux. The relationship for the laser source term (Equation (14.1)) is based on the assumption of a spatially and temporally Gaussian beam that satisfies Beer's law of absorption. While the nominal optical absorption depth at $\lambda = 800$ nm is ~ 12 nm, a modified optical absorption depth $d_{abs} = 1$ nm was fitted to the computed near-field absorbed-energy-density profile in the target obtained from the FDTD near-field simulations:

$$Q_{ab}(r, z, t) = \sqrt{\frac{4\ln 2}{\pi}}\frac{(1 - R)F}{t_{pulse}d_{abs}(1 - e^{-d/d_{abs}})}$$
$$\times \exp\left[-\left(\frac{r}{r_{1/e}}\right)^2 - \frac{z}{d_{abs}} - 4\ln 2\left(\frac{t - t_{pk}}{t_{pulse}}\right)^2\right]. \tag{14.5}$$

Predictions of the heat transfer within the thin films for the nanostructuring experiments are given in Figure 14.2.

14.2.3   Experimental results on surface nanostructuring

Figure 14.3 shows the effect of the laser pulse energy on the dimensions of grid lines produced on the surfaces of 25-nm-thick Au films. It can be seen that both the depth and the width of the line features increase linearly with the laser fluence. Wider and deeper features were produced by successively increasing the pulse energy. To form continuous line features, the laser was operated at 1 kHz while the sample surface was scanned at a speed of 1 μm/s. The lateral dimension of the minimum feature size in these experiments was less than 10 nm. Figure 14.4 shows nanostructuring of 15- and 25-nm-thick Au films. Features with complex contours can be produced, as demonstrated in Figures 14.4(c) and 14.4(d).

14.2.4   The mechanism of surface nanostructuring

Several investigators have attempted to explain the physical phenomena responsible for surface nanostructuring (Ukraintsev and Yates, 1996; Kawata *et al.*, 1999; Bachelot

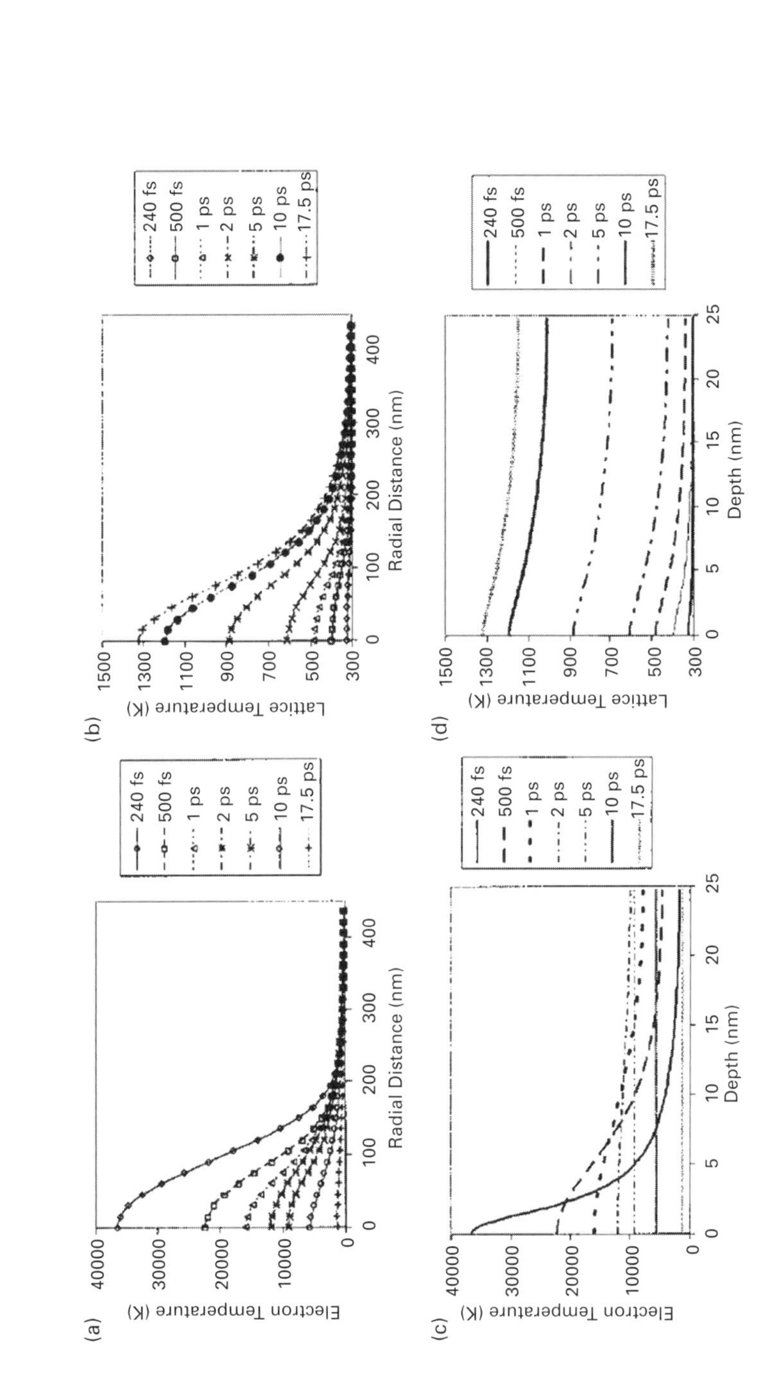

Figure 14.2. Electron and lattice temporal temperature distributions in the radial and depth directions in Au film for a laser-beam profile with $1/e^2$ radius $\sim$141 nm. (The time of maximum electron temperature and the time taken for the center node at the film surface to reach the melting temperature are 240 fs and 17.5 ps, respectively.) From Chimmalgi et al. (2005a), reproduced with permission by the American Institute of Physics.

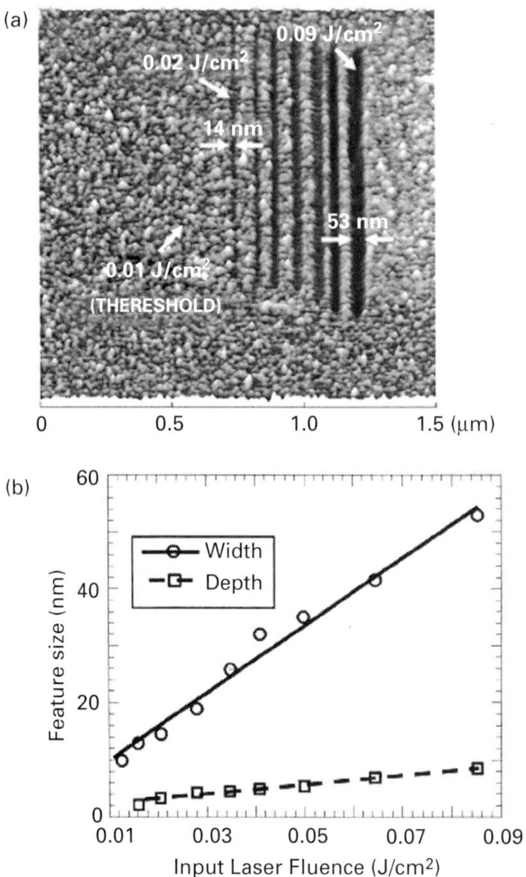

Figure 14.3. (a) Surface nanostructuring of 25-nm-thick Au film for laser fluence between 0.012 and 0.086 J/cm². (b) The dependence of lateral feature size on laser fluence. From Chimmalgi *et al.* (2005a), reproduced with permission from the American Institute of Physics.

et al., 2003). In general, material removal by ablation can be associated with photothermal, photomechanical, and photochemical effects (Hecht *et al.*, 2000). Significant theoretical and experimental effort has been expended to study the thermal expansion of the tip and the treated sample. Indentation of the sample surface due to thermal expansion of the tip has been proposed as the main reason for the formation of nanoscopic features (Boneberg *et al.*, 1998; Huber *et al.*, 1998). Several apertureless schemes have been developed to overcome resolution limitations and significant transmission losses in aperture near-field scanning microscopy. Ultrahigh-resolution fluorescence microscopy utilizing two-photon excitation and sharp metal tips for spectroscopy of single molecules has been demonstrated using similar configurations (Sanchez *et al.*, 1999). These schemes produce a highly localized and enhanced field underneath a sharp tip irradiated with a laser beam. Direct observations of the enhanced field using photosensitive azobenzene-containing films have confirmed the nanoscopic spatial confinement of the

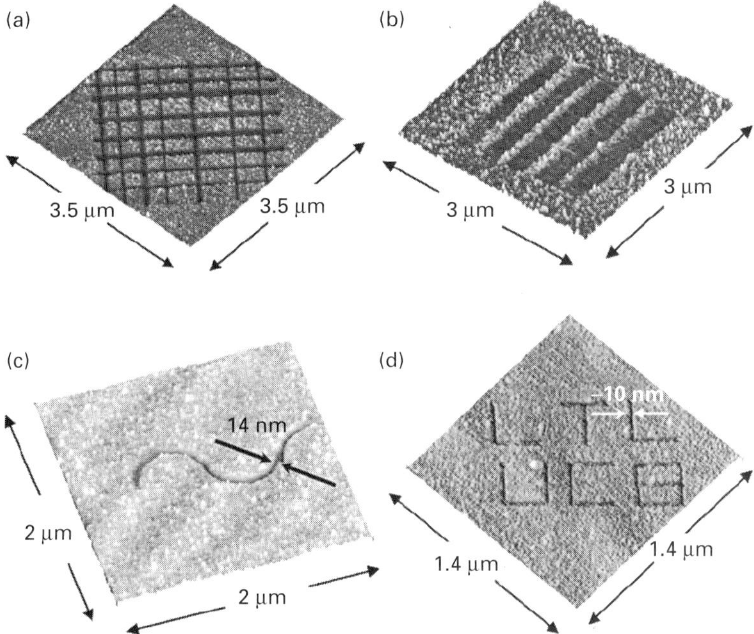

Figure 14.4. Surface nanostructuring of ~25-nm-thick Au film, (a), (c), and (d), and 15-nm-thick Au film, (b), for laser fluence 0.069 J/cm^2. From Chimmalgi *et al.* (2005a), reproduced with permission from the American Institute of Physics.

electric field, hence excluding the possibility of surface modification due to expansion of the tip (Bachelot *et al.*, 2003).

14.3 Apertured NSOM nanomachining

In the most common embodiment of apertured NSOMs, the optical fiber is heated locally, usually via a CO_2 laser to form a taper. A cladding layer, typically aluminum because of its high reflectance at visible wavelengths, is coated onto the outside of the fiber. The transmission efficiency through the fiber probe is defined as the ratio of the output energy to the input coupled energy and is roughly proportional to $(d_{ap}/\lambda)^n$ (n is around 2–4, depending on the probe design (Ebbesen *et al.*, 1998), and d_{ap} is the aperture diameter). Because considerable input laser pulse energy is required in order to offset losses from apertures of decreasing size, the trade-off between minimum aperture size and minimum required light throughput should be considered in materials-processing applications. Figure 14.5 shows FDTD simulations of the EM field emitted from tapered tips of various apertures. As the aperture size shrinks, the output distribution becomes bimodal. The throughput of the NSOM chiefly depends on the geometry of the tapered region, where the transmission efficiency drops as the fiber diameter becomes smaller than the laser wavelength.

Figure 14.6 shows a schematic diagram of the apertured NSOM nanomachining setup (Hwang *et al.*, 2006a). The processing laser beam is coupled to a bent cantilevered,

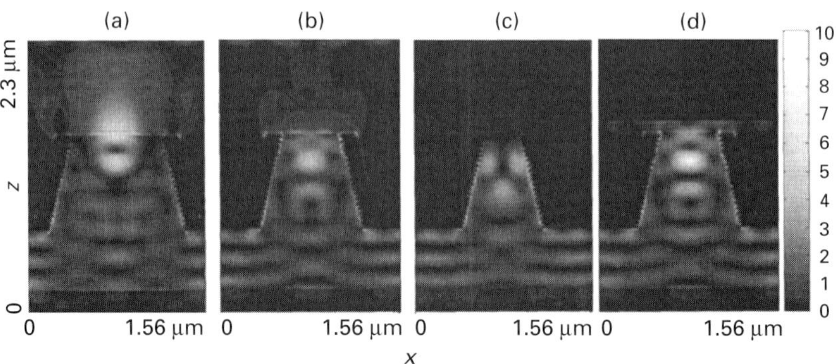

Figure 14.5. Contours of the intensity distribution of $\lambda = 532$ nm laser light transmitted through 800-, 500-, and 300-nm apertures without a sample, (a)–(c), and correspondingly with a gold sample (d) for the 500-nm aperture. The planes parallel to the electric-field-polarization directions are shown. From Hwang *et al.* (2006a), reproduced with permission by the American Institute of Physics.

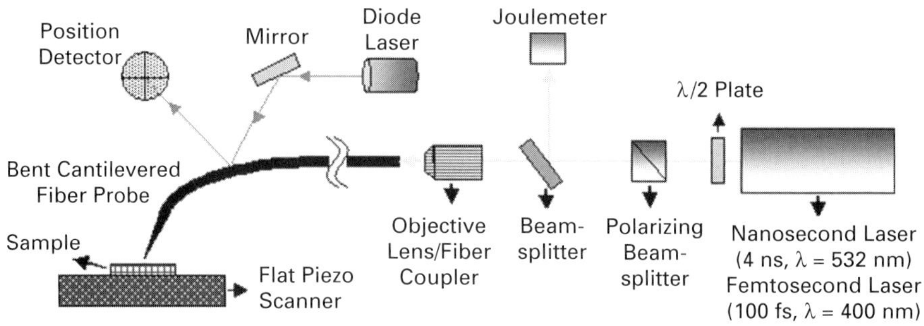

Figure 14.6. A schematic diagram of the experimental setup for processing utilizing a cantilevered NSOM system. From Hwang *et al.* (2006a), reproduced with permission by the American Institute of Physics.

metal-coated optical fiber and delivered onto the sample surface. Craters ablated by femtosecond-laser pulses (400 nm wavelength and $\sim$80 fs pulse width) are shown in Figure 14.7(a). In comparison with craters formed by nanosecond-laser processing, the craters ablated by femtosecond-laser pulses are of higher aspect ratio. Ablation craters with feature width less than 200 nm that penetrated through the thickness of the metal film were obtained.

The spectral bandwidth of ultra-short laser pulses is relatively wide. When femtosecond-laser pulses propagate through the optical-fiber core, they can be broadened by several mechanisms related to optical dispersion. If the optical fiber is a single-mode fiber, the femtosecond pulses primarily experience group velocity dispersion that typically broadens $\sim$80-fs pulses at the rate of roughly one picosecond per meter. Because the fiber probes used in the aforementioned experiment were drawn from multimode fiber, the dominant dispersion mechanism is modal dispersion. The degree of pulse broadening by modal dispersion depends on the refractive index of the glass medium as a function of the laser wavelength and core size. The modal dispersion for multimode

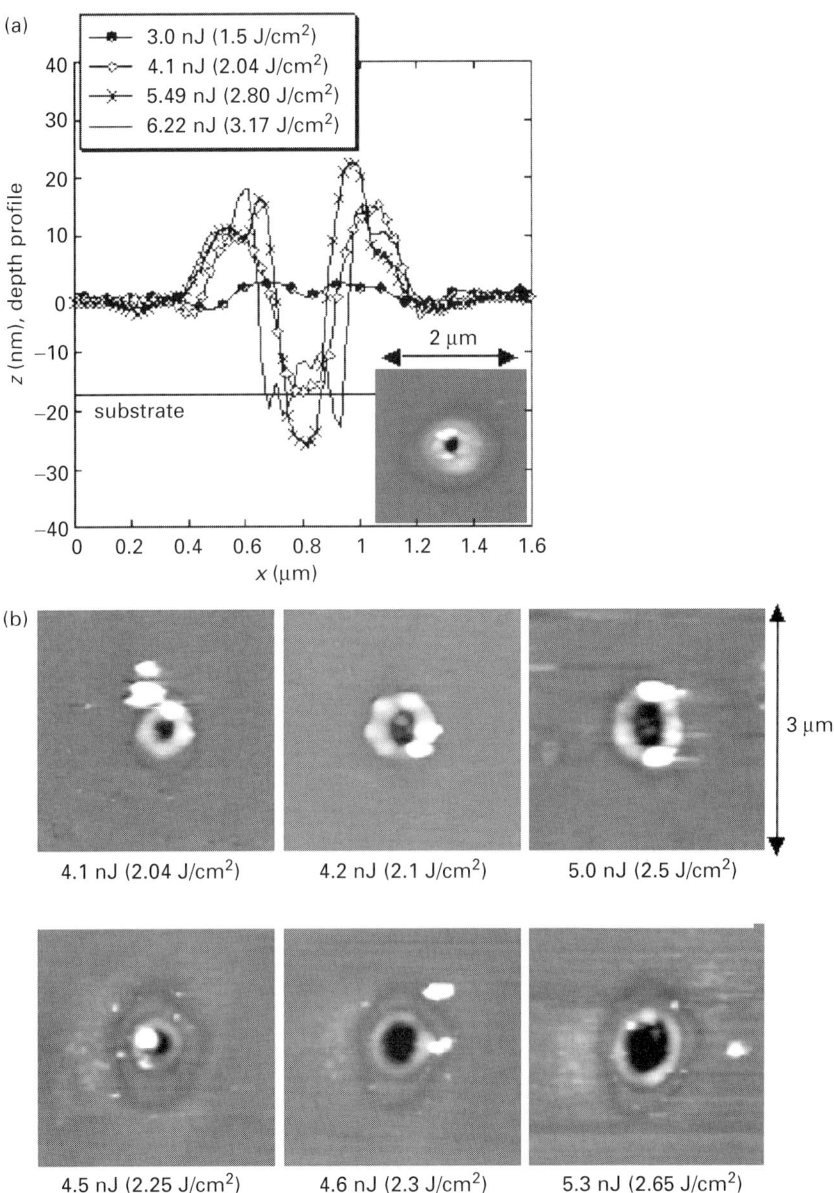

Figure 14.7. (a) Ablation-crater profiles produced by a femtosecond laser (500-nm probe, wavelength 400 nm) as laser energy varies. (b) A comparison of craters ablated in ambient air (left) and in moderate vacuum (right). From Hwang *et al.* (2006a), reproduced with permission by the American Institute of Physics.

fiber of core diameter ~ 100 µm is in the range of tens to hundreds of picoseconds. Hence, even though femtosecond-laser pulses were coupled to the fiber probe, the output laser pulse was much wider. In order to maintain ultra-short laser pulses, single-mode-fiber-based probes may be chosen, compromising light throughput, a drawback that is in part alleviated by the lower ablation threshold for shorter pulses. Special care needs to be taken to compensate for the group velocity dispersion. Nevertheless, in spite of the pulse broadening, the pulse width is still a couple of orders of magnitude shorter than nanoseconds.

Melting is difficult to avoid in laser ablation of metals, even in femtosecond-laser processing, for which thermal diffusion is limited. The melt phase tends to produce elevated rims at the periphery of the ablation craters, hence increasing the overall feature size. This is not an issue for applications such as ablation-based mass spectroscopy (Stöckle *et al.*, 2001).

When a solid sample is ablated using conventional lens-focusing optics, the final shape of the ablated features is sensitive to the environmental conditions. For example, when ultra-short laser light is focused onto the solid sample, electrons are first excited and emitted almost instantaneously (within $\sim$100 fs) due to their high mobility and small heat capacity. If the sample is exposed to air, the emitted electrons collide with air molecules, triggering the formation of a surface-electron-initiated air plasma by ionization. Once the plasma has been formed, it can further absorb the laser pulse energy for the remaining period of the pulse duration by inverse Bremsstrahlung, giving rise to a hot and dense state. The energy stored in the electron subsystem in the target solid is transferred to the lattice on a time scale of $\sim$0.1–10 ps due to the much higher heat capacity of the lattice compared with that of the electrons. Therefore, the material-ejection process commences after an elapsed time of about 10–100 ps. Upon interacting with the air plasma, the ejecta can bounce back and compress the melt. This can explain why much cleaner features can be fabricated in vacuum or nonreactive gas environments such as helium (Sun and Longtin, 2001). In nanosecond ablation, the incident beam interacts with the ejecta over a substantial fraction of the pulse duration. Consequently, the ablation threshold is not very sharp because of the more complex plasma-shielding effect compared with the case of femtosecond-laser ablation. In the case of ablation through NSOM fiber probes, the laser passes through a narrow, roughly nanometer scale gap before reaching the sample. The laser interaction with the plasma is therefore expected to be weaker than that in the far-field lens-focusing scheme. This argument is supported by the clear dependence of the feature size on the laser pulse energy applied for both nanosecond- and femtosecond-laser pulses as shown above. However, Figure 14.7(b) shows that the topography of features ablated in ambient air exhibits a qualitative difference from that of features ablated in vacuum.

14.4   Nanoscale melting and crystallization

In erasable phase-change optical data-storage applications, a "bit" of information is written on a crystallized region of the disk using a sufficiently strong laser pulse to induce local melting and rapid cooling leading to the formation of an amorphous phase (Yamada *et al.*, 1991). This bit can be read by using a low-energy laser pulse and noting the difference in optical response relative to the surrounding crystalline region. Furthermore, to erase this bit one can use an intermediate laser-pulse energy to raise the temperature of the film to a value above the crystallization temperature but below the melting temperature, thus converting the amorphous region back into its corresponding crystalline phase to erase optical contrast. Understanding the nanoscale amorphization and crystallization phenomena using optical near-field techniques is essential for ultrahigh-density data-storage applications.

For the study reported by Chimmalgi *et al.* (2005b), two different near-field schemes were employed to investigate the nanoscale melting and crystallization phenomena in a-Si thin films. In the fiber-coupled scheme, the nanosecond-laser beam was spatially confined by a cantilevered NSOM fiber tip. A second approach entailed local field enhancement in the near field of a SPM probe tip irradiated with nanosecond-laser pulses in an apertureless arrangement. The experimental setups for the two schemes have already been discussed in earlier sections (refer to Figures 14.1 and 14.6). A frequency-doubled ($\lambda = 532$ nm) Q-switched Nd : YAG pulsed nanosecond laser operating at a maximum repetition frequency of 10 Hz and pulse duration of 4 ns was used in the nanocrystallization studies. Bent cantilevered multimode NSOM fibers, of aperture size 800 nm, fitted onto a commercial NSOM system were used in the current experiment. The samples consisted of $\sim$20-nm-thick a-Si films that had been deposited onto fused quartz substrates by low-pressure CVD.

The results of a study employing single nanosecond-laser pulses of varying fluence, using the fiber-coupled probe, are detailed in Figure 14.8. Interesting observations about the melting transformation and the subsequent nucleation and crystallization processes were made with the gradual decrease of the fluence. These results can be explained by considering previous findings for thin-film a-Si crystallized by microscale laser pulses (Lee *et al.*, 2000).

For the high-fluence case of $\sim$5.9 J/cm^2 (Figure 14.8(a)), a clear central ablation regime with a narrow surrounding melt region can be seen. With a gradual decrease in energy, but at a still high enough fluence, the central region melts completely, whereas the adjacent surrounding area is partially melted. Therefore, nucleation occurs first in the outer region, as evidenced by the formation of small polycrystals. Grains launched from these seeds grow laterally toward the central and completely molten region until their growth is arrested by impingement on lateral grains growing from the opposite direction or by spontaneous nucleation triggered in the severely super-cooled molten-silicon pool (Lee *et al.*, 2000). Figure 14.8(b), corresponding to a laser fluence of $\sim$3.3 J/cm^2, shows such an occurrence wherein the progress of the nucleation front is inhibited by the development of an ultrafine microstructure of a polycrystalline/fine-amorphous phase forming a central spot structure. Considering the solid–liquid interfacial kinetics (Stolk *et al.*, 1993) and experimental evidence from time-resolved imaging (Lee *et al.*, 2001), the lateral crystal-growth speed is around 10 m/s. Consequently, the elapsed time for the lateral crystal growth would be approximately 20–30 ns. It is estimated that the three-dimensional conductive loss to the substrate that is not participating in nucleation drives quenching rates of the order of 10^{11} K/s at the center of the molten spot. As a result the undercooled liquid spontaneously freezes upon reaching the nucleation temperature (which is about 250 K below the nominal melting point), thereby arresting the inward propagating crystals as shown in Figure 14.8(b).

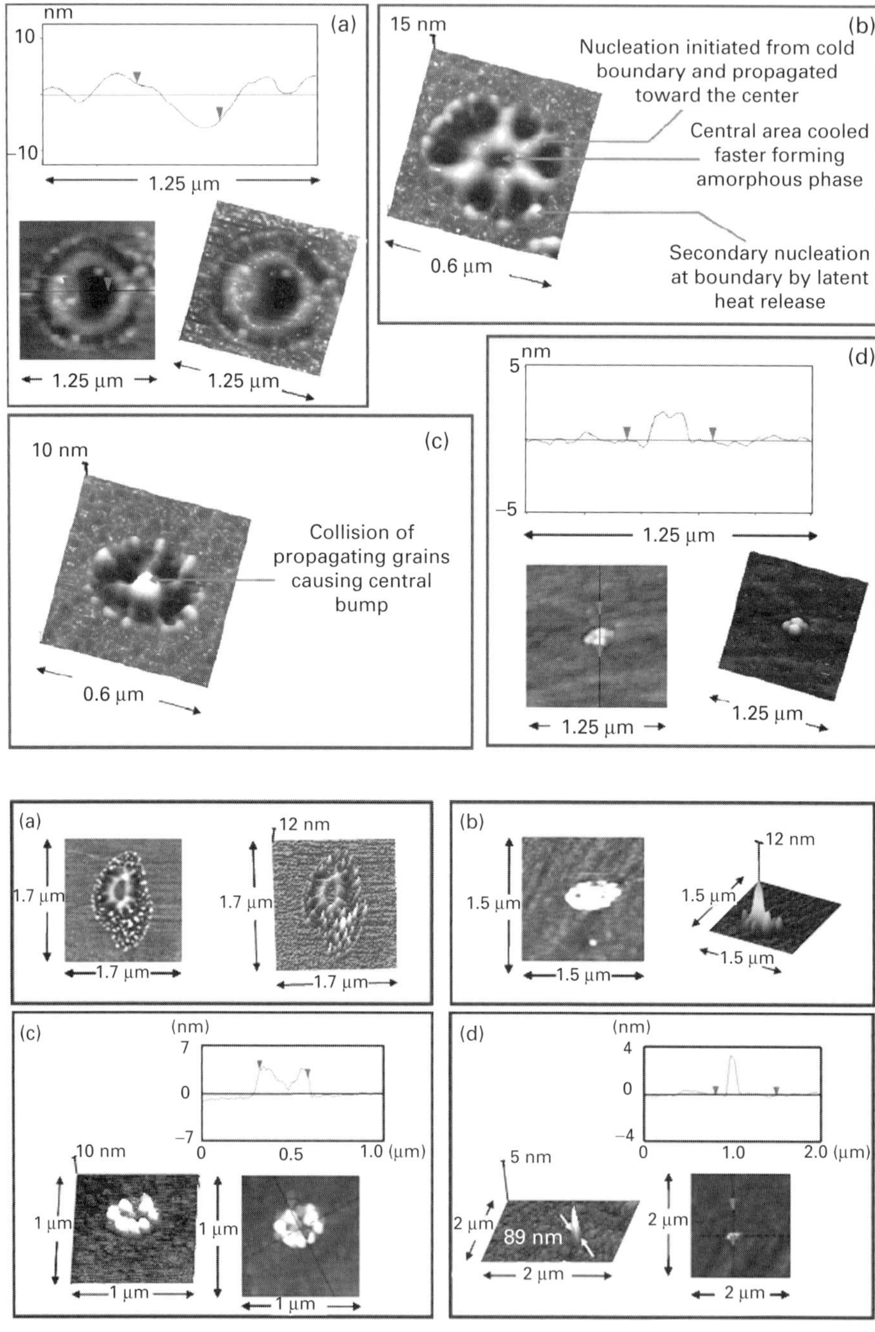

Figure 14.8. The upper panels depict limiting cases of a nanocrystallization study of 20-nm-thick a-Si films involving apertured NSOM for laser fluences (a) 5.9 J/cm^2, (b) 3.3 J/cm^2, (c) 3.0 J/cm^2, and (d) 2.0 J/cm^2. The part below shows corresponding results obtained by use of the apertureless NSOM scheme. The fluence in this case corresponds to the beam incident on the Si tip. From Chimmalgi *et al.* (2005b), reproduced with permission by the American Chemical Society.

The protrusions observed can be explained on the basis of volume expansion accompanying the quenching of liquid silicon that is trapped and squeezed by the propagating crystals. For a lower fluence of $\sim$3.0 J/cm^2, collision of the crystalline silicon grains led to the formation of a central bump structure (Figure 14.8(c)). When the fluence was further decreased to $\sim$2.0 J/cm^2 (Figure 14.8(d)), another regime, characterized by the formation of a set of individual grains, was encountered. The dimensions of these nanostructures were typically in the range of 80 nm. The same phase-transformation regimes could be identified with the apertureless configuration. In that case, a decrease of the input laser fluence to 0.09 J/cm^2 resulted in the formation of a single nanostructure with a lateral dimension of $\sim$90 nm. The lateral dimensions of these nanostructures could be further confined by using even lower fluence values and sharper tips, although limitations due to thermal diffusion would still apply.

14.5   Laser-assisted NSOM chemical processing

Various laser-assisted wet and dry etching schemes have been demonstrated; Bäuerle (2000) gives a comprehensive review. High-resolution etching of Si in a Cl$_2$ atmosphere could be done by direct writing, achieving sub-100-nm features and three-dimensional patterning. For low laser fluences the etching is photochemical in nature and driven by the production of free radicals in the chlorine phase and electron–hole pairs in the Si surface. At intermediate fluences that cannot quite induce melting, thermal processes are important, although photogenerated Cl radicals are still required. At laser energy densities surpassing the melting threshold, the etching is mainly thermally activated. Both regimes have been investigated utilizing an NSOM processing scheme shown in Figure 14.9 (Wysocki *et al.*, 2004). The minimum feature size obtained via a photophysical process was 30 nm (Figure 14.10).

Laser chemical vapor deposition (LCVD) is an extremely versatile materials-synthesis technique that enables the formation of technologically attractive microstructures of well-defined dimensions in a single-step maskless process (Bäuerle, 1998). Various applications have been pursued, including the fabrication of contacts, circuit lines, and interconnects, and the repair of lithographic masks. It also offers unique capabilities for the formation, coating, and patterning of nonplanar, three-dimensional objects. For example, LCVD was employed for the fabrication of micrometer-scale, photonic band-gap structures and the rapid prototyping of micro-mechanical actuators such as microtweezers and micromotors (Wanke *et al.*, 1997).

Recent experiments have yielded successful results for the pyrolytic NSOM–LCVD deposition of Si dots onto crystalline Si by the decomposition of SiH$_4$. Figure 14.11 shows an AFM image of the Si dots demonstrating that sub-100-nm features can be synthesized. Fabrication of three-dimensional objects of nanometric dimensions and arbitrary shape can be accomplished by extending the deposition and crystal growth through precise motion of the substrate. This nanoscale LCVD offers the possibility of selectively controlling nucleation and then manipulating the crystal-growth kinetics to

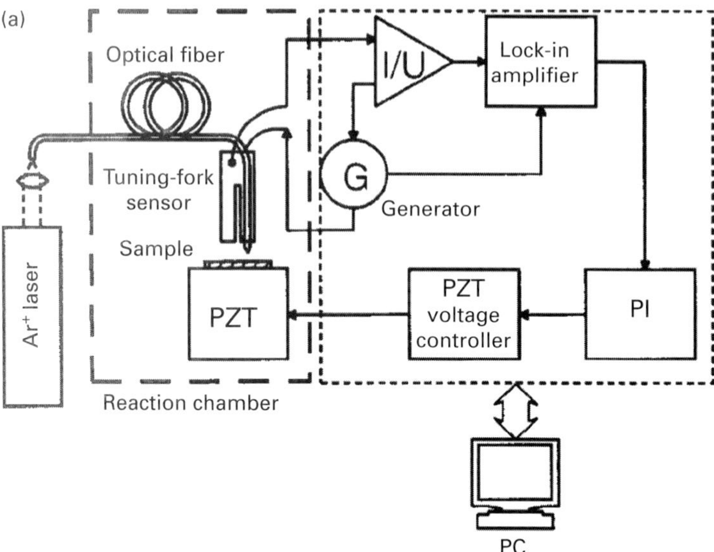

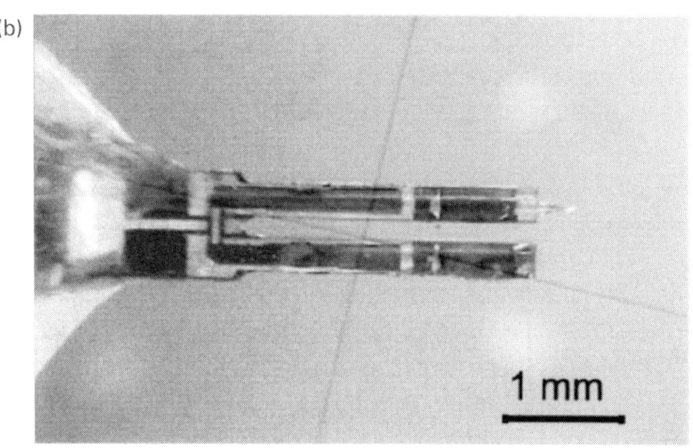

Figure 14.9. A schematic picture of the experimental setup used for laser-induced chemical etching of Si in Cl_2 atmosphere (Wysocki *et al.*, 2004). Courtesy of Professor Dieter Bäuerle. Reproduced with permission by the American Institute of Physics.

drive three-dimensional nanostructure fabrication and patterning to produce complex architectures.

Features written onto the metal films can be transferred to the underlying substrate via chemical etching. In a first study, the smallest crater profile obtained had a FWHM lateral dimension of ~90 nm (Figure 14.12, from Chimmalgi *et al.*, 2007). Hydrogen fluoride-based wet etching, which is an isotropic method, effectively enlarges the lateral dimensions of the final obtained crater. However, the issue of resolution loss could be remedied by using alternative dry-etching techniques such as deep reactive-ion etching (DRIE), which allow straight transfer of the pattern from the mask onto the underlying substrate. Using NSOM probes of smaller aperture dimensions in conjunction with

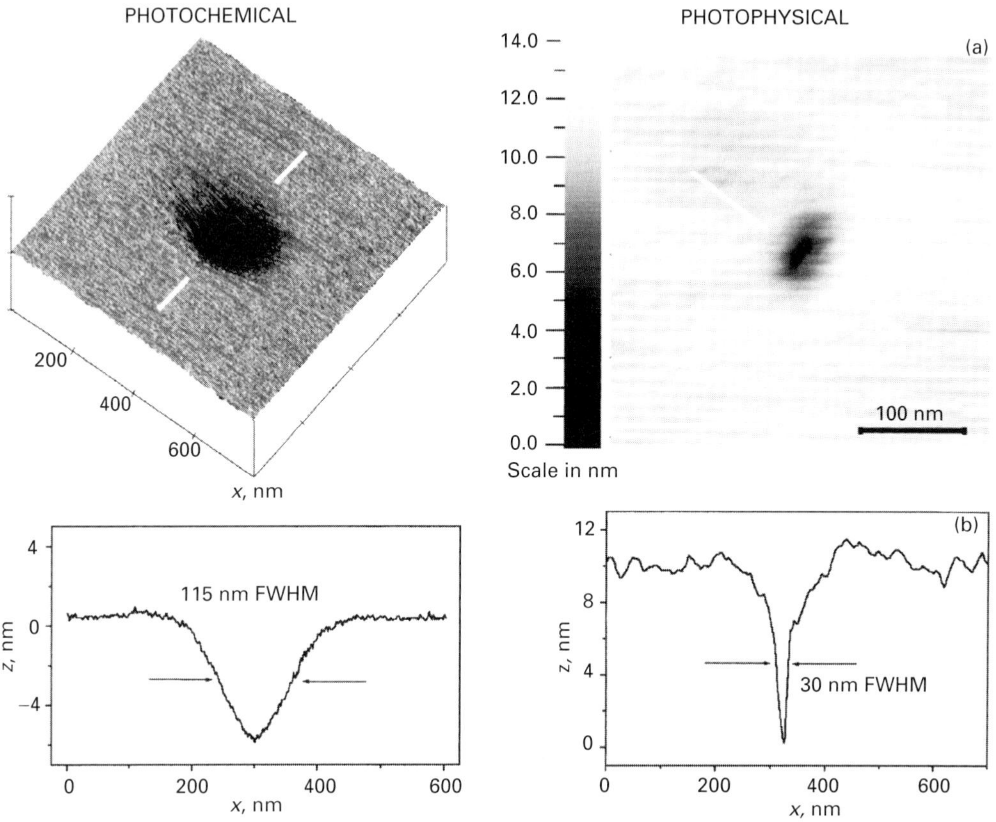

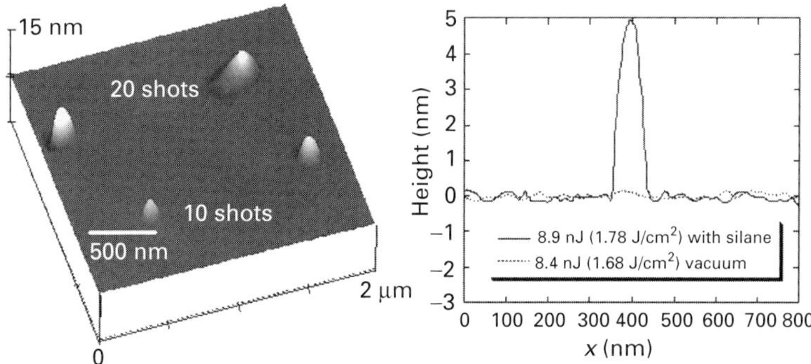

Figure 14.10. The picture on the left shows an AFM scan and a cross-sectional view of a hole in a Si surface etched in Cl_2 atmosphere using a tapered fiber tip ($\lambda = 351$ nm, $P_{tip} \sim 10$ mW, $t_{pulse} = 5$ s). The white lines show the scan direction of the cross section. The AFM and cross-sectional scan shown in the picture on the right were produced via a photophysical process driven by $\lambda = 514.5$ nm, $P_{tip} \sim 2.5$ mW, and $t_{pulse} = 10$ s (Wysocki *et al.*, 2004). Courtesy of Professor Dieter Bäuerle. Reproduced with permission by the American Institute of Physics.

Figure 14.11. Nanodeposition of Si dots from silane using single nanosecond-laser shots at $\lambda = 532$ nm through the apertured NSOM tip. The base diameter of the nanodeposit is ~ 100 nm.

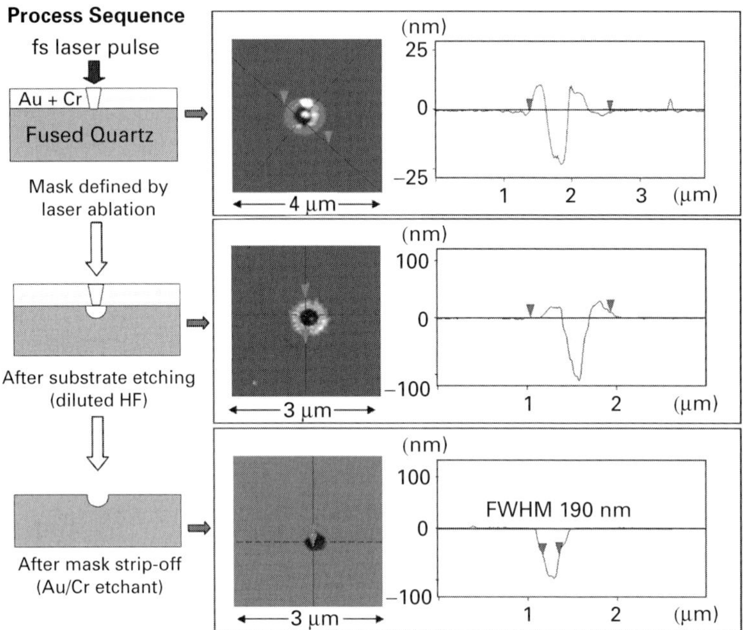

Figure 14.12. The process sequence for NSOM-based nanomachining of thin metal films and subsequent feature transfer onto a transparent quartz substrate. Also shown in the figure are the AFM scans corresponding to each of the steps in the process sequence, for a single case of an input laser fluence of 3.2 J/cm^2. From Chimmalgi *et al.* (2007), reproduced with permission by the Institute of Physics.

thinner metal films could also further reduce the lateral dimensions of the features. By employing probes of even smaller aperture sizes (corresponding to lower throughput), it should still be feasible to go through the metal-film layer and expose smaller regions on the quartz substrate. Furthermore, by choosing suitable etchants, the same process could be used for a variety of film–substrate combinations not just limited to transparent substrates.

14.6   Plasmas formed by near-field laser ablation

An intrinsic question in optical near-field-based ablation is that of the detailed expulsion process of the ablated material through the narrow probe-tip–target gap. Figure 14.13 shows a schematic diagram of the optical near-field-based ablation and plasma-emission measurement setup (Hwang *et al.*, 2006b). Figure 14.14 shows side-view emission imaging of the optical near-field-based ablation process. The symmetric traces toward the left in Figure 14.14(a) are mirror reflections of the ablation-induced emission from the sample surface. A bright spot appears near the probe–sample gap and jet-like material expulsion is observed away from the gap region and around the probe tip in a conical fashion. The bright emission near the gap evolves in a manner almost synchronized with

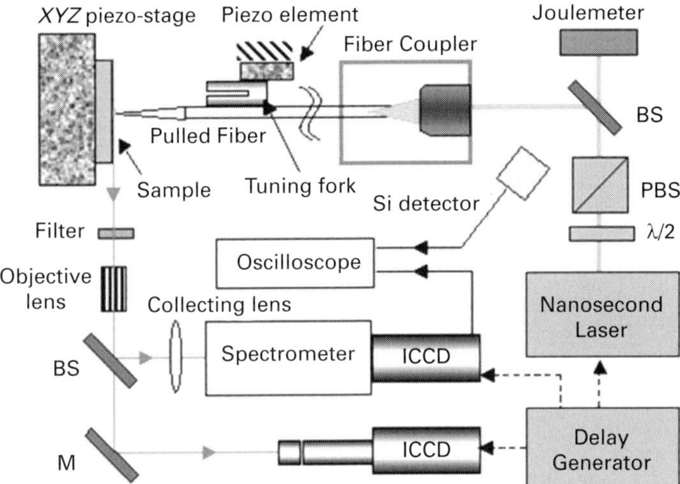

Figure 14.13. A schematic diagram of the optical-near-field-based ablation and plasma-emission-measurement setup. From Hwang *et al.* (2006b), reproduced with permission by the American Institute of Physics.

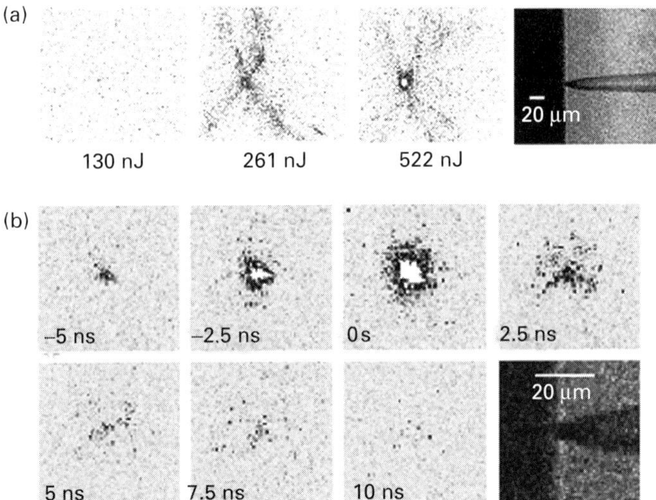

Figure 14.14. Side-view emission imaging of the optical-near-field-based ablation process. (a) The entire lifetime ($\sim$10 μs) measurement for various pulse energies. Measured output pulse energy is marked in the figure. (b) Time-resolved imaging with 2 ns exposure time for the case of pulse energy 522 nJ. The delay time is marked in each time step. The time-zero corresponds to the peak intensity timing of the temporally Gaussian-shaped nanosecond-laser pulse. Material ejection continues for $\sim$10 μs after this timing, showing the jet-like material-expulsion trajectories. From Hwang *et al.* (2006b), reproduced with permission by the American Institute of Physics.

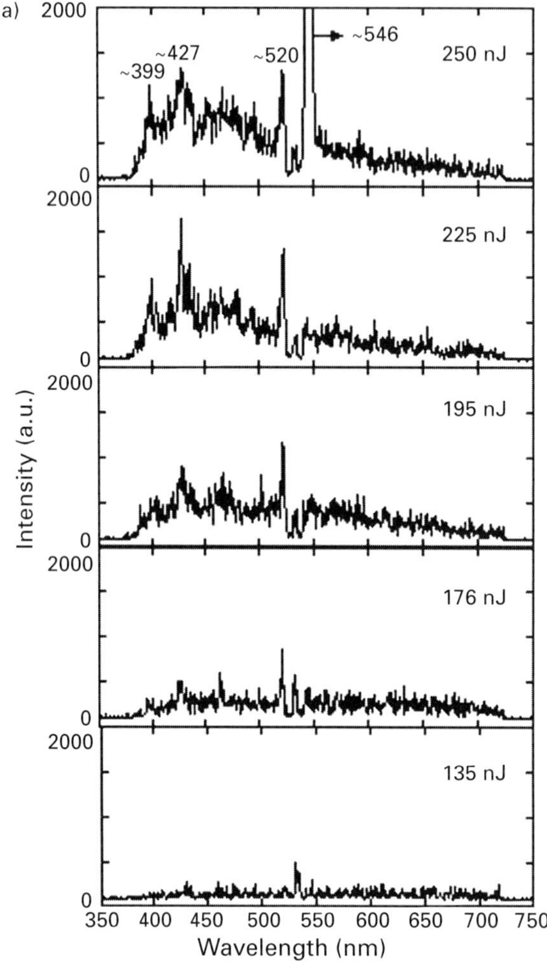

Figure 14.15. (a) Measured spectra in the optical-near-field ablation process for several pulse energies marked in the figure for the entire lifetime ($\sim$10 µs). (b) Measured AFM images of ablation craters are compared. The observed peaks correspond to Cr spectral lines due to electronic transitions (Meggers *et al.*, 1961). From Hwang *et al.* (2006b), reproduced with permission by the American Institute of Physics.

the laser pulse and then rapidly disperses away from the sample–probe gap, decaying within several nanoseconds of the laser pulse (Figure 14.14(b)). Material expansion continues for several microseconds, depending on the pulse energy applied.

Time-resolved spectra were also measured and compared with the ablation craters as displayed in Figure 14.15. These results demonstrate the possible use of optical near-field ablation in the context of laser-induced breakdown spectroscopy (LIBS) for chemical-species analysis with very high spatial resolution and minimal destruction/consumption of the target. Time-resolved spectral-emission measurement with 2-ns temporal resolution is shown in Figure 14.16 for pulse energy $\sim$195 nJ. This signal exhibits a very similar

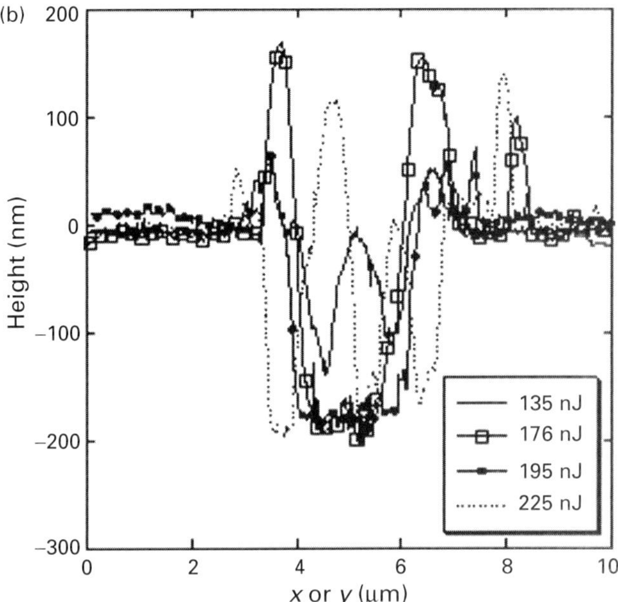

Figure 14.15. (*cont.*)

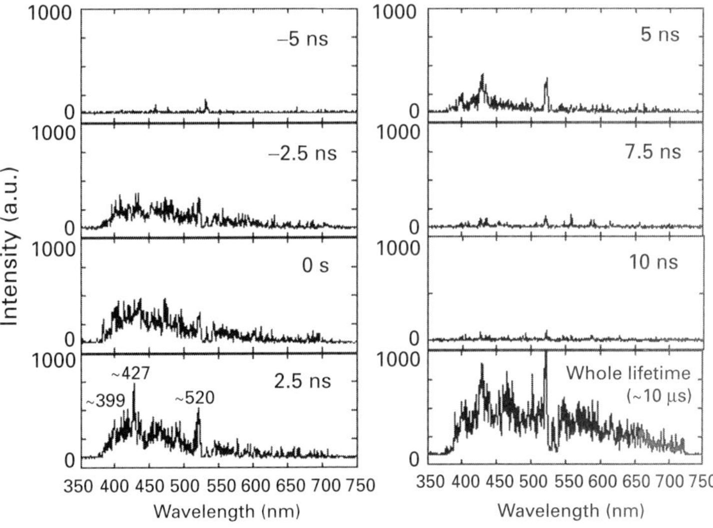

Figure 14.16. Measured time-resolved spectra in optical-near-field ablation with 2 ns exposure time for the case of pulse energy ∼195 nJ presented in Figure 14.15. The random, broad emission spectrum persisted for ∼10 μs after this timing. The collected signal for the entire lifetime is shown on the same data scale for comparison. From Hwang *et al.* (2006b), reproduced with permission by the American Institute of Physics.

trend to that of the bright-spot emission near the probe–sample gap as shown before in Figure 14.14(b). In the far-field laser-ablation configuration, the typical plasma lifetime is as long as 1–10 µs, whereas the LIBS signal appears about 100 ns after the laser pulse and is maintained over a longer time. The distinct deviation of plasma behavior is possibly due to the presence of the sharp probe tip at a gap distance of just ~10 nm from the sample and the much smaller plasma scale. Statistically very few collisional events of ejected electrons/material with air molecules occur in such a confined domain. Most ejected matter quickly moves away from the tiny volume underneath the tip and then collides with air molecules, forming a strong plasma. The laser coupling into the ablated plasma and the resulting reheating are minimal in the near-field configuration. This situation is analogous to ultrafast-laser LIBS, in which the laser pulse terminates before the initial plasma formation (Margetic *et al.*, 2000), although optical near-field-based LIBS is largely influenced by geometry constraints. Stronger initial electronic excitation by ultra-short laser pulses or UV laser pulses would be beneficial for achieving a higher-energy-level plasma state. Ultra-short pulsed radiation is expected to produce a smaller crater size and can provide higher excitation with less consumption of material.

14.7   Outlook

Using arrays of scanning-microprobe tips with integrated actuator and sensor mechanisms could lead to increased throughput of near-field optical surface-nanostructuring schemes. Furthermore, by incorporating improved probe designs with dedicated waveguide structures or using switching devices such as digital micromirror arrays, precise optical delivery schemes could be devised for coupling the beam with the microprobe tips. Possible applications of these nanostructuring processes are envisioned in high-resolution nanolithography, controlled nanodeposition, ultrahigh-density data storage, mask repair, nanoelectronics, nanophotonics, various nanobiotechnology applications, and nanoscale chemical analysis and imaging.

References

Ash, E. A., and Nicholls, G., 1972, "Super-Resolution Aperture Scanning Microscope," *Nature*, **237**, 510–512.

Bachelot, R., H'Dhili, F., Barchiesi, D., 2003, "Apertureless Near-Field Optical Microscopy: A Study of the Local Tip Field Enhancement Using Photosensitive Azobenzene-Containing Films," *J. Appl. Phys.*, **94**, 2060–2072.

Bäuerle, D., 1998, "Laser-Induced Fabrication and Processing of Semiconductors: Recent Developments," *Phys. Status Solidi* (a), **166**, 543–554.

2000, *Laser Processing and Chemistry*, 3rd edn, Heidelberg, Springer-Verlag.

Betzig, E., and Trautman, J. K., 1992, "Near-Field Optics: Microscopy, Spectroscopy and Surface Modification Beyond the Diffraction Limit," *Science*, **257**, 189–195.

Betzig, E., Trautman, J. K., Harris, T. D., Weiner, J. S., and Kostelak, R. L., 1991, "Breaking the Diffraction Barrier: Optical Microscopy at Nanometric Scale," *Science*, **251**, 1468–1470.

Boneberg, J., Munzer, H. J., Tresp, M., Ochmann, M., and Leiderer, P., 1998, "The Mechanism of Nanostructuring upon Nanosecond Laser Irradiation of a STM Tip," *Appl. Phys. A*, **67**, 381–384.

Chen, G., and Hui, P., 1999, "Thermal Conductivities of Evaporated Gold Films on Silicon and Glass," *Appl. Phys. Lett.*, **74**, 2942–2944.

Chimmalgi, A., Choi, T.-Y., Grigoropoulos, C. P., and Komvopoulos, K., 2003, "Femtosecond Laser Apertureless Near-Field Nanomachining of Metals Assisted by Scanning Probe Microscopy," *Appl. Phys. Lett.*, **82**, 1146–1148.

Chimmalgi, A., Grigoropoulos, C. P., and Komvopoulos, K., 2005a, "Surface Nanostructuring by Nano-/Femtosecond Laser-Assisted Scanning Force Microscopy," *J. Appl. Phys.*, **97**, 104319(1)–104319(12).

Chimmalgi, A., Hwang, D. J., and Grigoropoulos, C. P., 2005b, "Nanoscale Rapid Melting and Crystallization of Semiconductor Thin Films," *Nano Lett.*, **5**, 1924–1930.

Chimmalgi, A., Hwang, D. J., and Grigoropoulos, C. P., 2007, "Near-Field Scanning Optical Microscopy Based Nanolithography Using Thin Metal Film Masks," *J. Phys. D, Conf. Ser.*, **59**, 285–288.

Dickmann, K., Jersch, J., and Demming, F., 1997, "Focusing of Laser Radiation in the Near-Field of a Tip (FOLANT) for Applications in Nanostructuring," *Surf. Interface Anal.*, **25**, 500–504.

Ebbesen, T. W., Lezec, H. J., Ghaemi, H. F., Thio, T., and Wolff, P. A., 1998, "Extraordinary Optical Transmission through Sub-wavelength Hole Arrays," *Nature*, **391**, 667–669.

Gorbunov, A. A., and Pompe, W., 1994, "Thin Film Nanoprocessing by Laser/STM Combination," *Phys. Status Solidi A*, **145**, 333–338.

Hecht, B., Sick, B., Wild, U. P., 2000, "Scanning Near-Field Optical Microscopy with Aperture Probes: Fundamentals and Applications," *J. Chem. Phys.*, **112**, 7761–7774.

Huang, S. M., Hong, M. H., Lu, Y. F., 2002, "Pulsed-Laser Assisted Nanopatterning of Metallic Layers Combined with Atomic Force Microscopy," *J. Appl. Phys.*, **91**, 3268–3274.

Huber, R., Koch, M., and Feldmann, J., 1998, "Laser-Induced Thermal Expansion of a Scanning Tunneling Microscope Tip Measured with an Atomic Force Microscope Cantilever," *Appl. Phys. Lett.*, **73**, 2521–2523.

Hwang, D. J., 2005, *Pulsed Laser Processing of Electronic Materials in Micro/Nanoscale*, Ph.D. Dissertation, University of California, Berkeley.

Hwang, D. J., Chimmalgi, A., and Grigoropoulos, C. P., 2006a, "Ablation of Thin Metal Films by Short-Pulsed Lasers Coupled through Near-Field Scanning Optical Microscopy Probes," *J. Appl. Phys.*, **99**, 044905.

Hwang, D. J., Grigoropoulos, C. P., Yoo, J., and Russo, R. E., 2006b, "Optical Near-Field Ablation Induced Plasma Characteristics," *Appl. Phys. Lett.*, **89**, 254101.

Jersch, J., and Dickmann, K., 1996, "Nanostructure Fabrication Using Laser Field Enhancement in the Near Field of a Scanning Tunneling Microscope Tip," *Appl. Phys. Lett.*, **68**, 868–870.

Jersch, J., Demming, F., and Dickmann, K., 1997, "Nanostructuring with Laser Radiation in the Nearfield of a Tip from a Scanning Force Microscope," *Appl. Phys. A*, **64**, 29–32.

Kawata, Y., Xu, C., and Denk, W., 1999, "Feasibility of Molecular-Resolution Fluorescence Near-Field Microscopy Using Multi-photon Absorption and Field Enhancement near a Sharp Tip," *J. Appl. Phys.*, **85**, 1294–1301.

Lee, M., Moon, S., Hatano, M., Suzuki, K., and Grigoropoulos, C. P., 2000, "Relationship between Fluence Gradient and Lateral Grain Growth in Spatially Controlled Excimer Laser Crystallization of Amorphous Silicon Films," *J. Appl. Phys.*, **88**, 4994–4999.

Lee, M., Moon, S., and Grigoropoulos, C. P., 2001, "*In situ* Visualization of Interface Dynamics during the Double Laser Recrystallization of Amorphous Silicon Thin Films," *J. Cryst. Growth*, **226**, 8–10.

Lewis, A., Isaacson, M., Harootunian, A., and Muray, A., 1984, "Development of a 500 Å Spatial Resolution Light Microscope I. Light Is Efficiently Transmitted through Lambda/16 Diameter Apertures," *Ultramicroscopy*, **13**, 227–231.

Lieberman, K., Shani, Y., Melnik, I., Yoffe, S., and Sharon, Y., 1999, "Near-Field Optical Photomask Repair with a Femtosecond Laser," *J. Microscopy*, **194**, 537–541.

Lu, Y. F., Hu, B., Mai, Z. H. *et al.*, 2001, "Laser-Scanning Probe Microscope Based Nanoprocessing of Electronics Materials," *Jap. J. Appl. Phys.*, **40**, 4395–4398.

Lu, Y. F., Mai, Z. H., Qiu, G., and Chim, W. K., 1999, "Laser-induced Nano-oxidation on Hydrogen-Passivated Ge (100) Surfaces under a Scanning Tunneling Microscope Tip," *Appl. Phys. Lett.*, **75**, 2359–2361.

Margetic, V., Pakulev, A., Stockhaus, A. *et al.*, 2000, "A Comparison of Nanosecond and Femtosecond Laser-Induced Plasma Spectroscopy of Brass Samples," *Spectrochim. Acta B – Atomic Spectrosc.*, **55**, 1771–1785.

Meggers, W. F., Corliss, C. H., and Scribner, B. F., 1961, *Table of Spectral-Line Transitions, Part I*, Washington D.C., National Bureau of Standards.

Müller, R., and Lienau, C., 2001, "Three-Dimensional Analysis of Light Propagation through Uncoated Near-Field Fibre Probes," *J. Microscopy*, **202**, 339–346.

Nolte, S., Chichkov, B. N., Welling, H. *et al.*, 1999, "Nanostructuring with Spatially Localized Femtosecond Laser Pulses," *Opt. Lett.*, **24**, 914–916.

Ohtsu, M., 1998, *Near-Field Nano/Atom Optics and Technology*, Tokyo, Springer-Verlag.

Pistor, T. V., 2001, *Electromagnetic Simulation and Modeling with Applications in Lithography*, Memorandum No. UCB/ERL M01/19, Berkeley, University of California.

Pohl, D. W., Denk, W., and Lanz, M., 1984, "Optical Stethoscopy: Image Recording with Resolution λ/20, *Appl. Phys. Lett.*, **44**, 651–653.

Quate, C. F., 1997, "Scanning Probes as a Lithography Tool for Nanostructures," *Surf. Sci.*, **386**, 259–264.

Sanchez, J., Novotny, L., and Xie, X. S., 1999, "Near-Field Fluorescence Microscopy Based on Two-Photon Excitation with Metal Tips," *Phys. Rev. Lett.*, **82**, 4014–4017.

Smith, A. N., Hostetler, J. L., and Norris, P. M., 1999, "Nonequilibrium Heating in Metal Films: An Analytical and Numerical Analysis," *Num. Heat Transfer A*, **35**, 859–873.

Stöckle, R., Setz, P., Deckert, V., 2001, "Nanoscale Atmospheric Pressure Laser Ablation–Mass Spectrometry, *Anal. Chem.*, **73**, 1399–1402.

Stolk, P. A., Polman, A., and Sinke, W. C., 1993, "Experimental Test of Kinetic Theories for Heterogeneous Freezing in Silicon," *Phys. Rev. B*, **47**, 5–13.

Sun, J., and Longtin, J. P., 2001, "Inert Gas Beam Delivery for Ultrafast Laser Micromachining at Ambient Pressure," *J. Appl. Phys.*, **89**, 8219–8224.

Synge, E. H., 1928, "A Suggested Method for Extending Microscopic Resolution into the Ultramicroscopic Region," *Phil. Mag.*, **6**, 356–362.

Ukraintsev, V. A., and Yates, J. T., 1996, "Nanosecond Laser Induced Single Atom Deposition with Nanometer Spatial Resolution using a STM," *J. Appl. Phys.*, **80**, 2561–2571.

Wanke, M. C., Lehmann, O., Muller, K., Qingzhe, W., and Stuke, M., 1997, "Laser Rapid Prototyping of Photonic Band-Gap Microstructures," *Science*, **275**, 1284–1286.

Wegscheider, S., Kirsch, A., Mlynek, J., and Krausch, G., 1995, "Scanning Near-Field Optical Lithography," *Thin Solid Films*, **264**, 264–267.

Wysocki, G., Heitz, J., and Bäuerle, D., 2004, "Near-Field Optical Nanopatterning of Crystalline Silicon," *Appl. Phys. Lett.*, **84**, 2025–2027.

Yamada, N., Ohno, E., Nishiuchi, K., Akahira, N., and Takao, M., 1991, "Rapid-Phase Transitions of GeTe–Sb$_2$Te$_3$ Pseudobinary Amorphous Thin-Films for an Optical Disk Memory," *J. Appl. Phys.*, **69**, 2849–2856.

Index